Introduction to Geology

Second Edition

Daria Nikitina
West Chester University

Cover image Yellowstone National Park, photo by D.Nikitina

Kendall Hunt
publishing company

www.kendallhunt.com
Send all inquiries to:
4050 Westmark Drive
Dubuque, IA 52004-1840

ISBN 978-1-4652-6078-9

Printed in the United States of America

C o n t e n t s

Introduction

This textbook is prepared for an Introductory Geology course and aimed for a broad audience of students not majoring in Earth Science. Students without much of a science background should be comfortable with the material presented in this textbook. Additionally, it should be used as a comprehensive introduction to the discipline for Geology majors before stepping into more in-depth geology courses. The textbook includes chapters explaining how Earth works as a planet. The planet Earth is a dynamic, ever-changing system that includes four main components: lithosphere (solid Earth), atmosphere (gaseous outer sphere), hydrosphere (oceans, rivers, lakes, glaciers and groundwater) and biosphere (all living forms from primitive to human being). All components of the system are constantly interacting with each other to shape our planet.

Earth, along with other planets of the solar system, was formed about 4.6 billion years ago as the result of an orbiting cloud of gas and particles of ice and rocks called a **nebula** (Figure 1a).

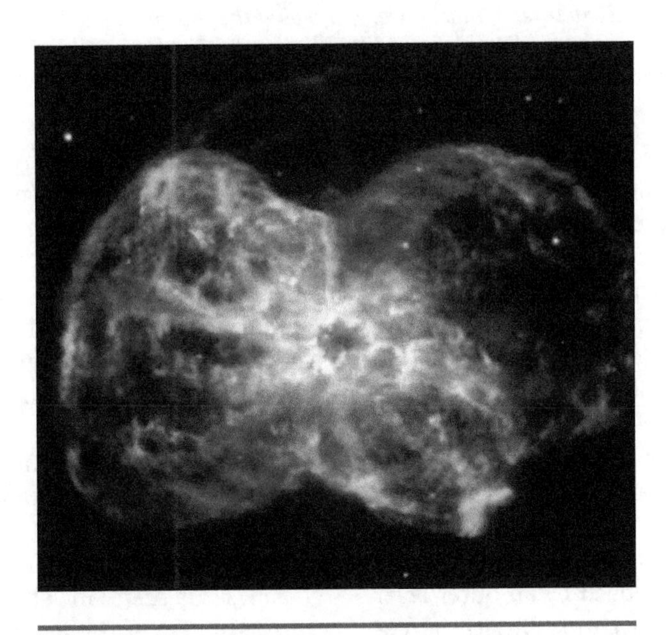

FIGURE 1a *The Nebula*

During its early stages of formation, Earth underwent a separation into layers which resulted in the formation of a **core, mantle,** and **crust** (Figure 1b). The density of the materials decreases systematically from the core to the crust with the crust having the lowest average density of 2.9 gm/cm^3, the mantle having an intermediate density of 5.7 gm/m^3 and the core having the highest density of 12.9 gm/m^3.

Geologists believe that the young planet was subject to widespread volcanism. Great volumes of water vapor, carbon dioxide, sulfur dioxide and nitrogen were released from the interior of the Earth from volcanoes, forming Earth's atmosphere. Because Earth's surface was too hot for liquid water to exist, the water vapor collected in the atmosphere. As the planet cooled, the water in the atmosphere condensed—forming clouds—and fell as rain, filling the ocean basin. The large amount of water on Earth can never have been produced by volcanism and degassing

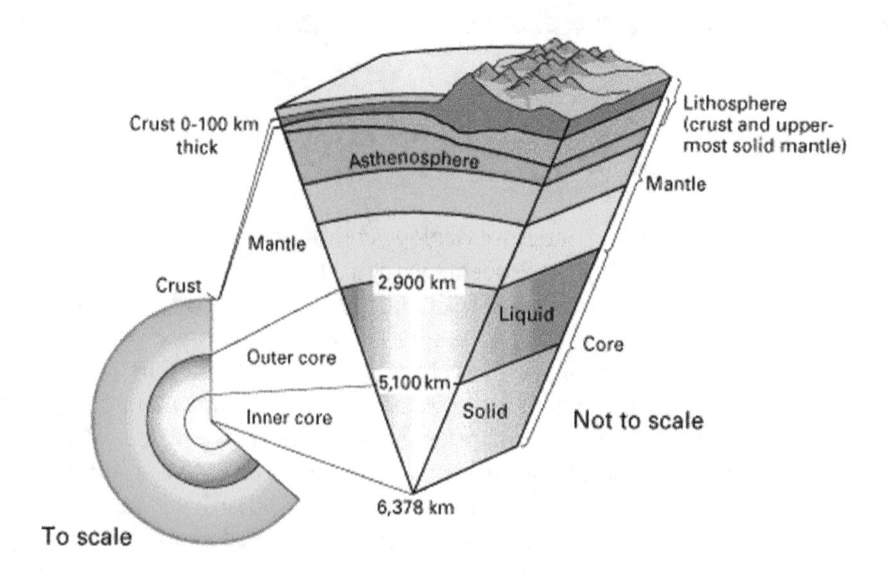

FIGURE 1b *Earth's Internal Structure: cutaway views showing the internal structure of the Earth. Below: This view drawn to scale demonstrates that the Earth's crust literally is only skin deep. Below right: A view not drawn to scale to show the Earth's three main layers (crust, mantle, and core). (http://pubs.usgs.gov/gip/dynamic/inside.html).*

alone. It is assumed that the water was derived from impacting comets containing ice. Recent evidence suggests the oceans may have begun forming 4.2 billion years ago.

The most abundant chemical elements that form Earth's interior are oxygen (O), silicon (Si), aluminum (Al), iron (Fe), calcium (Ca), sodium (Na), magnesium (Mg) and potassium (K). The rest of the known chemical elements are certainly present in the Earth materials, but are far less quantities. During the process of compositional differentiation, heavy elements, such as iron and nickel, melted and sank to the center of the planet, forming the core. The solid inner core of Earth is surrounded by an outer layer of molten iron (Earth's outer core). Physical forces within the liquid outer core cause molten metal to flow, forming currents responsible for Earth's magnetic field. The mantle that accounts for almost 82% of Earth's volume is formed by a mixture of metallic oxides and less dense magnesium- and iron-rich silicates (minerals olivine and pyroxene). The mantle is mostly solid, with the exception of its upper portion called the ***asthenosphere*** (from Greek asthenes 'weak' + sphere) where material is partially molten and able to flow. The Earth's crust consists of the relatively thin **oceanic crust** (4-6 km thick) and thicker (up to 70 km) **continental crust.** Most of our knowledge about geology comes from studying rocks of the Earth's crust. There are three different types of rocks: ***igneous***, ***sedimentary***, and ***metamorphic***.

Igneous rocks form from the cooling and crystallization of molten rock (**magma**). Pockets of magma form within Earth's mantle and crust. Molten rock material is hotter, and therefore less dense, than surrounding solid rocks, giving it the ability to rise. Magma extruded through the crust onto Earth's surface is called **lava.** Lava erupts from volcanoes, cooling and crystallizing into **extrusive** igneous rocks or **volcanic rocks.** Volcanic rocks accumulate around a vent to create volcanic mountains such as the Andes and Cascades in South and North Americas respectably, on the margins of the western and northern Pacific Ocean forming a chain of islands, such as the Japanese or Aleutian Islands, and Hawaiian Islands in the middle of the Pacific Ocean. Violent volcanic eruptions are associated w ith magmas of high viscosity (very resistant to flow). These eruptions are often explosive and result in ejection of pulverized magma and rock fragments into the air, where it solidifies in flight and falls around volcanic vents as **volcanic ash.** The solid fragments ejected from an eruptive volcano are collectively called **tephra** or **pyroclastic material** (from Greek pyro 'fire' + clastic 'broken bits and pieces of rocks'). Pyroclastic material larger than ash (>1/16 inch) accumulates in the immediate vicinity of the volcano. The fine-sized tephra smaller than 1/16 inch (volcanic ash and dust) from major volcanic eruptions can be carried long distances and even around the globe by the prevailing winds.

Less viscous magmas flow easily through fractures often long distances from a vent accumulating in flat sheet-like lava flows. Eruptions of magma of low viscosity are rarely accompanied by explosions or ejection of large volumes of ash into the atmosphere. Therefore, risk of such eruptions is easier to mitigate even though they may last for a long time, like the ongoing eruption of Kilauea, main volcano of Hawaii Islands. When low-viscosity magmas solidify, they form igneous volcanic rock called **basalt.** Most of basalt magmas erupt from rift zones along **mid-oceanic ridges** that present in all world oceans.

When magma plumes move slowly through the Earth crust they cool and crystallize before reaching the surface creating large bodies of **intrusive** igneous rocks called **plutons.** The most common **plutonic rock** is called **granite.** Intrusions of plutons add great volumes of rock material to the crust. Thick continental crust (~40 km) is mainly built of granite. Under the major mountain ranges, such as the Himalayan Mountains, the thickness of continental crust reaches 70 km. With time, erosion may strip away the overlying rocks, exposing intrusive rocks at the surface. For example, intrusive rocks that originally formed deep within the North American continent are now exposed at the surface throughout the Sierra Nevada, California. The Sierra Nevada Mountains were formed by volcanic activity at the margin of the North American continent (similar to present-day Andes and Cascade Mountains). Huge amounts of magma never reached the surface forming granitic intrusions at the roots of the volcanic mountains. Over millions of years, the volcanic mountains were removed by erosion, exposing granitic rocks that now form the Sierra Nevada.

The crust and the outermost portion of mantle above the asthenoshere form a rigid rocky layer called **lithosphere.** Under the oceans, the lithosphere is thinner and extends to the depths of approximately 50 km while under the major mountain ranges continental lithosphere may extends to the depths of more than 100 km. The difference between oceanic and continental lithospheres is not only in thickness and composition but also in density. Density of granitic continental crust is 2.6 gm/cm^3 while density of basaltic oceanic crust is 3.0 gm/cm^2. Because of high concentration of iron-rich minerals basalt is a relatively dense rock, while granite is mostly made of silica and other light elements. Both types of lithosphere float on top of partially melted asthenosphere. Since basalt is denser, the oceanic crust forms basins on earth's surface now filled with water. Due to the low density and great thickness of the continental rocks, the continents rise higher than the ocean basins. The average elevation of the continents is ~0.8 km above sea level while the average depth of the ocean floor is 3.7 km below sea level.

THREE TYPES OF PLATE BOUNDARY

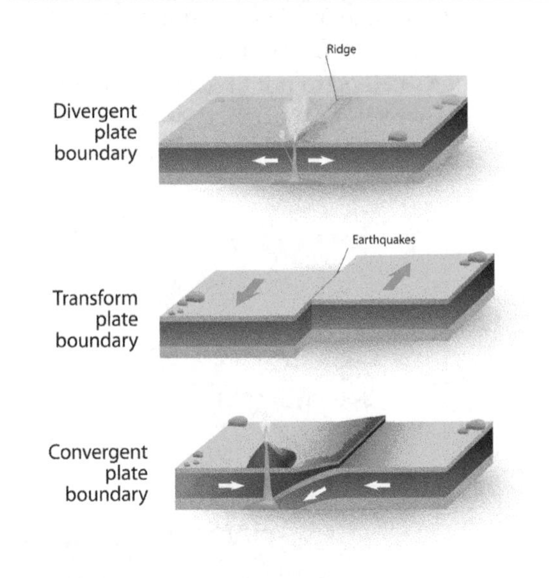

FIGURE 1c *Three type of plate motion along divergent, convergent and transform plate boundaries. Image © Designua, 2014. Used under license from Shutterstock, Inc.*

The rigid lithosphere is broken into series of **plates** that move relative to each other. The energy responsible for plate movement or **plate tectonics** is coming from the Earth interior. At some locations plates are moving away from each other or ***diverging*** while at others, plates are ***converging*** (moving towards each other) or sliding passing each other. At **diverging plate boundaries** a **rift** or fracture extends down to the top of the asthenosphere where molten rock (magma) begins to rise, then cools, and solidifies constantly forming new lithospheric rocks. Most of basalt magmas upwell and erupt along the crest of mid-oceanic ridges pushing away blocks of the oceanic lithosphere (Figure 1c (a)). Large blocks of the continental crust rift apart resulting in formation of **rift zone** like in eastern Africa where a divergent plate boundary is being formed. The rift is a narrow zone in which the African Plate is in the process of splitting into two new tectonic plates called the Somali Plate and the Nubian Plate. (Figure 1d) The East African Rift Zone includes a number of active volcanoes such as Mount Kilimanjaro and Mount Kenya. The East African Rift Zone is part of the larger Great Rift Valley. **Rift valleys** extend to hundreds and thousands of kilometers where the continental crust stretches and breaks forming linear depressions often filled with water forming long narrow lakes. As the plates move apart along the rift valley, **linear oceans** such as the Red Sea begin to open. Eruption of basaltic magmas along the mid-oceanic ridge within the linear ocean results in the continuous formation of oceanic crust, expansion and development of the **ocean basin** such as the Atlantic and Indian oceans.

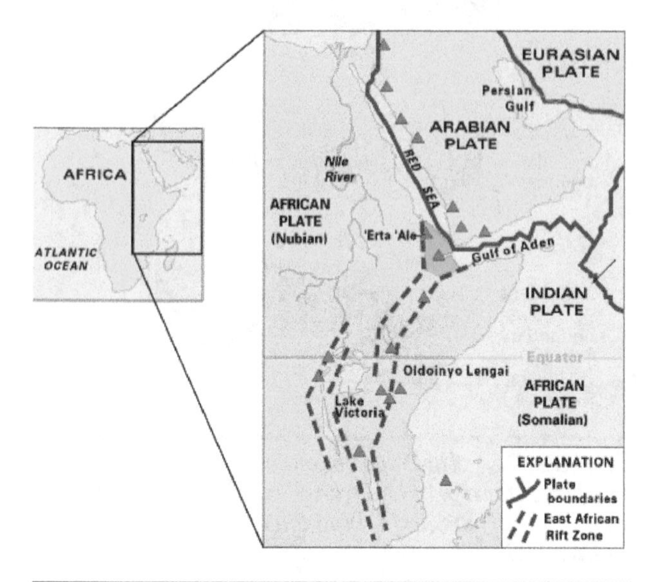

FIGURE 1d *East African Rift where the Arabian Plate, and the two parts of the African Plate (the Nubian and the Somalian) splitting along the East African Rift Zone. (http://pubs.usgs.gov/gip/dynamic/East_Africa.html).*

At **converging plate boundaries,** where two plates move towards each other, the more dense oceanic lithospheric plate descends, or **subducts,** beneath the less dense continental lithospheric plate or beneath another oceanic plate (Figure 1e). Water drawn down into the **subduction zone** lowers the melting point of mantle rocks, producing plumes of the molten material that rises to Earth's surface. This molten material intrudes through both the continental and oceanic crust of the overriding plate to build volcanic mountain chains, such as the Andes, Cascades, and Aleutian islands.

Over time, convergent plate movement along with the recycling of the oceanic lithosphere in the subduction zones cause continents to collide, forming large land masses often called **super continents.** Since the continental lithosphere has the same density and similar composition, compressive forces cause continental rocks to fold and thrust along the **collisional plate boundary,** forming mountain belts

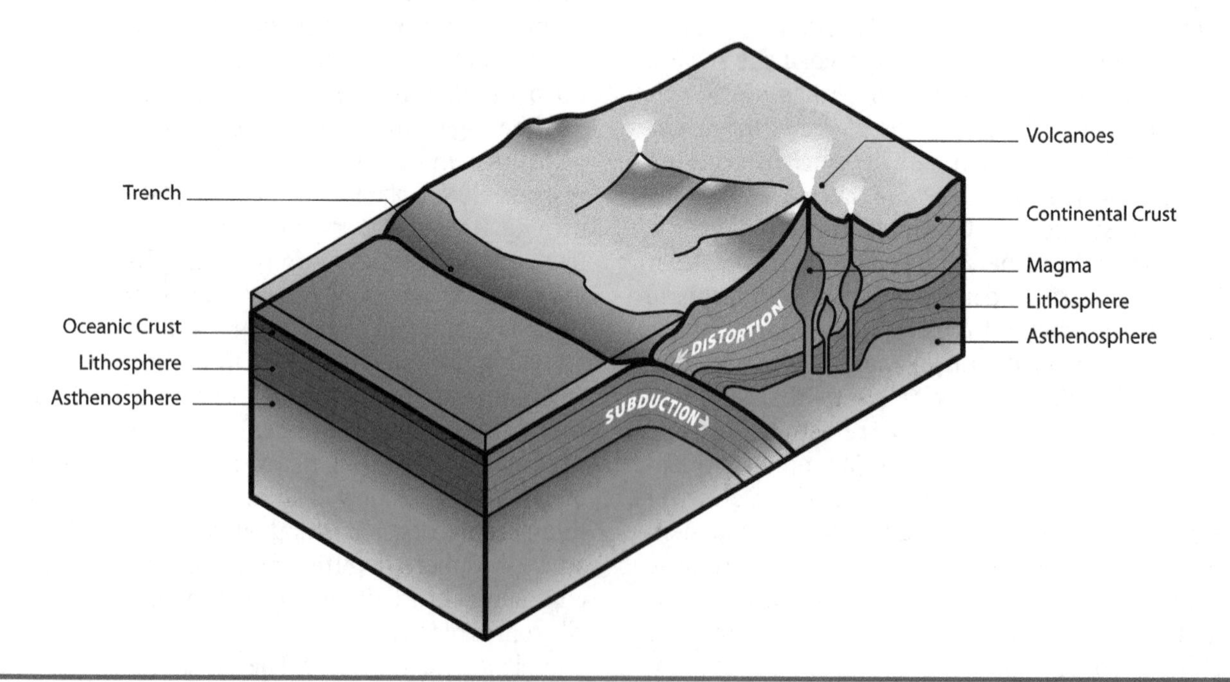

FIGURE 1e *Convergent plate boundary with plate subduction zone. Image © Zern Liew, 2014. Used under license from Shutterstock, Inc.*

such as the Himalayan Mountains (Figure 1f). The Himalayas are the result of the collision between India and Asia that began about 45 million years ago. As a result of the crustal thickening caused by the collision, The Himalayas are still rising at the rate of about 5 cm (2 inches) per year.

Because of Earth's spheroidal shape, other fractures called transform faults develop, allowing for lithospheric plates to move laterally along Earth's surface (Figure 1c (b)). Short segments of **transform fault boundaries** typically form perpendicular to the trend of the mid-oceanic ridges. However, extended over hundreds of kilometers, transform faults may also separate portions of the continental lithosphere, as seen along the San Andreas Fault in California (Figure 1g).

Rocks of oceanic and continental crust are covered with a veer of sedimentary and metamorphic rocks. Both types are formed from preexisting rocks. When igneous or any other rocks are exposed at Earth's surface, they are subject to processes of weathering that result in physical and chemical changes (disintegration and decomposition). Products of **physical weathering** are collectively called **sediments** or loose, solid particles of broken rocks that are usually moved by running water, glaciers, waves, wind and gravity to basins where the sediments are being deposited, commonly in layers. Such **sedimentary basins** or **environments of deposition** may be stream channels, floodplains, lakes, ponds, and swamps on land. However, the ultimate destination of nearly all sediments stripped from the continents and transported by agents mentioned previously is the ocean floor. **Chemical weathering** involves decomposition or chemical alteration of rocks by subtracting or adding various substances. For example, some soluble minerals dissolve within a solution and are transported in the dissolved physical state. Materials taken into solution may be precipitated out by evaporation or by organic secretion (by marine or fresh water organisms). Whether accumulated on land or in the ocean, each layer of sediment or precipitated minerals may be buried beneath the newly formed layers and undergo **lithification** (the transition from loose mass of sediments to consolidated rock layer) to form a **sedimentary rock.** Lithification involves **compaction** and **cementation.** Compaction is a mechanical process by which most of the open spaces between sediment particles initially filled with air or water are being eliminated because of the weight of overlying sediments. Cementation is the process by which sediments are converted into sedimentary rock by the precipitation of mineral cement (such as silica or calcite) between the sediment grains. As mineral cement holds sediment grains together, it becomes an integral part of the rock. Because the ocean basins are the major depositional environments for material derived from the continents, most of the sedimentary

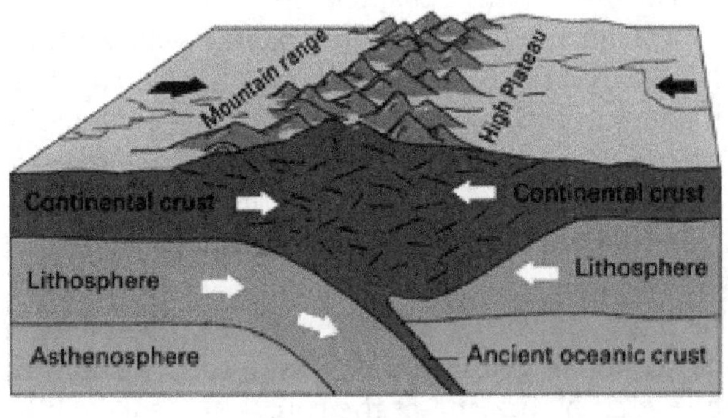

FIGURE 1f *Continental-continental collision. (http://pubs.usgs.gov/gip/dynamic/understanding.html).*

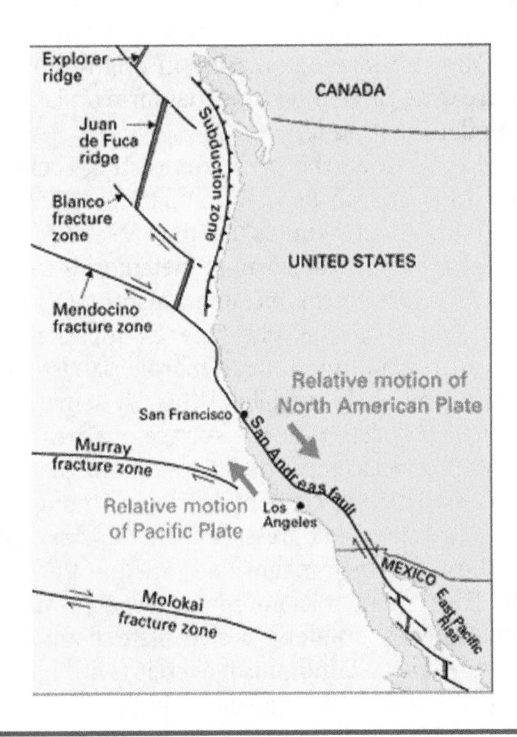

FIGURE 1g *The Blanco, Mendocino, Murray, and Molokai fracture zones are some of the many transform faults. The San Andreas fault, which is about 1,300 km long and in places tens of kilometers wide, separates Pacific Plate from the North American Plate. (http://pubs.usgs.gov/gip/dynamic/understanding.html)*

rocks are **marine** in origin. Sedimentary rocks made from material accumulated on land are called **continental rocks.** The sedimentary rocks may be uplifted from the bottom of the basins to higher elevations during mountain-building episodes common along converging plate boundaries or exposed by a change in sea level and thus being subjected again to the processes of weathering and erosion. The most abundant sedimentary rocks are **shale, sandstone,** and **limestone.** Shale is a fine-grained rock composed of mud that is a mix of flakes of clay minerals and tiny fragments of other minerals, especially quartz and calcite that accumulate on floodplains, wetlands, and on the ocean floor. Sandstone is composed mainly of sand-sized minerals or rock grains. Most sandstone is composed of quartz derived from previously existing rocks. Because it is resistant to chemical attack, quartz survives long after other minerals have decomposed. Limestone is composed mainly of interlocking crystals of calcite ($CaCO_3$) that have been precipitated by inorganic or organic means. Certain organisms, both animals and plants, remove $CaCO_3$ from water to make tests and shells. When organisms die, their hard remains accumulate at the bottom of the basins and undergo lithification to form limestone. Most limestone is marine in origin; however there are varieties of limestone that form on land and even under the ground.

Deep within Earth's crust, preexisting rocks are subject to alteration by high temperature, pressure and chemically active fluids that may not cause them to melt, but rather transform to **metamorphic rocks.** The transformation, or **metamorphism,** takes place while the rock is in the solid state. Examples are the transformation of limestone to **marble,** sandstone to **quartzite,** and shale to **slate.** Although some metamorphic rocks form near the surface of Earth, most metamorphic rocks form under conditions that exist only deep within the cores of developing mountains. High temperature and pressure conditions are similar to those in which magma is formed in the Earth crust. Thus many metamorphic rocks form near magma sources or along active plate boundaries. Often, extreme metamorphism overlaps with igneous activity. In such environments, highly metamorphosed and partially melted rock may form. These rocks exhibit characteristics of both igneous and metamorphic rocks and are called **migmatites.**

From the above discussion we may conclude that the main rock types are tied into the **rock cycle,** the dynamic transition that each type of rocks experiences or undergoes through geologic time to change from one type into another. Each type of rock is altered or destroyed when it is forced out of its equilibrium conditions, the place in the environment in which it was originally formed. An igneous rock, such as basalt, may break down and dissolve when exposed to the atmosphere or melt as it is subducted under a continent. Due to the driving forces of plate tectonics or interaction with atmosphere, hydrosphere and living matter, rocks do not remain in equilibrium and are forced to change as they encounter new environments.

Further chapters and practical laboratories presented in this textbook will introduce you to the basic concepts of geology in greater details. We will start our journey through planet Earth with the fundamental concept of plate tectonics that logically explains formation of magma, igneous activity, metamorphism, distribution of earthquakes and volcanoes and why most of Earth's major mountains are located where we find them today. We will spend a great deal of time learning about rocks. The rocks are assemblages of minerals, thus we will start our study with an introduction to minerals and their properties. Understanding chemical processes and conditions under which chemical elements bond together to form minerals will help us with understanding the origin of different types of rocks and processes contained in the rock cycle. We will discuss internal processes that are responsible for the formation of Earth's surface features and how the interaction with atmosphere, hydrosphere and biosphere is shaping them. Continuous changes occurring inside and at the surface of our planet will be discussed in the chronological sequences of geological events introducing students to the concept of **geologic time.**

The intent of this chapter was to introduce you to basic terminology and to illustrate how all geologic phenomena are related to each other. Please note that all topics in this textbook are tied together. All of the geologic features we will discuss in this course are related to real geographic localities; therefore students will work with maps and satellite images and will interpret photographs of real geologic features. The basic knowledge of the world, North American and regional geography is assumed. The author hopes that by the end of the course students will be able to relate geologic concepts to real places and find it fascinating to learn about physical environment they live in.

CHAPTER 1

Plate Tectonics

INTRODUCTION For each science, one concept can be identified that, more than any other, has changed how that science visualizes its domain. For geology, *plate tectonics* is that concept. Plate tectonics has been called the *"unifying principle of geology."* Its impact on geology has been comparable to the effects of Newton's laws of motion and universal gravitation on physics, Darwin's concept of organic evolution on biology, and Copernicus's view of the Solar System on astronomy.

Before the concept of plate tectonics was proposed in the mid-1960s, we could not logically (or correctly) explain the distribution of earthquakes and volcanoes, the source of energy for the creation of mountains, or why most of Earth's major mountains were located along the margins of continents while a few, such as the Urals of Russia, were not. The origin of the deep-sea trenches was a mystery as were the origins of the most extensive mountain ranges of Earth, the oceanic ridges. Although many competing theories attempted to explain the origin of the ocean basins, none of them took into account the fact that ocean basins are born, grow in size, and then close, all within a period of a few hundred million years.

Many of Earth's surface features, as well as its surface and internal processes, remained mysteries until the formulation of the concept of plate tectonics. Looking back now with the clarity and wisdom of hindsight, we may wonder why the concept of plate tectonics was not proposed earlier. Certainly, some of the evidence that supports it had been known for decades. The answer perhaps is that the concept required the combination of modern techniques that looked more critically at Earth and enabled geologists to reevaluate old data and scientific insight that recognized the single thread that tied the various kinds of data together.

HISTORICAL DEVELOPMENT

Ever since cartographers produced maps with a sufficient degree of accuracy to portray the shapes of the continents realistically, individuals have noticed the correspondence in the shapes of coastlines on opposite sides of the Atlantic Ocean, in particular, the similarity between the eastern coastline of South America and the western coastline of Africa (Figure 1.1). As early as 1620, *Francis Bacon* commented on the similarity of the Atlantic coastlines of South America and Africa. Almost certainly, at least a few of those who made the same observation must have wondered whether the similarity was just coincidental or meant that the two continents had once been joined like two pieces of a gigantic jigsaw puzzle. In 1668, *François Placet* proposed that the continents of North and South America were once joined to Europe and Africa by islands or by a continent that foundered. Based on the similarity of present-day animals and plants, the *Compte de Buffon,* a French naturalist, suggested in 1750 that North America and Europe were once joined. In 1858, *Antonio Snider-Pellegrini* commented on the similarity of the fossil plants associated with coal-rich rocks on opposite sides of the Atlantic. In the late 1800s, *Edward Suess* pointed out the nearly identical geological records of the continents in the Southern Hemisphere. Suess proposed that all of the southern continents were once joined together in a supercontinent he called **Gondwana** (Figure 1.2).

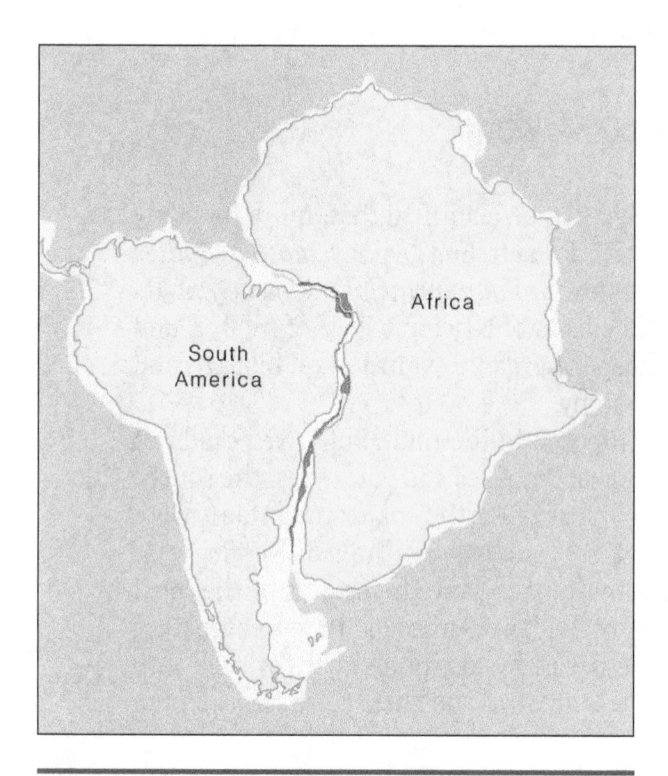

FIGURE 1.1 *The marked similarity between the Atlantic coastlines of South America and Africa led many to wonder whether the two continents had once been joined. (After A. G. Smith, "Continental Drift." In* Understanding the Earth, *edited by I. G. Gass. Courtesy of Artemis Press.)*

Although it was intriguing, few pursued Suess's suggestion that the continents had been joined. First of all, it went against all that had been taught about Earth. From the beginnings of the science of geology in the late eighteenth century, the size, shape, and location of the continents as we see them now were considered to have been established during the early days of Earth's creation when the brittle outer layer we call the crust was first formed. Secondly, if the continents were indeed once connected, some mechanism must have caused them to migrate to their present positions. What conceivable force could tear a continent apart and move it? As a result of these difficulties, the obvious similarity in the shapes of the Atlantic coastlines remained little more than an interesting and puzzling observation.

F. B. Taylor

It was inevitable that someone would take on the intellectual challenge of fitting and moving the continents. The first to suggest that the continents had moved over time was *F. B. Taylor* who, in 1908, argued that the location of young, folded mountains long the margins of certain continents was strong evidence that the continents had moved laterally. He theorized that the mountains had been created when the leading edge of the moving continent encountered the immobile rocks of the ocean basin in much the same way that the front end of a car crumples when it encounters an immovable wall. Taylor even suggested a possible source of energy to drive the movement, namely, strong lunar tides that existed during a theoretical period when the Moon's orbit was closer to Earth. His ideas were not widely accepted by the professional community.

H. B. Baker

The first geologist to actually suggest that the continents had once been joined together, had broken apart, and had moved with time was *H. B. Baker* in 1911. Impressed by the similarities in the shapes of the continental margins, Baker proposed the existence of a supercontinent made up of the joined masses of the present continents. In support, however, he offered little evidence other than the similarity in coastal shapes. He also failed to suggest a mechanism by which the proposed supercontinent fragmented and the resultant continents ultimately reached their present locations.

Alfred Wegener

Alfred Wegener, a German meteorologist, had also noted the similarities in the coastal outlines, especially the outlines of the opposing coastlines of South America and Africa. Picking up on Baker's idea, he proposed that all the continents were joined together in a supercontinent called **Pangaea** (Figure 1.2). Wegener, however, went further than any of his predecessors and presented evidence for the supercontinent. He showed not only that the continents fit together physically, but that mountain ranges, rock types, fossils, glacial deposits, directions of glacial striations, and other features from the eastern margin of South America matched those of the western margin of Africa (Figure 1.3). In short, he presented what would be accepted today as strong, if not irrefutable, evidence that the continents, now separated by an ocean, were once connected.

Wegener first published his hypothesis on continental drift in 1912. A few years later, while recuperating from wounds suffered during World War I, he wrote his famous work *The Origin of Continents and Oceans,* which was published in 1915. In this paper,

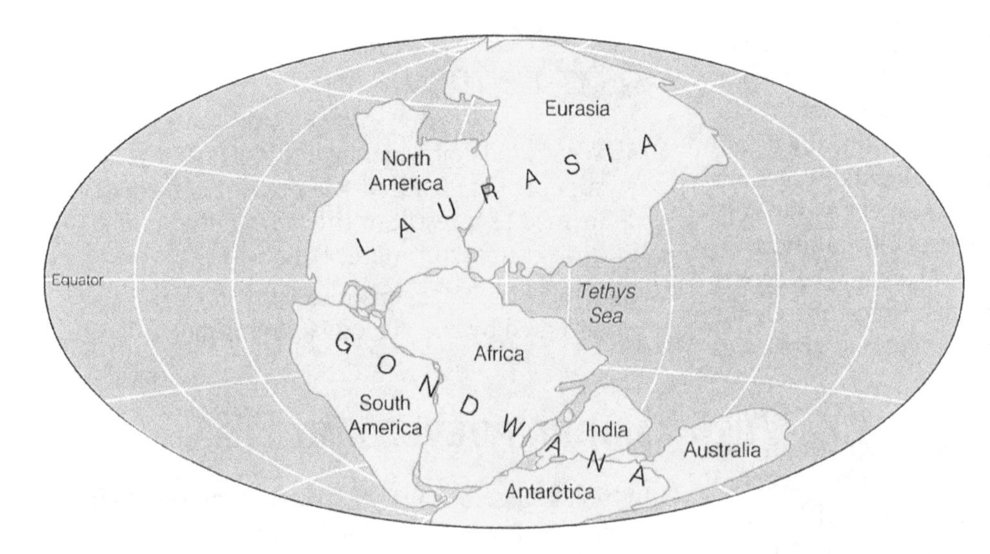

FIGURE 1.2 *Alfred Wegener proposed that all of the present continents were once joined into a supercontinent he called* Pangaea.

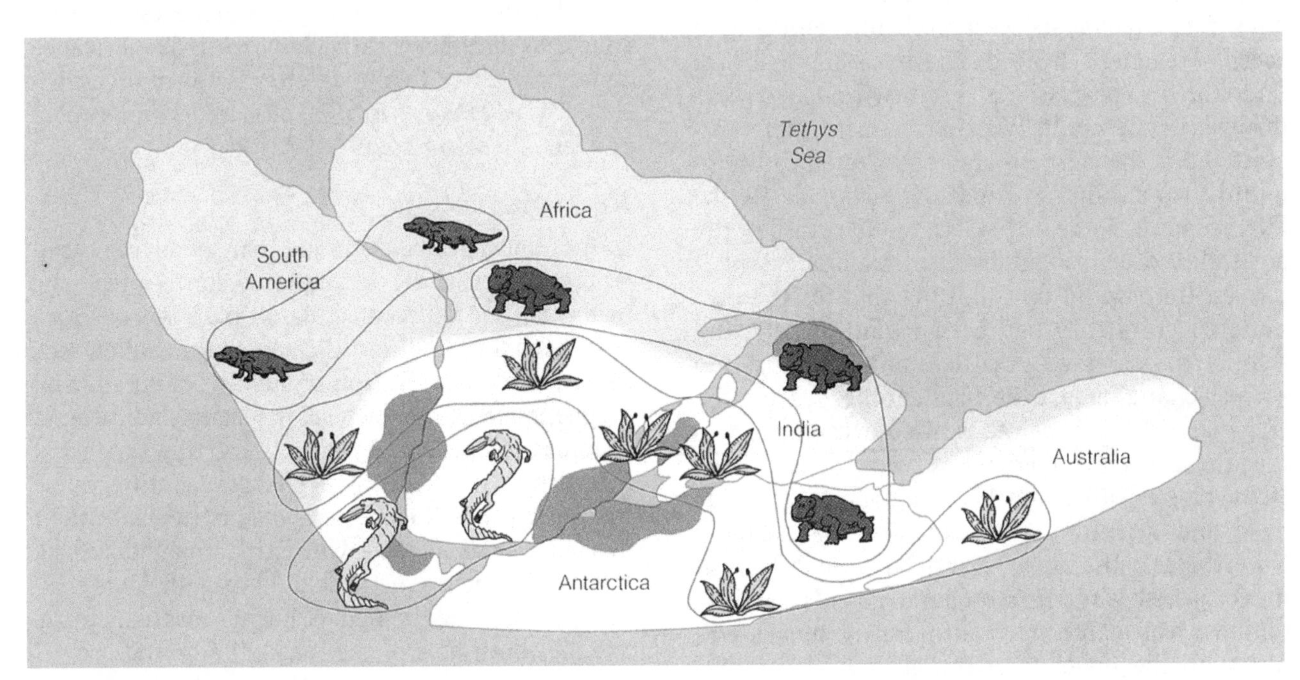

Wegener noted that fossils of *Mesosaurus* were found in Argentina and Africa but nowhere else in the world.

Fossil evidence tells us that *Cynognathus*, a Triassic reptile lived in Brazil and Africa.

Remains of *Lystrosaurus* were found in Africa, Antarctica, and India.

Fossil ferns, *Glossopteris* were found in all the southern land masses.

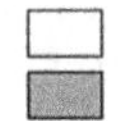

Distribution of late Paleozoic glaciers

Locations of major plateau basalts

FIGURE 1.3 *Various kinds of data exist to support the idea that the southern continents were once joined into a continent called* Gondwana, *some 200 million years ago.*

he postulated the existence of the supercontinent he called Pangaea, which based upon fossil evidence, began to break up about 200 million years ago. However, he could not identify a scientifically defensible mechanism to move the continents. Wegener's best attempt at explaining the movement was to suggest that the lower-density masses of granitic continental rocks somehow plowed through the denser basaltic rocks of the ocean basins. Unfortunately for Wegener, no deformation of the oceanic crust was ever found to support his proposal.

Like the proposals of his predecessors, Wegener's ideas found a mixed reception in the scientific community. Though widely accepted by South American and African geologists, they were ridiculed by most British and American geophysicists. Geologists were probably reluctant to accept Wegener's hypothesis for several reasons. Undoubtedly, professional jealousy and ego were involved. How could Wegener, a mere meteorologist, have the nerve to tell the geological community something so fundamental about Earth? Furthermore, if geologists were to consider the possibility of continental drift, let alone accept it as fact, they would be admitting that for the past hundred years they had been wrong in believing that Earth possessed a brittle, immobile crust and a never-changing face. Perhaps another reason why Wegener's thesis received limited acceptance was that most of his evidence was from the Southern Hemisphere while most geologists of the day lived and worked in the Northern Hemisphere. Nevertheless, the only scientifically sound argument against Wegener's proposal was his inability to find a reasonable mechanism for the movement. Although the majority of the geological community refused to accept Wegener's theory, it was not rejected by everyone. Like any radical new thesis, it found a few disciples willing to keep the idea alive.

Unfortunately, Wegener did not live to see his hypothesis vindicated. He died on a scientific expedition to the Greenland ice cap, presumably from a heart attack brought on by a strenuous march. From Wegener's death in 1930 until the late 1950s, the idea of continental drift remained in limbo. Except for geologists from the Southern Hemisphere and paleontologists (geologists who study fossils), relatively few geologists were willing to espouse the concept. Consequently, classroom instructors gave the topic little time or attention. All the while, the continents continued to drift.

SPOT REVIEW

1. What kinds of evidence did Wegener use to support his idea that the present continents formed as a result of the rifting of a much larger supercontinent, Pangaea?
2. Why were Wegener's ideas not universally accepted by the geologists of his day?

MODERN DEVELOPMENTS

In the 1950s, as bits of evidence appeared from various sources, continental drift was resurrected, and ultimately, Wegener and his predecessors were shown to have been basically correct. The evidence came from three totally different areas of investigation: (1) *rock magnetism,* (2) *ocean bottom topography,* and (3) *seismology.*

Rock Magnetism

In the 1950s, geologists found that solidifying basaltic lavas recorded the orientation of Earth's magnetic field when the lava cooled. Basalt is an igneous rock rich in magnesium-iron silicates. As the molten rock solidifies, some of the iron crystallizes in the mineral *magnetite* (Fe_3O_4), which, as its name indicates, is ferromagnetic (ferro = iron). As the tiny crystals of magnetite cool below a certain temperature called the **Curie point,** they become magnetized with their magnetic fields aligned with Earth's magnetic field. Not only do the magnetic fields of the individual crystals line up along the field of Earth's magnetic force, but the south pole of the magnetite crystals points toward Earth's north magnetic pole. As a result, the magnetite crystals "lock in" and preserve a record of Earth's magnetic field at the time of solidification. The orientation of the crystals' magnetic field acts both as a *compass* and a *dipmeter.* In a compass, the needle moves in a horizontal plane and determines the direction to the magnetic north pole (Figure 1.4). In a dipmeter, the needle moves in a vertical plane and determines the **magnetic inclination,** which is the vertical angle between the direction of the magnetic field and the horizontal (Figure 1.5). The magnetic inclination is used to calculate the latitude of the rock's location by the simple relationship:

$$\tan \lambda = 1/2 \tan \Theta$$

where λ = the latitude
Θ = the inclination

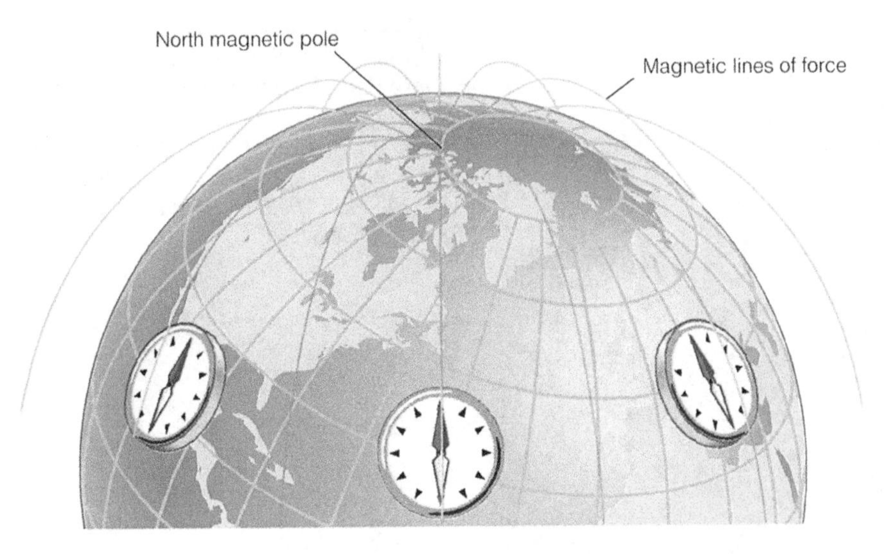

FIGURE 1.4 *The horizontally oriented needle of the* compass *aligns parallel to Earth's magnetic lines of force and thereby points toward the magnetic poles.*

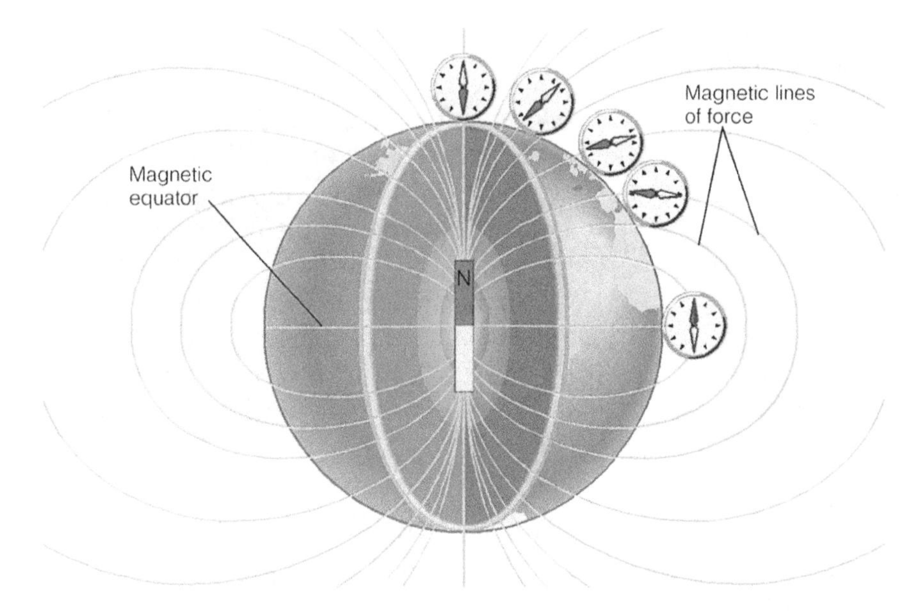

FIGURE 1.5 *The needle of the* dipmeter *also aligns parallel to Earth's magnetic lines of force, but in a* vertical *plane. Note that at the magnetic equator, the needle is parallel to Earth's surface (no dip angle). At locations closer to the magnetic poles, the angle between the needle and Earth's surface increases with increasing latitude until, over the magnetic pole, the needle is vertical (90° angle of dip).*

In this way, the magnetite crystals in the basalt record not only the orientation of the magnetic field but also the direction and distance to the magnetic poles. Both of these orientations can be determined in an oriented rock specimen with a sensitive instrument called a *magnetometer.* Because basaltic lavas are the most abundant kind of lava extruded onto Earth's surface, a large database of magnetic data can readily be amassed on any continent.

Magnetic Reversals

In establishing the magnetic orientation in successive lava flows (Figure 1.6), early paleomagnetists (workers who investigate rock magnetism) found that Earth's magnetic field had been **reversed** at times; that is, the north magnetic pole became the south magnetic pole and vice versa. The investigators found that the periods of reversed magnetism lasted a few hundred thousand years or more in duration and that the time interval between any two reversals was essentially random.

To this day, we do not understand the exact process by which the reversals take place. However, the fact that they *do* take place and are recorded simultaneously in the basaltic lavas around Earth was to play an important role in proving the existence of plate movement.

FIGURE 1.6 Stacked lava flows, *such as those that characterize the Columbia Plateau, record the orientation of Earth's magnetic field at the time of solidification and allow geologists to determine changes in the magnetic field over time. (Courtesy of P. Weis/USGS)*

Polar Wandering

Another important discovery made by researchers in rock magnetism was **"polar wandering."** It is generally accepted that Earth's magnetic field is the result of movements within Earth's molten iron core and that these movements are generated by Earth's rotation. We also know that the geographic locations of the present north and south magnetic poles are not constant. Although the magnetic poles do migrate over geologic time, they have always been located near the rotational poles. As far as we know, the magnetic reversals are not related to the movements within the liquid portion of the core thought to be responsible for Earth's magnetic field.

As worldwide paleomagnetic data began to accumulate, researchers began to plot the locations of Earth's north magnetic pole over time for the individual continents. The maps showed that the north magnetic pole for each continent had followed a *different* pathway (Figure 1.7). Note that if the continents had remained fixed in location while the magnetic poles wandered, each continent would have shown the *same* polar track. On the other hand, if the locations of the magnetic poles had remained relatively fixed over time, the different polar pathways shown by the data from the individual continents could only be due to the movement of the continents. The conclusion was that most of the observed polar wandering was the result of continental movements. The "wandering" of the magnetic poles was only an *apparent* movement; most of the movement was actually due to changes in the relative positions of the continents. Interest in continental drift was suddenly revived.

SPOT REVIEW

1. How do basaltic rocks record the magnetic field that exists at the time of their formation?
2. How can a compass and a dipmeter be used to locate the position of the north magnetic pole?
3. How can magnetic data from different continents indicating different tracks of polar movement be used as evidence that the continents moved and not the magnetic poles?

Topography of the Ocean Bottom

Beginning with military investigations of the 1940s and continuing throughout the 1950s and 1960s, studies of the ocean bottom began to reveal unexpected features. Although limited **bathymetry** (Greek *bathus* = deep + metry) (topography of the ocean bottom) of the Atlantic Ocean was known by the late 1800s, until the 1950s, many geologists had taught, for no good reason, that the ocean bottom was a broad featureless plain extending from continent to continent. With the exception of a few volcanic peaks that mysteriously rose from the ocean floor, the ocean bottom was viewed as being flat as a billiard table. That picture was soon to change. As a result of echo-sounding studies originally begun by the U.S. Navy during World War II, an entirely different picture of the ocean bottom evolved. The data showed that the ocean bottom was hardly a flat, featureless surface (Figure 1.8).

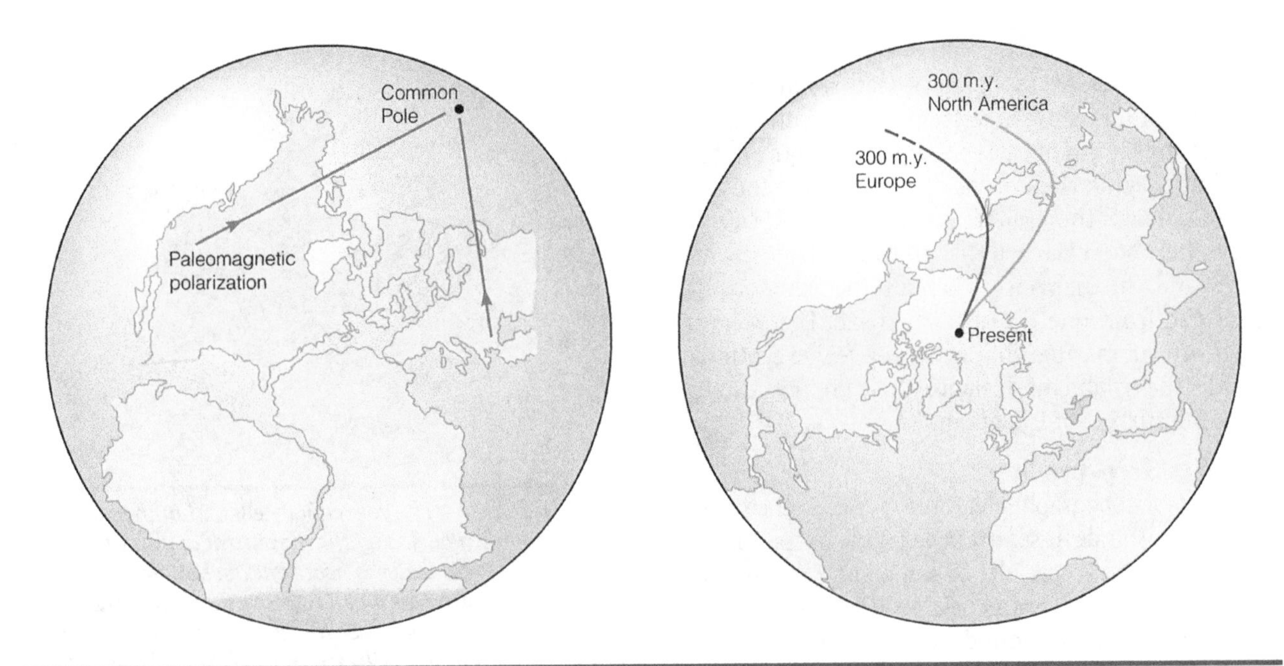

FIGURE 1.7 *Paleomagnetic data for individual continents indicate that the magnetic pole has had different locations over time. The patterns of apparent movement differ for individual continents. Possible explanations include the following: (1) the continents remained fixed in position as the magnetic poles moved, (2) the magnetic poles remained fixed in position as the continents moved, or (3) both the magnetic poles and the continents moved with time. Because the magnetic pole is believed to have always been located close to the geographic pole, the second option is the most likely.*

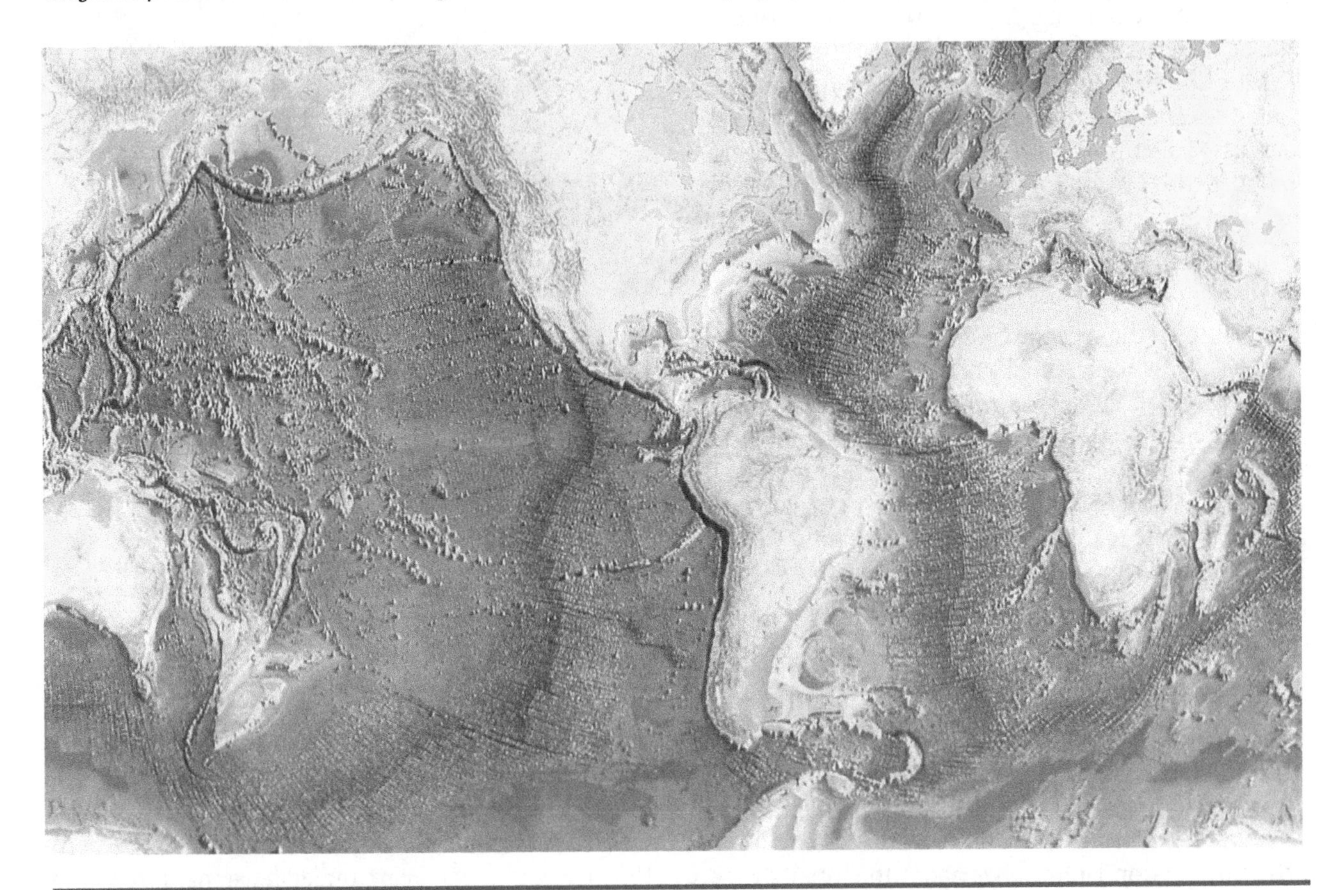

FIGURE 1.8 *Data largely amassed since the invention of* echo sounding *have revealed a varied landscape for the ocean bottom. (World Ocean Floor by Bruce C. Heezen and Marie Tharp, 1977 © Marie Tharp 1977. Reproduced by permission of Marie Tharp, 1 Washington Avenue, South Nyack, NY 10960.)*

Oceanic Ridges

The most impressive topographic features discovered by echo sounding were volcanic mountain ranges that ran the length of each ocean basin (Figure 1.18). The first **oceanic ridge** to be discovered was the one running down the middle of the Atlantic Ocean. Researchers soon learned that all oceans had oceanic ridges and, furthermore, that all the ridges were connected from one ocean to the next. The oceanic ridges are submarine; that is, with a few exceptions, notably the island of Iceland, they do not extend above the surface of the ocean.

Deep-Sea Trenches

Other major topographic features found on the ocean bottom are the **deep-sea trenches**, very deep, narrow troughs that trend parallel to some continental margins and chains of volcanic islands (Figure 1.8). Most of Earth's trenches are found around the margin of the Pacific Ocean. Invariably, the trenches are associated with a volcanic chain, either on the adjoining continental margin, such as the Andes Mountains along the western margin of South America or the Cascade Mountains of the Pacific Northwest, or just offshore of the continental margin, such as the Japanese Islands off the eastern margin of the Asian mainland (Figure 1.8). In the late 1950s and early 1960s, no one had an acceptable explanation for these trenches and their associated volcanism, at least not an explanation that met with the approval of the majority of the geological community.

Mantle Convection Cells

In 1960, *Harry H. Hess,* proposed a mechanism to explain both the oceanic ridges and the deep-sea trenches. Hess resurrected a hypothesis that had first been presented by *Arthur Holmes* in 1928. Holmes had proposed that convection cells in the mantle were the source of energy that drives the process of continental drift. According to Holmes, the convection cells in the mantle were driven by heat flow from Earth's interior much as currents are generated in a pot of boiling water (Figure 1.9). He postulated that as heat was conducted outward from Earth's center, convection cells caused a slow movement in the rocks within the mantle. He further postulated that if a convection cell became positioned beneath a continent, the lateral movement of the underlying mantle rocks would break the continent into fragments and drive them apart (Figure 1.10). Considering how little information Holmes had about the inner structure of Earth, his suggestion is amazingly close to the most popular mechanism that we now have.

FIGURE 1.9 *The* convection cells *within the mantle are commonly modeled by the movement of water in a heated pan. As the water is heated at the bottom of the pan, the water expands and its density decreases. The more buoyant hot water rises over the heat source and spreads out along the surface of the water where it cools. As the water cools and its density increases, the water sinks along the sides of the pan to complete the convection cell.*

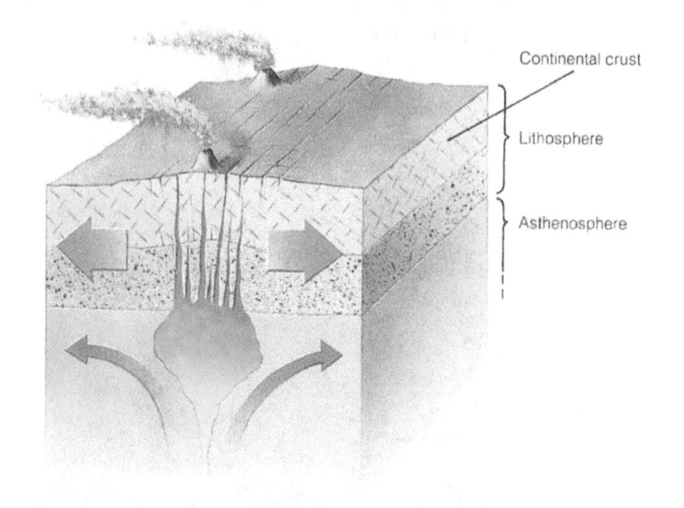

FIGURE 1.10 *Tensional forces that develop in the lithosphere above the rising portion of a mantle plume cause the continental lithosphere to fracture or rift. Molten rock forms at the top of the asthenosphere and moves upward along the fractures to produce the volcanism that characterizes a* rift zone.

Hess proposed that the oceanic ridges represented the rising portion of Holme's convection cell. He hypothesized that, as the mantle rock moved upward and outward beneath a continent, the continent would break apart under tensional forces and separate, just as Holmes had envisioned. As the two continental masses drifted apart, a rift zone would form and, in time, develop into a rift valley that would eventually open to the sea. The valley would then begin to flood, and as the continental masses continued to separate, a new ocean basin would be born (Figure 1.11). According to Hess's hypothesis, basaltic magmas welling up along fractures would continuously create new crust for the new, opening ocean basin. Once solidified, the newly formed oceanic crust moves laterally away from the oceanic ridge, and the continents continue to recede, a process called **sea-floor spreading.**

Hess further reasoned that with new oceanic rock continuously being created at the oceanic ridges, one of two events had to occur: (1) Earth had to expand to accommodate the volume of new oceanic crust, or (2) somewhere, an equal volume of old oceanic crust had to be consumed. He considered the idea of an expanding Earth, but dropped that possibility because there was no evidence to indicate that Earth had been much smaller in the past and had expanded to the extent necessary to accommodate all the oceanic crust generated over geologic time. The most

FIGURE 1.11 *An ocean basin begins with the formation of a continental rift zone and develops through rift valleys and linear oceans to the point where the landmasses finally separate completely to form the ocean basin.*

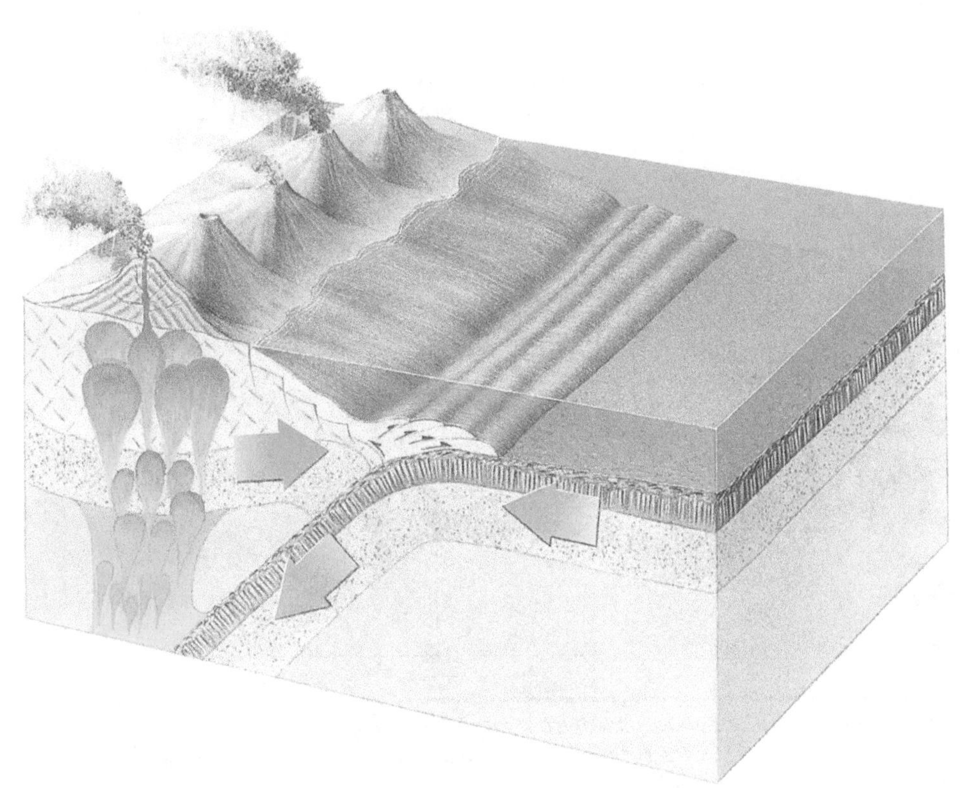

FIGURE 1.12 *A deep-sea trench forms at a convergent plate boundary where the accretionary wedge at the edge of the overriding plate encounters the downwarping oceanic plate.*

obvious candidate for a location where oceanic crust was likely to be consumed was the deep-sea trench. Hess proposed that the trenches are located over the downward-moving portion of a mantle convection cell (Figure 1.14). As the crust is compressed by the movement of the mantle rocks and breaks, the oceanic crust sinks into the mantle and is assimilated. Hess therefore hypothesized that a balance exists between the creation of new oceanic crust at the oceanic ridges and the destruction of old oceanic crust at the trenches. According to this idea, the continents are carried as passive passengers as the oceanic crust moves from the oceanic ridges to the trenches. Could this be a mechanism that Wegener had so desperately sought? We now know that it is.

SPOT REVIEW

Why must there be a balance between the volume of crustal rocks formed at the oceanic ridges and the volume of crustal rocks consumed at the trenches?

Magnetic Zonation of the Oceanic Crust Within a few years, other geologists published research that confirmed Hess's hypothesis. In 1963, *F. Vine* and *D. H. Mattews* published a paper describing their finding of symmetrical bands of varying width and magnetic intensity in the rocks of the ocean floor (Figure 1.13). *L. Morely* published a similar, but independently developed, explanation of the magnetic striping on the ocean floor. Although submitted for publication earlier than Vine and Matthew's paper, Morely's paper had at first been rejected as too speculative. Some now suggest that the concept should be referred to as the *Vine-Matthews-Morely* hypothesis.

In an attempt to explain what they had found, Vine and Matthews utilized Hess's idea of a spreading center along the oceanic ridge. Consider the drawing in Figure 1.14. Basaltic magma wells up through fractures parallel to the axis of the oceanic ridge. We have already discussed how the magnetite grains record the existing magnetic field of the Earth as the molten rock solidifies. Once formed, the rocks split and move laterally in opposite directions to make way for new basaltic magma. In time, an everwidening band of basalt forms, recording both the orientation of the existing magnetic field and the direction to the north magnetic pole.

Consider what would happen if a magnetic reversal were to occur. The lavas welling up along the oceanic ridge would record the magnetic reversal by reversing the poles of the magnetite grains. In other words, the magnetic fields of the magnetite grains would flip end for end. As long as the magnetic reversal remained in effect, newly formed basaltic rocks would continue to record the magnetic field as a new band parallel to the previously generated band. Subsequent reversals would be recorded in similar fashion. The result would be a set of symmetrical bands with the *width* of each band reflecting the *duration* of the respective reversal episode and the *spreading rate* of the oceanic crust.

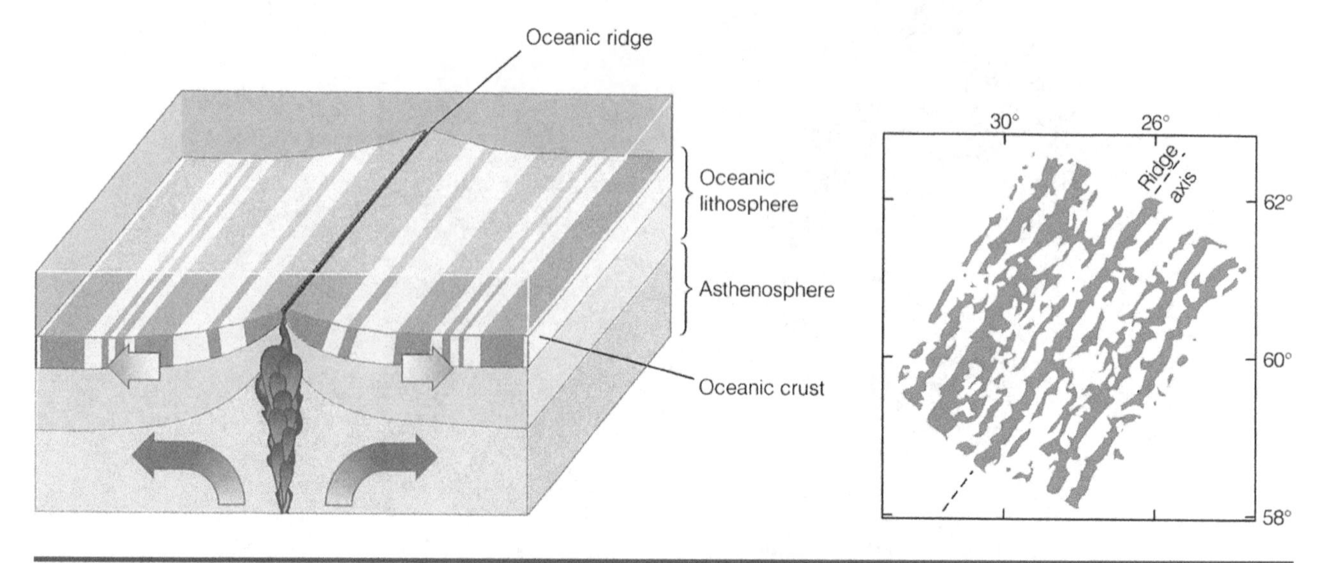

FIGURE 1.13 *As newly formed oceanic lithosphere moves away from an oceanic ridge in opposite directions, magnetic reversals are recorded in the oceanic basalts in the form of symmetrically disposed pairs of stripes of varying magnetic intensity oriented parallel to the trend of the ridge.*

If magnetic measurements were made in a direction *perpendicular* to the axis of the oceanic ridge, they would show a variation in the **magnetic intensity** of the rocks. The magnetic intensity of a rock depends primarily on the strength of the original magnetic field and the magnetite content of the rock. The maximum magnetic intensity would be recorded for rocks where the popular orientation when the lava cooled was the *same* as the polar orientation at the time of measurement, that is, when the directions of the north magnetic pole were the same. The minimum magnetic intensity would occur when, because of magnetic reversals, polar orientations with the rocks were *opposite* to the polarity of the magnetic field at the time of the measurement (Figure 1.15). Note that the patterns of the magnetic bands on opposite

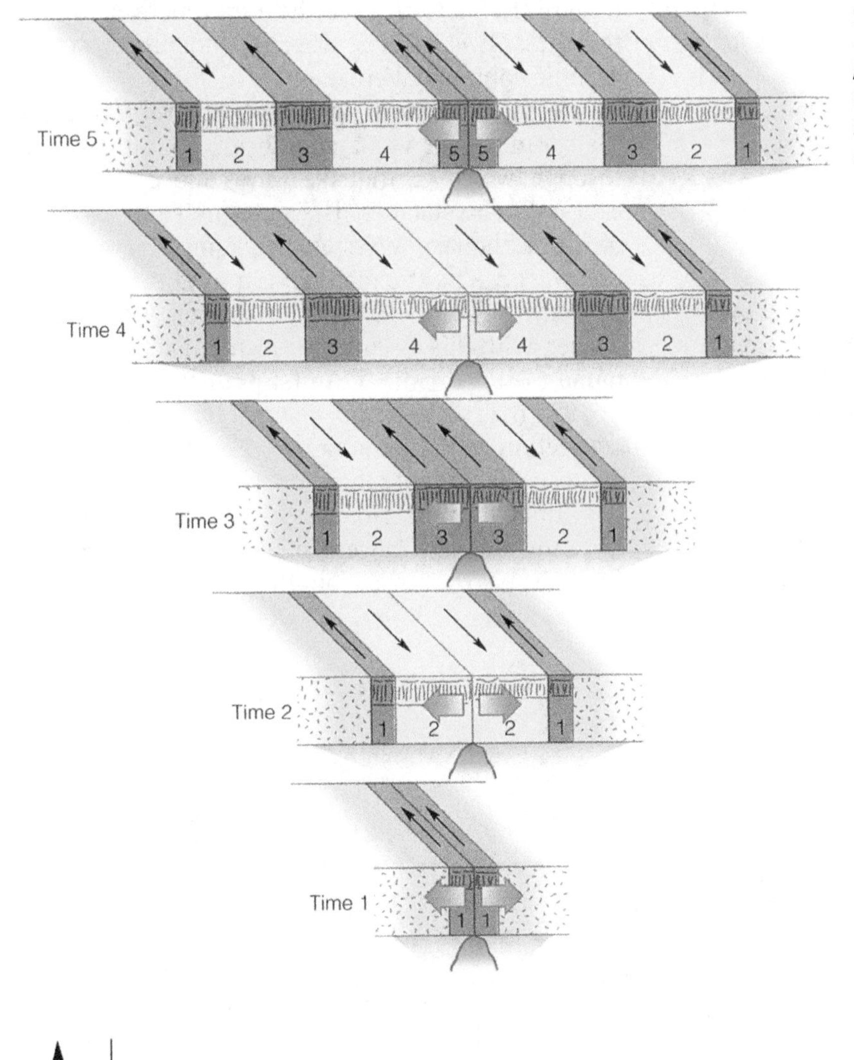

FIGURE 1.14 *The width of each pair of stripes depends on the length of time from one magnetic reversal to the next and on the rate of formation of new oceanic lithosphere.*

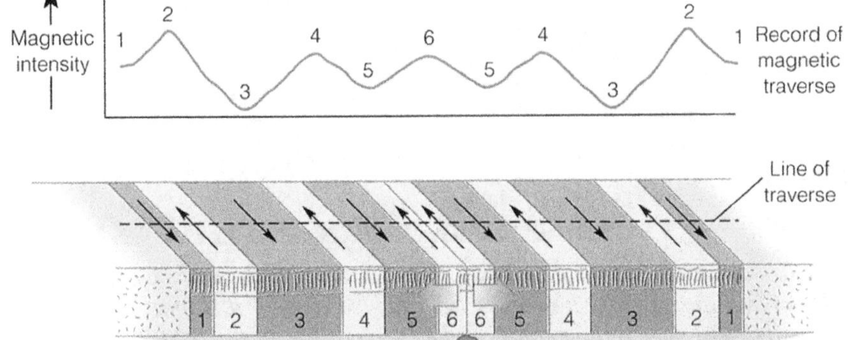

FIGURE 1.15 *A traverse perpendicular to the trend of the oceanic ridge shows a systematic variation in the magnetic intensity of the magnetic stripes.*

sides of the oceanic ridge would be symmetrical. After the presentation of the Vine and Matthews paper, researchers found that the magnetic bands associated with the Atlantic Oceanic Ridge showed the predicted symmetrical arrangement, a discovery that was strong evidence in support of continental drift.

Soon researchers showed that the patterns of magnetic intensity in bands from different ocean basins were similar; the bands differed only in width, reflecting different rates of sea-floor spreading. Analysis of the bands showed that the Atlantic and Indian oceans are spreading at rates of 0.5 to 1.5 inches (1–4 cm) per year while the Pacific Ocean is spreading at about 4 inches (10 cm) per year.

SPOT REVIEW

Why do the rocks of the oceanic crust show parallel bands of reversed magnetism? Why do the widths of the bands differ, and why are they symmetrical on opposite sides of an oceanic ridge?

Seismic Investigation of Earth's Interior

With the magnetic data clearly indicating that the oceanic crust is moving at a measurable rate, geologists began to look for evidence to support the convection cell theory. In the 1960s, seismic (earthquake) data began to provide a new picture of Earth's internal structure that was somewhat more complex than the older core-mantle-crust model. Specifically, seismologists discovered that the combined oceanic and continental crust and the outermost portion of the mantle acted together as a single brittle layer they called the **lithosphere** (Figure 1.16). Note that the lithosphere varies in thickness. It is thickest where it contains the thick granitic continental crust.

Perhaps the most significant discovery made by the seismologists was a zone below the lithosphere through which earthquake waves move at significantly slower velocities. They interpreted the slowing of the earthquake waves as indicating that the layer was *plastic;* a plastic material is a solid that deforms by *flowing* like the ice at the bottom of a moving glacier. With this discovery, evidence had finally been found to support the idea first presented by Holmes in 1928 that rocks within the mantle could move as convection cells. The plastic layer within the mantle is called the **asthenosphere** (Figure 1.16.)

Although most geologists agree that the driving force for the plate movements is mantle convection, there is no consensus on the convection mecha-

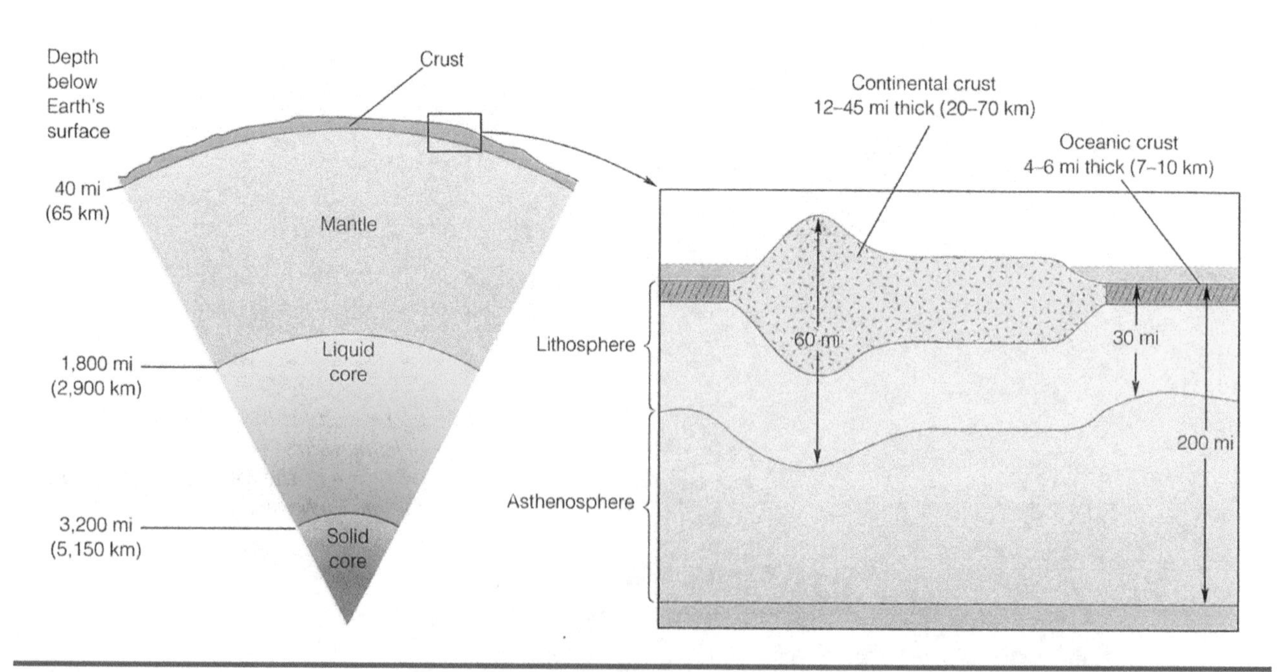

FIGURE 1.16 *Earthquake data have shown that the outer portion of Earth is divided into a brittle layer called the* lithosphere, *made up of the crust and the outermost part of the mantle, and an underlying layer called the* asthenosphere *within which the rocks show liquidlike properties.*

nism. One model suggests that the convection cells are located within the asthenosphere (Figure 1.17). According to this model, the rising portion of a convection cell induces tensional forces in the overlying lithosphere. The tensional forces cause the lithosphere to break along fractures that eventually extend to the surface, producing **rift zones** and **rift valleys** that may eventually open to form ocean basins (Figure 1.18). Molten rock from the asthenosphere fills the fractures, solidifies, and adds to the lithosphere, which moves off in opposite directions from the zone of rifting. According to this model, the major force causing the lateral motion of the lithosphere away from the zone of rifting, called **mantle drag,** is generated by the movement of the asthenospheric rocks beneath the lithosphere. With time, the rocks of the asthenosphere cool, increase in density, and descend to form the downward-moving portion of the convection cell. At this point, forces generated by the downward-moving portion of the convection cell break the overlying lithosphere parallel to the continental margin. The portion of the lithosphere bearing the higher-density oceanic plate is driven downward beneath the edge of the lighter continent by a combination of compressional forces and descending mantle drag (Figure 1.19). The downward-moving portion of the convection cell drags the edge of the lithosphere into the asthenosphere where it is assimilated. The location of the downward-moving lithosphere is marked on the ocean floor by the presence of a deep-sea trench. The volume of lithosphere converted to asthenosphere at the descending portion of the convection cell compensates for the volume of asthenosphere converted to lithosphere at the rising portion of the convection cell. In this way, constant volumes of asthenospheric and lithospheric rocks are maintained. To summarize, in this model, the lithosphere is passively carried or dragged along by the convective movement of the underlying asthenosphere with the major force being supplied by mantle drag.

Others disagree with this model, arguing that the degree of cohesion between the asthenosphere and the lithosphere is insufficient to drag the lithosphere laterally. These geologists maintain that the lithosphere is actually the uppermost, cooling portion of the convection cell and that the convection cells are not limited to the asthenosphere but involve a large portion of the mantle (Figure 1.20). According to this scenario, as the heated mantle rocks rise, they undergo changes in both composition and physical properties as the conditions of heat and pressure change. Beneath rift zones, rift valleys, and oceanic spreading centers, the dense rocks of the asthenosphere melt to form the lower-density lithospheric rocks. As the new lithospheric rocks form and move laterally, tensional forces are generated that cause fractures to develop within the newly formed lithosphere. Molten rock moving up from below continuously fills the fractures and adds to the lithosphere. The elevated mass of the new lithospheric rocks formed at the rift zone, rift valley, or oceanic ridge exerts a gravitational force, called **ridge push,** on the older, adjacent lithosphere that causes the lithosphere to move laterally over the underlying plastic asthenosphere. In this model, mantle drag resulting from the movement of the underlying asthenospheric rocks may aid in the lateral movement of the lithosphere, but this force is not thought to be a major source of energy for plate movement.

The buoyancy of the hot, lower-density rocks rising beneath the oceanic ridge lifts the ridge higher

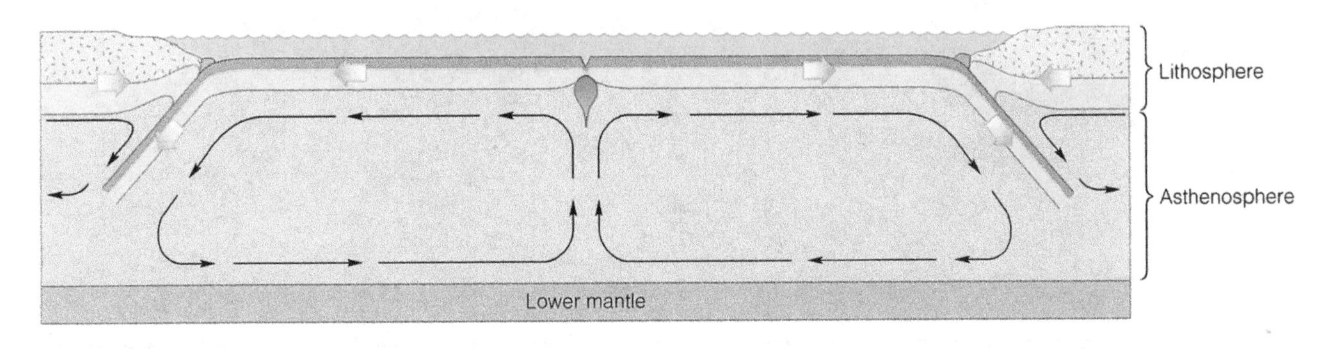

FIGURE 1.17 *Some geologists believe that the convection cells associated with plate tectonics are located within the asthenosphere.*

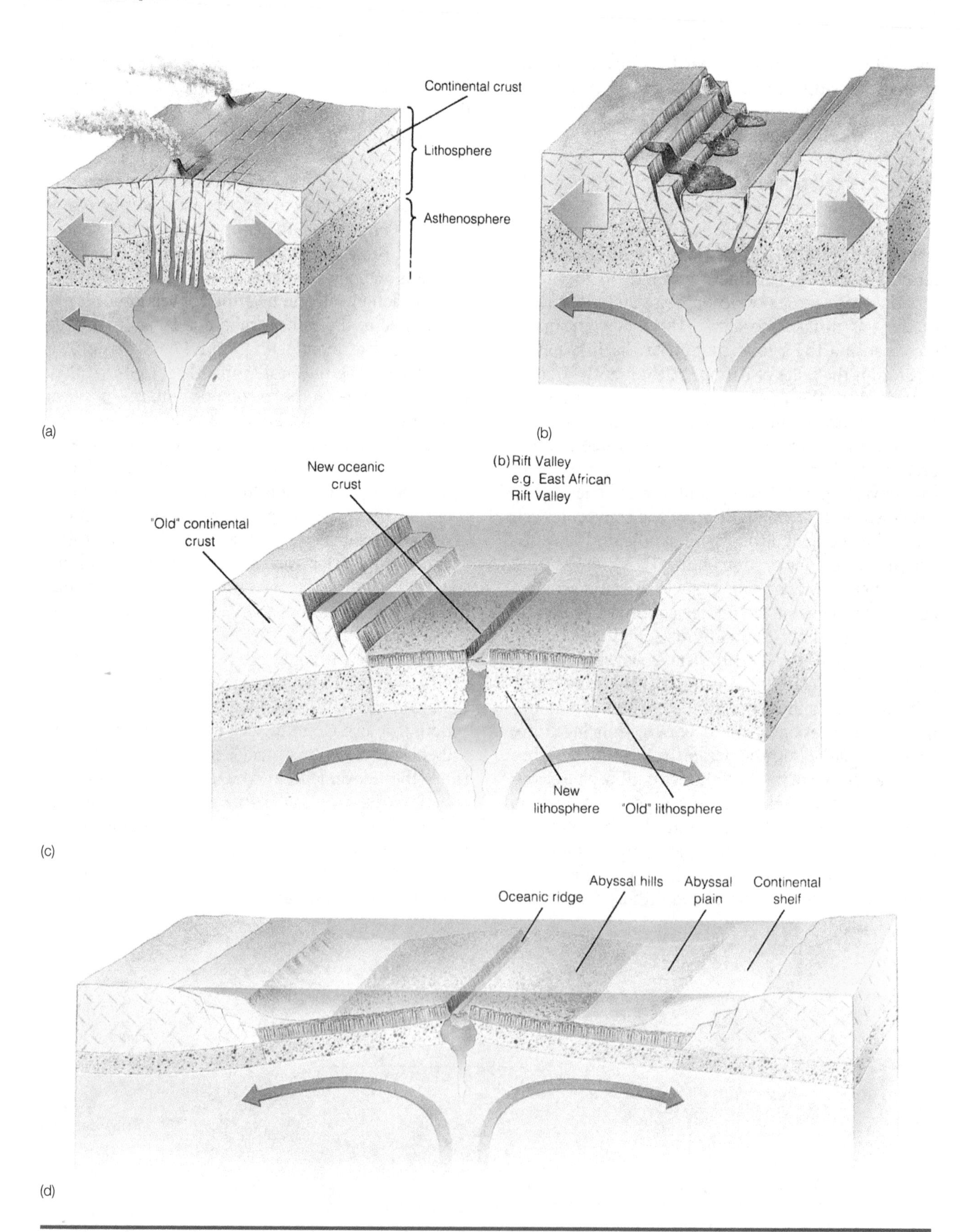

FIGURE 1.18 *Some geologists believe that the movement of the asthenospheric rocks* away *from the top of the* rising *portion of the mantle convection cell generating the tensional forces within the overlying lithosphere that cause it to break into plates, which then move away from each other.*

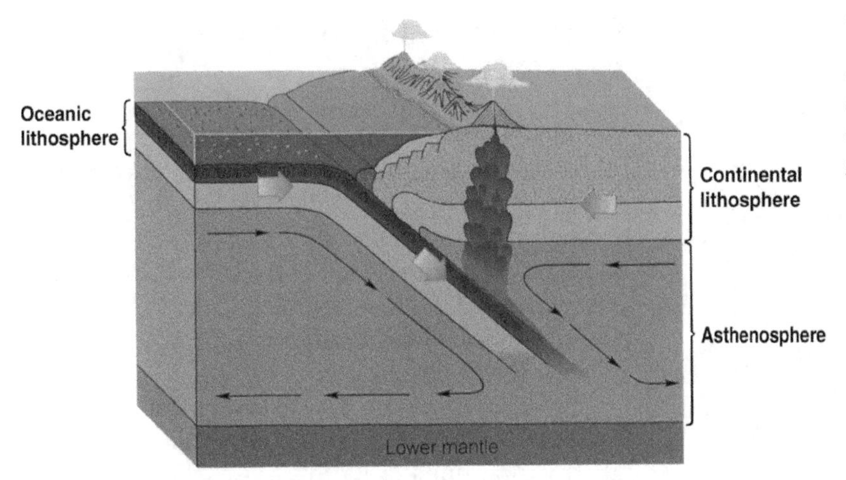

FIGURE 1.19 *The forces generated at convergent plate margins that result in the formation of a zone of subduction are associated with the* descending *portion of the mantle convection cell.*

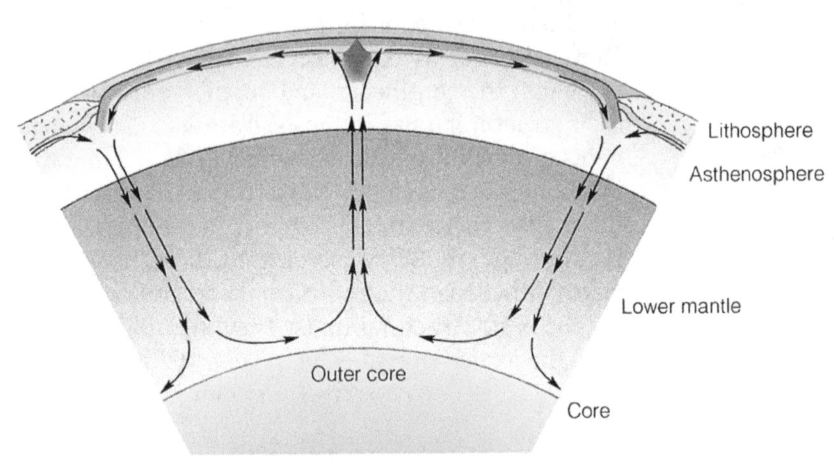

FIGURE 1.20 *Some geologists believe that the convection cells associated with plate tectonics are not restricted to the asthenosphere but rather involve the entire mantle. In this model, the lithosphere is the uppermost portion of the cell. Rather than simply responding to the movement of underlying rocks, the lateral movement of the lithosphere is the result of a combination of forces, which are produced as buoyant, molten rock is injected at the oceanic ridge and the cooled lithosphere sinks at the zone of subduction.*

than the adjacent cooler, denser oceanic lithosphere (Figure 1.11). Because of this difference in elevation, some geologists have suggested that both the lateral motion of the lithosphere and ridge push may result, at least in part, from *gravitational sliding* of the more rigid lithosphere down and away from the ridge over the more plastic asthenosphere.

Eventually, as the spreading oceanic lithosphere cools and increases in density, it breaks and begins to sink, generating a force called **slab pull** that draws the lithosphere downward into the mantle. Proponents of this model consider slab pull the most important of the three sources of energy involved in the movement of the lithosphere and regard mantle drag and ridge push as less significant. A major difference between this model and the model discussed previously is that here the lithosphere is part of the convection cell and both ridge push and slab pull *contribute* to the convective option, whereas in the first model the convection cell is restricted to the asthenosphere and mantle drag is the primary force.

The most recent data generated by three-dimensional seismic **tomography**, a seismic technique comparable to three-dimensional X-ray scans (CAT scans) of the human body, suggest that movements within the mantle are far more complex than any of the existing models portray (Figure 1.21). We have known about plate tectonics for fewer than 25 years, and it is apparent that much remains to be learned about Earth's interior.

SPOT REVIEW

1. How is convection within the mantle thought to be involved in the movement of the lithospheric plates?
2. What are the various mechanisms that have been proposed to explain how mantle convection results in the movement of the plates?

FIGURE 1.21 *The most recent information produced by a three-dimensional seismic technique called* tomography *reveals that Earth's interior is far more complex than the simplistic crust-mantle-core model would indicate. Such data will play an increasingly important role as our understanding of the driving forces behind plate tectonics evolves over coming decades. (Andrea Morelli et al. 1986, Geophys. Res. Lett. 13, 1545–1548).*

THE PRINCIPLE OF PLATE TECTONICS

A comprehensive principle to explain all that had been observed was the result of the combined efforts of four geologists, *D. P. McKenzie* and *R. L. Parker* of Cambridge University, *W. J. Morgan* of Princeton University, and *X. LePichon* of the Lamont Geological Observatory. Their theory, which became known as **plate tectonics,** views Earth's surface in a new and different way.

Earlier workers such as Wegener had regarded the continents and the ocean basins as the fundamental structural units of Earth's crust. They attempted to explain how the continents could first break up and then move about in this "sea" of oceanic basalt. McKenzie and his co-workers considered the possibility that the continental and oceanic crusts are not basic structural units, but instead are actually components of the lithosphere. They then proposed that the lithosphere is broken into about a dozen pieces, which they called **plates,** that fit together like a huge jigsaw puzzle (Figure 1.22). According to their theory, the plates rather than the continental and oceanic crusts are the basic structural units. They theorized further that the plates are moving relative to each other (Figure 1.23). As a result of this movement, certain

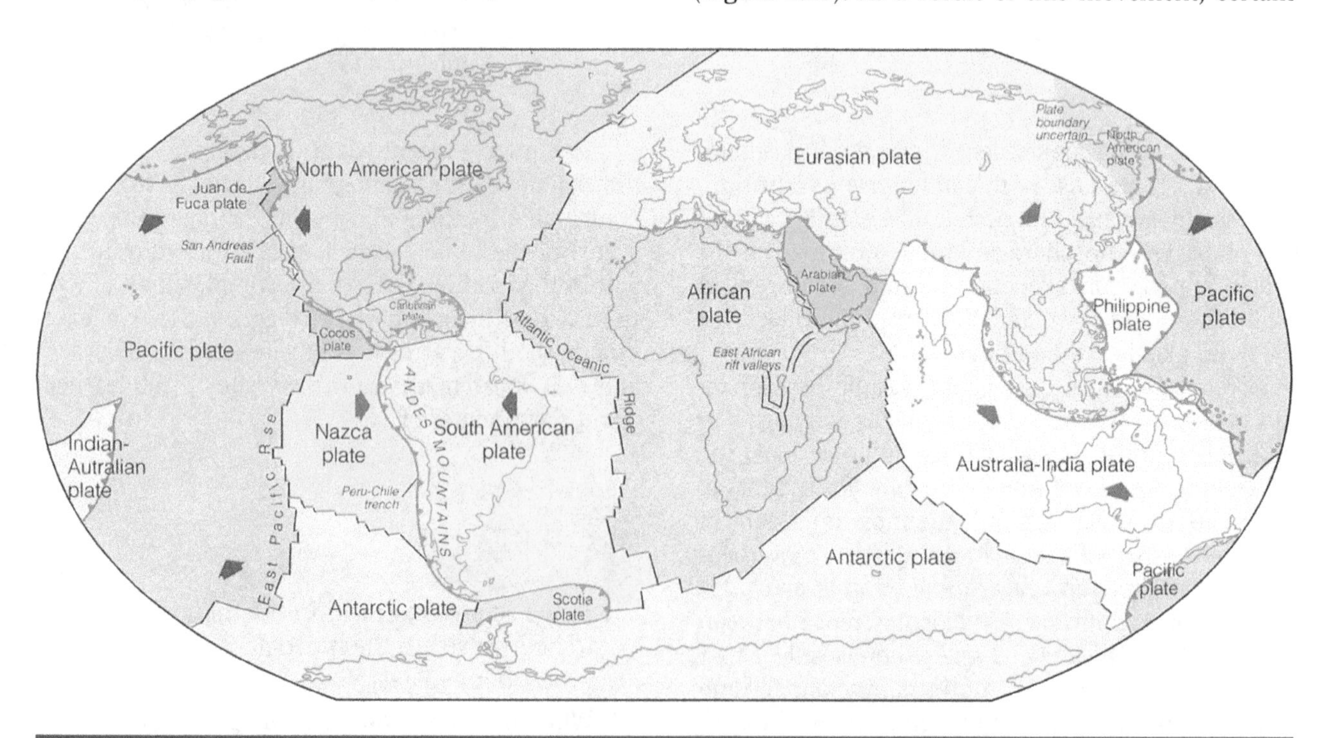

FIGURE 1.22 *The lithosphere is broken into about a dozen plates. The number of plates changes with time as large plates break apart and smaller plates collide and weld together. Sawtooth lines indicate the convergent boundary; the double color lines indicate divergent boundaries; and transform boundaries are shown with single color lines.*

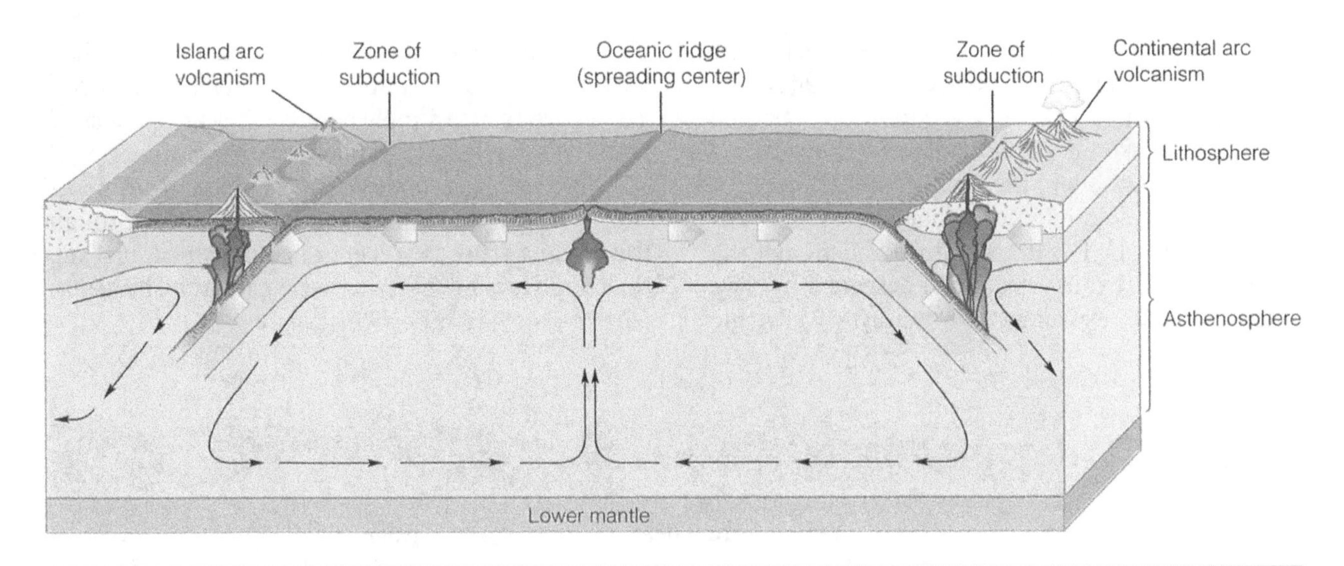

FIGURE 1.23 *We now know that oceans are born, open, and close, and that the continents move as continental crust is carried as the passive component of a lithospheric plate. Although most geologists agree that the driving force behind plate tectonics is mantle convection, the depths from which the convection cells rise is still a matter of debate.*

plates are moving away from each other. The margins of these plates, called **divergent margins** or **spreading centers,** are the sites of the rift zones, rift valleys, and the oceanic ridges where new oceanic lithosphere is being formed. At other margins, called **convergent margins** or **zones of subduction,** the plates are moving toward each other. Along these margins, the denser oceanic plate is driven or pulled downward beneath the lower-density continental plate or the edge of the opposing oceanic plate. These plate margins are the locations of the deep-sea trenches. Volcanic mountains located either on land, such as the Andes and the Cascades, or just seaward of the continental margins, such as the Japanese Islands and the Aleutians, mark the convergent boundaries.

In order to allow the plates to move along the surface of a sphere, other fractures called **transform faults** develop perpendicular to the linear trend of the oceanic ridges (Figure 1.24). The plates move laterally along the transform faults. No volcanism is associated with the transform boundaries, but later on we will see that the transform faults are the locale of shallow earthquakes. When we study earthquakes we will see that, with few exceptions, the locations of most of Earth's earthquakes coincide with the plate margins and that the most powerful earthquakes are concentrated along the zonesof subduction.

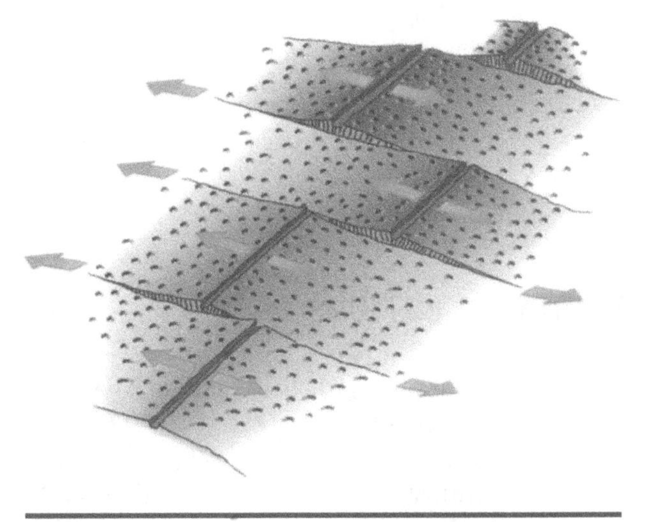

FIGURE 1.24 *Fractures called* transform faults, *which develop perpendicular to the trends of the oceanic ridges, allow the plates to move across the spherical surface of Earth.*

The introduction of plate tectonics in the late 1960s revolutionized the way we view Earth. Many geologic processes that had defied explanation now became understandable. For many years, the coincidence in the distribution of volcanoes and earthquakes had given rise to many a heated discussion over which was the cause and which was the effect.

Some had maintained that "quite obviously" the direct association of these phenomena indicated that either the onset of a volcanic eruption was initiated by an earthquake or that volcanic eruptions caused earthquakes. Although the rise of molten rock to Earth's surface does initiate earthquakes, such earthquakes are usually low in magnitude and are not the earthquakes that cause death and destruction. Plate tectonics finally explained why these two geologic phenomena are so intimately associated. Rather than a cause and effect relationship, earthquakes and volcanoes are related simply because they are associated with the same feature, namely, plate margins.

Now we also understand how the continents "*wander.*" We know that they do not plow through the oceanic crust as Wegener had imagined. Instead, the continents are passive passengers, transported as part of a moving lithospheric plate.

CONCEPTS AND TERMS TO REMEMBER

Gondwana
Pangaea
Rock magnetism
- Curie point
- magnetic inclination
- magnetic reversals
- polar wandering

Topography of the ocean floor
- bathymetry
- oceanic ridges
- deep-sea trenches
- sea-floor spreading

Magnetism of the oceanic crust
- magnetic intensity

Seismological data
- lithosphere
- asthenosphere
- rift zone
- rift valley
- driving mechanisms
  - mantle drag
  - ridge push
  - slab pull
  - tomography

Principle of plate tectonics
- plates
- divergent plate margins
- spreading centers
- convergent plate margins
- zones of subduction
- transform faults

REVIEW QUESTIONS

1. Ocean basins open at the rate of a few centimeters per
   a. year.
   b. century.
   c. 1,000 years.
   d. 1 million years.
2. Which of the following Earth features is an "ocean in the making"?
   a. the Grand Canyon
   b. the East African Rift Valley
   c. the Mediterranean Sea
   d. the Great Lakes
3. Which of the following landmasses is an exposed portion of a mid-oceanic ridge?
   a. Hawaii
   b. the Aleutian Islands
   c. Iceland
   d. Greenland
4. Oceanic trenches are associated with
   a. convergent plate margins.
   b. divergent plate margins.
   c. transform plate margins.
   d. both convergent and divergent plate margins.
5. The convection cells that drive the plates are thought to be located within the
   a. lithosphere.
   b. continental crust.
   c. oceanic crust.
   d. asthenosphere.
6. The most distinctive characteristic of the asthenosphere is its
   a. thickness.
   b. composition.
   c. lack of rigidity.
   d. content of molten rock.

REVIEW QUESTIONS *continued*

7. Why are basaltic rocks primarily involved in recording Earth's magnetic field?
8. How are the magnetic data recorded in the rocks of the ocean floor used to determine the rates of plate movement?
9. In what way is the asthenosphere involved in the movement of the plates?
10. What is the basic difference between the plate boundaries of the eastern and western Pacific Ocean basin?
11. How do you explain that while continental rocks more than 3 billion years old have been found, no oceanic rocks older than about 250 million years have been found?
12. Explain why basaltic rocks of the same age on different continents indicate different locations for the magnetic poles.
13. How can rocks of the asthenosphere constantly be converted into lithospheric rocks without depleting the asthenosphere?

THOUGHT PROBLEMS

1. Much of the data used to prove that the continents move relative to each other came from studies of the Earth's magnetic field, in particular, the phenomenon known as "polar wandering." A hypothetical polar wandering diagram is shown in Figure 1.25. The drawing illustrates the "apparent" wandering of the north magnetic pole based on magnetic data locked into the rocks of the continent at 10 different times (t_1 through t_{10}). How are magnetic data used to determine the direction and distance to the north magnetic pole at any time t? The drawing assumes a stationary continent and a moving pole. Prepare a second drawing to illustrate how the same magnetic data can be used to prove continental wandering. (*Hint:* Assume a stationary pole and a moving continent. Note that in the original drawing, the direction to

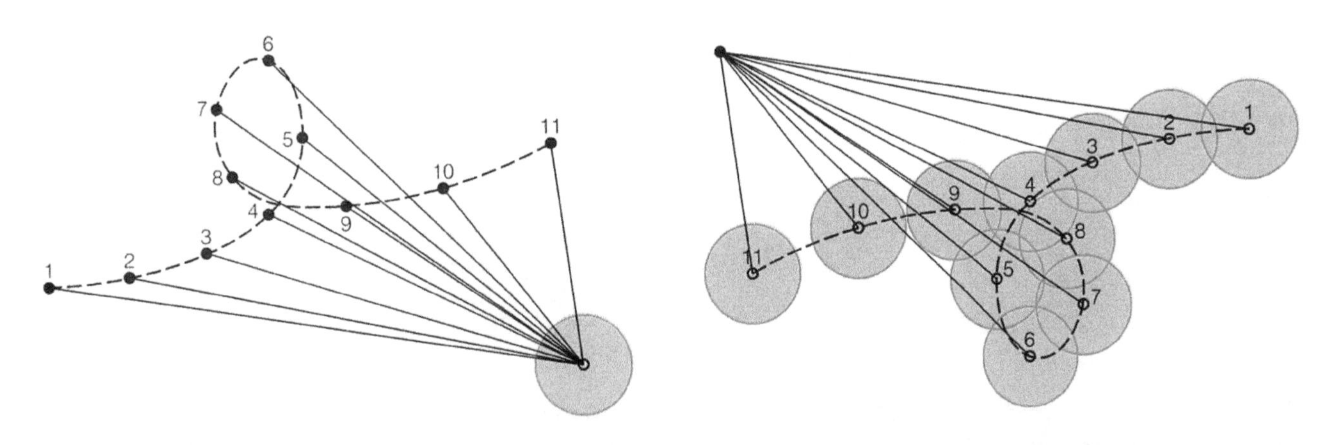

FIGURE 1.25 *Magnetic data interpreted on (a) fixed continent—moving magnetic pole, and (b) fixed magnetic pole, moving continent.*

Note: The lines between the continent and the magnetic pole are parallel and are of equal length but the relative positions of pole and continent for each time are opposite.

THOUGHT PROBLEMS *continued*

the north magnetic pole and the distance between the continent and the pole at any time *t* were determined by the orientation of the Earth's magnetic field at that time.)

When you have completed your solution to the problem, compare the pattern of "polar wandering" in the original drawing to the pattern of "continental wandering" in your drawing. How do they compare?

2. Assuming that the rifting of Pangaea that produced the North Atlantic Ocean began approximately 200 million years ago, what is the approximate rate of opening of the North Atlantic Ocean in centimeters per year?
3. If the Earth's major mountain ranges are the result of the collision of two continents, such as the recent (45 million years ago) collision of India and Asia that produced the Himalayas, what two continents are most likely to be involved in the next collision? Following that collision, what two continents do you predict will be next to collide and where will the mountains rise?

FOR YOUR NOTEBOOK

The news media increasingly are using illustrations of plate movements to explain earthquakes and volcanic eruptions to their readers and viewers. A perusal of recent newspapers and weekly news magazines in your library will provide you with excellent examples of plate tectonics in action. Including the stories or summaries of the stories will help you understand more fully the profound effect that the movement of plates has on our everyday lives.

CHAPTER 2

Minerals

INTRODUCTION The rocks of Earth are assemblages of **minerals.** If we understand minerals, we can better understand Earth, how it was created, how both internal and external forces are constantly changing its face, how its history has been recorded throughout the eons of time, and how the history of eons yet to come will be preserved. Minerals in igneous rocks give us information about the temperature, composition, and perhaps the source of the original magma. The mineral assemblage of a metamorphic rock provides insight into the pressures and temperatures that existed when and where the rock formed. Sedimentary rocks contain clues as to the kinds of rocks that were weathered to produce sediments. The relief of the land reflects, at least in part, the relative resistance of the minerals in rocks to the relentless processes of weathering and erosion. All this information and more is engraved in the kinds and abundances of minerals that make up the rocks of Earth.

MINERALS DEFINED

A mineral is defined as *"a naturally occurring inorganic element or compound having an orderly internal structure of atoms and characteristic element composition, crystal form, and physical properties."*

The requirement that a mineral be naturally occurring precludes the use of mineral names to describe any number of synthetic materials, such as synthetic rubies or diamonds, which may be exact copies of their natural counterparts. The requirements that minerals be inorganic and have an orderly internal or solid structure eliminate a number of naturally occurring materials commonly referred to as "mineral" resources such as coal, oil, and water. There is no question that coal, oil, and water are *natural resources,* but by definition, they cannot be considered *mineral resources.* It should be pointed out that not all geologists agree that minerals should be restricted to inorganic compounds. Some geologists argue that natural occurring organic compounds such as calcium oxalate (CaC_2O_4), a common ingredient in plants, should be considered minerals.

The most important parts of the definition are the orderly internal structure and the characteristic elemental composition. A mineral's chemical composition and crystal structure are products of its environment of formation and also influence how the mineral will react at Earth's surface.

CHEMICAL COMPOSITION OF MINERALS

Chemists define an **atom** as the smallest subdivision of matter that retains the chemical characteristics of that particular material. The atom consists of three major subatomic components: (1) **protons,** which are positively charged particles with a mass of 1 AMU (atomic mass unit, equal to $^1/_{12}$ the mass of a carbon atom), (2) **electrons,** which are negatively charged particles with infinitesimally small size and mass, and (3) **neutrons,** which are an electrically neutral combination of a proton and an electron with a mass of 1 AMU (Figure 2.1).

The atom is often pictured as a miniature solar system with most of the mass concentrated in a tiny central nucleus composed of protons and neutrons. The electrons circle the nucleus at specific distances within layers called *shells* (Figure 2.2). The maximum number of shells within the naturally occurring elements is seven. The shells are designated with sequential letters of the alphabet starting with K for the innermost shell. The high orbital speeds of the negatively charged electrons keep them from being pulled into the positively charged nucleus.

The number of protons in the nucleus determines the **atomic number** of the atom, which in turn dictates the *identity* of the element. Of the first 92 elements (hydrogen to uranium), element

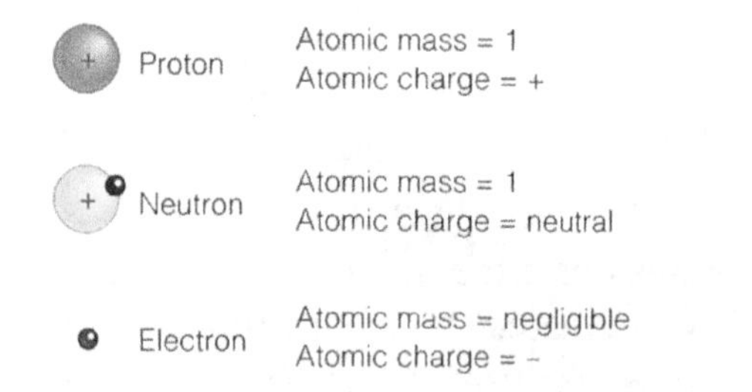

FIGURE 2.1 *The basic components of the atom are the* proton, *the* neutron, *and the* electron. *Note that a neutron is a combination of a proton and an electron.*

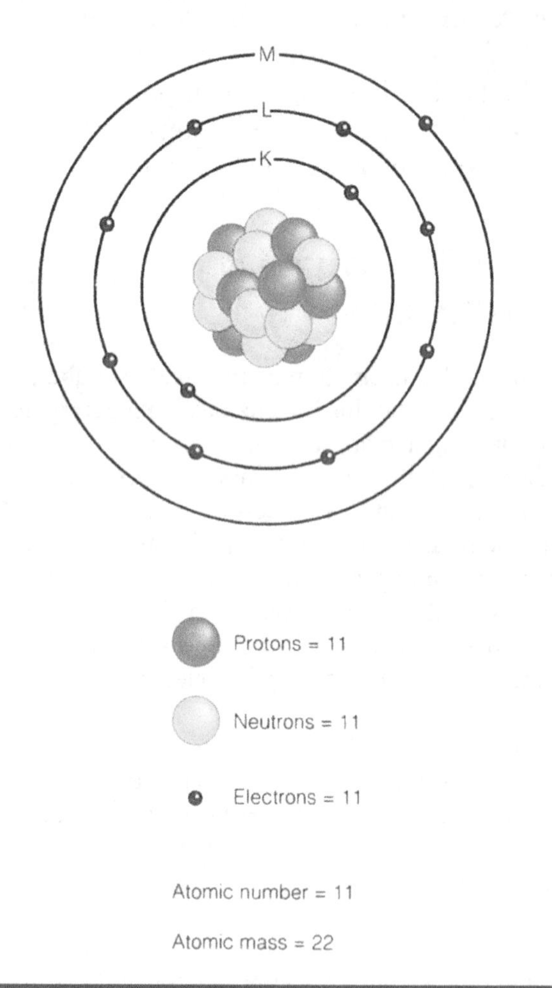

FIGURE 2.2 *The schematic of the sodium atom. Note that the outermost (M) shell has only one electron.*

61 (promethium) does not occur naturally, and elements 43 (technetium), 85 (astatine), and 87 (francium) undergo very rapid radioactive decay and have only been identified in stellar spectra. Elements with atomic numbers greater than 92 do not occur naturally, but several have been made in atomic reactors. They are very unstable (radioactive) and disintegrate with time to atoms with atomic numbers less than 92.

The sum of the protons and neutrons in the nucleus is the **atomic mass** of the atom. The number of neutrons in atoms of the same element can vary, resulting in **isotopes** of the same element with differing atomic masses. For example, most atoms of hydrogen have only one proton in the nucleus and consequently have an atomic mass of 1, but the hydrogen isotopes deuterium and tritium have one and two neutrons, respectively, resulting in atomic masses of 2 and 3. Although the isotopes of a particular element differ in atomic mass, they exhibit most of the same chemical properties.

As protons are added to the nucleus, electrons are added to the electron shells, generally filling the shells from the innermost shell outward. The number of electrons allowed in any shell is limited. The two lightest elements, hydrogen and helium, possess only one orbital shell, which can hold two electrons. Other elements possess additional shells that may contain up to 32 electrons. However, the *outermost* shell of elements with higher atomic numbers can contain a maximum of eight electrons. With increasing atomic number, the atom adds electrons either by forming a new outer shell or by adding electrons to an existing inner shell (Figure 2.3).

The number of electrons in the outermost orbital shell is very important in determining the chemical reactivity of atoms. The **octet rule** states that when the outermost shell is filled with eight electrons, the atom becomes quite stable and does not readily participate in chemical reactions. The elements referred to as the noble or **inert gases** have an outermost shell filled with either two or eight electrons. All the other elements have fewer than the maximum allowable electron complement in the outermost shell are chemically reactive, although some elements such as gold and platinum react very slowly.

SPOT REVIEW

1. Prepare a list of naturally occurring materials that would not fit the definition of a mineral and indicate what property or properties preclude each from being considered a mineral.
2. Explain the effect that each of the following would have on the atomic number and atomic mass of an atom:
   a. Gaining a proton in the nucleus.
   b. Losing a neutron from the nucleus.
   d. Losing an orbiting electron.

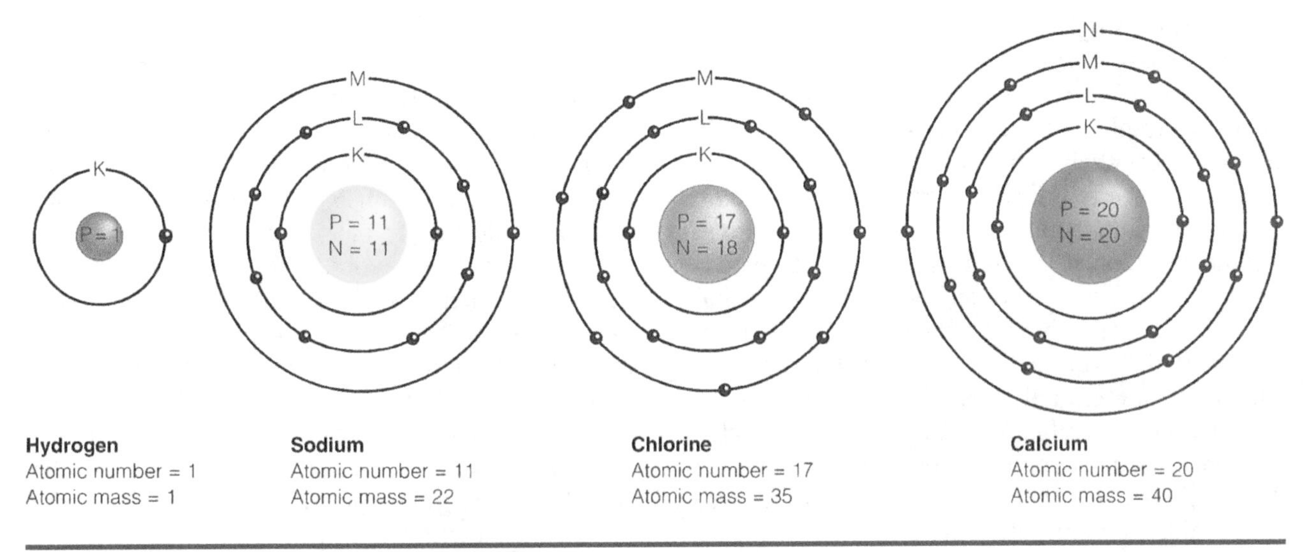

FIGURE 2.3 *The identity of an atom is determined by the number of protons in the nucleus; its mass is equal to the sum of the number of protons and neutrons.*

RADIOACTIVE DECAY OF ATOMS

The elements were created in the fiery cores of stars and once formed, most remain unchanged. However, some isotopes of certain elements are unstable and disintegrate by emitting *alpha* and *beta* particles from their nuclei. An alpha particle consists of two protons and two neutrons. Each emitted alpha particle decreased the atomic number of the element by two and the atomic mass by four. The beta particle is an electron released from a neutron. The emission of a beta particle therefore changes a neutron into a proton with a subsequent increase of one in the atomic number of the element and no change in the atomic mass. Isotopes that undergo these nuclear decays are the **radioactive** isotopes.

In addition to losing protons, neutrons, and electrons from the atom, the nucleus of the potassium isotope ^{40}K (atomic number 19) can be changed into argon (atomic number 18) if one of its orbital electons is captured by a proton in the nucleus, a process called **electron capture.**

Compounds

Atoms of elements react with each other to form **compounds.** A chemical reaction results in each atom filling its outermost shell. The formation of a compound is therefore the mechanism whereby individual atoms can become chemically stable. When some atoms react with each other and lose or gain outer electrons, they are converted into ions. Electron loss converts an atom to a positively charged ion called a **cation** while electron gain produces a negatively charged ion called an **anion.**

An atom will lose or gain electrons depending upon the number of electrons in its outermost shell. Atoms with 1, 2, or 3 electrons in the outermost shell tend to *lose* electrons while those with 4 or more electrons tend to *gain* electrons. For example, the sodium atom (atomic number 11) has 11 electrons: 2 in the innermost shell, 8 in the second shell, and 1 in the outermost shell (refer to Figure 2.2). Should the sodium atom lost the lone outermost electron, its outermost shell would be filled with 8 electrons. In losing the electron, the atom would become a cation with a single positive charge (Na^{1+}) because it would possess one more proton in the nucleus than electrons in its electron shells. The atom of chlorine (atomic number 17) has 17 electrons: 2 in the innermost shell, 8 in the second shell, and 7 in the outermost shell. Chlorine can accept one more electron to fill its outermost shell and in the process become an anion (Cl^{1-}). Ions with opposite charges strongly attract each other and, if other criteria that we will discuss shortly are fulfilled, can unite to form a compound (Figure 2.4). Sodium (Na^{1+}) and chlorine (Cl^{1-}), for example, combine to form sodium chloride, the mineral *halite* (NaCl) or common table salt. Once the ions combine, the compound will be electrically neutral; that is, the total positive charge will equal the total negative charge.

The element that account for 99% of Earth's crust are listed in Table 2.1. Of the 90 available natural elements, only 8 make up almost all of the rocks and minerals of Earth's crust. Note that of the elements in Table 2.2, only oxygen is prone to become an anion (O^{2-}). The remainder will form cations. Given the large number of elements that prefer to form cations, the number of elements that will ionize as anions seems insufficient to balance the cations and form the variety of minerals found in nature. The answer to this seemingly anomalous situation is that certain elements, although themselves prone to become cations, will combine with oxygen to form a multielement anion that chemists call a *functional group.* Silicone (Si), for example, readily joins with oxygen to form the silicate functional group, $(SiO_4)^{4-}$, a major component of a group of minerals called the *silicates,* which make up most of the rocks of the mantle and crust.

But even with the inclusion of the silicate functional group, there still appear to be only two important anions, limiting the kinds of minerals to oxides

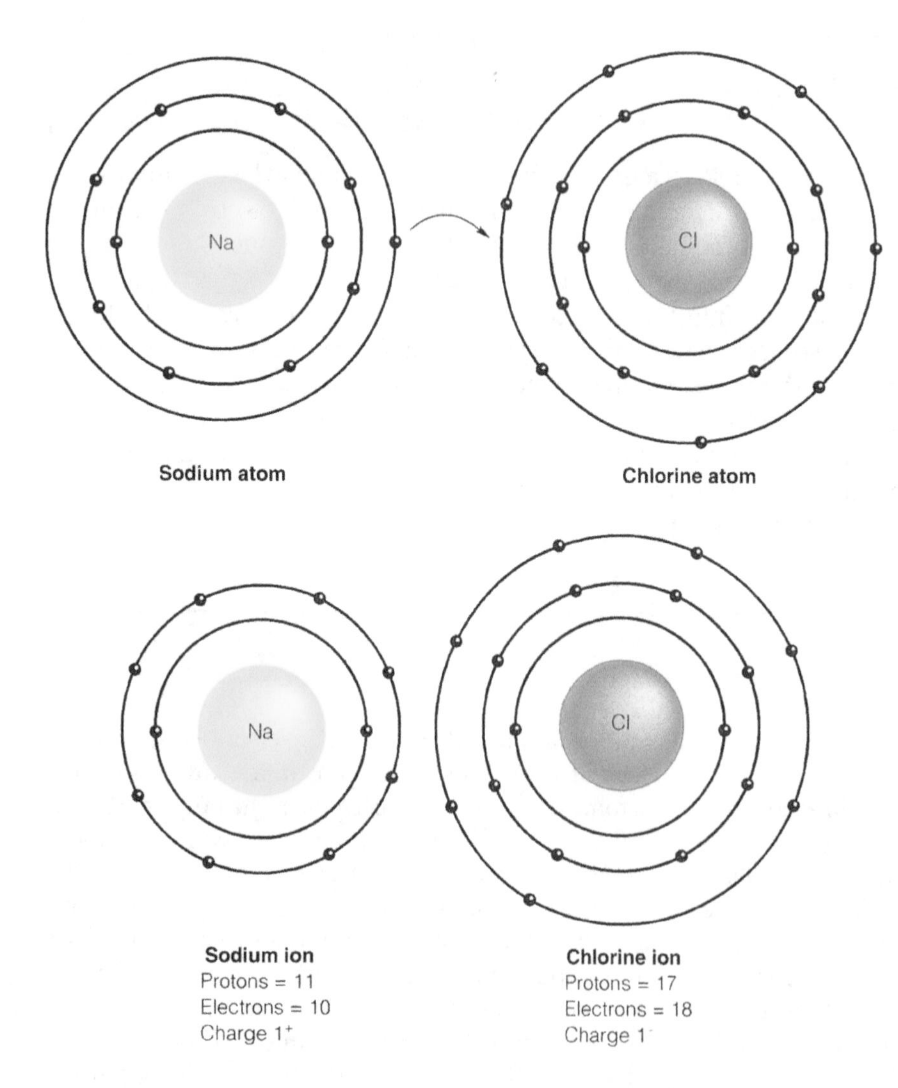

FIGURE 2.4 *The sodium and chloride ions in halite are held together by* ionic bonding *with the electron exchange fulfilling the octet rule for each atom. When the sodium atom donates the single electron from its M shell to chlorine, its L shell, containing 8 electrons, becomes the outermost shell. At the same time, the donated electron fills the outermost shell of chlorine with 8 electrons.*

and silicates. If, however, we add *carbon* (*C*), *nitrogen* (*N*), *phosphorus* (*P*), *sulfur* (*S*), and *chlorine* (*Cl*) to the original list of eight, the situation improves significantly (Table 2.3). The first four elements—carbon, nitrogen, phosphorus, and sulfur—will also combine with oxygen to form multielement functional groups: the *carbonate* $(CO_3)^{2-}$, *nitrate* $(NO_3)^{1-}$, *phosphate* $(PO_4)^{3-}$, and *sulfate* $(SO_4)^{2-}$ functional groups. In addition to forming the sulfate functional group, sulfur can form two other anions, the *sulfide* (S^{2-}) and *disulfide* ($S^{2}_{\bar{2}}$) ions. Chlorine will readily accept electrons to become the *chloride* (Cl^{1-}) ion. With our expanded list of elements, not only is the division of positive and negative ions more equable, but we can account for nearly all the common minerals and, in most cases, the not-so-common minerals.

In order for a mineral to form, the necessary ions must first exist, usually in a solution. An important exception to this generalization is the formation of metamorphic minerals, many of which may form without the intervention of a solution. Once in solution, conditions, which we will discuss shortly, come into play that result in the union of the various ions to form a *crystal lattice*. Once the lattice grows to sufficient size, a crystal of the mineral will begin to precipitate from solution.

The role of the solvent is to mobilize the ions so that they may be recombined into compounds. In nature, water is a powerful solvent. Water may contain a wide variety of ions ranging in concentration from the salt-rich waters of a perennial desert lake to the pristine waters of a mountain stream containing very low concentrations of relatively few kinds of ions.

Many minerals consist of ions that were at one time contained in a water solution; calcite ($CaCO_3$) and halite (NaCl) are two examples. The ions that formed most of the *silicate* minerals, however, were not provided from water solution but rather crystallized as molten rock (magma or lava) cooled.

Once in solution, ions move at random through the solvent with a limited tendency to form permanent combinations. A number of conditions promote the union of cations and anions or functional groups to precipitate a compound from a solution.

A basic requirement for precipitation to occur is that the concentration of the ions be sufficient to ensure that they come in contact within the solution. Ionic concentration and subsequent ion contact can be increased either by adding more ions to the solution or by reducing the volume of the solvent.

TABLE 2.1

AVERAGE ABUNDANCE OF THE MAJOR ELEMENTS IN EARTH'S CRUST

Element	Symbol	Percentage (By Weight)
Oxygen	O	46.4
Silicon	Si	28.2
Aluminum	Al	8.2
Iron	Fe	5.6
Calcium	Ca	4.1
Sodium	Na	2.4
Magnesium	Mg	2.3
Potassium	K	2.1

TABLE 2.2

ION AFFINITIES OF THE MAJOR ELEMENTS IN EARTH'S CRUST

Element	Cation	Anion
Oxygen		O^{2-}
Silicon	Si^{4+}	
Aluminum	Al^{3+}	
Iron	Fe^{2+}, Fe^{3+}	
Calcium	Ca^{2+}	
Sodium	Na^{1+}	
Magnesium	Mg^{2+}	
Potassium	K^{1+}	

TABLE 2.3

EXPANDED LIST OF CRUSTAL ELEMENTS SHOWING THE CATION-ANION AFFINITIES

Element	Cation	Anion
Oxygen		O^{2-}
Silicon	Si^{4+}	$(SiO_4)^{4-}$
Aluminum	Al^{3+}	
Iron	Fe^{2+}, Fe^{3+}	
Calcium	Ca^{2+}	
Sodium	Na^{1+}	
Magnesium	Mg^{2+}	
Potassium	K^{1+}	
Carbon	C^{4+}	$(CO_3)^{2-}$
Nitrogen		$(NO_3)^{1-}$
Phosphorus		$(PO_4)^{3-}$
Sulfur	S^{4+}	S^{2-}, $S^{2}_{\bar{2}}$, $(SO_4)^{2-}$
Chlorine		Cl^{1-}

Precipitation also depends on the temperature of the solution. The higher the temperature, the faster the ions move and the less tendency they have to combine to make a compound. This is why most materials are more soluble in hot water than in cold. As the movement of the ions in solution slows, the chances of a union improve.

In summary, precipitation of compounds from solution is favored by (1) increasing the concentration of ions either by adding ions to the solution or by reducing the volume of the solvent and (2) by decreasing the temperature.

Another requirement for the union of ions of dissimilar charge is the establishment of **electric neutrality** within the compound. In order for a union to be made, the total positive and negative charges must balance. Electric neutrality must be maintained at all times.

Lastly, the ions must be able to combine and form a three-dimensional crystal lattice. Ion size is an important constraint in building crystal lattices. For example, a specific ion with all the proper electrical qualifications may be excluded from a particular lattice because the diameter of the ion is too large to fit into the available space. To illustrate with a simple example, imagine a table top covered with tennis balls where each tennis ball is in contact with all adjacent tennis balls (Figure 2.5). Can you replace certain tennis balls with baseballs? Yes, because the two balls are about the same size. Can you replace some tennis balls with 1/4-inch ball bearings? You can, but when you do, there will be nothing to support the tennis balls adjoining the site occupied by the ball bearing, and they will begin to move from their original positions. Thus, substituting ball bearings for tennis balls will disrupt the pattern. In this case, the intended replacement is too small. How about replacing some tennis balls with basketballs? It can be done, but the basketball will make the layer excessively thick. As a rule of thumb, in order for an ion to occupy an available space within a lattice, the diameter of the ion and the diameter of the available space must not differ by more than 15%.

SPOT REVIEW

1. What determines whether an atom will tend to become a cation or an anion?
2. What conditions promote the combining of ions and the subsequent precipitation of a compound from solution?

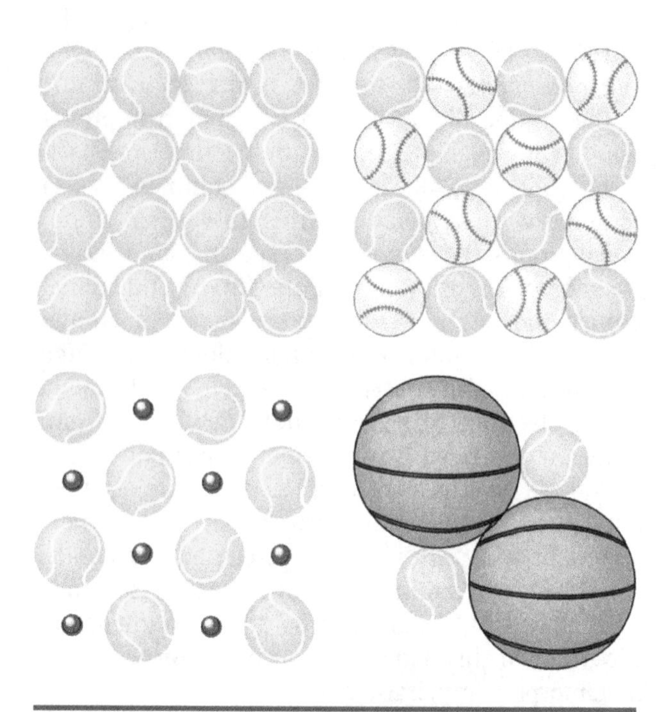

FIGURE 2.5 *Assuming all other criteria are acceptable, ions with ionic diameter within 15% of each other may readily interchange, just as baseballs and tennis balls may be substituted for each other. If the ionic radius is too small, subsequent lattice deformation may prevent substitution from occurring. Substitution of very large ions would result in unacceptable lattice distortions.*

Chemical Bonding

Once the concentration of ions needed for precipitation has been exceeded, compounds can form by the **bonding** of positive and negative ions into an orderly three-dimensional structure. Bonding results from the attempt of each atom to maintain the maximum number of electrons in its outermost orbital shell. There are four basic kinds of bonds: (1) *ionic,* (2) *covalent,* (3) *metallic,* and (4) *van der Waals.*

During **ionic bonding** (ionization), the outermost electrons of one atom or functional group are relinquished and given to another. The transfer of electrons results in each atom in the compound achieving the desired number of eight electrons in the outermost shell. Let's once again take the example of halite (NaCl), common table salt. As shown in Figure 2.4, the electron donated by the sodium atom fills the outermost shell of chlorine with eight electrons and at the same time leaves sodium with eight electrons in its outer shell. Both atoms now have a filled outermost shell. Once the two ions are present in sufficient concentration, they attract each other and create a crystal lattice, producing a compound with the required state of electric neutrality. Compounds in which atoms are joined by ionic bonding are usually water soluble.

Covalent bonding forms the strongest bonds. In true covalent bonding, the outermost electrons of both atoms are *shared.* True covalent bonds can only be formed by two atoms of the same element. For example, carbon with four electrons in the outermost shell joins with other carbon atoms to form *diamond* and *graphite* (Figure 2.6) by the mutual sharing of their outermost electrons. The union produces an extremely strong bond by establishing the desired eight electrons in the outer shell of each atom. Because of its strength, covalent bonding produces some of the most chemically resistant compounds, including most minerals. Covalently bonded compounds are only slightly soluble in water, and more severe chemical conditions are required to return them to solution. Such conditions can be provided by a mass of molten rock, for example.

A continuum exists between ionic bonding where electrons are *completely exchanged* and covalent bonding where the electrons are *equally shared.* Whether a bond will have an ionic or covalent character depends on the degree to which the electrons are attracted to one of the two atoms. As the ability of the atoms to attract electrons becomes more equal, the covalent character of the bond increases.

Metallic bonding, as the name implies, is the bonding occurring in metals. In metallic bonding, the atoms are locked into place within a rigid lattice, but the outer electrons roam easily throughout the crystal structure. The characteristic properties of metals such as high electrical and thermal conductivity are due to this unique bonding style.

Van der Waals bonding differs from the other three types of bonding in that it does not involve electron transfer or sharing but rather a weak attraction between subunits *within* the main crystal structure. In the mineral graphite, for example, carbon atoms are joined by strong covalent bonding to form sheets. The sheets are then stacked to form the structure of graphite (Figure 2.6b). Within the sheets, the orientation of the electron fields around the carbon atoms causes positive and negative charges to develop,

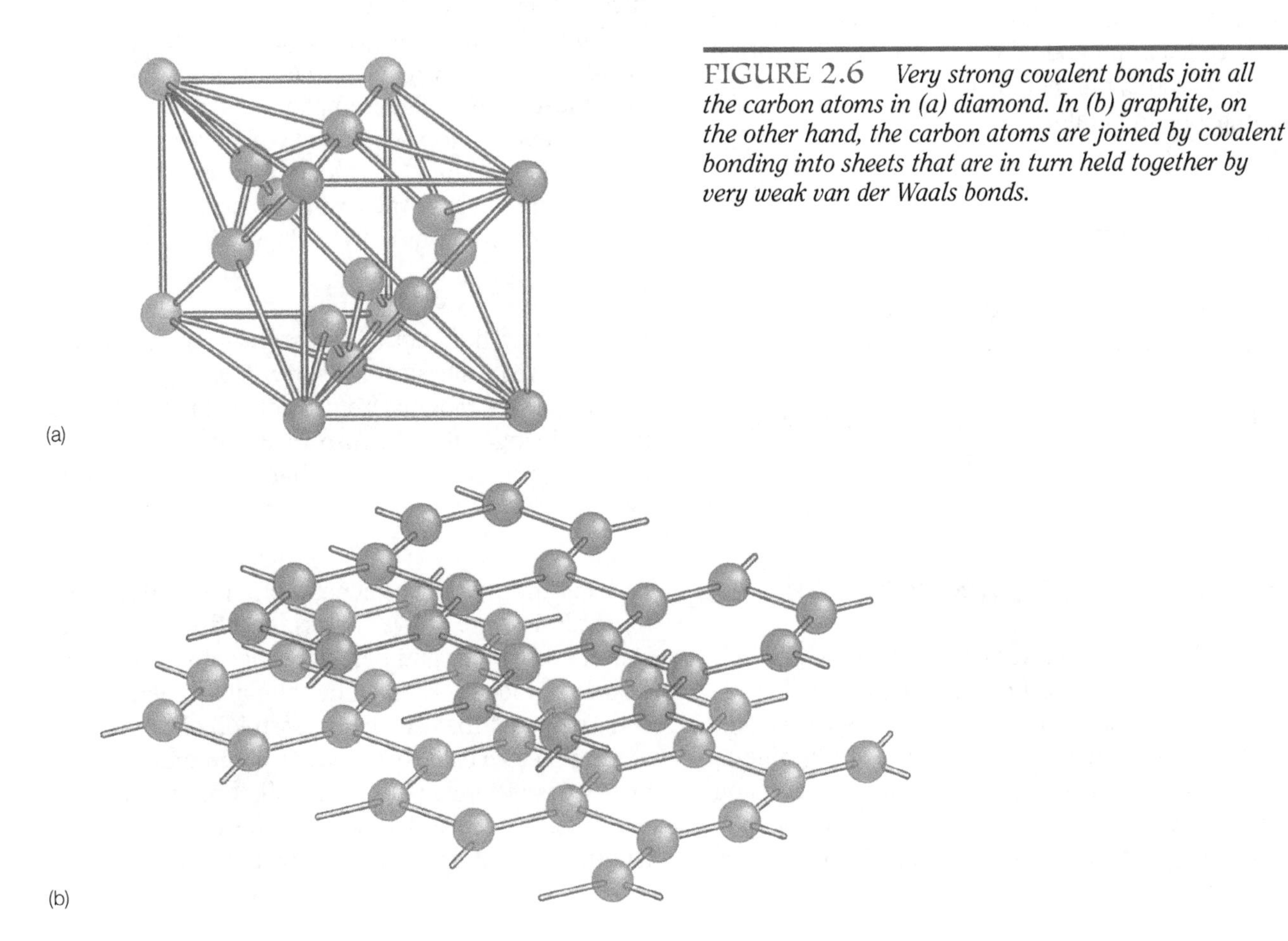

FIGURE 2.6 *Very strong covalent bonds join all the carbon atoms in (a) diamond. In (b) graphite, on the other hand, the carbon atoms are joined by covalent bonding into sheets that are in turn held together by very weak van der Waals bonds.*

generating an electrostatic attraction between the sheets. This attraction is an example of van der Waals bonding. Most van der Waals bonds are very weak. The bond in graphite is so weak it can easily be broken with the application of the smallest force generated between your thumb and forefinger. When you rub graphite or soft pencil lead, it feels greasy because sheets of carbon atoms are slipping over each other as you break the van der Waals bonding. Because of the weak van der Waals bonds, graphite can be used as a solid lubricant in door locks, electrical equipment, and similar instances when liquid lubricants are inappropriate.

Crystal Structure

Solids can form in two ways. The individual atoms can join together into an orderly three-dimensional array called a **crystal lattice** in which each atom, ion, or functional group is located at a specific site (Figure 2.7). Materials that form in such orderly arrangements are said to be **crystalline,** and if they form naturally, they can be called minerals. Atoms can also join together to form a solid where the individual atoms are arranged somewhat randomly with no specific order or assigned locations within the solid. The material is then said to be **amorphous** and is called a **mineraloid** or **glass.** *Opal* ($SiO_2 \cdot nH_2O$ where n is the number of water molecules) is an example of a mineraloid.

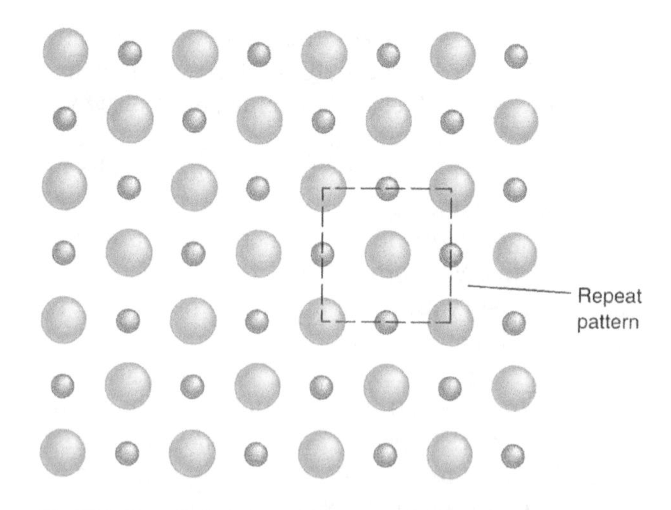

FIGURE 2.7 *Within every crystal lattice, the atoms or ions are arranged in a symmetrical, three-dimensional collection called a* unit cell, *here represented in two dimensions. The* crystal structure *is constructed by repeating the unit cell in the three crystallographic directions.*

SPOT REVIEW

1. What is meant by the statement that a continuum exists between ionic and covalent bonding?
2. In what ways does the kind of bonding affect the chemical and physical characteristics of a compound?

MINERAL IDENTIFICATION

Minerals have orderly internal structures, characteristic composition, typical crystal forms, and physical properties. Any of these parameters may be used in identification. Sometimes, one can identify mineral grains with the unaided eye or a low-powered hand lens. At other times, mineral grains or crystals are too small to be observed and tested with a hand lens, and their identification requires other instruments such as powerful microscopes or X-ray analyzers.

When the mineral grains can be seen with the unaided eye or with low-power magnification, the identification technique employed is referred to as *visual identification.* With a low-powered hand lens and some experience, a geologist can usually identify most common mineral grains or crystals from an examination of one or more of its **physical properties.**

Physical Properties

Any property that can be determined by the senses is considered a physical property. A variety of physical properties ranging from color or taste are used in identifying minerals. Also included among these physical properties are such characteristics as hardness, specific gravity, and luster.

Color

Perhaps the most conspicuous physical property of minerals is *color.* Nevertheless, the first thing you learn as a beginning mineralogy student is not to rely on color for identification. The color of a mineral is the end product of the degree of absorption of white light as it is reflected back to the eye. Most minerals contain impurities of foreign atoms. Small concentrations of impurities can have significant effects upon the wavelengths of light absorbed and will therefore have a

FIGURE 2.8 *As certain minerals are drawn across a piece of unglazed tile, small (powder-sized) particles of the mineral break off and are left behind as a* streak. *The color of the streak is diagnostic of the mineral.*

major effect on the color. Nevertheless, color is very useful for distinguishing between the dark-colored ferromagnesian (iron- and magnesium-rich) minerals such as augite and the light-colored nonferromagnesian minerals such as albite.

Streak The color of the powdered mineral (streak) is relatively diagnostic of some minerals, particularly metallic minerals such as *hematite* (Fe_2O_3). Experience has shown that the color of the powder is fairly constant. It is called streak because the most common way of obtaining a powder is by dragging the mineral across the surface of a plate of unglazed ceramic tile, thereby producing a line or streak (Figure 2.8).

Hardness An often-used physical property is **hardness.** Hardness, the resistance of a mineral to scratching, is a result of the strength of the bonds within the crystal structure. A geologist named *Fredrick Mohs* studied the relative hardness of minerals nearly two centuries ago (1822). When his study was completed, he selected 10 common or well-known minerals to represent the entire range of hardness. For the softest mineral, he selected *talc,* and for the top of his scale, he used the hardest mineral known, *diamond.* The minerals in **Mohs's Scale of Hardness** are listed in Table 2.4. The hardness of an unknown mineral is determined by finding the pair of minerals in Mohs's list, the harder of which will scratch the surface of the unknown mineral, and the softer of which will be scratched *by* the unknown. A geologist can carry a set of hardness mineral standards or use a few common items to estimate hardness.

TABLE 2.4

MOHS'S SCALE OF HARDNESS

Hardness	Mineral	Composition
10	Diamond	C
9	Corundum	Al_2O_3
8	Topaz	$Al_2SiO_4(F,OH)_2$
7	Quartz	SiO_2
6	Orthoclase	$KAlSi_3O_8$
5	Apatite	$Ca_5(PO_4,CO_3)_3(F,OH,Cl)$
4	Fluorite	CaF_2
3	Calcite	$CaCO_3$
2	Gypsum	$CaSO_4{\cdot}2H_2O$
1	Talc	$Mg_3Si_4O_{10}(OH)_2$

Before we leave the topic of hardness, perhaps some comments on a few of the minerals in Mohs's scale are in order. *Talc* is used to make baby powder and talcum powder while *gypsum* is widely used in the construction industry as plasterboard and dry wall. *Calcite* is the mineral that makes up the shells of animals such as clams, oysters, and snails, while *apatite* is used in the construction of bones and teeth, including those of humans. *Orthoclase,* one of the potassium feldspars, with a hardness of 6, is the abrasive used in most household cleansers. Because the abrasive has a hardness of 6, there is no chance of scratching such things as tile tubs and washbasins, which have a hardness of about 7. In fact, the logo of one cleanser includes the statement "hasn't scratched yet" encircling a baby chick just emerging from its shell. Orthoclase is also used as the abrasive agent in many toothpastes. Remember, the enamel of your teeth is made of apatite, which has a hardness of 5.

Quartz is the hardest of all the common minerals and has long been used as an abrasive. Antique grinding stones and wheels were usually carved from blocks of sandstone, which is made up primarily of quartz grains. Sandblasting, the method commonly used to clean the surfaces of stone or metal, often utilizes loose quartz sand grains blown at the surface by compressed air. At one time, quartz sand similar to the sand that makes up your favorite beach was glued to paper to make "sandpaper." Today, *corundum,* which is harder, has nearly replaced quartz as the major abrasive in stones, wheels, emery cloth, and sandpaper. Another common use of quartz is as the timing device in quartz watches.

It may come as a surprise that most *diamonds* are not used as gemstones, but as abrasives. The edges of saw blades used to cut rocks are embedded with diamonds. Drilling bits used to drill through extremely hard rock are encrusted with diamonds, and you can purchase diamond "paste" (powdered diamonds in an oil matrix) for fine polishing.

Crystal Form The **form** of a mineral is the geometric shape of its crystal as defined by the angle between faces. The form is the outer expression of the crystal structure. If a crystal were able to grow without interference, it would have a distinctive crystal form. Many minerals have diagnostic forms that may be used in their identification (Figure 2.9). However, most mineral grains grow in restricted space, and the diagnostic crystal forms do not develop.

Mineral Cleavage **Mineral cleavage** is a very diagnostic physical property of some minerals. Cleavage describes the tendency of certain minerals to break along specific planes within the crystal structure. These planes or surfaces develop where the bonding across some crystallographic planes is weaker than others. If stressed, the chemical bonds between these planes will break preferentially. An excellent example of cleavage is the ability of the *micas* to be peeled apart into sheets, giving rise to the term "books" of mica. Mica is made up of sheets in which atoms and ions are held together by strong bonding. The bonds that hold the sheets together are much weaker. In fact, they are so weak that they can easily be broken by prying with a fingernail. Other minerals with a single dominant cleavage plane due to weak van der Waals bonding are graphite and talc.

Many minerals have multiple cleavage planes. The mineral *halite* (NaCl), for example, has three mutually perpendicular cleavage planes, which explains why crushed salt grains have a cubic form (Figure 2.10). *Calcite* also has three sets of cleavage planes, but because they are not mutually perpendicular, the crystal form is rhombic. Other common minerals with multiple cleavage planes include *feldspar, augite,* and *hornblende,* each with two planes, and *fluorite* with four.

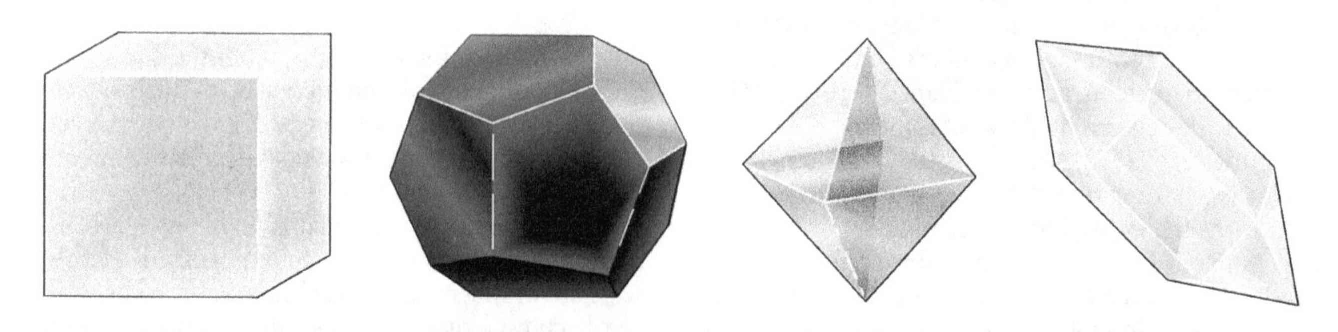

FIGURE 2.9 *The outer appearance of a crystal, called the crystal form, is the external expression of the systematic internal arrangement of atoms or ions within the crystal lattice.*

FIGURE 2.10 *The cubic arrangement of the atoms within halite results in its characteristic* cubic form. *(M. Claye/Jacana, Photo Researchers)*

Fracture Not all minerals show cleavage. Minerals such as quartz show no cleavage because the interatomic bonds are all of equal strength. If you shatter a piece of quartz, it will break or **fracture** along random, usually curved, surfaces like the breaking of glass rather than along planes of cleavage. Such curved fracture is called *conchoidal fracture.* Minerals such as the feldspars that have two planes of cleavage will show fracture in non-cleavage directions.

Specific Gravity The **specific gravity** is a useful physical property that can be estimated by "hefting" the mineral. Specific gravity is the ratio of the mass of a solid to the mass of an equal volume of pure water at 4°C. Specific gravity is determined by the kinds of atoms in the crystal structure and the closeness with which they are packed. Most rocks and minerals have specific gravities from 2.5 to 3. Specific gravity increases with increasing atomic number. For example, the specific gravity of *quartz* (SiO_2), is 2.65, *pyrite* (FeS_2) 5.02, *chalcocite* (Cu_2S) 5.8, *galena* (PbS) 7.6, and *gold* (Au) 19.3. The *specific gravity* and **density** of a substance are numerically equal. The only difference is that because the density of water at standard conditions of temperature and pressure is 1.0 gm/cm^3, specific gravity is dimensionless whereas density is mass per volume and has the units gm/cm^3.

Luster The **luster** of a mineral is the appearance of its surface under reflected light. Luster can be described as *metallic* or *nonmetallic.* Nonmetallic luster is further subdivided into descriptions that are self-explanatory such as *pearly, silky, glassy* or *vitreous,* and *earthy* and into others that are not so obvious such as *adamantine* (sparkles like a diamond). Metallic luster, such as that shown by galena, refers to the metallic appearance of the mineral.

Acid Reaction Certain carbonate minerals, especially calcite, can be readily identified by the fizz that results when a few drops of dilute acid are applied to the surface. The generation of the gas is explained by the following reaction:

$$CaCO_3 + 2HCl \rightarrow CO_2 + H_2O + Ca^{2+} + 2Cl^{1-}$$

calcite + acid →
gas + water + in solution + in solution

Few geologists would go into the field without a small bottle of dilute hydrochloric acid. Acid must always be used with care, however.

Taste **Taste** can be used (also with care) to identify some minerals. Examples include the minerals *halite* (NaCl) and *sylvite* (KCl). Halite is common table salt while sylvite is sold as a substitute for halite. Halite is identified by its familiar salty taste while sylvite, although salty, has a distinctive bitter taste.

SPOT REVIEW

1. List the mineral characteristics that are most responsible for each physical property.
2. How are crystal form, density, cleavage, and fracture related?

MINERAL CLASSIFICATION

The more than 3,200 known rock-forming minerals are classified into two groups: (1) **silicates** and (2) **nonsilicates** (Table 2.5), a classification that once again emphasizes the importance of the silicate functional group. Depending on the *iron* (Fe) and *magnesium* (Mg) contents, the silicate minerals may be further subdivided by composition into two mineral subgroups: (1) the *ferromagnesian* and (2) the

TABLE 2.5

CLASSIFICATION OF MINERALS

Common Silicate Minerals	
Ferromagnesium Minerals	
Olivine group	
Fayalite Fe_2SiO_4	
Forsterite Mg_2SiO_4	
Pyroxene group	
Augite $(Ca,Na)(Mg,Fe^{3+}, Fe^{2+}, Al)_2(Si, Al)_2O_6$	
Amphibole group	
Hornblende $(Ca,Na)_{2-3}(Mg,Fe^{2+},Fe^{3+},Al)_5(Al,Si)_8O_{22}(OH)_2$	
Mica group	
Biotite $K(Mg,Fe^{2+})_3(Al,Fe^{3+})Si_3O_{10}(OH)_2$	
Nonferromagnesian Minerals	
Feldspar group	
Plagioclase feldspar subgroup	
Anorthite $CaAl_2Si_2O_8$	
Albite $NaAlSi_3O_8$	
Orthoclase $KAlSi_3O_8$	
Mica group	
Muscovite $KAl_2(Al,Si_3)O_{10}(OH)_2$	
Quartz SiO_2	
Common Nonsilicate Minerals	
Oxides	Nitrates
Hematite Fe_2O_3	Niter KNO_3
Corundum Al_2O_3	Soda niter $NaNO_3$
Carbonates	Phosphates
Calcite $CaCO_3$	Fluorapatite $Ca_5(PO_4)_3F$
Dolomite $(Ca,Mg)CO_3$	Chlorapatite $Ca_5(PO_4)_3Cl$
Sulfates	Halides
Gypsum $CaSO_4{\cdot}2H_2O$	Halite NaCl
Anhydrite $CaSO_4$	Sylvite KCl
Sulfides	Fluorite CaF_2
Galena PbS	Native Elements
Sphalerite ZnS	Graphite C
Pyrrhotite FeS	Sulfur S
Disulfides	Gold Au
Pyrite FeS_2	Silver Ag
Chalcopyrite $CuFeS_2$	Copper Cu
	Platinum Pt

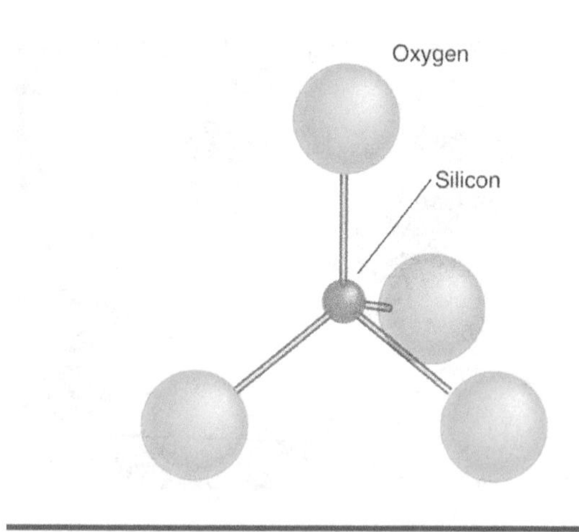

FIGURE 2.11 *The basic building block of all silicate minerals is the* silicon-oxygen tetrahedron.

nonferromagnesian minerals. Subdivisions of the silicate subgroups are based primarily on structure.

The nonsilicate minerals are subdivided into groups based on the anion or functional group. A common mineral representative is included in Table 2.5 for each anion or functional group.

Silicate Crystal Structures

Silicate minerals are the primary building blocks of rocks. The basic building block of all the silicate minerals is the **silicon-oxygen tetrahedron** (Figure 2.11), which consists of four oxygen atoms arranged at the corners of a tetrahedron around a central silicon atom. One of the negative charges of each of the four oxygen atoms is balanced by the 4 positive charges of the silicon, resulting in a net charge 4– charge for the silicate functional group.

The silicon-oxygen (silica) tetrahedra can be joined into five major structural arrangements: (1) *isolated tetrahedra,* (2) *single chains,* (3) *double chains,* (4) *sheets,* and (5) *framework structures.* Each structural type is illustrated in Figure 2.12.

Isolated Tetrahedral Structure

A crystal structure can be envisioned as a three-dimensional arrangement of anions or functional groups, *glued* together by cations. In the **isolated tetrahedral structure,** the silica tetrahedra are *isolated* from each other and held (glued) together by the cations. The silicon/oxygen ratio in these minerals is 1:4. An example is the **olivine group,** $(Fe,Mg)_2SiO_4$. The olivine group is an example of a **solid-solution series,** which is a group of minerals with a common crystal structure

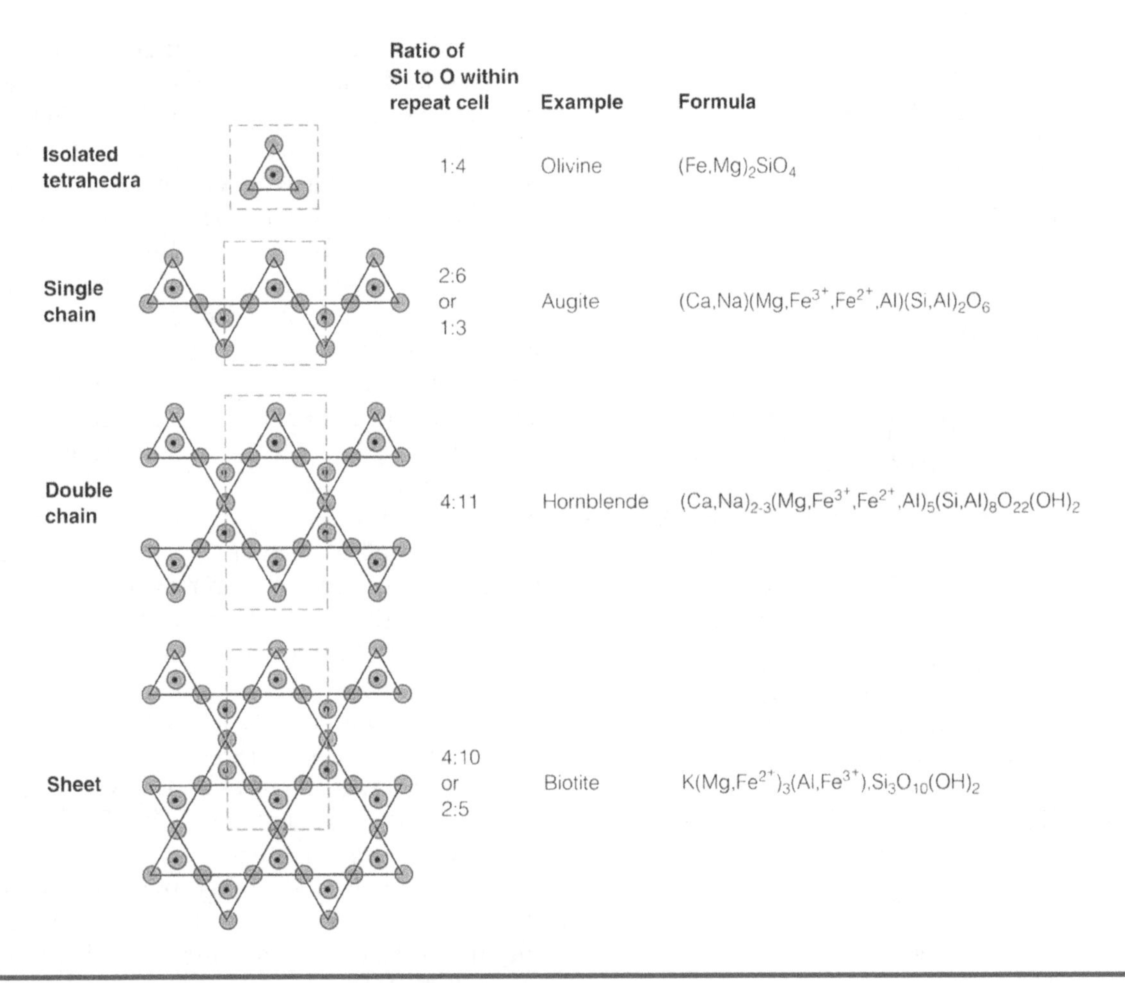

FIGURE 2.12 *Five different arrangements of silicon-oxygen tetrahedra, held together by cations, make up most of the important rock-forming silicate minerals.*

in which two positive ions of nearly the same size can substitute interchangeably. A solid-solution consists of two end-member minerals in which all the potential cation sites are occupied by one ion or the other.

Single-Chain Structure In the **single-chain structures,** the silicon-oxygen tetrahedra are joined into a chain in which two corners of each tetrahedron are shared, producing a unit with a silicon/oxygen ratio of 1:3. The single-chain structure is represented by the **pyroxene group** minerals. *Augite,* $(Ca,Na)(Mg,Fe^{3+},Fe^{2+},Al)_2(Si,Al)_2O_6$, is the most common representative. The single-chain minerals can be envisioned as stacked cordwood or logs with the long dimensions of the chains parallel to each other. Again, the cations bond the chains, producing a single mineral structure. The weaker bonding of adjacent chains by the positive ions as compared to the strong bonding within the chains results in the characteristic near–90° cleavage of the pyroxene group (Figure 2.13). Note also that as the crystal structure becomes more complex, the compositional makeup of the minerals becomes more diverse.

Double-Chain Structure **Double-chain structures** are made by joining two single chains, once again by a sharing of oxygen atoms. This time, alternate third corners of tetrahedra are linked. The double-chain minerals are represented by the **amphibole group** of silicate minerals, of which *hornblende,* $(NaCa)_{2-3}(Mg,Fe^{2+},Fe^{3+},Al)_5(Si,Al)_8O_{22}(OH)_2$, is the most common representative.

The double-chain units are like planks with the crystal structure built up by overlapping the planks in successive layers, again *glued* together with cations. Note, once again, that compositional complexity increases with structural complexity. As in the single-chain minerals, the weak bonding of adjacent double chains by the positive ions results in the characteristic cleavage of the amphibole minerals (Figure 2.14).

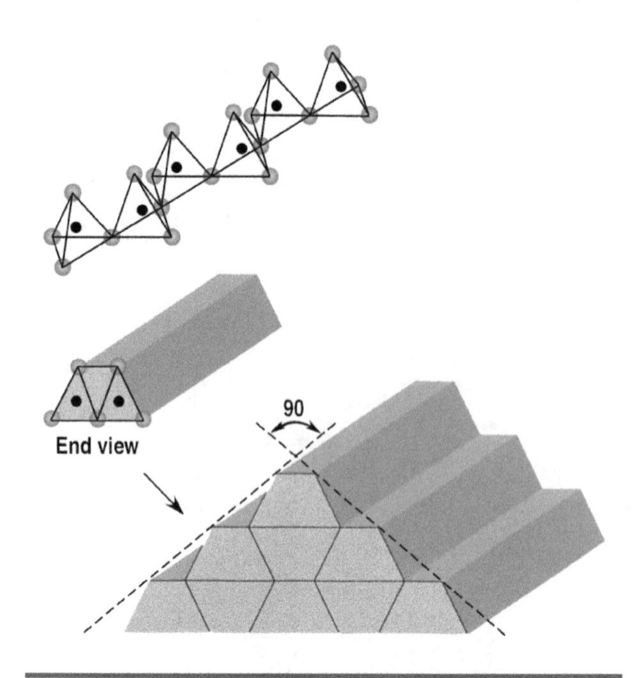

FIGURE 2.13 *The pyroxene minerals consist of parallel stacked, single chains held together by cations including iron and magnesium. The combination of the cross-sectional shape of the individual chains and the stacking arrangement results in two well developed cleavage planes that intersect at a 90° angle.*

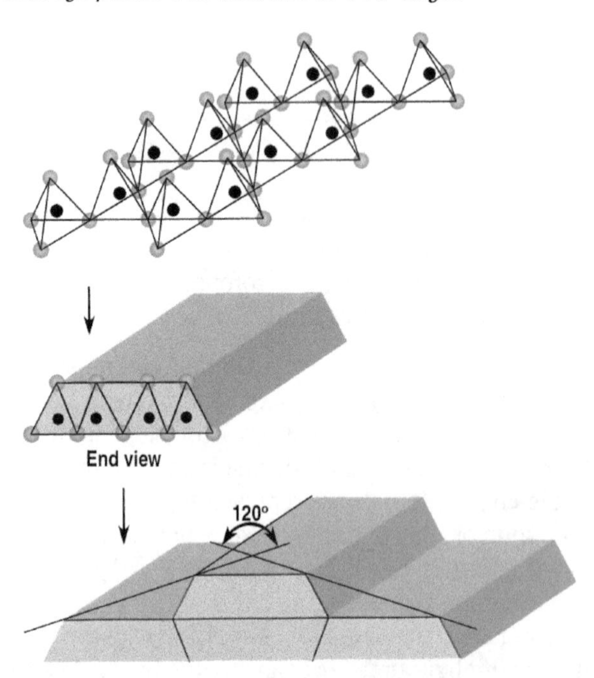

FIGURE 2.14 *The crystal structure of the amphibole minerals is similar to that of the pyroxenes except that the structural unit is the double chain rather than the single chain. The amphiboles also have two well developed cleave directions, but, because of the width of the double chain structure, the angle between the cleavage planes is 120°.*

Sheet Structure In the **sheet structures,** the silican tetrahedra are joined into sheets with a symmetrical hexagonal (six-sided) pattern (Figure 2.15). Three oxygens of each tetrahedron are shared, leaving the one unshared oxygen pointing away from the sheet. The (silicon + aluminum)/oxygen ratio of the structure is now 2:5. Most sheet minerals are constructed by joining two sheets of tetrahedra together with cations bridging the negatively charged oxygen surfaces. The bond between the joined tetrahedral sheets is very strong. The joined sheets are stacked like sandwiches to create the mineral structure. A representative mineral group is the **mica group.** In the ferromagnesian mica, *biotite,* $K(Mg,Fe^{2+})_3(Al,Fe^{3+})Si_3O_{10}(OH)_2$, the two tetrahedral sheets are bonded together with iron and/or magnesium whereas in the nonferromagnesian mica, *muscovite,* $KAl_2(AlSi_3)O_{10}(OH)_2$, aluminum links the two sheets (Figure 2.16). In both cases, the silicate sandwiches are bonded relatively weakly to each other by the potassium ion. The characteristic cleavage of mica is due to the weakness of the potassium bonds between adjacent structural sheets. Note once again the substitution of aluminum (Al^{3+}) silicon for (Si^{4+}) in the tetrahedral structure.

Clay minerals are another important example of the sheet silicate structure. These minerals form by the chemical alteration of most of the silicate minerals, including the micas, and are a major constituent of many soils.

Framework Structure The **framework structure** is a three-dimensional arrangement of tetrahedra in which each oxygen is shared by adjoining tetrahedra giving a silicon/oxygen ration of 1:2 (Figure 2.17). The *feldspars,* the major rock-forming minerals, and *quartz* have this type of structure.

The **feldspar group** is subdivided into two subgroups, the **plagioclase** feldspars and the **alkali** feldspars. The plagioclase group is a solid-solution series in which composition varies from a *calcium-rich* (Ca^{2+}) end-member, *anorthite,* $CaAl_2Si_2O_8$, to a *sodium-rich* (Na^{1+}) end-member, *alabite,* $NaAlSi_3O_8$. The alkali feldspars include *orthoclase, microcline,* and *sanidine.* Orthoclase and microcline are identical in composition ($KAlSi_3O_8$), and sanidine differs in that sodium may substitute for potassium, $(K,Na)\ AlSi_3O_8$. All three minerals differ in crystal structure.

Quartz (SiO_2), is the second most abundant silicate mineral in Earth's crust after feldspar. In the crystal structure of quartz, each oxygen atom in the silicon-oxygen tetrahedra contributes a negative

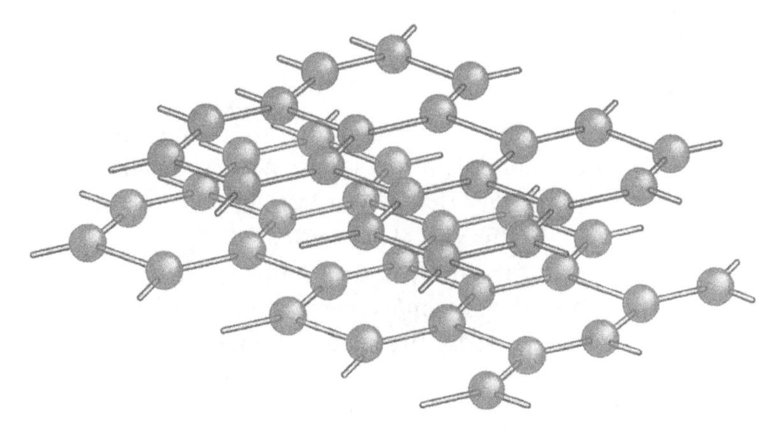

FIGURE 2.15 *The silicon-oxygen tetrahedra can also be joined in a two-dimensional hexagonal pattern, forming sheets. Pairs of sheets are then held together like a sandwich by a variety of ions including iron, magnesium, and aluminum. The sandwiches are then stacked and held together by ions such as potassium, calcium, and sodium to form the micas. The characteristic cleavage of the micas is due to the weakness of the bonds that join the sandwiches. The kinds of cations that join the sheets into sandwiches and hold the sandwiches together determine the identity of the individual mica mineral.*

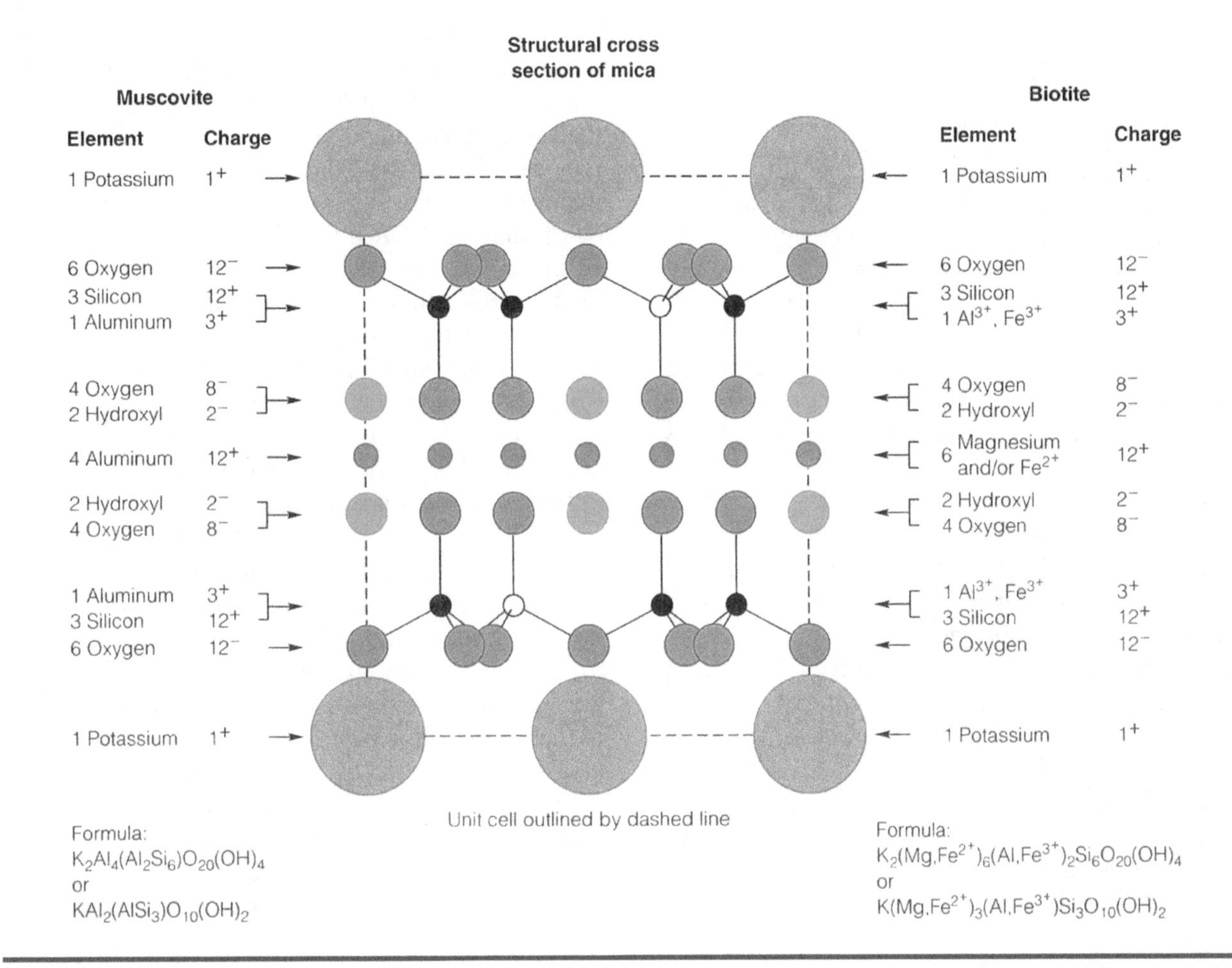

FIGURE 2.16 *The difference between biotite and muscovite is largely determined by the kinds of cations contained between the sheets. Biotite, the ferromagnesian mica, contains iron and/or magnesium while muscovite, the nonferromagnesian mica, contains aluminum. Note, however, that in both cases, the total cation charge is equal to the total anion charge; the lattice must be electrically neutral.*

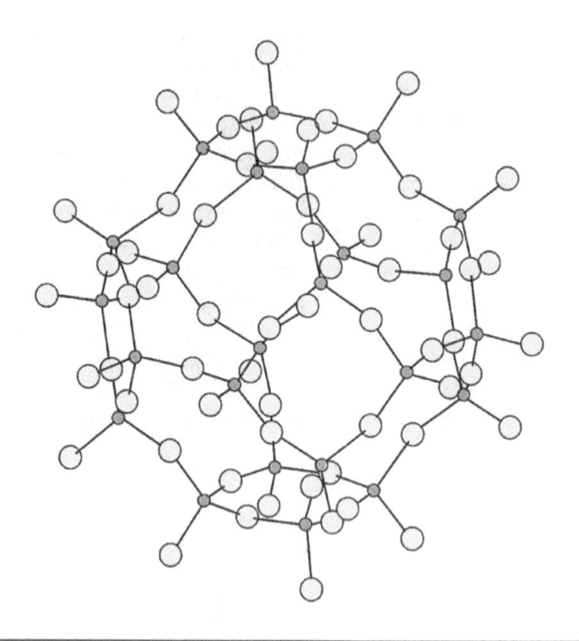

FIGURE 2.17 *Some silicate minerals, such as quartz, have a three-dimensional lattice where each silicon-oxygen tetrahedron is bonded to four other tetrahedra forming a very strong structure that is extremely resistant to chemical attack.*

charge to two Si^{4+} atoms; consequently, no other positive ions are needed to attain electric neutrality. The structural formula of quartz, SiO_2, is therefore the same as the silicon/oxygen ratio of the network structure, 1:2.

The compositional subdivision of the silicates into ferromagnesian and nonferromagnesian minerals also separates the minerals in terms of a number of basic physical properties. Because the iron in crystal lattices is highly absorptive of light, ferromagnesian sililcates are all dark colored—black, dark green, and brown. The nonferromagnesian silicates, lacking iron, are light in color—colorless, white, or pink. Because of their iron content, the ferromagnesian minerals have a higher specific gravity than the nonferromagnesians. By utilizing these basic differences, mineral representatives of the two groups can usually be easily distinguished.

SPOT REVIEW

1. Many of the silicate minerals are members of a solid-solution series. What is a solid-solution series? Give examples of minerals that represent such a series.
2. Using the drawings in the text, demonstrate how the silicon/oxygen ratio for each of the silicate structures is determined.

Nonsilicate Minerals

Although thus far we have emphasized the silicate minerals, we cannot end a discussion of minerals without placing the nonsilicate minerals in proper perspective. Although the nonsilicate minerals make up less than 10% of Earth's crust, they are nevertheless of great importance. *Hematite,* Fe_2O_3, the representative of the oxide group in Table 2.5, is the major ore of iron. Where would modern civilization be without iron? *Calcite,* $CaCO_3$, is the major mineral of limestone. Limestone is the raw material used for the production of Portland cement, an essential material for the construction industry. *Soda niter,* $NaNO_3$, is an industrial mineral that serves as the raw material for a wide range of products including fertilizer. *Apatite,* $Ca_5(PO_4)_3(OH,F,Cl)$, is used in the production of phosphoric acid, which is a basic chemical ingredient in the manufacture of fertilizers. *Gypsum,* $CaSO_4 \cdot 2H_2O$, is used to make wallboard for home and building construction. The various crystalline and amorphous forms of Al_2O_3 found in *bauxite* are the source of aluminum, which is used for everything from aircraft bodies to beverage cans and aluminum foil. The sulfide minerals are major sources of a range of metals including copper, zinc, and lead. A number of elements, known as *native elements,* appear in nature uncombined in a solid or liquid state. Native elements can be either nonmetallic such as *carbon* and *sulfur,* semimetallic such as *arsenic* and *bismuth,* or metallic such as *gold, copper, mercury,* and *silver.*

CONCEPTS AND TERMS TO REMEMBER

Minerals
Atomic structure
 proton
 electron
 neutron
atomic number
atomic mass
isotope
radioactive
electron capture
octet rule
inert gas
Combining of Atoms
 compound
 cation

CONCEPTS AND TERMS TO REMEMBER *continued*

- anion
- electric neutrality

Bonding
- ionic bonding
- covalent bonding
- metallic bonding
- van der Walls bonding

Crystal structure
- lattice
- crystalline
- amorphous
- mineraloid (glass)

Physical properties
- streak
- hardness
  - Mohs's Scale of Hardness
- form
- mineral cleavage
- fracture
- specific gravity
- density
- luster

Instrumental analysis
- light microscopy
- X-ray diffraction

Classification
- silicates
- nonsilicates

Silicate crystal structure
- silicon-oxygen tetrahedron
- isolated tetrahedral structure
- single-chain structure
- double-chain structure
- sheet structure
- framework structure

Solid-solution series

Mineral groups
- olivine group
- pyroxene group
- amphibole group
- mica group
- feldspar group
  - plagioclase feldspar subgroup
  - alkali feldspar subgroup

REVIEW QUESTIONS

1. The tendency for a mineral to break along a plane of weakness is called
   a. cleavage. c. hardness.
   b. fracture. d. streak.
2. The loss of an electron from an atom will produce a(an)
   a. cation. c. anion.
   b. isotope. d. neutron.
3. The hardness of quartz on Mohs's Scale of Hardness is
   a. 4. c. 7.
   b. 5. d. 9.
4. An atom has 13 protons and 14 neutrons in its nucleus. Which of the following is true?
   a. The atomic weight is 13.
   b. The atomic number is 13.
   c. The atomic number is 27.
   d. The nucleus is surrounded by 27 electrons.
5. The most abundant element in Earth's crust is
   a. silicon. c. oxygen.
   b. iron. d. aluminum.
6. The silicate mineral group that is characterized by the single-chain structure is
   a. pyroxene. c. plagioclase feldspar.
   b. mica. d. amphibole.
7. The physical property known as "fracture" is shown by
   a. all minerals.
   b. minerals that do not show cleavage.
   c. silicate minerals with sheet-type structures.
   d. minerals with hardness in excess of 7.
8. The physical property that would most readily distinguish a pyroxene from a feldspar is
   a. density. c. form.
   b. color. d. hardness.

REVIEW QUESTIONS *continued*

9. The most abundant group of minerals in Earth's crust are
   a. oxides. c. ferromagnesians.
   b. feldspars. d. carbonates.
10. Why does the proper identification of any object usually require information about both its composition and its structure? Give some examples where using only compositional or only structural information might result in an erroneous identification.
11. Several naturally occurring substances that are commonly referred to as "mineral resources" are by definition not minerals and therefore should be referred to as natural resources, not mineral resources. Give some examples and explain why each technically would not qualify as a mineral resource.
12. Why must atoms be converted into ions before they can be joined together into compounds?
13. Discuss the relative importance of composition and crystal structure in determining each of the physical properties used to identify minerals.
14. Of the feldspar minerals, anorthite and albite represent the end-members of a series of minerals referred to as the plagioclase feldspars. In what way are the plagioclase feldspars similar? In what way are they different? Why is orthoclase feldspar excluded from the plagioclase series?

THOUGHT PROBLEMS

1. Assume you have the following items: (a) a beaker graduated in tenths of milliliters, (b) a balance capable of determining mass to a tenth of a gram, and (c) a mineral specimen. How would you determine the density of the mineral?
2. What determines whether a particular atom will be prone to form a cation or an anion?
3. Using diagrams of the basic structures, explain why halite has three mutually perpendicular sets of cleavage planes, the pyroxene and amphibole minerals have two sets, and the micas have only one set.

FOR YOUR NOTEBOOK

Most of us come in contact with minerals each day. We have already commented on the use of feldspars in many household cleansers. It might be interesting to consider your home room by room to determine which, if any, minerals can be found. Don't forget certain health additives such as dolomite tablets. Many of us wear jewelry made of gold or silver, much of which is adorned with various kinds of gemstones. Although the names of many of the more common gemstones are familiar, their compositions and modes of origin are not. Using a list of birthstones as a starting point, collect information about the composition, mode of occurrence, and common sources of the gems. In addition to the information that may be available in a mineralogy textbook, a visit to a local jeweler or a rock and mineral shop will provide some more practical information such as their relative values and how they are most likely to be found as settings.

KEY TERMS		
Anhedral Crystal	Effervescence	Octahedron
Basal Cleavage	Euhedral Crystal	Rhombohedron
Cleavage	Fracture	Specific Gravity
Cube	Luster	Striations
Double Refraction	Mohs Hardness Scale	Subhedral Crystal

IDENTIFICATION AND PROPERTIES

Minerals are arguably the most important aspect of your Physical Geology laboratory experience. They are found in every aspect of our lives and make up all rocks. Once you understand how minerals form and how they make up the earth's materials, your textbook readings will become much more meaningful.

Glance around your laboratory to confirm the importance of minerals. The floor, walls, ceiling, lighting . . . and even your clothes are products of natural minerals obtained and refined.

Some minerals are constructed of atoms and molecules that form beautiful gems like diamonds, emeralds, and rubies . . . but much more useful minerals occur as ores of metals, materials used in wallboard, paper, and concrete. More than 3000 minerals are now known to scientists, and the number continues to grow. The majority are obscure and seldom seen, so we will concentrate on those minerals most common and useful to society.

The definition of a mineral is fundamental. Geologists generally agree that a mineral must:

1. **Be naturally occurring.** No materials known only from the laboratory or from industry will qualify. Example: There are a few man-made materials that are believed to be slightly harder than diamonds, but they are manufactured and therefore do not qualify as minerals.
2. **Be inorganic.** Minerals are normally considered NOT to have been the product of some living organism or composed largely of compounds of carbon. Example: Coal made from partially decomposed plant materials cannot be a mineral.
3. **Have a definite chemical composition.** Each mineral has a defined and distinctive chemical composition. Example: NaCl is the chemical formula for ordinary table salt, the mineral halite.

   If we were able to prepare a hypothetical compound whose formula could be written as Na_2Cl (bisodium chloride), it would almost certainly have different properties such as taste, color, crystal shape, etc., and would no longer be halite. (But such a compound is impossible due to the limits of atomic size and chemical bonding.)
4. **Have distinctive and definitive physical properties.** Mineral appearance and character is controlled by the arrangement of atoms and molecules. Example: Cleavage of halite is controlled by the packing order and atomic size of sodium and chlorine atoms. These atoms fit so that halite crystals can only break along planes at right angles to each other, forming ***cubes*** (see later discussion about cleavage).
5. **Be a crystalline solid or have characteristic internal structures** (atomic arrangement). Minerals could be identified solely from texture, mass, color, and general appearance, but, if you used this method, you would never be sure whether the mineral you discovered on your vacation was one you already knew or one that you had never seen before. It is better to study diagnostic features of minerals and to learn to distinguish them through the use of identification keys.

This chapter presents a list of characters known to be most useful in identifying minerals and will teach you to quickly and accurately determine their names.

The tests are all simple, because you won't have laboratory balances, hardness points, or other apparatus while standing on the side of that tall mountain with an unknown mineral specimen in your hand.

To acquire these skills, you need the following materials:

- A tray with 25 to 35 mineral specimens
- A dropper bottle containing 3 to 5 percent hydrochloric acid

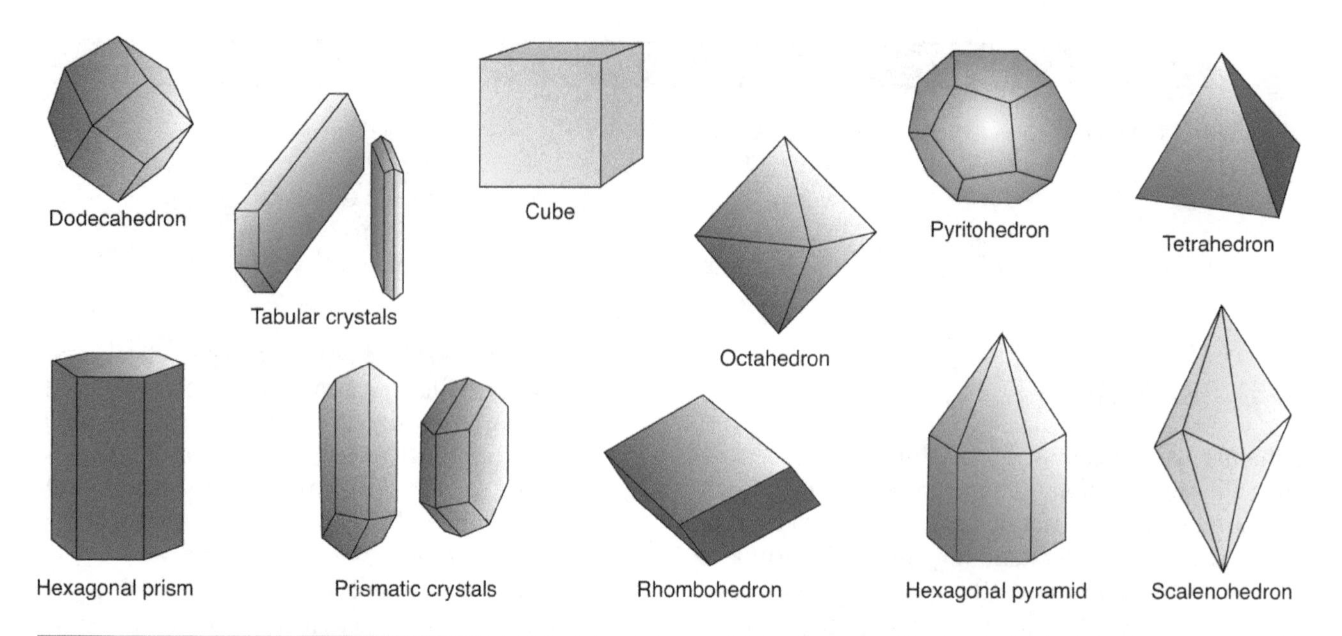

FIGURE 2.18 *Common crystal habits.*

- A copper penny
- An ordinary wire carpenter's nail
- A hardened masonry nail
- A glass plate
- An unglazed porcelain streak plate
- A magnet
- A pencil
- An eraser

We first need to build a vocabulary of useful terms. Ask your instructor to help apply these concepts as you sort minerals by the major character features described below such as streak, luster, hardness, or specific gravity. When you learn to recognize these features, you will quickly develop the ability to distinguish between even very similar minerals.

Mineral Features Useful for Identification

Crystal Shape

Crystals are naturally occurring structures that grow by the addition of atoms on smooth faces of minerals. These surfaces (faces) that are produced give a regular shape that is often characteristic for a given type of mineral. For instance, quartz crystals are elongated, six-sided shapes ***(hexagonal),*** while calcite crystals have a distinctive "leaning" rectangular shape known as a ***rhombohedron*** (see Figure 2.18). One of the most important identification tools a mineralogist can use is the fact that the angles between any two adjacent faces of a particular mineral are always the same.

When crystals meet the precise angular rule above, we say that the crystal is ***euhedral***—that is, extremely regular and uniform in shape. (See Figure 2.19.)

Crystals often lack the open space within a rock that allows them to grow into perfect euhedral

FIGURE 2.19 *It is important to remember that the same crystal shape can occur in more than one kind of mineral. In this photograph of the mineral galena, note that the crystal in the upper left corner is an octohedron, while others are cubes (upper right), and still more show combinations in between the two shapes. All of these dark gray crystals have the same chemical formula and are the same mineral.*

shapes. The crystal may have one or several of the sides pressed against other crystals, and those sides are then suppressed or distorted. This is a much more common condition, and these crystals are called ***subhedral.*** (See Figure 2.20.)

If mineral grains grow in compact masses so that there is little space for individual crystal faces to develop, then the crystals will lack well-defined euhedral faces and the grain will be irregular in shape. This condition is called ***anhedral*** and is the most common condition of all.

Fracture and Cleavage

Some minerals have internal atomic order that prevents regular breaking—that is, if you break these minerals, the resulting pieces are irregular. An example would be the fragments produced by dropping a glass bottle onto a hard concrete surface. No two pieces of the broken glass would have exactly the same shape. This manner of breakage is termed ***fracture.*** Quartz and obsidian are two examples of substances that fracture into odd-shaped pieces (see Figure 2.21), even though crystals of quartz may have beautifully regular shapes.

Cleavage is a very useful feature, because minerals that cleave will break along predetermined planes of weakness generated by the internal atomic arrangement. A piece of the mineral halite has chlorine and sodium atoms arranged to produce three planes of weakness at right angles to each other.

Test this by placing a knife blade against a piece of halite parallel to one of the flat sides and tap the blade with a hammer. Halite breaks smoothly, producing a flat surface. Move the blade in a little toward the center of the halite piece, parallel to the flat surface just produced, and tap it again. The halite will again break, producing a new surface perfectly parallel to the first one.

Cleavages can produce different numbers of cleavage patterns, and cleavage planes often have different angular relationships. Figure 2.22 shows the common types you will work with in this laboratory session.

Micas (biotite and muscovite) are examples of minerals with a single perfect cleavage, a type often called ***basal cleavage,*** causing them to split into thin parallel sheets. In working with basal cleavage, think of a stack of paper sheets. Each sheet is parallel to the others, and each will lift smoothly, but the only way to divide the sheets across will result in an irregular torn edge. This type of cleavage can also be referred to as "basal pinacoid."

Other minerals cleave into two sets of planes (four parallel faces) with the other two sides of the mineral rough and irregular. Some minerals you test will have two planes at right angles, while others will have a pair of planes at some angle clearly NOT at right angles.

A few minerals have three cleavage planes (six parallel faces) and cleave into cubes, rectangles, or parallelograms called ***rhombohedrons. Octahedral*** minerals with four cleavage planes (eight parallel faces) are more rare, while those with more than four are seldom seen among the common minerals. The number and arrangement of cleavage planes is one of the best identification clues. (See Figure 2.22.)

FIGURE 2.20 *Photograph of anhedral (left), subhedral (center), and euhedral crystals (right).*

FIGURE 2.21 *Fractured obsidian and cleaved halite.*

Number of Cleavage Directions	Shapes that Crystal Breaks Into	Sketch	Illustration of Cleavage Directions
0 No cleavage, only fracture	Irregular masses with no flat surfaces	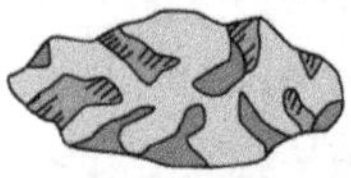	None
1	"Books" that split apart along flat sheets		
2 at 90°	Elongated form with rectangular cross sections (prisms) and parts of such forms	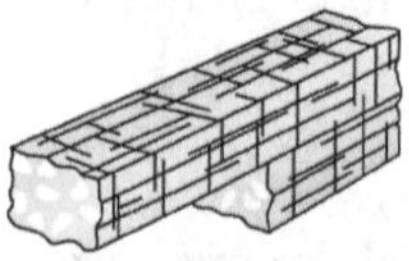	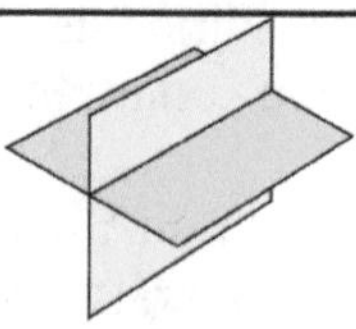
2 not at 90°	Elongated form with parallelogram cross sections (prisms) and parts of such forms	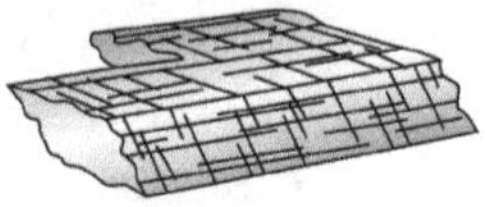	
3 at 90°	Shapes made of cubes and parts of cubes		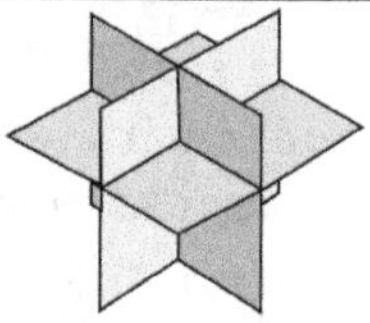
3 not at 90°	Shapes made of rhombohedrons and parts of rhombohedrons	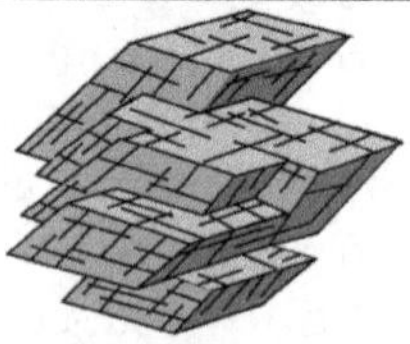	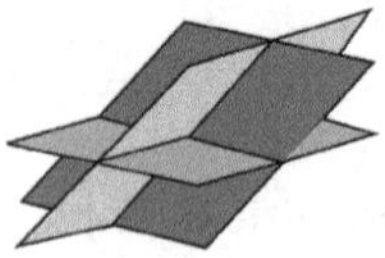
4	Shapes made of octahedrons and parts of octahedrons	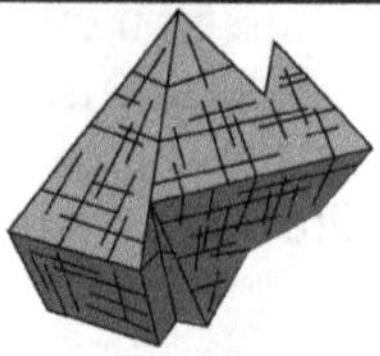	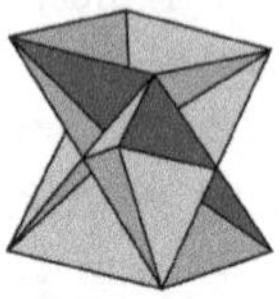
6	Shapes made of dodecahedrons and parts of dodecahedrons		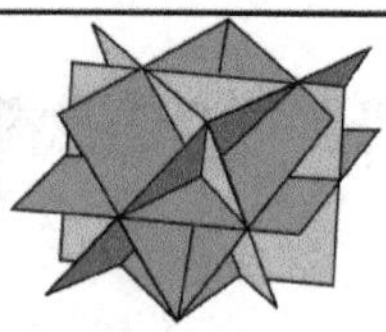

FIGURE 2.22 *Common cleavage patterns in minerals. From Laboratory Manual in Physical Geology, 3rd edition by Busch and Tasa, MacMillan Publishers 1996.*

It is important to note that some minerals, such as clays, DO TECHNICALLY HAVE cleavage, but their mineral grains and crystals are so small that cleavages can only be seen with electron microscopes.

Identification keys later in the chapter will help guide you to the right answer in these unusual cases.

One way to determine whether cleavage is present is to hold the specimen in your hands and rotate it with a beam of light reflecting from the surface. A bright evident flash of reflected light will disclose a smooth surface, often a cleavage surface. Then, rotate the specimen carefully through 180° (rotate it so that

INSIGHTS

If a mineral displays cleavage, each broken surface will be parallel to those surfaces already present. These parallel broken surfaces are called cleavage planes. Therefore, cleavage planes on opposite sides of the mineral sample should be parallel to each other, or nearly so. If the mineral is a crystal, this may not be true. Place the mineral on the table so it rests on one of the flat surfaces. Is there another flat surface parallel to the table on the top surface of the mineral? If so, you may have found a pair of cleavages that represent one of the cleavage planes. If not, you may be working with a subhedral crystal.

Be aware that crystals may well have paired faces, too, but, in the important examples, the broken ends of the crystal will generally show fractured surfaces.

(NOTE: Two parallel cleaved surfaces in a mineral specimen represent ONE plane of cleavage. Example: A cleaved halite cube has six surfaces representing three pairs of parallel surfaces and therefore has three planes of cleavage.)

it turns over and the opposite side is upwards), and if you get the bright reflection repeated, you are probably looking at the same cleavage plane.

Surfaces will reflect light differently if they are cleaved than they will if they are fractured. Some minerals will cleave with broad smooth surfaces (Figure 2.23) while others may split so that there are many parallel mini-faces that are at different heights. But in this second case, all the parallel steps will still reflect together (Figure 2.24).

Fractured faces will scatter light and never return good clean reflections (Figure 2.25). (Caution: Some materials like quartz have smooth crystal faces that may resemble cleavage surfaces, but broken surfaces are always irregular.)

Note that in Figure 2.23, parallel light rays arrive at a smooth plane surface and the reflected rays all leave at a common angle. This type of cleavage is found in many minerals and is essentially the same effect as that seen when light is reflected off a sheet of glass.

In Figure 2.24, parallel light rays arrive at a surface, but the smooth plane surface is broken up into many flat surfaces at different heights, but which are

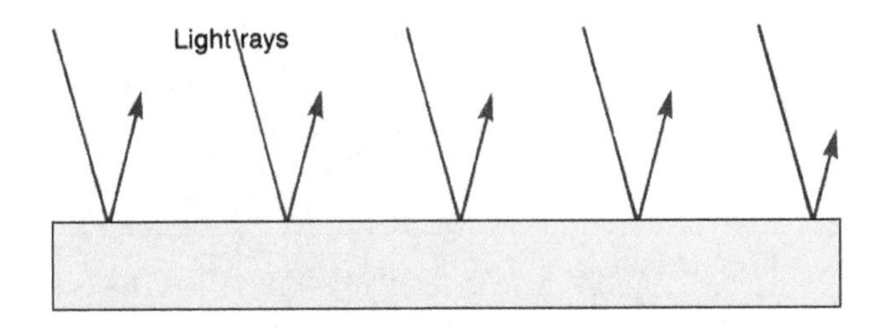

FIGURE 2.23 *Smooth cleavage faces reflect light evenly.*

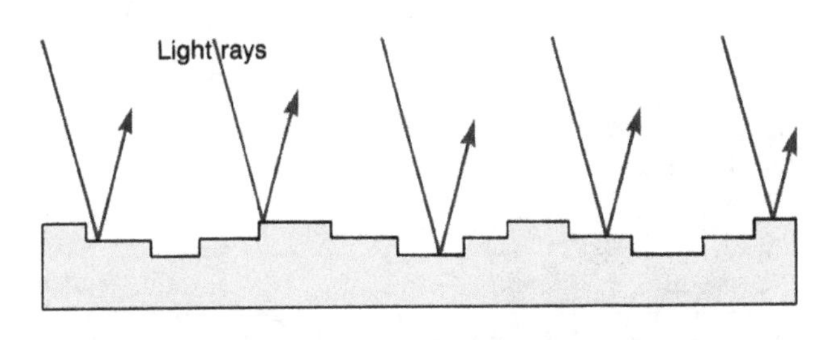

FIGURE 2.24 *Some cleavage faces reflect light evenly, but from surfaces made of parallel faces they may appear irregular.*

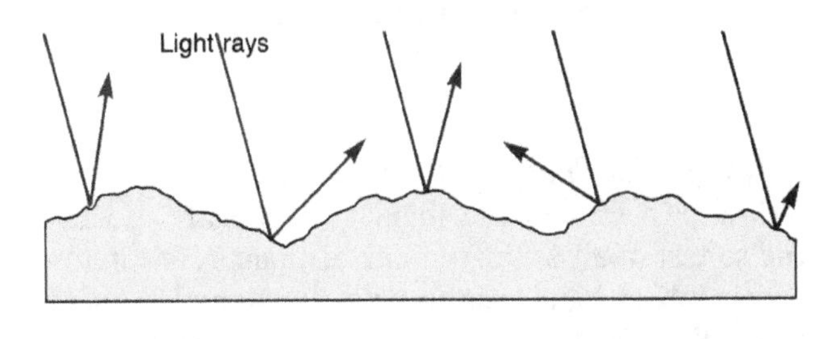

FIGURE 2.25 *Fractured faces of minerals reflect light unevenly.*

still parallel to each other. The arriving light rays will STILL be reflected at a common angle. This type of reflection effect is often seen in minerals such as hornblende.

In Figure 2.25, parallel arriving light rays strike an irregular surface. The reflected rays must then leave the surface in many different directions. This is called "scattered" light, or disordered light, and is the condition seen in minerals that lack cleavage.

Hardness

Hardness of minerals is also important and is easily determined thanks to the Mohs Hardness Scale (Figure 2.26). This scale compares the hardness of all known minerals against each other.

Mohs Scale of Relative Hardness	Useful Aids
10 Diamond	
9 Corundum	
8 Topaz	
7 Quartz	
6 Orthoclase	
	5.5 Glass, Masonry Nail
5 Apatite	
	4.5 Carpenter's Nail
4 Fluorite	
	3.0–3.5 Copper Penny
3 Calcite	
	2.5 Fingernail
2 Gypsum	
1 Talc	

FIGURE 2.26 *Mohs hardness scale.*

Mohs Hardness Scale uses ten common or well-known minerals as memory aids, starting with the softest of all ordinary minerals, talc. A few items commonly at hand help us with the Mohs Hardness Scale. Your fingernails are hardness 2.5; a REAL copper penny is about 3.5 (modern composite pennies can be softer); a standard carpenter's nail is 4.5; and a hardened masonry nail is 5.5. (Be careful, many masonry nails today are much softer than you might have expected.) Another 5.5 hardness tool can be a piece of common glass.

Luster is a measure of the way minerals reflect light. Reflection styles include: metallic, nonmetallic, vitreous (glassy), earthy, waxy, glossy, and greasy. Practice determining these surface features with the minerals in your kit.

Specific gravity determines how heavy (dense) a mineral is compared to the density of water. Water has a specific gravity of 1.0, and when we learn that a mineral has a specific gravity of 2.5, we know that, volume for volume, the mineral is 2.5 times as heavy as water.

Determining specific gravity is easily learned by using your hands as scales, a technique known as hefting. Place an unknown mineral in one hand, a known mineral in the other, and while holding your arms slightly away from your body, gently bounce the minerals three or four times in your hands. Then, quickly switch hands and do it again. One mineral should now feel substantially heavier than the other, and the general specific gravity of the unknown mineral can be estimated. Make sure the unknown and known mineral samples are about the same size.

Minerals generally resist breaking, a feature known as **tenacity**. Some are **brittle**, like glass, shattering into many pieces, while others are **malleable**, like many metals, and can be easily deformed. A few are **flexible** (like the easily bent micas), and even more rarely flexible minerals are also **elastic** (they return to their original shapes after the stress is released, like a metal spring).

Striations are shallow, straight, near-microscopic grooves on cleavage surfaces. Orient the specimen so that the flat surfaces reflect light, and striations will become evident. This is a very useful feature to distinguish between the two major groups of feldspars.

INSIGHTS

SUMMARY: Minerals that display cleavage break into regular shapes, while those that exhibit fracture break into irregular shapes.

INSIGHTS

Hardness and strength are NOT the same thing. Some very hard minerals are fragile and can be easily crushed or broken. (Diamond is such a mineral.)

Because some minerals are brittle, it is often useful to run the hardness test both ways. Press a carpenter's nail into a piece of granular olivine, and because olivine is so brittle and fragile, it will show a clear scratch, even though olivine is hardness 8 and the nail is 4.5. Take olivine and rub it against the nail, and you will see bright shiny steel. The olivine easily scored the metal, demonstrating how much harder it was than the nail. Granular olivine crystals are poorly bonded and come apart when scratched, falsely suggesting that the nail is harder.

Remember, minerals with higher Mohs numbers always scratch those with lower numbers.

> **INSIGHTS**
>
> Lines resembling striations will appear on the external surfaces of some crystals. These lines are often the result of pauses in the growth of those crystals and do not appear on the cleavage surfaces.

FIGURE 2.27 *Labradorite feldspar with striated surface.*

Plagioclase feldspars (generally white, cream, gray, or dark in color) show well-developed **striations** on some cleavage surfaces. The striations are very perfectly developed and are precisely spaced or in groups of parallel sets. These grooves will be parallel to the other cleavage face (Figure 2.27).

Orthoclase (potassium) feldspars are generally brighter in color than plagioclase types (pink, orange, blue, and green) and lack striations (Figure 2.28).

Effervescence (fizzing) results when a weak solution (3–5 percent) of hydrochloric acid is dropped onto the surface of some carbonates. Calcite, the main mineral present in limestone, reacts well (see Figure 2.29). Dolomite, another carbonate, only fizzes well when the mineral is powdered. When testing for dolomite, the hydrochloric acid will effervesce on ordinary surfaces, but the bubbles will be produced very slowly, while the fizzing on calcite is immediate.

The acid should be supplied in dropper bottles and when used as directed, will not be dangerous. Try not to get it on clothing (it might change the color of

FIGURE 2.28 *Nonstriated orthoclase feldspar.*

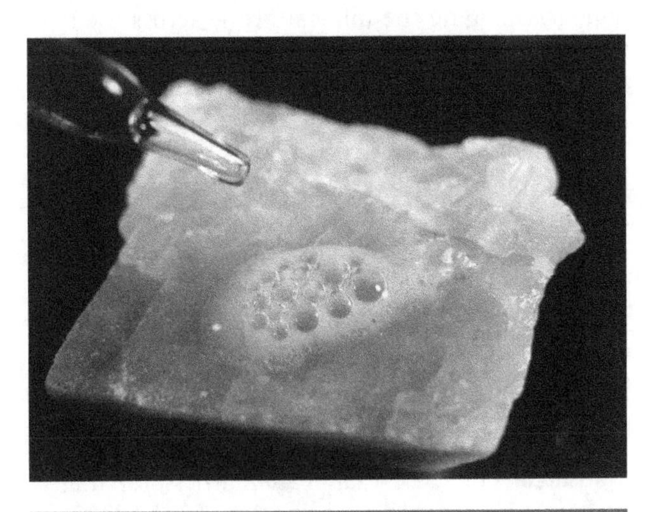

FIGURE 2.29 *Effervescence of hydrochloric acid on calcite.*

FIGURE 2.30 *Color varieties of calcite.*

dyes) and keep it off your lab manual (it will degrade the paper). If spilled, just wipe it up. Good manners dictate that you should always wash acid off the minerals you have tested.

Mineral color would seem to be one of the very best tools in identification. Instead, it is one of the most treacherous for the beginner. Some common minerals such as quartz occur in a wide variety of colors. Color is often the result of trace amounts of chemicals within the main mineral. While the different colored varieties of quartz have been used as semiprecious gemstones, they are still only quartz. Calcite and fluorite display a similar wide range of colors (Figures 2.30 and 2.31).

Streak color is much more reliable in identifying minerals and is one of your best tools. Streaking is done by dragging the mineral firmly across the surface of a piece of unglazed porcelain (a **streak plate**). Blow off the excess powder and small broken grains and examine the color produced (Figure 2.32).

The color on the streak plate is the actual color of the mineral. The way your eye perceives color is a complex effect. Colors you see are a combination of reflected and absorbed light frequencies, plus the texture of an object will also affect the way you see the color. By rubbing the mineral on a streak plate, the mineral is ground into small particles of uniform size, and therefore all varieties of the same mineral will have identical colors when viewed after streaking.

In some cases, such as with the mineral hematite, the color on the streak plate may be quite different than the apparent color of the unstreaked mineral.

Magnetism occurs naturally in a very few minerals, especially the iron mineral magnetite. This form of the mineral is called a lodestone and will attract iron filings, paper clips, and other magnets. This feature, plus the black streak, is definitive for the mineral. See Figure 2.33.

Double refraction is a condition where the atomic structure causes light to be broken into two rays that pass through the mineral in different pathways. If such a mineral is transparent, you will see a pair of images when an object is viewed through it. See Figure 2.34.

Some minerals will be **flexible** in thin sheets. Flexible minerals will bend, but after being bent, will not return to their original shape when the bending force is removed. Gypsum is actually flexible in THIN sheets (but the sheets must be very thin and bent very slowly). Other minerals can be both **flexible** and **elastic**, such as micas, which bend and then return to their original flat shapes when the force is removed.

INSIGHTS

Orthoclase feldspars sometimes appear to be striated, but close examination will show that these false striations result from parallel light- and dark-colored lamellae (planes) within the crystal itself and are NOT surface features (See Figure 2.28).

INSIGHTS

Reserve the use of color until you are nearly finished with the identification process and employ it mainly as a confirming feature, never as a primary character.

FIGURE 2.31 *Color varieties of fluorite.*

FIGURE 2.32 *Streak plate and color varieties of hematite.*

Streak color is a powerful tool. Which samples:	
SHOW Colored Streaks	SHOW White Streaks
Sample Number	Sample Number

A few minerals will separate easily into fibers such as asbestos. Due to the current controversy over the potential health hazards of this mineral, you will not work with it in your kit, but perhaps your instructor will demonstrate this interesting feature while taking proper safety steps (exhibit fibrous asbestos only in a vented fume hood).

INSIGHTS

Be sure to press the mineral FIRMLY against the streak plate. If the mineral is only timidly pressed against the porcelain, it may produce a streak so weak that the true color will not be apparent. Remember, STREAK the mineral firmly onto the streak plate, but don't STRIKE the plate.

FIGURE 2.33 *Magnetite showing natural magnetism.*

FIGURE 2.34 *Double refraction of calcite.*

ESS 101

Lab 1: Mineral Properties and Mineral Identification

Student Name: ______________________________

The goal of this Laboratory activity is to acquire the skills necessary to identify common minerals based on their properties.

Part 1. Define a mineral:

____

____

____

____

____

Part 2. Mineral Properties

1. Luster Metallic vs Non-metallic
2. Use Mohs Scale of Hardness to estimate the hardness of the following materials:

   Finger nail ______________________________

   Copper penny ______________________________

   Wire (iron) nail ______________________________

   Glass ______________________________
3. Define cleavage:

Part 3. Mineral Identification

Recognize and record mineral's properties. Note that every single mineral has luster, color, and hardness. Some minerals have a tendency to break along planes of weakness between parallel layers of atoms in crystal (show cleavage), while others have no cleavage. Properties such as reaction with hydrochloric acid, striations, magnetism, feel, odor, or double reflection are specific to certain minerals only.

Use mineral identification charts to identify mineral samples.

Sample #	Luster	Hardness	Streak	Cleavage	Other Properties	Mineral Name	Mineral Composition	Mineral classification (based on chemical composition)

Part 4. Mineral Classification.

Known rock-forming minerals are classified based on their chemical composition. First there are two major groups:

1. Silicates:
   a. Ferromagnesian silicates, classified based on iron (Fe) and magnesium (Mg) content
   b. Non-ferromagnesian silicates

2. Non-silicates
   a. Oxides
   b. Carbonates
   c. Sulfates
   d. Sulfides
   e. Halides
   f. Native Elements

Once you correctly identified mineral samples, refer to the mineral's chemical composition listed under the mineral name in the tables to classify minerals into the groups listed.

IDENTIFICATION KEY TO MINERALS

Methods:

(Use the accompanying data sheets to record your observations.)

Identification of minerals is easy when keys are used. The keys presented here (Figures 2.35 and 2.36) use the techniques and physical characteristics described earlier.

> **INSIGHTS**
>
> First, sort the minerals into similar groups; for example, find all the metallic minerals first or perhaps all those that show cleavage. You might want to determine the hardness of the minerals before beginning to precisely identify your specimens.

STEP ONE: If your unknown mineral shows cleavage, use the first key (Figure 2.36). Start by trying to produce a colored streak, and if you are successful, compare the streak with the colors indicated on the key.

STEP TWO: If there is no streak produced (or if the streak is white), determine the number of cleavages present and the angles between adjacent cleavage planes to find the correct name of the mineral.

STEP THREE: If your unknown mineral does not show cleavage, move to key two (Figure 2.36) and follow the same steps. First, check for streak color and, if unsuccessful, move to the lower part of the key where miner tiated by their hardness.

STEP FOUR: After you feel the mineral has been identified, check Tables 2.6 through 2.10 and see if the detailed clues also match. These tables are arranged by characters: luster, hardness, streak, general properties, and chemical formulae.

Your instructor will provide you with the basic tools for mineral identification:

- Set of 15 to 25 common mineral specimens
- Dropper bottle with 5 percent hydrochloric acid
- Copper penny
- Carpenter's nail
- Masonry nail
- Sheet of glass
- Streak plate

Work in teams of two to four students and use the keys on the following pages to identify the mineral specimens.

NOTE: Be alert to the results of each of your test steps. If you have made an error and are on the wrong chart, you will quickly begin getting nonsense results. If you believe this is happening, turn to the other chart. To a remarkable degree, the chart is self correcting . . . if you pay attention!

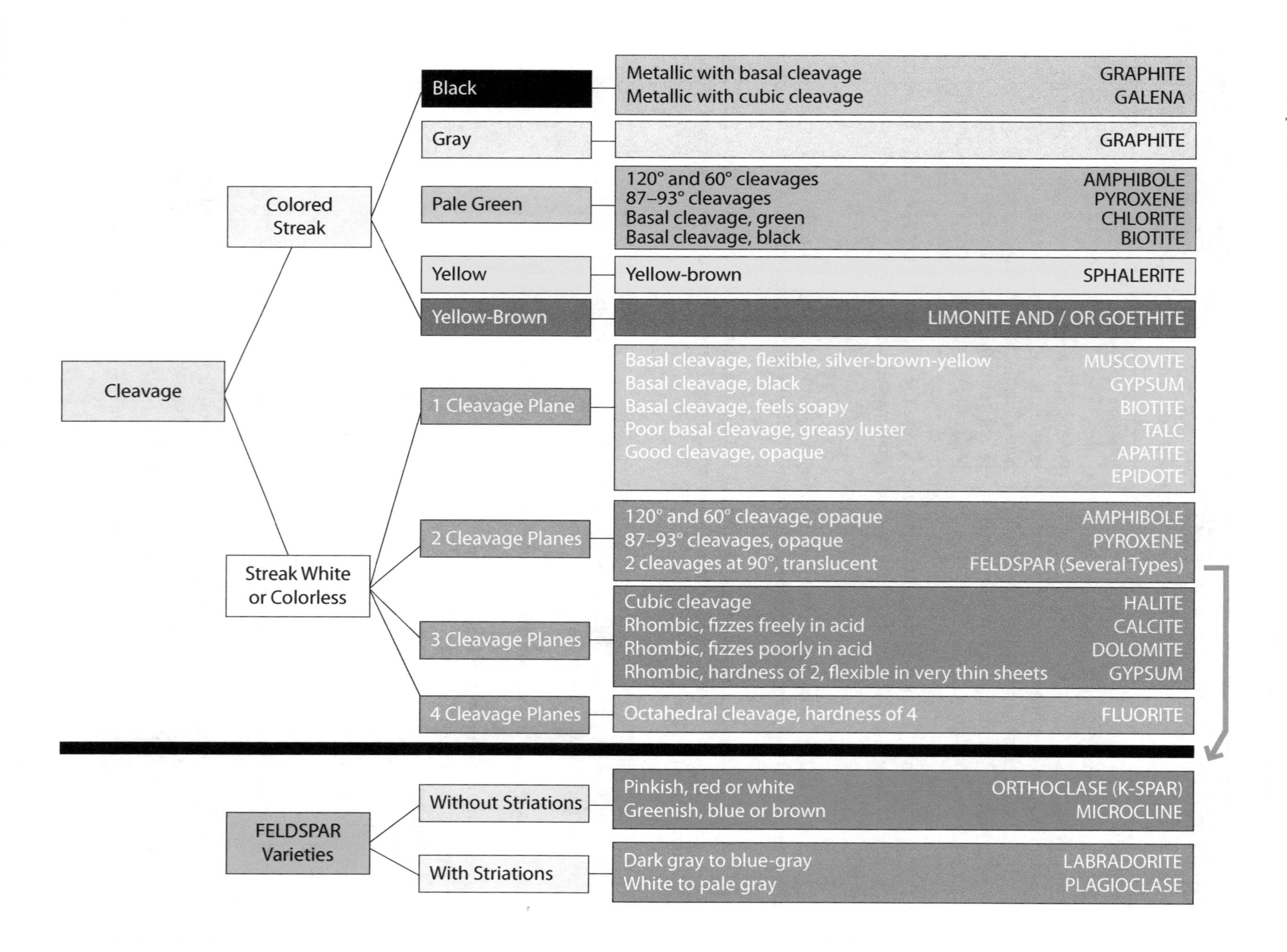

FIGURE 2.35 *Minerals WITH Cleavage.*

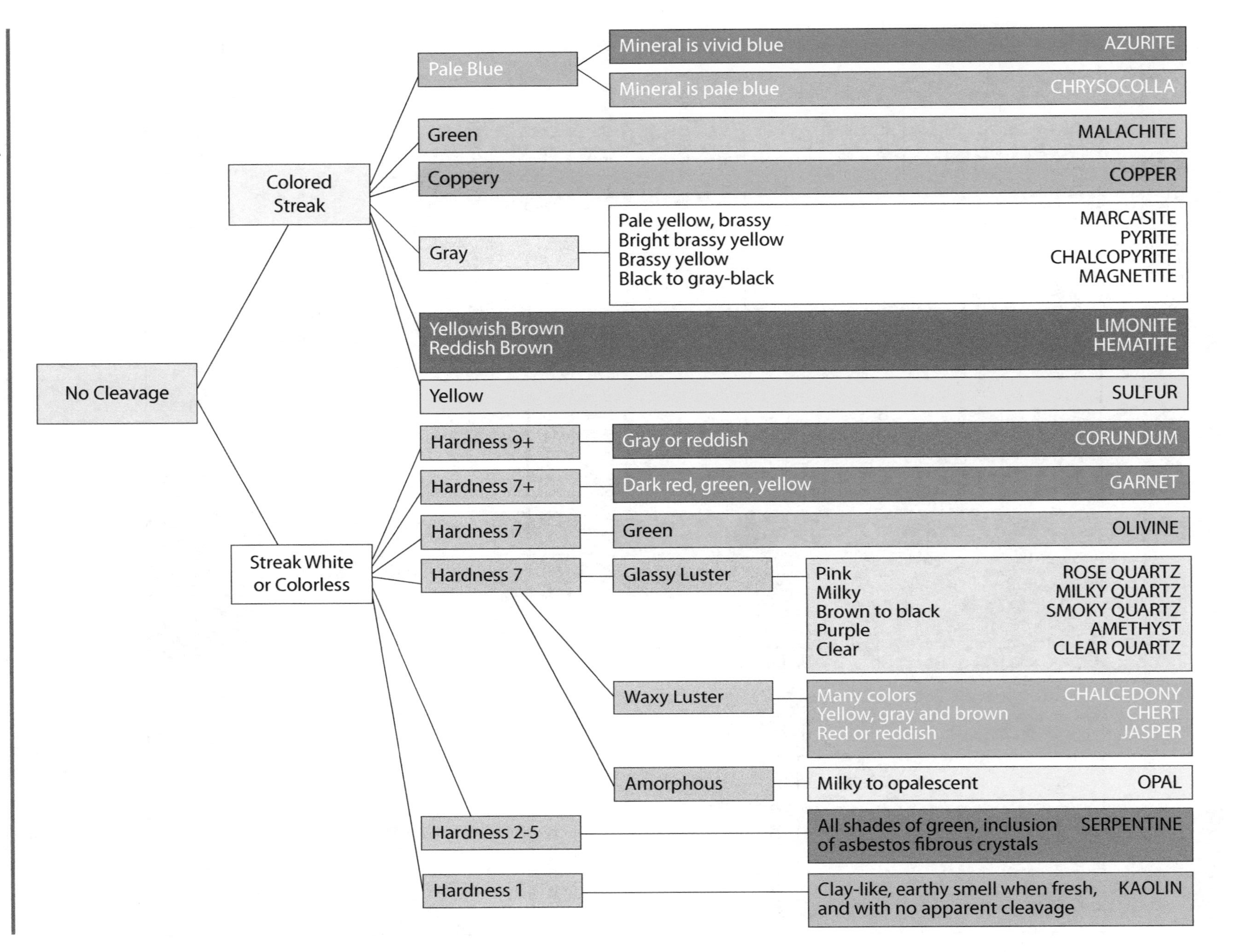

FIGURE 2.36 *Minerals WITHOUT Cleavage.*

TABLE 2.6

	METALLIC MINERALS			
	Hardness	Streak	Other Properties	Mineral
Not scratched by steel nail or knife	6.5-6	Dark gray	brass yellow, may tarnish brown; brittle, no cleavage, cubic crystals common, S.G. = 5.0.	PYRITE ("fool's gold") FeS_2 iron sulfide
	6.5-6	Dark gray	pale brass yellow to whitish gold; brittle, no cleavage, radiating masses and "cockscombs," S.G. = 4.9.	MARCASITE FeS_2 iron sulfide
	6	Dark gray	dark gray to black; magnetic, no obvious cleavage, S.G. = 5.2.	MAGNETITE Fe_3O_4 iron oxide
Scratched by steel nail or knife	6.5-5	Red to red-brown	silver to gray, may be tiny glittery flakes may tarnish red, S.G. = 4.9-5.3.	HEMATITE Fe_2O_3 iron oxide
	5.5-5	Yellow-brown	dark brown to black, in radiating layers, S.G. = 4.3.	GOETHITE FeO(OH) hydrous iron oxide
	5.5-5	Yellow-brown	yellow-brown to dark brown; amorphous, but may be pseudomorphic after pyrite, S.G. = 4.1-4.3.	LIMONITE $Fe_2O_3 \cdot nH_2O$ hydrous iron oxide
Scratched by wire nail	4-3.5	Dark gray	golden yellow, may tarnish purple; brittle, no cleavage, S.G. = 4.1-4.3.	CHALCOPYRITE $CuFeS_2$ copper-iron sulfide
	4-3.5	White to yellow-brown	brown to yellow, or black; submetallic, dodecahedral cleavage, S.G. = 3.9-4.0.	SPHALERITE ZnS Zinc sulfide
Scratched by penny	3-2.5	Copper	copper to dark brown, may oxidize green; malleable, S.G. = 8.8-8.9.	NATIVE COPPER Cu copper
	2.5	Gray to dark gray	silvery gray, tarnishes dull gray; cubic cleavage, not scratched by fingernail, S.G. = 7.4-7.6.	GALENA PbS lead sulfide
Scratched by fingernail	1	Dark gray	gray to black, marks paper easily; greasy feel, S.G. = 2.1-2.3.	GRAPHITE C carbon

TABLE 2.7

	Hardness	Streak	Other Properties	Mineral
	NONMETALLIC MINERALS			
Not scratched by steel nail or knife	9	White	gray, red, brown, blue; greasy luster, commonly in six-sided crystals with striated flat ends; no cleavage, S.G. = 3.9-4.1.	CORUNDUM Al_2O_3 aluminum oxide
	8	White	colorless, yellow, blue, or brown; one perfect cleavage, crystal faces often striated, S.G. = 3.5-3.6.	TOPAZ $Al_2O_4(OH,F)_2$ hydrous fluoro-aluminum silicate
	7.5-7	White	green, yellow, pink, blue, brown, or black slender crystals with rounded triangular cross sections; striated crystal faces, no cleavage, S.G. = 3.0-3.2.	TOURMALINE complex silicate
	7	White	any color to colorless, transparent to translucent, greasy luster, no cleavage, conchoidal fracture, S.G. = 2.7.	QUARTZ SiO_2 silicon dioxide
	7	White	white, light colors; waxy luster, translucent, often banded masses cryptocrystalline, S.G. = 2.5-2.8.	CHALCEDONY SiO_2 cryptocrystalline quartz
	7	White	black, cryptocrystalline, waxy, conchoidal fracture, translucent to opaque, S.G. = 2.5-2.8.	FLINT SiO_2 cryptocrystalline quartz
	7	White	gray, brown, yellow; cryptocrystalline, waxy luster, opaque, conchoidal fracture, S.G. = 2.5-2.8.	CHERT SiO_2 cryptocrystalline quartz
	7	White	red, opaque, waxy luster, cryptocrystalline, conchoidal fracture, S.G. = 2.5-2.8.	JASPER SiO_2 cryptocrystalline quartz
	7	White	green, black, or yellow; conchoidal fracture, no cleavage, S.G. = 3.3-3.4.	OLIVINE $(Fe,Mg)_2SiO_4$ ferromagnesian sillicate
	7	White	dark red, brown, pink, green, or yellow; transparent to translucent, no cleavage, S.G. = 3.4-4.3.	GARNET complex silicate

TABLE 2.8

	NONMETALLIC MINERALS			
	Hardness	Streak	Other Properties	Mineral
Not scratched by steel nail, knife	7-6	White	green to yellow-green, striated crystals or dull granular masses, one cleavage, S.G. = 3.3-3.5.	EPIDOTE complex silicate
	6	White	blue-gray, black, or white; striations on some cleavage planes; two cleavages at nearly 90°, S.G. = 2.6-2.8.	PLAGIOCLASE FELDSPAR $NaAlSi_3O_8$ to $CaAl_2Si_2O_8$ calcium-sodium aluminum silicate
	6	White	white, pink, brown, green; exsolution lamellae are present and subparallel, two cleavages at 90°, S.G. = 2.6.	POTASSIUM FELDSPAR $KAlSi_3O_8$ potassium aluminum silicate
	6	White	colorless, white, orange, gray, yellow, green red, blue; may have play of colors (opalescence), amorphous, greasy luster to earthy luster; conchoidal fracture, S.G. = 1.9-2.3.	OPAL $SiO_2 \bullet nH_2O$ hydrated silicon dioxide
Scratched by steel nail, knife	5.5	White	green to black, dull, stout crystals; two cleavage directions that intersect at about 87° and 93°, S.G. = 3.2-3.5.	PYROXENE (AUGITE) calcium ferromagnesian silicate
	5.5	White	green to black, opaque, two cleavage directions at 60° and 120°, slender crystals, may be splintery or fibrous, S.G. = 3.0-3.3.	AMPHIBOLE (HORNBLENDE) calcium ferromagnesian silicate
	5	White	brown, green, blue, yellow, purple or black; one poor cleavage, common as six-sided crystals, S.G. = 3.1-3.2.	APATITE $Sa_5F(PO_4)_3$ calcium fluorophosphate
	5-2	White	green, yellow, gray, or variegated green, gray, and brown; dull masses or asbestos fibrous crystals, no cleavage, S.G. = 2.2-2.6.	SERPENTINE $Mg_6Si_4O_{10}(OH_8)$ hydrous magnesian silcate
	5.5-1.5	Red to red-brown	red, opaque, earthy luster, S.G. = 4.9-5.3.	HEMATITE Fe_2O_3 iron oxide
	5.5-1.5	Yellow-brown	yellow-brown to dark brown, amorphous, but may be pseudomophic after pyrite, S.G. = 3.6-4.0.	LIMONITE $Fe_2O_3 \bullet nH_2O$ hydrous iron oxide

TABLE 2.9

	NONMETALLIC MINERALS			
	Hardness	Streak	Other Properties	Mineral
Scratched by wire nail	4	White	colorless purple, blue, yellow, or green; dioctahedral cleavage, crystals usually cubic, S.G. = 3.0-3.3.	FLUORITE CaF_2 calcium flouride
	4-3.5	Light blue	vivid royal blue, earthy masses or tiny crystals, effervesces in dilute HCl, S.G. = 3.7-3.8.	AZURITE $Cu_3(CO_3)_2(OH)_2$ hydrous copper carbonate
	4-3.5	Green	green to gray-green laminated crusts or masses of tiny, granular crystals; effervesces in dilute HCl, S.G. = 3.9-4.0.	MALACHITE $Cu_2CO_3(OH)_2$ hydrous copper carbonate
	4-2.0	Very light blue	pale blue to blue-green crusts or massive, amorphous, conchoidal fracture, S.G. = 2.0-2.4.	CHRYSOCOLLA $CuSiO_3 \cdot 2H_2O$ hydrated copper silicate
	4-3.5	White	white, gray, pink, or brown; opaque; rhombohedral cleavage; effervesces in dilute HCl only if powdered, S.G. = 2.8-2.9.	DOLOMITE $CaMg(CO_3)$ magnesian calcium carbonate
Scratched by penny	3	White	colorless, white, yellow, gray, green, brown, red, blue; transparent to translucent, rhombohedral cleavage; effervesces in dilute HCl, S.G. = 2.7.	CALCITE $CaCO_3$ calcium carbonate
	3	White	colorless, white, red, brown, yellow, blue; platy crystals, massive, or in rose-like shapes; three cleavages, one perfect and at right angles to others; very heavy, S.G. = 4.5.	BARITE $BaSO_4$ barium sulfate
	3	Gray-brown	very dark brown to black, one perfect cleavage; flexible, very thin sheets, S.G. = 2.7-3.1.	BIOTITE MICA ferromagnesian potassium, hydrous aluminum silcate
	2.5	White	colorless, white, yellow, red, blue, brown; cubic crystals and cubic cleavage, salty taste, S.G. = 2.1-2.6.	HALITE NaCl sodium chloride

TABLE 2.10

	NONMETALLIC MINERALS			
	Hardness	Streak	Other Properties	Mineral
Scratched by fingernail	2.5-1.5	Pale Yellow	yellow to red, bright crystals or earthy masses, brittle, no cleavage, conchoidal fracture, S.G. = 2.1.	NATIVE SULFUR S sulfur
	2.5-2	White	colorless, yellow, brown, red-brown; one perfect cleavage; flexible, elastic sheets, S.G. = 2.7-3.0.	MUSCOVITE MICA potassium hydrous aluminum silicate
	2	White	dark green, one perfect cleavage, S.G. = 2.6-3.0.	CHLORITE ferromagnesian aluminum silicate
	2	White	one good cleavage (two poor cleavages); nonelastic sheets, colorless to white; H = 2, easily scratched with fingernail, S.G. = 2.0-2.4.	GYPSUM $CaSO_4 \cdot 2H_2O$ calcium sulfate
	2-1	White	white to very light brown, one perfect cleavage, common as earthy, microcrystalline masses, S.G. = 2.6.	KAOLINITE $Al_4(Si_4O_{10})(OH)_8$ hydrous aluminum silicate
	1	White	white, gray, green, pink, brown, yellow; soapy feel, pearly to greasy luster, massive or foliated, S.G. = 2.7-2.8.	TALC $Mg_3Si_4O_{10}(OH)_2$ silicate

ESS 101

Lab 2: Visit to Geology Museum

Student Name: ______________________________

Museum is located in Schmucker Science Center LINK, Room 145
Museum is open Mondays–Thursdays, 10 A.M.–2 P.M.

Enter the Geology Museum and start with the display cases along the curved wall to your left. Answer the questions below. The title of each display case is underlined below.

Geothermal Energy

1. What is the Temperature of the rock below the earth's surface?
   By how much will WCU reduce its heating costs by using geothermal energy?

   ____

Asbestos

2. Name three different minerals that can be called asbestos because they grow in long fibers:

   ____

3. What properties make asbestos minerals useful?

   ____

Electrical Properties of Minerals (it's fun to play with the buttons and switches!)

4. Is copper metal an electrical conductor or an electrical insulator? (Underline the correct answer).
5. What mineral is used to keep time in digital watches because of its atomic vibrations?

   ____

Earth Minerals in Everyday Products

6. Write the name of the correct mineral in front of its use on the list below.

____________ glass	____________ dry wall
____________ pencils	____________ paint
____________ electric batteries	____________ watches
____________ cosmetics	____________ de-icing

Silicate Minerals

7. The two chemical elements that build up this group of minerals are ________ and ________ .
8. What is a basic structural unit of this group of minerals? ______________________

Draw a sketch and label the atoms:

The Many Mines of Chester County and their Minerals

9. Where was the location of the first chromite mine in the U.S.?

____

   What is the main use of chromium?

____

10. The mineral Chalcopyrite was mined in Phoenixville for what element?

____

11. What mineral do we eat with Oreo cookies?

____

12. The mineral rutile belongs to this group of minerals ________________ .
    (see chemical formula of Rutile)

Minerals of Southeast Pennsylvania

13. List minerals familiar to you from Mineral Identification lab that are mined in Southeastern PA:

____

Fluorescent Minerals

14. What type of light (radiation) is used to produce fluorescence in minerals?

____

15. How fluorescent minerals are being used by humans?

____

Minerals in Meteorites

16. Name two minerals that are found in meteorites:

____

Quartz

17. What is the chemical formula of quartz?

____

18. Quartz belongs to this group of minerals ________________ .

19. List varieties of quartz that do not include word "quartz" in their name:

____

Copper

20. List physical properties of copper:

____

21. How old is the oldest copper artifact?

____

22. How much copper will be used by every American born in 2008?

____

Marvelous Micas

23. Micas belong to this group of minerals ______________ .

24. List four different minerals that are called mica:

____

25. What is the major difference between muscovite and biotite micas?

Minerals and Biology

26. List minerals that replace hard parts of organism in fossils?

____

27. What mineral was found in brains of certain animals, including humans?

____

28. How is this mineral being used by an organism?

____

Caves and Caverns

29. Caves and caverns are formed in these rocks ______________ .
30. Why? (explain the process of cave formation)

____

31. What mineral formations are formed only in caves?

____

Minerals in Fireworks

32. Write the name of the correct mineral in front of the color.

____________	yellow	____________	red
____________	orange	____________	white
____________	green	____________	blue

QUESTIONS FOR FURTHER THOUGHT

1. What minerals do you commonly use daily in their natural form?
2. What common minerals do you recognize in your home?
3. What common minerals were used to make your automobile?

EXTRA PROJECTS

Visit local museums. Many of you will be fortunate to have natural history museums where you live. These museums will have exceptionally fine specimens showing the range of variation and beauty of perfect crystals. Make a visit and see how many of the minerals in the displays are the same as those you studied in this chapter.

Rock and mineral shows and sales. Amateur collectors of mineral specimens and gemstones have formed clubs all over the world. These groups often hold combination events where you can view fine rocks and minerals, plus purchase your own samples. Another benefit is that you can meet collectors to learn where and how you might find your own minerals.

Field Trips. Minerals are found everywhere. Although some parts of the world are richer in unusual or beautiful minerals, your instructor can be helpful in suggesting localities you might visit to begin making your own collection.

Always remember to be sure you are welcome at these sites by getting permission BEFORE entering and collecting.

Internet Links. Use Internet sources to find information on useful, rare, and beautiful minerals. Possibly check out minerals that are common but may also be dangerous in the environment.

Check the Internet for national and local gem and mineral societies. Use the Internet browsers to search using these terms: rock, gem, mineral, "rockhound," etc.

CHAPTER 3

Igneous Rocks

INTRODUCTION Igneous rocks form from the solidification of molten rock that originated within Earth. Molten rock below the surface is called **magma.** Once formed, magma rises because it expands and is of lower density and therefore more buoyant than the surrounding rock. Some magmas solidify within Earth's crust and form **intrusive** igneous rocks. In other instances, the magma may move upward along cracks, eventually reaching the surface where the molten rock is extruded as **lava.** The solidification of lava forms **extrusive** igneous rocks. Other extrusive igneous rocks form by the accumulation of bits and pieces of lava or solid rock fragments blasted into the air during volcanic eruptions.

The major constituents of igneous rocks are silicate minerals. At this point, a review of the silicate minerals in Table 3.1 will help prepare you for the discussion that follows.

MELTING AND CRYSTALLIZATION OF SOLIDS

Every solid substance has a temperature above which it will melt and below which the melt will crystallize or solidify. This temperature is called either the *melting temperature* or the *crystallization (freezing) temperature,* depending upon the direction of the transformation. The melting-crystallization temperature is a physical property of every material and is dependent upon its composition and crystal structure. Because the minerals listed in Table 3.1 all differ in composition and/or crystal structure, it is reasonable to assume that they will have different melting-crystallization temperatures. Minerals with variable compositions melt and crystallize over a range of temperatures.

At temperatures above the melting-crystallization temperature, the amount of thermal energy (heat) available is sufficient to overcome the energy of the bonds that hold the crystal lattice together. As the bonds between ions and atoms within the crystal lattice are broken, the material melts. As long as the temperature is maintained above the melting point, the amount of thermal energy within the melt is sufficient to keep the ions and atoms from forming permanent bonds.

TABLE 3.1

THE MAJOR ROCK-FORMING SILICATE MINERALS

Ferromagnesian	NonFerromagnesian
Olivine group	Feldspar group
Fayalite	Plagioclase subgroup
Forsterite	Anorthite
Pyroxene group	Albite
Augite	Orthoclase
Amphibole group	Mica group
Hornblende	Muscovite
Mica group	Quartz
Biotite	

As the temperature drops below the melting-crystallization point, the attraction between the ions is sufficiently great to overcome the effects of the thermal energy. A crystal lattice starts to form (nucleate) as bonds begin to be established between the ions and atoms. As the crystal lattice grows, the mineral begins to precipitate from solution. Below the melting-crystallization temperature, the amount of thermal energy is insufficient to disrupt the chemical bonds, and the crystal lattice remains intact.

For a material of given composition, the melting-crystallization temperature can be affected by two external parameters: (1) *pressure* and (2) the presence of *volatile substances,* in particular, water. Volatile substances are those components that can be readily converted to gas. An increase in pressure increases the melting-crystallization temperature according to a basic precept in chemistry called LeChatelier's rule. Applied to this situation, this rule states that when a liquid-solid transformation is involved, increased pressure will favor the denser of the two phases (Figure 3.1). Therefore, an increase in pressure usually favors the solid phase. The melting temperature will rise because more thermal energy is needed to disrupt the chemical bonds. The common and important exception is water. Because water is denser than ice, increased pressure causes the melting point to decrease. As a result, in the presence of water, the melting temperature of ice decreases with increasing pressure.

One aspect of this relationship will be very important later in our discussions of magma formation. If a solid is already at high pressure and sufficiently high temperature, a rapid reduction in pressure without an associated decrease in temperature may cause the solid to melt (refer to Figure 3.1).

Volatiles, especially water, assist in breaking the chemical bonds, thereby reducing the amount of thermal energy needed for melting. In other words, the introduction of water lowers the melting temperature of a solid. Therefore, if a rock is dry and at a temperature just below the melting point, the addition of water may reduce the melting point and cause the rock to melt (Figure 3.2).

The Melting of Single Minerals of Constant Composition

A single mineral of a constant composition will begin to melt when the temperature reaches the melting point. Once melting begins, as long as heat is added, the temperature of the system will be maintained at the melting point of the mineral until it is entirely melted. When melting is complete, the temperature of the melt may then rise if heat continues to be added.

The Melting of Mixtures of Minerals and Minerals of Variable Composition

A mixture of minerals melts according to a quite different scenario. The melting of a mixture of minerals cannot be considered as simply the sequential melting of the individual minerals as their individual melting points are reached. First of all, with the exception perhaps of quartz, most of the minerals making up igneous rocks have variable compositions and therefore melt over a range of temperatures, rather than having specific melting points. In addition, because many of the minerals are members of solid solutions, melting points will vary depending on the specific composition of the mineral. Even combinations of minerals that are not members of a solid-solution series will melt over a range of temperatures depending on their relative concentrations. As melting progresses, the remaining solid and the highly reactive molten liquid become intermixed and continuously react with each other. As a result, as the rock melts, the compositions of both the melt and the remaining solid portion constantly change.

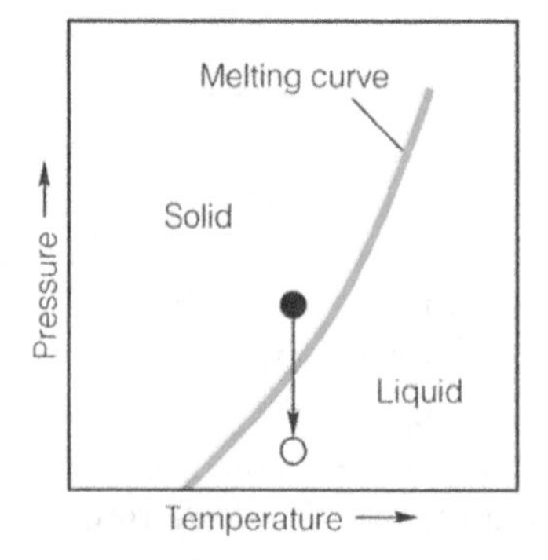

FIGURE 3.1 *When a solid is both under pressure and at high temperature, dropping the pressure may cause the solid to melt.*

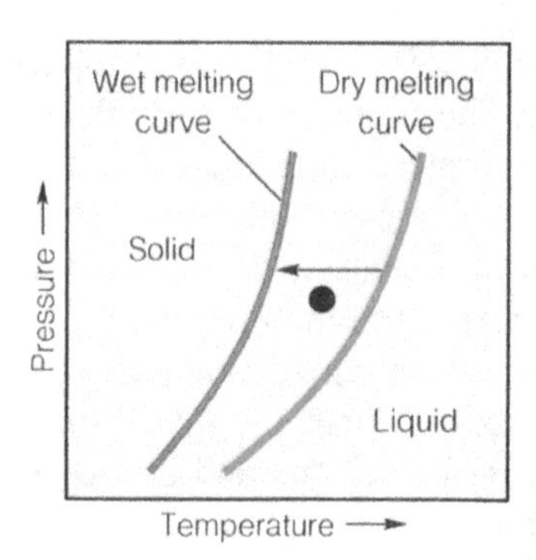

FIGURE 3.2 *When a solid is under both high temperature and pressure, the introduction of water may cause the solid to melt by providing an additional agent of bond-breaking.*

As heat is applied and the temperature of the mineral rises, the material melts sequentially, beginning with the components with the lowest melting points; the result is a **partial melt**. An important point to note is that the composition of the melt at any point during the melting process will be preferentially enriched by the early melting components and will therefore be *different* from the composition of either the original rock or the remaining solid. Partial melting may, for example, enrich a melt in potassium and sodium while iron, magnesium, and calcium remain in the solids. Only if the rock completely melts and the melt does not separate from the remaining solid during the melting process will the elemental composition of the melt be the same as that of the original solids.

Order of Mineral Crystallization

Igneous rocks form by the crystallization of silicate minerals from magma or lava. Crystallization is the opposite of melting. Just as there is an order of melting, there is also an order of crystallization or a sequence in which the major silicate minerals appear as a molten rock mass cools. Like melting, the actual process of crystallization is complex and for the same reasons. As the individual minerals form, they may react with the remaining melt and change in both composition and structure.

The order of crystallization of the minerals that comprise an igneous rock can be investigated using several different approaches. One possibility is to observe the rock under a microscope by preparing a thin section. By identifying the individual minerals and observing their shapes, sizes, and grain boundary relationships, the order of crystallization can often be determined. Because the early formed minerals grow within the melt with no obstructions, they are able to grow relatively large and have well-developed crystal faces. As the later formed minerals begin to crystallize, their growth is partially impeded by existing crystals. As a result, their shapes are not as perfect. The last minerals to form crystallize in whatever openings are available between the existing crystals and are therefore limited severely in both size and shape.

A second method that has been used to establish the order of crystallization is to collect samples of cooling magma. Much of this work has been done by petrologists (geologists interested in the origin, occurrence, structure, and history of rocks) of the U.S. Geological Survey at the Hawaiian Volcano Observatory, a research facility on the island of Hawaii. Following an eruption, scientists venture out onto the solidified surface of the lava lakes and drill through the frozen crust of lava into the liquid lava below, an extremely dangerous adventure to say the least (Figure 3.3). (Over the years, some scientists have been seriously burned when the lava crust gave way beneath their feet like the ice breaking on a pond.) Once a hole has been opened in the lava crust, probes are extended into the lava to different depths, the temperature is recorded, and a sample of lava is retrieved. The lava is quickly cooled, converting the molten portion of the lava into a glass that entraps the mineral crystals that have already formed. Thin sections are then prepared and described. The percentage of glass relative to mineral grains allows the

FIGURE 3.3 *Scientists at the Hawaiian Volcano Observatory on Hawaii study the crystallization of magma by taking samples of the molten lava for analysis. (Courtesy of R. T. Holcomb/USGS)*

scientists to determine the extent to which crystallization has progressed at any particular temperature. The kinds and abundances of the individual minerals encased within the glassy matrix are then identified. By describing a series of samples taken from various depths within the cooling lava representing different temperatures and stages of solidification, the sequence of mineral crystallization can be determined. The results of such a study are summarized in Figure 3.4.

A third method of establishing the order of crystallization of the silicate minerals utilizes synthetic "magmas" prepared by melting powdered samples of various kinds of igneous rock. Portions of the melts are then slowly cooled to various temperatures; at that point, they are rapidly cooled and analyzed according to the same procedure used for the Hawaiian magma samples.

As was the case with the Hawaiian studies, the amount of glass indicates the degree of solidification that the "magma" has experienced as it cools to a particular temperature, and the minerals identification provides the order of mineral crystallization. Whereas the Hawaiian studies are limited in both magma composition and potential rock types, the preparation of synthetic magmas from different kinds of igneous rocks allows scientists to study a wider range of magma compositions and rocks. An example of the results from this kind of study is shown in Figure 3.5.

The champion of the experimental investigation of silicate mineral crystallization was *N. L. Bowen.* In 1928, based upon experimental data and the microscopic study of mineral associations in synthetic melts, Bowen established a sequence of mineral crystallization. He demonstrated that the major silicate minerals crystallize from a cooling molten mass in a specific order (Table 3.2). Studying basaltic magmas, Bowen showed that olivine forms first with the remainder of the ferromagnesian minerals crystallizing sequentially from augite to biotite. Following the crystallization of olivine, and simultaneous with the crystallization of the remaining ferromagnesian minerals, the plagioclase feldspars crystallize progressively from the calcium-rich anorthite at higher

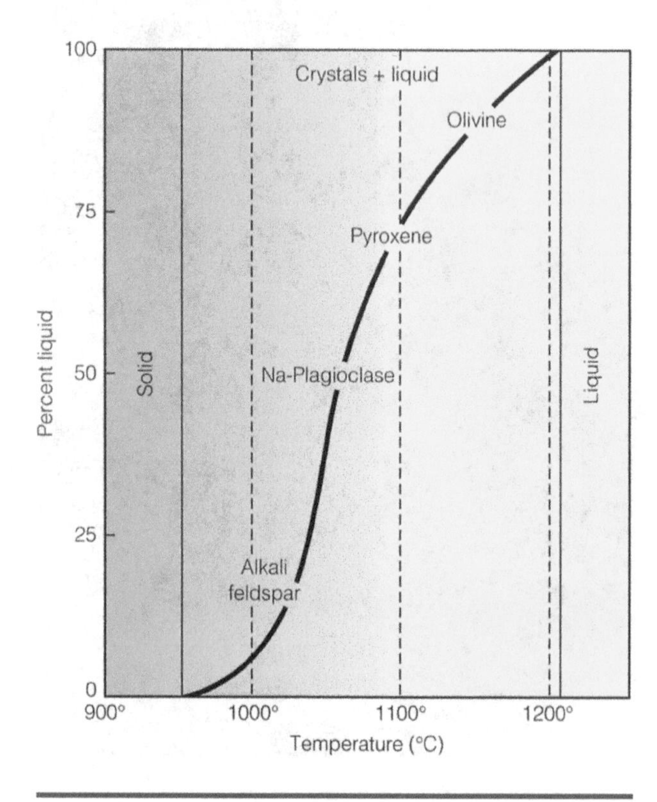

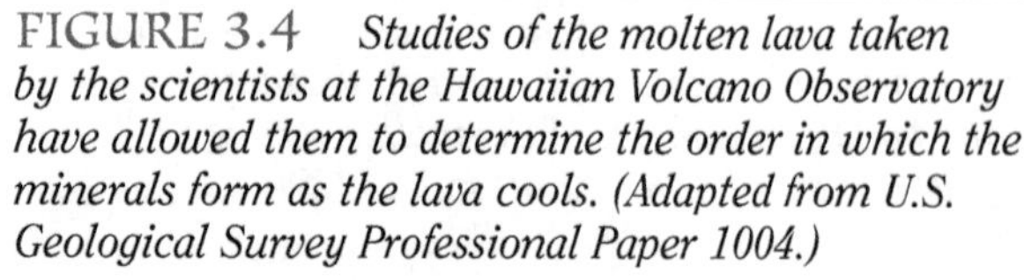

FIGURE 3.4 *Studies of the molten lava taken by the scientists at the Hawaiian Volcano Observatory have allowed them to determine the order in which the minerals form as the lava cools. (Adapted from U.S. Geological Survey Professional Paper 1004.)*

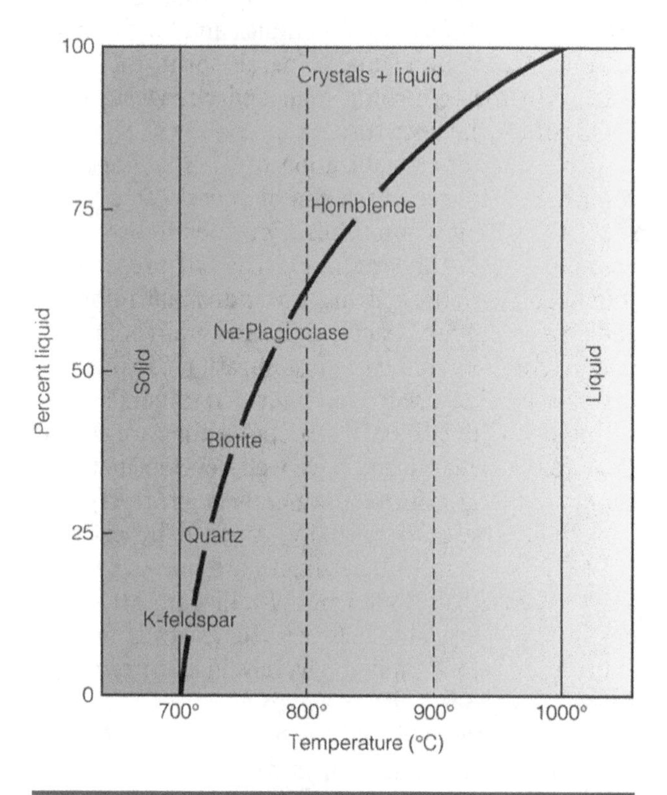

FIGURE 3.5 *Studies of synthetic magmas formed by melting various kinds of igneous rocks have verified the order in which minerals crystallize from cooling magmas. (Adapted from A. J. Piwinskii,* Journal of Geology *Vol. 76, University of Chicago © University of Chicago)*

temperatures to the sodium-dominated albite at lower temperatures.

The combined results of all these studies indicate that the major minerals that constitute igneous rocks crystallize in a specific order with minerals such as olivine crystallizing early at temperatures of 2,400°F to 2,200°F (1,300°C to 1,200°C) and minerals such as quartz and orthoclase feldspar forming last at temperatures of 1,300°F and 1,100°F (700°C to 600°C).

It is important to understand that the process of mineral crystallization is very complex. In his 1928 book, Bowen described his own work in a statement that is applicable to all experimental attempts to establish the order of crystallization of the silicate minerals: "An attempt is made . . . to arrange the minerals of the . . . rocks as reaction series. The matter is really too complex to be presented in such simple form. Nevertheless the simplicity, while somewhat misleading, may prove of service in presenting the subject in concrete form."

TABLE 3.2

BOWEN'S CRYSTALLIZATION SERIES

High Temperature	Olivine		First to crystallize
	Augite	Anorthite	
	Hornblende		
	Biotite	Albite	
	Orthoclase		
	Muscovite		
Low Temperature	Quartz		Last to crystallize

SPOT REVIEW

1. How can magmas of different compositions be formed during the cooling of a single magma?
2. How do pressure and the presence of water affect the melting point of rocks?
3. In what order do the major rock-forming silicate minerals form from a cooling magma?

ROCK TEXTURE

The **texture** of any rock refers to the size, shape, arrangement, and mutual interrelationships among the individual mineral grains.

The igneous rock textures are: (1) *phaneritic,* (2) *aphanitic,* (3) *glassy,* (4) *porphyritic,* and (5) *pyroclastic* (6) *vesicular,* (7) *pegmanitic* (Figure 3.6). In a rock with **phaneritic** texture, the individual grains are sufficiently large to be seen with the naked eye, usually larger than about 0.2 inches (5 mm) in diameter. **Aphanitic** texture indicates that the grains are too small to be resolved without the use of a hand lens or microscope, and **glassy** means that the rock has the amorphous (noncrystalline) structure of glass. **Porphyritic** texture describes a rock with phaneritic-sized crystals surrounded by either a finer phaneritic or an aphanitic matrix, and **pyroclastic** texture describes igneous rocks composed of rock fragments generated by volcanic eruptions.

The Rate of Cooling

The texture of igneous rocks is determined chiefly by the *rate of cooling* of the molten rock. Slow cooling favors the formation of fewer nucleation sites and the growth of larger crystals. Most intrusive magmas cool slowly because they are surrounded by rocks that conduct heat very poorly. As a result, most intrusive igneous rocks will be phaneritic (Figure 3.6a). Lavas, poured out onto Earth's surface and exposed to either the atmosphere or water, and sometimes to ice, cool rapidly and are invariably either aphanitic due to the formation of many nucleation sites or glassy if solidification occurs without the formation of crystals (Figure 3.6b).

Because the ions in a melt exhibit a certain degree of bonding, a melt can be readily converted to a glass if the cooling rate is exceptionally fast. Superfast cooling often occurs when lava is extruded underwater or beneath a glacier. In addition, it is not uncommon for the surface of basaltic lava flows to be glassy.

Porphyritic texture is a combination of two textures, usually phaneritic and aphanitic. The rock, referred to as a **porphyry,** exhibits larger crystals called **phenocrysts** surrounded by a finer-grained **groundmass** (Figure 3.7). Such a rock may form as a result of an interrupted cooling sequence. The phenocrysts form as the magma begins to cool slowly within Earth's crust or mantle. At some time before solidification is complete, the mixture of crystals and remaining melt is then injected into cooler rocks, or perhaps extruded onto Earth's surface, where solidification is completed by more rapid cooling, forming a finer-grained groundmass.

(a)

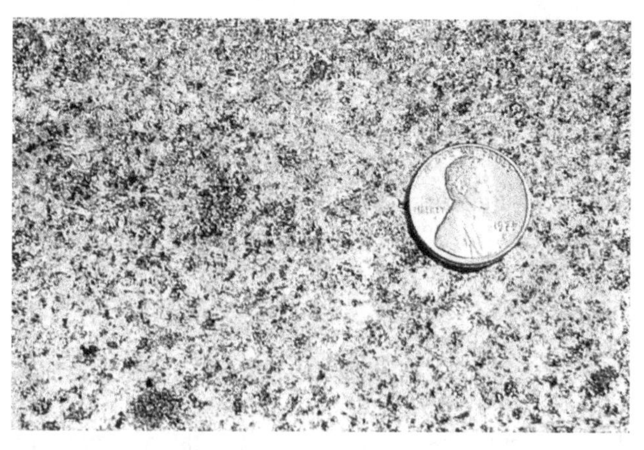

(b)

FIGURE 3.6 *Depending upon the rate at which molten rock cools, the texture of the rock that forms may be (a)* coarse-grained, *(b)* fine-grained, porphyritic *(a mixture of coarse and fine grains) or* glassy. *The texture of rocks that form from the fragments of rock produced by a volcanic eruption is called pyroclastic. (a N. K. Huber/USGS.)*

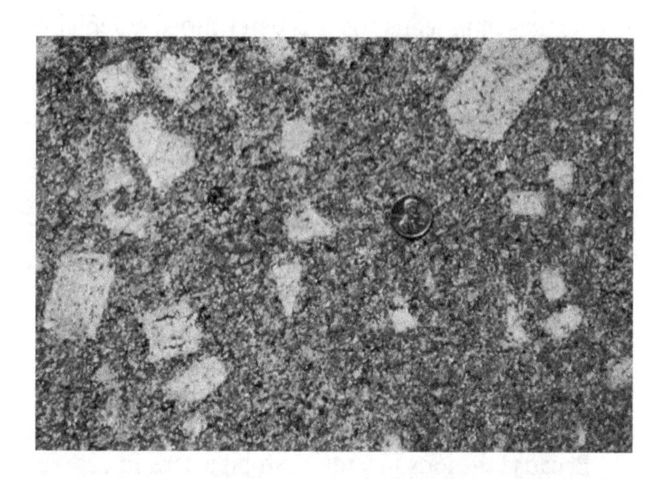

FIGURE 3.7 Porphyritic texture *is a* mixture *of coarse-grained crystals surrounded by fine-grained crystals. The large crystals formed when the magma was cooling slowly. At some point in time, the slow cooling was interrupted and the remaining molten rock was cooled rapidly, possibly by being extruded and cooled by air or water or by being injected into narrow fractures where it was cooled rapidly by the surrounding, cold rocks. (Courtesy of N. K. Huber/USGS)*

SPOT REVIEW

1. How does the cooling rate affect the grain size of igneous rocks?
2. What is the significance of porphyritic texture?

CLASSIFICATION OF IGNEOUS ROCKS

Igneous rocks are named and classified on the basis of (1) mineral *composition* and (2) *texture.* The names of the major igneous rocks are summarized and illustrated graphically in Figure 3.8. The rocks are arranged horizontally in rows based on texture (cooling rate) and vertically in columns based on mineral composition.

Proceeding from right to left, the mineral fields in the lower portion of the diagram depict the order of crystallization of the major silicate minerals. The rocks on the right-hand side of the chart contain minerals that crystallize at relatively high temperatures while rocks on the left are made from minerals that crystallize at lower temperatures. The diagram also illustrates from right to left the progressive change in the composition of the plagioclase feldspars and the ferromagnesian minerals, reflecting the decreased temperatures of crystallization.

The terms **mafic** and **felsic**, which appear at the top of the chart, refer to the compositions of igneous rocks, the silicate minerals, or the melts from which they form. Mafic rocks are rich in calcium plagioclase feldspars and in the ferromagnesian minerals olivine and pyroxene and have no minerals such as orthoclase feldspar and quartz. Rocks such as basalt or gabbro are mafic igneous rocks. On the other hand, igneous rocks enriched in minerals such as orthoclase feldspars, quartz, and sodium plagioclase but

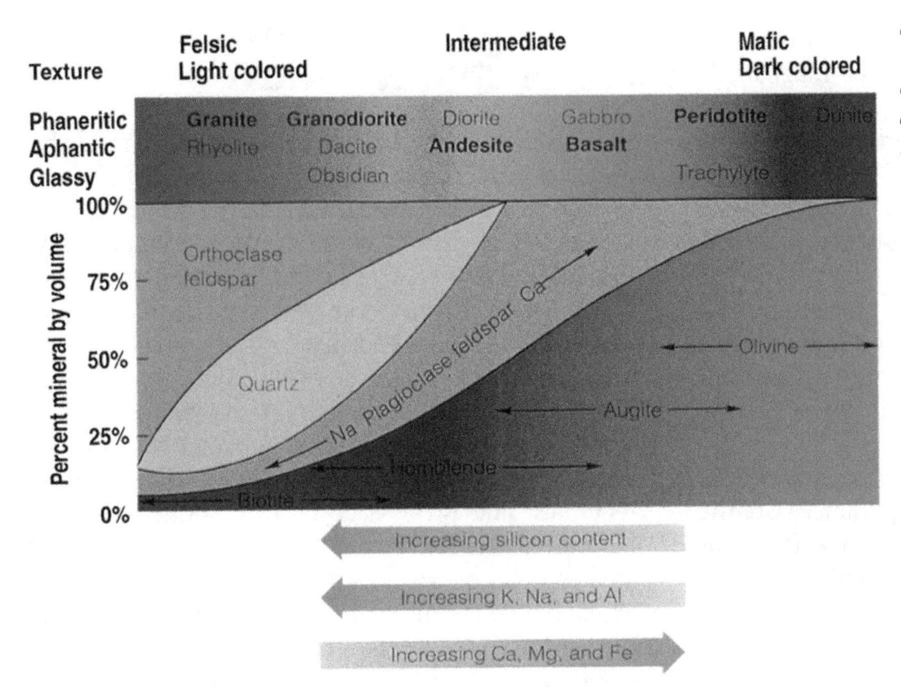

FIGURE 3.8 *The igneous rock chart summarizes the composition and cooling history of the major types of igneous rocks.*

with few ferromagnesian minerals are felsic. Rocks, such as peridotite (especially the variety called **dunite**), that contain no felsic minerals and are composed almost entirely of olivine are said to be **ultramafic.**

These terms are also used to describe the compositions of magmas and lavas. A mafic magma is a molten mass that will crystallize to produce mafic minerals and rocks. Similarly, a felsic magma or lava will produce felsic minerals and rocks when it solidifies. These two terms are quite useful as a quick way to describe the composition of igneous materials, be they rocks, minerals, or molten materials.

The chart also graphically portrays the compositions of the various igneous rocks. To determine the mineralogy of any rock in the upper part of the chart, drop a line down from the rock type into the portion of the chart containing the minerals and observe the relative proportions of the various mineral fields cut by the line. A gabbro or basalt, for example, is made up of about 50% ferromagnesian minerals and 50% plagioclase feldspar, with no quartz or orthoclase feldspar. Thus, as the chart shows, one would expect to find calcium-rich plagioclase feldspar in basalt along with the ferromagnesian minerals augite and olivine. In contrast, a granite is composed primarily of orthoclase feldspar and quartz, with lesser amounts of sodium plagioclase and ferromagnesian minerals; the latter would most likely be biotite with possibly a small amount of hornblende.

Rock combinations such as granite-rhyolite or gabbro-basalt that appear above and below each other are identical in composition and differ *only* in texture. Glasses can be of any composition. Felsic glasses are called **obsidian** while mafic glasses are called **tachylyte.**

Variations in elemental components are indicated at the bottom of the chart. One of the most important components is the *silicon* (Si) content, which increases in the direction of the more felsic rocks and magmas. The silicon content is important for its effect on the *viscosity* of the magma. Viscosity, a physical property of liquids, is a measure of the liquid's resistance to flow. Low-viscosity liquids such as water flow readily while more viscous liquids such as syrup flow more slowly. The viscosity of magma increases with increasing silicon content. As silicon becomes more available within the magma, more strong silicon-oxygen bonds are formed, causing the viscosity of the melt to increase.

Because of their low silicon content, mafic magmas are generally less viscous than felsic magmas. This explains, at least in part, why basalt is the most common extrusive rock type. The viscosity of basaltic magma is low enough to allow the molten mass to move to the surface along narrow fractures with minimal resistance to flow. More felsic magmas, in contrast, are commonly too viscous to flow through these fractures and may solidify within Earth before reaching the surface.

Another compositional parameter that affects viscosity is water content. Because water inhibits the formation of silicon-oxygen bonds, an increase in the abundance of water dissolved in the magma causes the viscosity of the magma to decrease. For this reason, it is possible for some *hydrous* (water-rich) felsic magmas to have viscosities as low as basaltic magmas.

Several rock names are in boldface type on the chart because of their singular importance. **Granite** and **granodiorite** are referred to collectively as *granitic* rocks, the most common intrusive igneous rock in the continental crust. **Andesite** is the dominant extrusive rock type associated with subduction zone volcanism. **Basalt**, the most abundant extrusive igneous rock, makes up much of the oceanic crust and is associated with rift zones and the volcanism associated with most oceanic and continental hot spots. Finally, **peridotite** makes up Earth's upper mantle and is the source for most of the magmas that intrude Earth's crust at rift zones, rift valleys, and oceanic ridges.

Komatiite, a rare fine-grained rock named after the Komati River in the Transvaal of South Africa, is the extrusive equivalent of peridotite or dunite. Because of the low silicon content, one would expect the magmas that form komatiite to be even less viscous than basaltic magmas and therefore to rise to Earth's surface with greater ease and produce more voluminous lavas than basalt. Such is not the case, however, presumably because the mineral constituents of komatiite crystallize at such high temperatures that the magma solidifies before reaching the surface. When the komatiites now exposed in South Africa were emplaced 2 billion years ago, crustal and mantle rocks were hotter than they are now, allowing the magmas to rise to Earth's surface without solidifying.

SPOT REVIEW

1. What factors affect the viscosity of magma?
2. Why are mafic lavas more common than felsic lavas?

THE ORIGIN OF MAGMAS AND PLATE TECTONICS

Any discussion of the origin of magmas must be prefaced with the statement that theories about their formation are speculative. Because the origin of magmas cannot be directly observed, their exact modes of formation will always be somewhat uncertain.

Since no one has ever observed the natural formation of a magma and likely no one ever will, our ideas about the origin of magmas are based upon (1) laboratory experiments, (2) observed distributions of the various kinds of igneous rocks at or near Earth's surface, (3) our understanding of plate tectonics, and (4) a large dose of sound scientific reasoning. We saw earlier that magmas are associated with four crustal areas: (1) *continental rifts,* which herald the initial breakup of continental crust; (2) *oceanic ridges,* where new oceanic crust is constantly being formed; (3) isolated *hot spots* beneath oceanic and continental crust; and (4) *zones of subduction,* where oceanic crust is being subducted and mountains are being created.

From our understanding of Earth's crust, magma appears to have three basic compositions: (1) *basaltic* magma, which provides the igneous rocks extruded at rift zones, rift valleys, oceanic ridges, and most hot spots (Figure 3.9 and 3.10); (2) *andesitic magma,* which is associated with subduction zone volcanic arcs; and (3) *granitic magma,* which accounts for most of the massive intrusive igneous rock bodies that are associated with the convergence of plates.

Most basaltic magmas are associated with zones of continental rifting and oceanic ridges, both of which are positioned above the rising portions of mantle convection cells (Figure 3.9 and Figure 1.11). The specific combination of temperature and pressure within the asthenosphere places the rocks just within the solid-liquid transition curve of Figure 3.1. As they move closer to Earth's surface, the pressure progressively decreases. By the time the rocks in the convection cell arrive at the boundary between the asthenosphere and the lithosphere, the decrease in pressure is sufficient for melting to be initiated. The partial melting of the peridotite produces a basaltic magma that, because of its buoyancy, rises through the lithosphere to the surface. The question is, how can the melting of peridotite generate a basaltic magma?

Referring once again to the rock chart in Figure 3.8, you will see that peridotite is more mafic

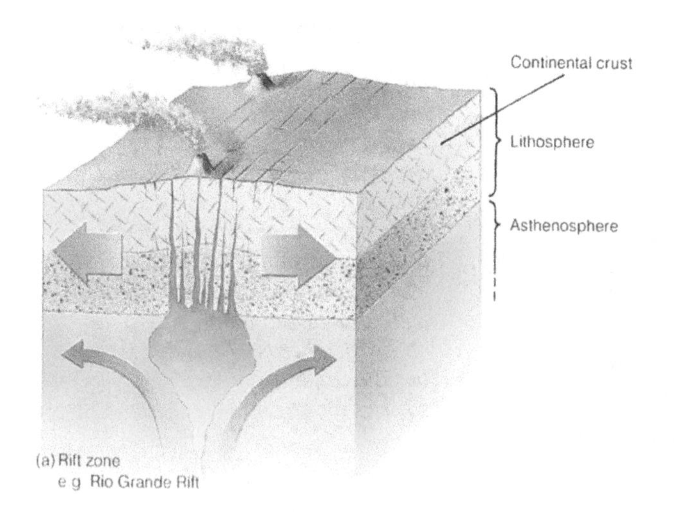

FIGURE 3.9 *At rift zones, basaltic magma, formed at the top of the asthenosphere, makes its way to the surface along fractures formed in the lithosphere. Some of the magma solidifies along the way to produce new lithospheric rocks. Other masses of the magma are extruded on the surface to produce the volcanic features associated with rift zones and rift valleys.*

than gabbro-basalt. As noted above, during the partial melting of multicomponent silicate rocks, the order of mineral melting is the reverse of the order of crystallization. As a result, the composition of the melt is always more *felsic* than the composition of the original rock. Referring to Figure 3.8, note that the next rock composition that is more felsic than peridotite (just to the left on the chart) is gabbro-basalt. Therefore, the melt formed by the partial melting of peridotite is gabbro-basaltic. In this case, because most of the molten rock makes its way to the surface to become lava and solidifies as an aphanitic rock, we refer to the magma as being basaltic. In summary, the basaltic magmas associated with rift zones, rift valleys, and oceanic ridges form from the partial melting of peridotite induced by a decrease in pressure.

Hot spots positioned below the lithospheric plate are the other major source of basaltic magmas. The hot spots are located at the tops of heat-driven plumes that, most evidence suggests, rise from deep within the mantle. Once again, the magmas form by the partial melting of peridotite as a result of a reduction in pressure. Once formed, the magmas rise due to their lower density and pierce the lithosphere at point sources to form seamounts and the spectacular shield volcanoes characteristic of oceanic volcanic islands such as the Hawaiian Islands (Figure 3.10).

FIGURE 3.10 *A well-known locality where one can see igneous rocks being made is Hawaii where basaltic magmas are being extruded and cooled to form the basaltic rocks that make up the Hawaiian Islands (Courtesy of USGS/HVO)*

The most complex magma-forming scenario involves the zones of subduction where we must explain the simultaneous generation of three kinds of magma (Figure 3.11): (1) the basaltic magmas that reach the surface along fractures that develop in the overriding plate behind the arc of volcanic mountains in what geologists call the back-arc basin; (2) the andesitic magmas that feed most of the arc volcanics associated with zones of subduction such as the Andes Mountains, the Aleutian Islands, and the Japanese Islands; and (3) the massive volumes of granitic magmas that account for the huge masses of igneous rocks that intrude into and add to the continental crust.

As the lithospheric plate bearing the oceanic crust subducts beneath an oncoming lithospheric plate bearing either an oceanic or continental plate, it is driven or pulled downward into the mantle. The major source of heat available within the zone of subduction is heat contained within Earth (refer to Figure 3.11).

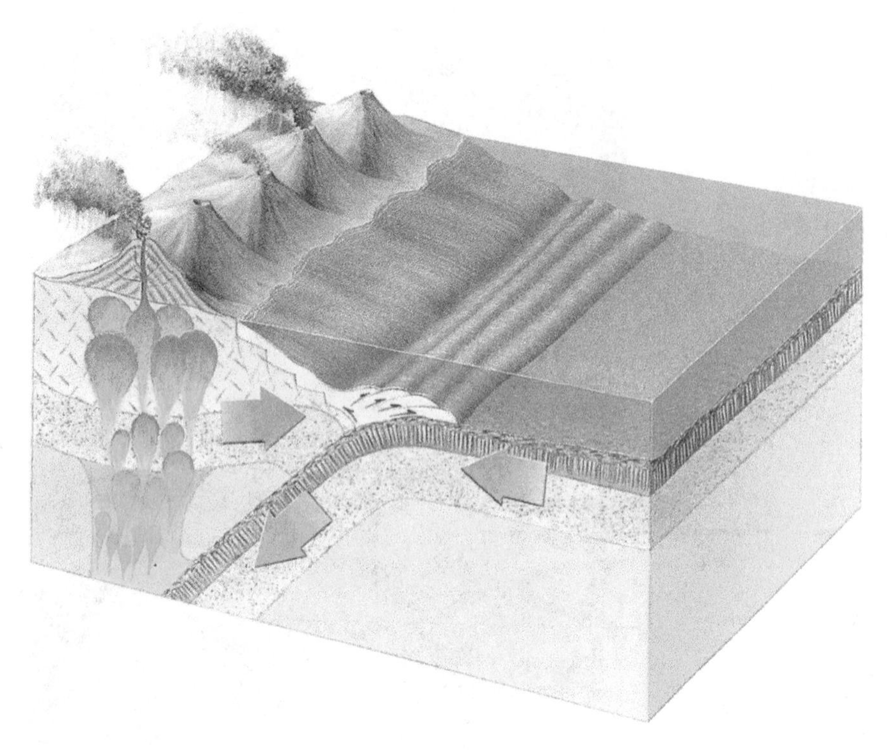

FIGURE 3.11 *Three compositions of magma are being formed within subduction zones: basaltic, andesitic, and granitic. The andesitic magmas are extruded and form the chains of volcanic mountains associated with converging plate margins. Most of the granitic magmas cool underground and add to the continental crust. Occasionally, basaltic magmas may reach the surface along fissure eruptions.*

Water is introduced into the mantle as rocks and sediments of the oceanic plate are subducted. As the hydrous minerals contained within the crustal rocks and oceanic sediments are heated, water is released and penetrates into the overlying mantle. With the combined effect of the heat provided at depth and the lowering of the rock melting points caused by the presence of water, massive quantities of basaltic magma are produced by partial melting of the mantle peridotites. Some of these basaltic magmas may rise to the surface and emerge along cracks in the crust to produce floods of basalt such as those that accumulated to form the Columbia Plateau of eastern Washington and Oregon (Figure 3.12) and the basaltic volcanics found throughout Nevada and adjoining states.

The silica-rich sediments commonly found on the deep-ocean bottom are derived from two major sources: (1) wind-transported dust from the continents, consisting primarily of clay minerals, quartz, and feldspar, and (2) the siliceous (silica-rich) shells of marine microorganisms such as diatoms and radiolarians. Some of these siliceous ocean bottom sediments are drawn down into the zone of subduction, melt, and become incorporated into the basaltic magmas, contributing to the increasingly felsic nature of the magma. Andesitic magmas may also form from the partial melting of subducted basaltic oceanic crust. Although andesitic magmas are more viscous than basaltic magmas due to their higher silica content, they can still rise along fractures to the surface. Because of the higher viscosity of the andesitic magmas, large volumes of gases remain dissolved in the magmas until they emerge onto the surface. Once the magmas are exposed to the atmosphere, the gases escape rapidly, producing the violent explosions characteristic of subduction zone volcanism. The large amounts of fragmental materials generated during the eruptions are responsible for the steep-sided volcanoes of the Japanese Islands, the Aleutians, the Cascades, and the Andes. We will discuss the various kinds of volcanoes in the next chapter.

The magmas that are the most difficult to explain are the huge masses of highly felsic granitic magmas that intrude into the overriding plate (refer to Figure 3.11). Partial melting of oceanic crustal basalt cannot generate granitic magmas. Even the intermixing of the silica available from the subducted oceanic sediments will not alone produce granitic magmas because the elements necessary to create the other required mineral constituents are not present. Two scenarios have been suggested for the production of the granitic magmas:

1. Intense compressional deformation within the subduction zone generates metamorphic rocks from a combination of lithospheric plate rocks and sedi-

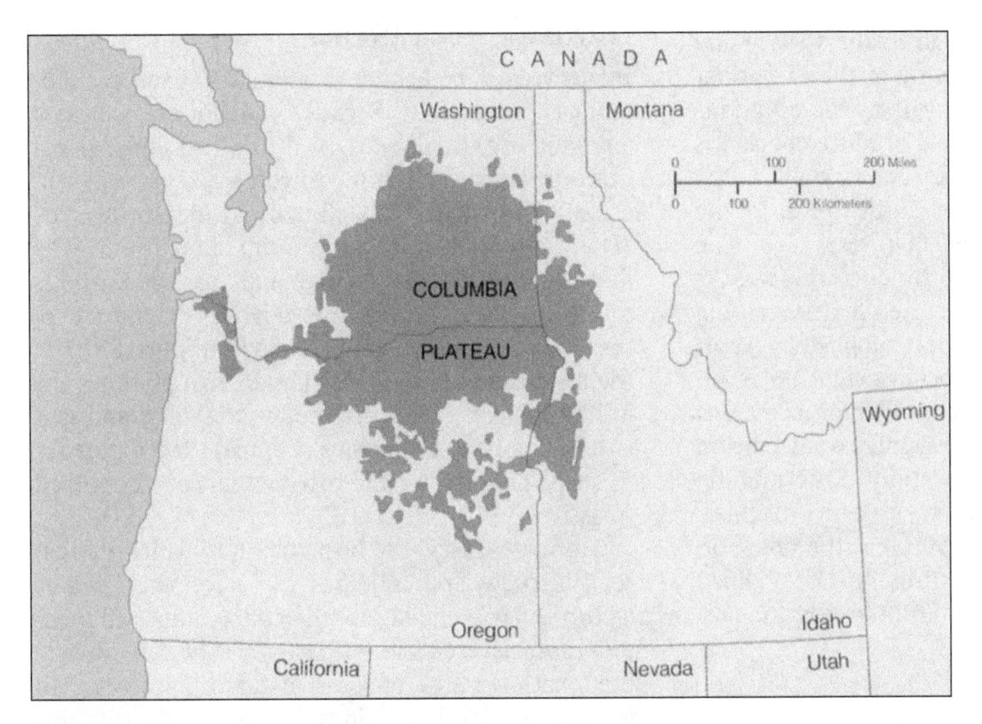

FIGURE 3.12 *Massive fissure eruptions of basaltic magma resulted in the formation of the layered basalt flows of the Columbia Plateau. (Courtesy of P. Weis/USGS)*

ments. Because the oceanic sediments are enriched in silica, the overall elemental composition of the metamorphic rock-oceanic sediments assemblage is dominantly felsic. The melting of this mixture may be the source of large volumes of granitic magma.

2. The source of the granitic magmas may be the granitic continental crust itself. As the basaltic magmas generated by the partial melting of the mantle peridotite rise toward the surface and encounter these more felsic (and therefore lower melting point) rocks, the basaltic magma pools below the crust because of the crust's lower density. As the basaltic magma accumulates, heat is transferred from the magmas to the granitic rocks and causes them to melt. As the heat is removed from the basaltic magmas and they begin to solidify, their upward movement is arrested within the crust. This may explain why basaltic lavas are not widely represented at the surface of the continents even though massive volumes of basaltic magmas are thought to be generated within zones of subduction.

Once generated, the low-density granitic magmas move upward and intrude into the overlying crust. Because of their high silica content, granitic magmas are usually highly viscous and tend to solidify within the crust, generating the huge masses of granitic rock that add to the growing mass of the continental crust. Upon occasion, however, these felsic magmas do break through to the surface and emerge as rhyolitic lavas. Because of the exceptionally high gas content of the magmas, the eruptions are typically highly explosive and produce enormous volumes of fragmental material. An example is the eruption that produced the rhyolitic rocks in the area of Yellowstone Park 600,000 years ago. The felsic magmas that erupted to produce these rocks are thought to have been generated by the melting of granitic crustal rocks with heat supplied by basaltic magmas rising from a subcontinental hot spot. The presence of the lower-density granitic magma prevented the denser basaltic magmas from moving upward. Following the solidification of the granitic magmas, it is believed that the basaltic magmas, having been continuously produced at the hot spot, rose to the surface along fractures in the solidified granities to emerge as the posteruption basaltic lava flows found in the area.

SPOT REVIEW

1. Why are the magmas associated with divergent plate boundaries and most hot spots dominantly mafic in composition while those associated with convergent plate boundaries range in composition from basaltic to granitic?
2. What are the two scenarios commonly presented for the formation of granitic magmas?

IGNEOUS ROCK BODIES

To fully understand how igneous rock masses are emplaced, we need to concern ourselves with the form of the igneous rock bodies. Three criteria are used to categorize igneous rock bodies: (1) where they are formed relative to Earth's surface, (2) their shape, and (3) their orientation relative to layered features in the surrounding rock.

We have already introduced the terms *intrusive* and *extrusive,* which categorize igneous rock bodies according to whether they are on or below Earth's surface. Magmas solidify to produce intrusive rock bodies while lavas and the fragmental materials produced by volcanic eruptions solidity to form extrusive rock bodies.

Intrusive Rock Bodies

Intrusive rock bodies are classified by shape as either **tabular** or **massive.** Because most rock bodies of any kind are usually observed in a two-dimensional exposure such as a cliff, canyon wall, or road cut, establishing the shape involves measuring the (1) maximum and (2) minimum dimensions. The dinstinction between tabular and massive shape is arbitrarily based upon the relative magnitude of these dimensions of the rock body (Figure 3.13). If the maximum dimension is larger than 10 times the minimum dimension, the body is tabular, and if it is less, the body is massive. Compare the difference in shape between a table top (tabular) and a football (massive).

In many cases, the host rock into which molten rock intrudes and solidifies will have some *planar feature* such as the layers that characterize sedimentary rocks, layered lava flows, pyroclastic deposits and the foliation planes of some metamorphic rocks. An igneous rock body is said to be **discordant** if its surface *cuts across* the layers, but if the body *parallels* the layers, it is **concordant** (Figure 3.14).

In summary, igneous rock bodies are classified as (1) *intrusive* or *extrusive* according to where they formed, (2) *tabular* or *massive* based on their shape, and (3) *discordant* or *concordant* based on their relationship to the surrounding rock.

Note that igneous rock bodies are *not* distinguished by *size.* Geologists are usually less concerned about the size of any an object or feature than they are about its origin. In only one instance, the designation of the *stock* and the *batholith* (to be discussed), is size a consideration in classifying an igneous rock body.

Tabular Intrusive Igneous Bodies

The two kinds of tabular intrusive rock bodies are illustrated in Figure 3.13a. A **dike** is defined as a "tabular discordant, intrusive igneous body" whereas a **sill** is a "tabular, concordant, intrusive igneous body." Again, it should be pointed out that there is no size restriction to these rock bodies. Except for the aphanitic portion of tabular rock bodies where they make contact with the host rock, most dikes and sills are phaneritic.

Massive Intrusive Igneous Bodies A **laccolith** develops when molten rock, fed by a feeder dike, is injected between layered rocks at a rate faster than the molten mass can move laterally between the layers of the intruded rocks. As the mass becomes thicker over the feeder site, the intrusion arches the

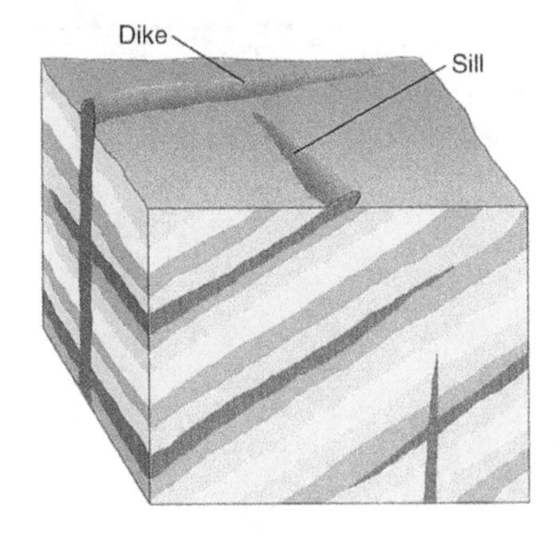

(a) Tabular igneous bodies

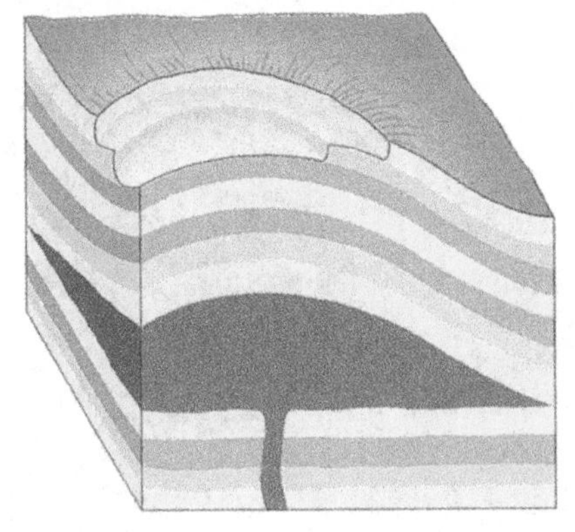

(b) Massive igneous body

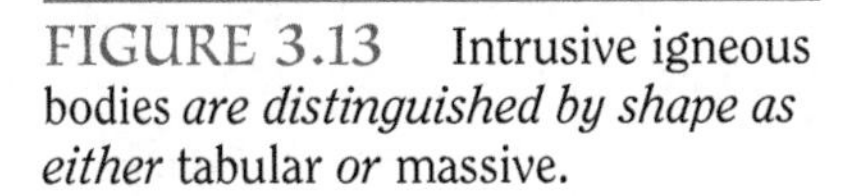

FIGURE 3.13 Intrusive igneous bodies *are distinguished by shape as either* tabular *or* massive.

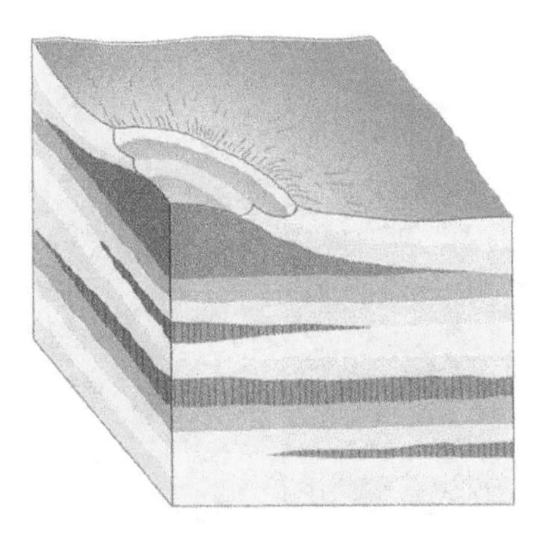

(a) Concordant contacts

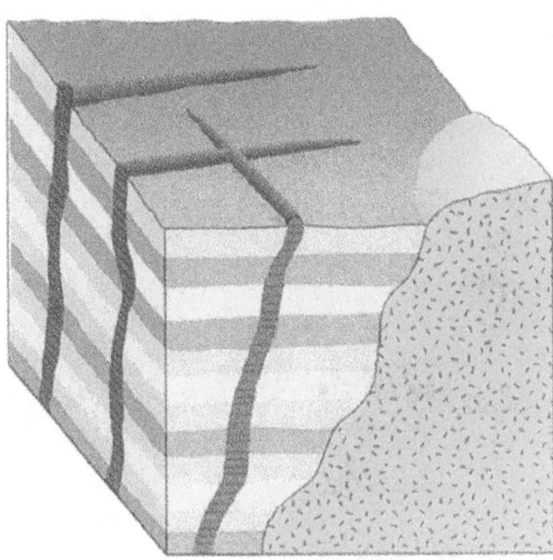

(b) Discordant contacts

FIGURE 3.14 Intrusive igneous bodies *are referred to as either* concordant *or* discordant *depending upon whether the boundary of an intruded igneous body* parallels *or* cuts across *layering in the host rock.*

overlying rocks and, at some point, changes in shape from concordant *tabular* to concordant *massive* (Figure 3.13b). Once the small dimension becomes more than one-tenth of the large dimension, the original sill turns into a laccolith. Once again, there is no size requirements. Laccoliths can be of any size. An example of exposed laccolithic intrusions that resulted from the removal of the overlying rock by erosion are the Henry Mountains of southern Utah (Figure 3.15). In a similar structure called a **lopolith,** the central portion of the body is sunken due to the sagging of the underlying rock (Figure 3.16).

The only two massive igneous bodies that are distinguished by size are the **stock** and the **batholith.** A stock is defined as a massive intrusive igneous body with a surface area of exposure of less than 40 square miles (100 km^2), whereas a batholith is a massive intrusive igneous body with a surface exposure greater than 40 square miles. In fact, the term *batholith* is synonymous with large size, often being defined as "a huge body of intrusive igneous rock." The concordant-discordant relationships of these massive bodies give an indication of the mode of intrusion. In some instances, the host rocks are deformed and arched over the top of the rock body, indicating a *forceful entry* of the molten mass. In other instances, the layering in the host rock is truncated at the body margin with little or no indication of deformation (Figure 3.17). In this case, the emplacement of the molten mass is interpreted as having involved **stoping,** a process where the host rock is detached and engulfed within the rising magma. These pieces of host rock either sink to the bottom of the magma chamber, become entrapped in the magma near the boundary of the igneous body as an **xenolith,** or are assimilated within the magma.

Perhaps the most spectacular, and familiar, intrusive igneous rock body is the **volcanic neck,** which is the remains of an extinct volcano. In time, every volcano becomes extinct as the heat source

FIGURE 3.15 *The core of the Henry Mountains in Utah is a* laccolith.

FIGURE 3.16 *A* lopolith *is a massive, concordant, igneous body whose center is downwarped due to the sagging of the underlying host rocks.*

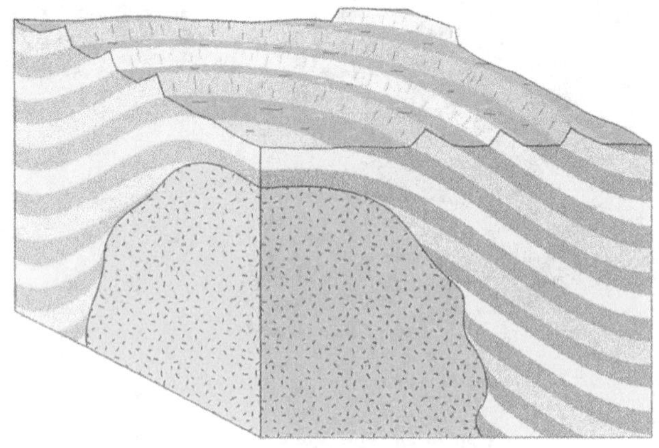

(a) Introduction by forceful entry

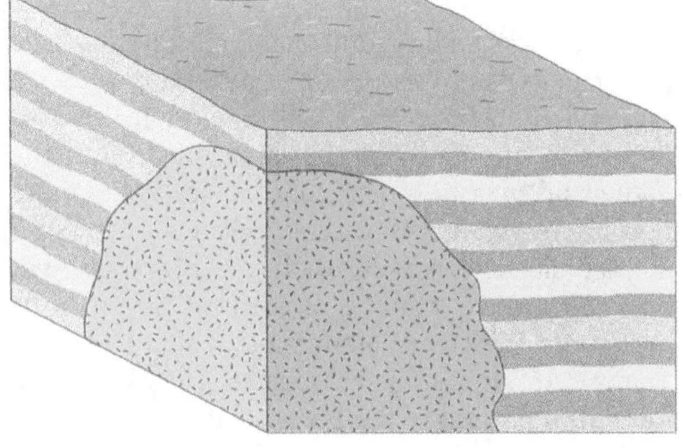

(b) Introduction by stoping

FIGURE 3.17 *Most plutons are* forcefully injected *into the host rocks as shown by the deformed layers of the host rocks. Some plutons, however, may be introduced by the* assimilation *or* stoping *of the host rock by the rising magma.*

maintaining the molten rock in the source magma chamber is exhausted. At this point, the constructive phase of the volcano ends. Erosion will then remove the outer conical structure of the volcano and expose the igneous rock in the core of the cone as a vertical spire. A well-known example of a volcanic neck is El Capitan, Arizona (Figure 3.18).

SPOT REVIEW

1. How can a sill be distinguished from a buried lava flow?
2. How can one determine whether an intrusive rock body has been emplaced forcefully or by stoping?

FIGURE 3.18 El Capitan *in Arizona is an example of a volcanic neck.*

CONCEPTS AND TERMS TO REMEMBER

- Kinds of molten rock
  - magma
    - intrusive
  - lava
    - extrusive
- Melting of solids
  - partial melting
  - magmatic fractionation
  - crystal settling
  - filter pressing
- Cooling of molten rock
  - texture
    - phaneritic
    - aphanitic
    - glassy
    - porphyritic
    - porphyry
    - phenocrysts
    - groundmass
  - pyroclastic
- Classification of igneous rocks
  - mafic
  - felsic
  - ultramafic
- Kinds of igneous rock
  - dunite
  - obsidian
  - tachylyte
  - granite
  - granodiorite
  - andesite
  - basalt
  - peridotite
  - komatiite
- Kinds of rock bodies
  - tabular
  - discordant
    - dike
  - concordant
    - sill
  - massive
    - laccolith
    - lopolith
    - stock
    - batholith
      - stoping
      - xenolith
    - volcanic neck

REVIEW QUESTIONS

1. The most common igneous rock found in intrusive igneous rock bodies is
   a. gabbro. c. basalt.
   b. granodiorite. d. peridotite.
2. According to Bowen's crystallization series, the first mineral to form from a crystallizing mafic magma would be
   a. quartz. c. hornblende.
   b. olivine. d. albite.
3. The igneous rock that makes up the oceanic crust is
   a. rhyolite. c. granite
   b. basalt. d. peridotite.
4. At which of the following localities is magma produced primarily by the pressure drop mechanism?
   a. zone of subduction (convergent boundary)
   b. hot spot
   c. mid-oceanic spreading center (divergent boundary)
   d. transform boundary
5. Which of the following statements will *always* be true?
   a. Intrusive igneous rocks are always coarse grained.
   b. Extrusive igneous rocks are never coarse grained.
   c. Intrusive igneous rocks can be glassy.
   d. Obsidian forms by the crystallizataion of either magma or lava.
6. Which of the following pairs of igneous rocks have the same composition and differ only in grain size?
   a. gabbro-basalt c. diorite-peridotite
   b. granite-andesite d. gabbro-rhyolite
7. Which of the following is defined as a "tabular, concordant, intrusive igneous body"?
   a. sill c. laccolith
   b. stock d. dike
8. The difference between magma and lava is that magma
   a. is hotter.
   b. exists within the Earth while lava exists on the Earth's surface.
   c. is more viscous.
   d. has a higher silica content.
9. If you were shown an igneous rock that was phaneritic and dark in color, which of the following could it be?
   a. granite c. granodiorite
   b. basalt d. gabbro
10. The composition of most lavas is
    a. andesitic. c. felsic.
    b. rhyolitic. d. basaltic.
11. If the asthenosphere is composed of the rock peridotite, how can the melting of the upper portion of the asthenosphere be a source of the basaltic magmas that rise to the surface at mid-oceanic ridges and hot spots?
12. If less viscous, silica-poor mafic magmas rise more easily to Earth's surface, why are komatiites not extensively exposed at Earth's surface?
13. Why does the rapid rate of crystallization of magma favor the formation of small crystallites while slow rates of crystallization promote the formation of large crystals?
14. Explain why the inner portion of a crystal of plagioclase feldspar is commonly a calcium-rich variety while the outer portion of the crystal is a more sodium-rich variety.
15. Why is the mineral quartz not found in mafic igneous rocks even though mafic magmas contain as much as 50% silica by weight?
16. Can you give a possible explanation why the crystal structures in the ferromagnesian minerals become more complex with decreasing crystallization temperatures?

THOUGHT PROBLEMS

1. What evidence would you look for to distinguish between a basaltic still injected between layers of sedimentary rocks and an ancient, buried basaltic lava flow?
2. Why does the silica content critically control the viscosity of molten rock?
3. How can igneous rocks of different compositions form from a single magma?

FOR YOUR NOTEBOOK

If you live or go to school in an area where igneous rocks are exposed, a field trip to observe and describe some of the local occurrences would certainly be in order. A geologic map that you can obtain from your state geological survey or from your instructor will outline the distribution of the various kinds of rocks as well as the most likely place to find them well exposed. You may want to collect some specimens for a rock collection, and you will want to indicate the effect that different rock types have on the topography with a series of sketches or photos.

For the great majority of us who do not live in areas where igneous rocks can be observed in outcrop, a field trip through your town or city will undoubtedly produce some examples of igneous rocks used for construction or trim. (As you make your survey, not the use of other types of stone so that you may include them in your notebook under the appropriate chapter.) You may want to seek the help of your instructor in identifying some of the rock types, and he or she may suggest some examples you will not want to overlook. In addition to noting the kind of rock, indicate how the rock was used; that is, whether it was used as a major construction stone or as a decorative trim. A conversation with a local architect would be a source of important information concerning the use of different kinds of rocks in your area.

KEY TERMS		
Aphanitic Texture	Mafic Igneous Rock	Pumice
Breccia, Volcanic	Pegmatitic	Pyroclastic
Extrusive Igneous Rock	Phaneritic	Scoria
Felsic Igneous Rock	Phenocryst	Tuff, Volcanic
Intermediate Igneous Rock	Porphyry	Ultramafic Igneous Rock
Intrusive Igneous Rock		

IGNEOUS ROCKS IDENTIFICATION AND PROPERTIES

Igneous rocks are the first of the major rock categories we will investigate. Geologists generally believe that the earth formed in such a way that the first rocks generated by earth processes were all igneous, with sedimentary and metamorphic rocks forming later. As the Rock Cycle demonstrates, all rocks are interrelated, and the concept of **Uniformitarianism** can explain the formation of any type of rock or mineral (Figure 3.19).

All rocks consist of one or more minerals. Now that you have learned to recognize common minerals, we can move on to identifying rocks.

Igneous rocks formed during the cooling of melted rock materials **(magma)** originally formed within the earth, melted by pressure and intense heat found there.

Sometimes these molten materials solidified beneath the surface; these rocks are termed ***intrusive (coarse grained) igneous rocks.*** Once molten magma flows out onto the surface of the earth, it is termed lava, and when crystallized, the resulting rocks will

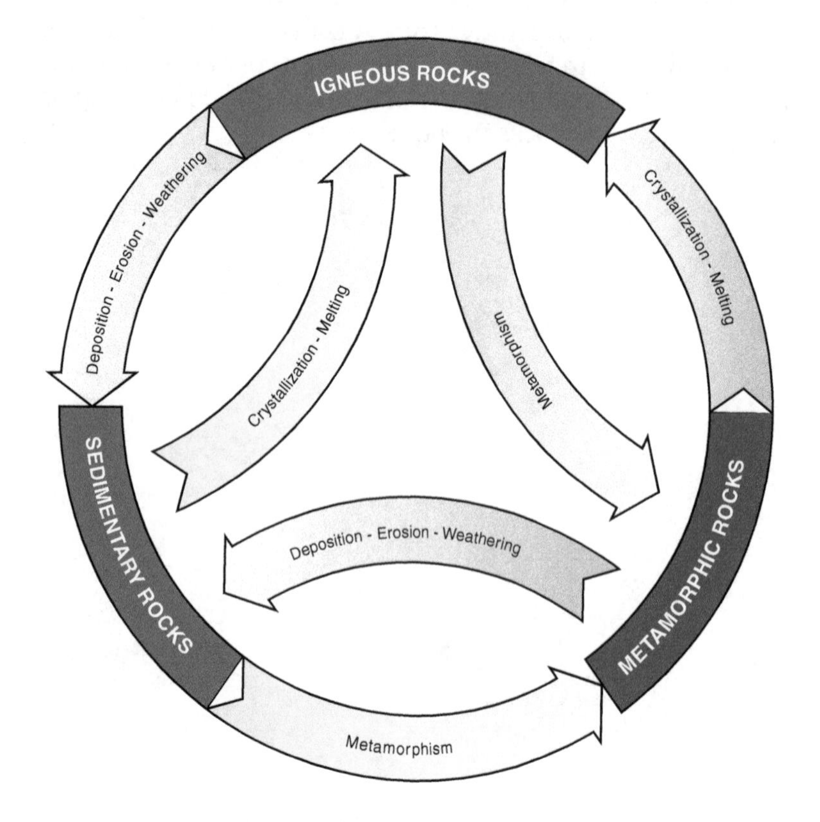

FIGURE 3.19 *The rock cycle demonstrates uniformitarianism.*

be considered ***extrusive (fine grained)***. Therefore, crystal size in the final rock is a function of how slowly or how quickly the rock cooled.

MATERIALS AND METHODS

- A set of the most common igneous rocks
- A ruler with metric measurements
- A low hand lens

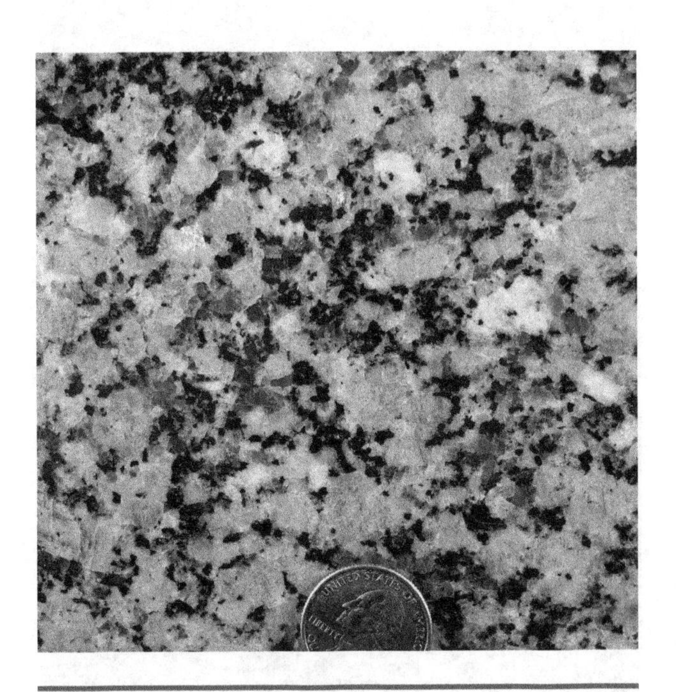

FIGURE 3.20 *Phaneritic rock.*

FIGURE 3.21 *Pegmatitic rock.*

Properties of Igneous Rocks

The primary classification of igneous rocks is based on **texture** and color. As stated, intrusive rocks, which cool slowly, are coarse grained, and, if the crystals of the rock are large enough to be seen easily with the unaided eye, the rock is said to be ***phaneritic*** (Figure 3.20). Sometimes the cooling rate will be unusually slow and the mineral crystals will grow to exceptionally large sizes (crystals on average exceeding one centimeter in size); such rocks are then called ***pegmatites*** (Figure 3.21).

Extrusive igneous rocks that cool rapidly produce crystals that are too small to be seen and identified by the unaided eye and are termed ***aphanitic*** (Figure 3.22). If cooled too quickly for crystals to

INSIGHTS

A down-to-earth visual aid might be to think of a chocolate chip cookie where the phenocrysts are the chocolate chips and the groundmass is the cookie dough. In naming porphyries, the "cookie dough" is named before the "chocolate chips"; that is, a porphyritic rock with the general mineral content of granite containing large feldspar phenocrysts would be called granite porphyry.

FIGURE 3.22 *Aphanitic rock.*

FIGURE 3.23 *Glassy rock.*

FIGURE 3.25 *Andesite porphyry.*

FIGURE 3.24 *Vesicular rock (scoria).*

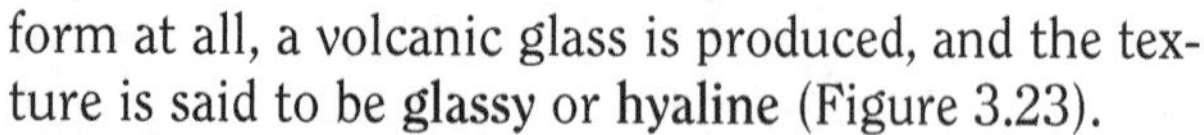

FIGURE 3.26 *Volcanic ash or tuff.*

form at all, a volcanic glass is produced, and the texture is said to be **glassy** or **hyaline** (Figure 3.23).

If the igneous rock is extrusive and gas rich, escaping gases form bubbles that may be trapped in the crystallizing rock. These bubbles are called vesicles, and the resulting rock will appear sponge-like, a condition termed vesicular or frothy (Figure 3.24).

A special case of texture results when a rock records two distinct patterns of cooling and is termed a ***porphyry*** (Figure 3.25). These rocks are primarily intrusive and begin cooling in a manner that produces relatively large crystals ***(phenocrysts)*** and seem destined to become phaneritic rocks. At some point before the rock has entirely crystallized, the temperature drops and the remaining rock mass is formed from aphanitic crystals **(groundmass)**. The end product is a rock with both large and small crystals.

Two general classes of igneous extrusive rocks are produced during the activity of volcanoes, and the materials produced are termed ***pyroclastic.*** Often these materials were blasted free from the liquid magma to produce lava that flowed from or was thrown violently out of the crater. If the fragments are very fine (2 mm or less in diameter), they are called ***volcanic tuff*** or **volcanic ash** (Figure 3.26), but if the grains are larger, they may be encountered as masses of irregularly shaped rock fragments and are then called ***volcanic breccias*** (Figure 3.27).

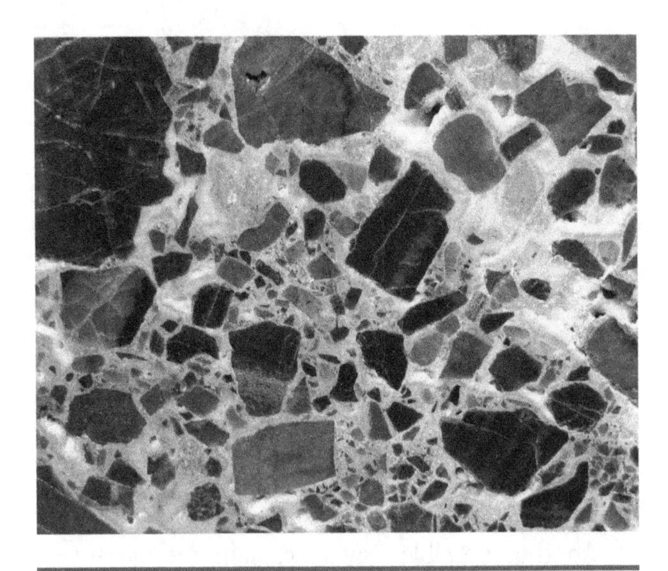

FIGURE 3.27 *Volcanic breccia.*

Particles of volcanic debris are named according to their general size classes. **Ash particles** are very small (up to 2 mm in diameter, **lapilli** range between 2–64 mm, and pieces of ejected materials larger than 64 mm in diameter are called **blocks** and ***bombs*** (these can be very large in the case of major volcanic explosion debris).

The other major aid for identification of igneous rocks is general color. Igneous rocks rich in quartz and orthoclase feldspars are light in color (brown, pink, tan, yellow, etc.) and are termed ***felsic,*** while igneous rocks rich in plagioclase feldspars and **ferromagnesian** minerals are called ***mafic*** (black, black-brown, greenish-black, etc.). Rocks falling

INSIGHTS

Volcanic tuffs are sometimes difficult to distinguish from rhyolites and andesites that cooled from pools of molten magma (see the Igneous Rock Identification Chart, Figure 3.29). Often, tuffs show layering or rows of tiny bubbles trapped along planes within the rock. Color banding may also be evident in tuffs. Rhyolites and andesites are seldom banded and are generally very uniform in texture.

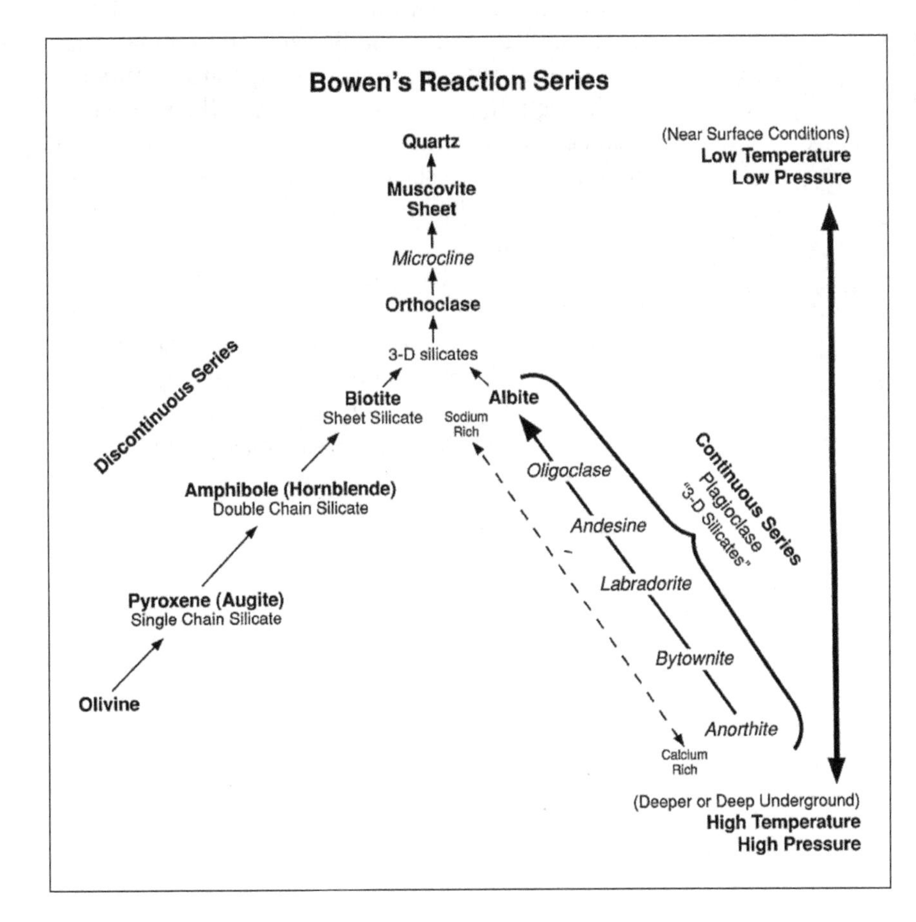

FIGURE 3.28 *Bowen's reaction series.*

between felsic and mafic commonly have a light and dark speckled appearance and are termed ***intermediate.*** A fourth class of rocks is the ***ultramafics*** that consist almost entirely of ferromagnesian minerals.

WHY ARE THERE SO MANY DIFFERENT IGNEOUS ROCKS?

Bowen's Reaction Series (Figure 3.28) is a very important aid to understanding how different igneous rocks are formed and what governs their chemical content.

Bowen was able to show in the laboratory that, as magmas cooled, two series of silicate rock-forming minerals were produced. One group was the so-called **discontinuous series** where each mineral was only stable under certain temperature and pressure conditions. As the temperature dropped, a series of discrete mineral types formed and disappeared. In the other series, a continuously changing ratio of calcium-rich to sodium-rich plagioclase feldspars formed as the temperature fell (**continuous series**). Near the top of Bowen's chart were the lowest temperature and pressure minerals.

Bowen's Reaction Series demonstrates that, if the magma was cooled while the minerals in the high temperature phase were present, you would get a gabbro (or basalt, gabbro's fine-grained equivalent), while if the series was allowed to run to completion, granite (or rhyolite, granite's fine-grained equivalent) would result. The intermediate case would produce diorite (or andesite, the fine-grained equivalent).

Bowen's chart also demonstrates why you don't find abundant olivine and pyroxene in granite, because these two minerals would not have been present during the formation temperatures that generate low-temperature granitic rocks.

Another possible way to explain the variety of observed igneous rock categories is found in process known as **crystal settling** (or **fractional crystallization**). In this process, a dense mineral formed early in the cooling process might settle out leaving the remaining magma depleted in that mineral. If this occurred, the resulting rock would have a mineral content different from the original magma. This is probably the main reason for the formation of intermediate and felsic magmas (Figure 3.29).

An igneous rock already formed might also be reheated. A glance at the Bowen's chart shows that, if this occurred, the lowest temperature minerals would begin to melt first. In this case, the rock would become partially melted, and the low-temperature minerals could be separated, leaving behind a much more mafic residual rock mass. This process is called **partial melting**.

Although perhaps rare, two magma chambers with different chemical formulations might be joined in the subsurface by means of faults or melting together. If this **magma mixing** occurred, the igneous rock ultimately produced would be different from that in either of the original magmas (Figure 3.30).

INSIGHTS

What is wrong with Figure 3.29? Nothing! This figure is upside down relative to the chart orientation printed in most textbooks. Bowen's Reaction Series chart needs to be seen as it appears below, however, because geologists always use Steno's Laws, the first of his Laws stating that geologic events are always shown in stratigraphic order. Therefore, hotter and higher pressure settings are to be found deep inside the earth, while lower temperature and pressure conditions are found near the surface.

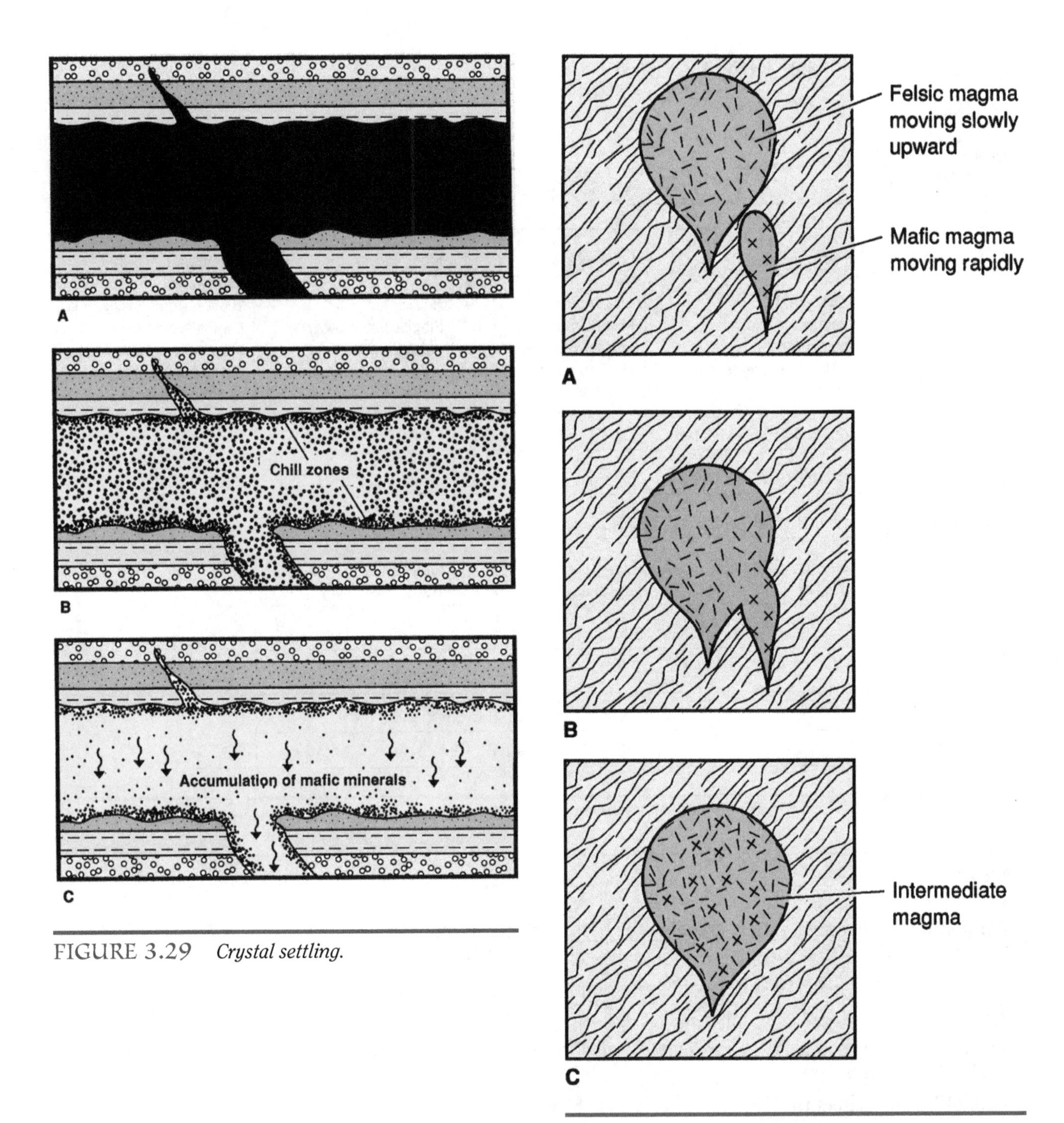

FIGURE 3.29 *Crystal settling.*

FIGURE 3.30 *Magma mixing. Two magmas of different composition merge (A), they intermingle (B), and they produce a new magma (C) with a composition different from either original body.*

Identification Procedures for Igneous Rocks
Using the Identification Key

Composition / Texture	Felsic (Light) ≥5% quartz. Potassium feldspar > plagioclase. ≤15% ferromagnesian minerals.	Intermediate <5% quartz. Plagioclase > potassium feldspar. 15–40% feromagnesian minerals.	Mafic (Dark) No quartz. Plagioclase ≤ 50%, no potassium feldspar, ≤40% ferromagnesian minerals.	Ultramafic Nearly 100% ferromagnesian minerals, olivine commonly present and minerals have a high specific gravity.
Pegmatite	GRANITE-PEGMATITE	DIORITE-PEGMATITE	GABBRO-PEGMATITE	
Phaneritic	GRANITE (SYENITE)	DIORITE	GABBRO (DIABASE)	PERIDOTITE
Aphanitic	RHYOLITE (TRACHYTE)	ANDESITE	BASALT	
Porphyry	Granite or Rhyolite Porphyry	Diorite or Andesite Porphyry	Gabbro or Basalt Porphyry	
Glassy	OBSIDIAN			
Frothy or cellular	PUMICE		SCORIA	
Pyroclastic or fragmental	VOLCANIC TUFF (Fragments ≤ 2 mm)			
	VOLCANIC BRECCIA (Fragments ≤ 2 mm)			

FIGURE 3.31 *Key for the identification of igneous rocks.*

INSIGHTS

For porphyritic textures, simply combine the basic rock name with the terms porphyry or porphyritic. For example, granite with phenocrysts can be called porphyritic granite, or granite porphyry.

Syenite is a name applied to felsic rocks resembling granite, but containing no quartz.

Trachyte is a name applied to felsic rocks that resemble rhyolite, but containing no quartz.

Diabase is a name applied to gabbros composed almost totally of plagioclase and pyroxene mineral crystals about 1.2 mm in size.

Peridotite consisting primarily of olivine is called a **dunite.**

All pale to white mineral grains are NOT quartz. Check with the microscope to make certain that light-colored grains do not show cleavage and striations, making them plagioclase feldspars.
If your rock is PHANERITIC, you can make use of the **‚ÄòBlack Grain Abundance Chart’** above the Identification Chart.

In this strip chart, black dots represent the relative abundance of BLACK grains (crystals) in the rock. Slide your rock sample along the strip chart and look for the relative abundance of truly BLACK grains, not grains that are gray, brown, green or any other color. The relative abundance of the black grains will often lead you directly to the correct color column on the chart.

NOTE: This will not work on aphanitic rocks, because you cannot clearly see how many black grains are present.

IDENTIFICATION KEY TO IGNEOUS ROCK

Methods:

(Use the accompanying data sheets to record your observations.)

STEP ONE: Texture of igneous rocks is a primary classification feature. Determine whether your sample is phaneritic (coarse grained), aphanitic (fine grained), glassy or exhibits some unusual texture.

Phaneritic Rocks: Decide whether your sample is just phaneritic or is coarse enough to be a **pegmatite,** or whether there are two size classes of mineral crystals present making it a **porphyry.**

Aphanitic Rocks: Decide whether your rock is simply fine grained, glassy, **vesicular,** or **pyroclastic.**

STEP TWO: Determine whether the sample is felsic (light in color), intermediate, or mafic (dark in color). Ultramafic rocks are not only dark colored but are generally much more dense than expected for their size.

Phaneritic Rocks: Check the minerals present to see if they support your color decision. For example, a felsic rock should have abundant quartz.

Aphanitic Rocks: You can't normally see which minerals are present here and must rely on general color alone.

Glassy: Color is of little use with these rocks. Rely on texture alone.

Frothy or Vesicular: Gray to white colored samples are normally considered ***pumice,*** while red to black samples are termed ***scoria.***

STEP THREE: Determine the relative abundance of **ferromagnesian** grains to all other mineral grains. Ferromagnesian minerals are dark-colored minerals. The strip chart accompanying your key shows black dots (ferromagnesians) in contrast to **ALL** grains of any other color (Figure 3.31). If you hold your rock sample against the chart and make a visual percentage estimate, it will be one more way to assure yourself that you are in the correct part of the chart.

Phaneritic Rocks:

Felsic types will have few to no ferromagnesian grains (0–15 percent).

Intermediate rocks will commonly have a light-to-dark "salt-and-pepper" appearance due to a mix of ferromagnesian and non-ferromagnesian grains (15–40 percent ferromagnesians).

Mafic rocks will have large percentages of ferromagnesian grains (40 percent or more) giving the rocks their characteristic dark colors.

Ultramafic rocks will consist almost entirely of ferromagnesian minerals.

Aphanitic Rocks: Since grain size is so small, you will not be able to determine which minerals are present without a microscope.

Results: If you carefully follow these steps, you realize that once you correctly check color and texture, the intersection of the color column and the texture class will provide the name of the rock being examined.

Using the identification key (Figure 3.32), record the data for each of your specimens on the included data pages. If you are careful and follow these instructions, you will find it easy and enjoyable to identify your igneous rock samples.

CHAPTER 4

Volcanism

INTRODUCTION During April 11 and 12, 1815, Earth experienced the largest single volcanic eruption in recorded time. Tambora, a volcano on the island of Sumbawa in Indonesia, erupted, blasting an estimated 36 cubic miles (150 km^3) of material into the air. The ash content of the atmosphere was so dense that for three days total darkness settled over an area of more than 900 square miles (2,500 km^2). The blast and the resultant seawave that inundated the surrounding islands killed an estimated 50,000 people with thousands more dying during the famine that followed. The winds of the upper atmosphere carried fine volcanic ash around the world, producing spectacular sunsets for several years. Unfortunately, the ash also filtered out the radiant heat of the Sun, and atmospheric temperatures dropped worldwide. The year 1816 became known as "the year without a summer." In Maine, frosts were recorded every month of 1816, and a snowstorm swept New England in June. Worse yet, around the world, at least 80,000 people starved as growing seasons were shortened and crops failed. Some blame the Tambora eruption for a famine in Ireland as well as for an outbreak of cholera in Asia that lasted until 1833 and claimed millions of lives. Can such a cataclysmic event happen again? The answer is yes. Where and when remain unresolved.

Before the advent of plate tectonics, the topic of volcanism was the center of a geologic debate. Geologists had long been aware that, for the most part, the distribution of active volcanoes coincided with zones of major earthquakes. The association sparked vigorous discussions. Some geologists argued that as rocks break under stress, they release energy in the form of earthquakes in much the same fashion as energy is released with the snap of a breaking stick. The earthquakes cause fractures to open in the lithosphere that allow molten rock to make its way to Earth's surface. In other words, earthquakes cause volcanic eruptions.

The other camp argued that volcanic eruptions cause earthquakes. As the molten rock forces its way to the surface, rocks are forced out of the way, stressed, and broken, giving rise to the release of energy responsible for the earthquake. It is true that the rise of magma does cause rocks to rupture, but the resultant earthquakes are all low magnitude. Major earthquakes are not caused by the ascent of magma.

We now know that the association of volcanism and major earthquakes is not a cause and effect relationship. Instead, both phenomena are associated with the same geologic setting, that is, hot spots and the edges of the lithospheric plates. Earthquakes occur where rocks are broken and moved as a result of being pulled apart or thrust together. Magma, generated beneath the plate edges, rises along the fractures that form as a result of these movements and emerges at the surface as a volcanic eruption. As the following discussion will explain, the composition of magma, the intensity of volcanic eruptions, and the magnitude of earthquakes associated with plate margins vary considerably, depending upon whether the margin is divergent or convergent.

Volcanoes are described as *active, dormant,* or *extinct*. Although the distinction between active and dormant is not precise, the term **active** is usually used to describe an erupting volcano. A **dormant** volcano, on the other hand, is one that is not now erupting but has erupted in historic time and is considered likely to erupt in the future. An **extinct** volcano is one that is not now erupting and is not likely to erupt in the future. At present, there are from 40 to 50 active and about 600 dormant volcanoes on Earth.

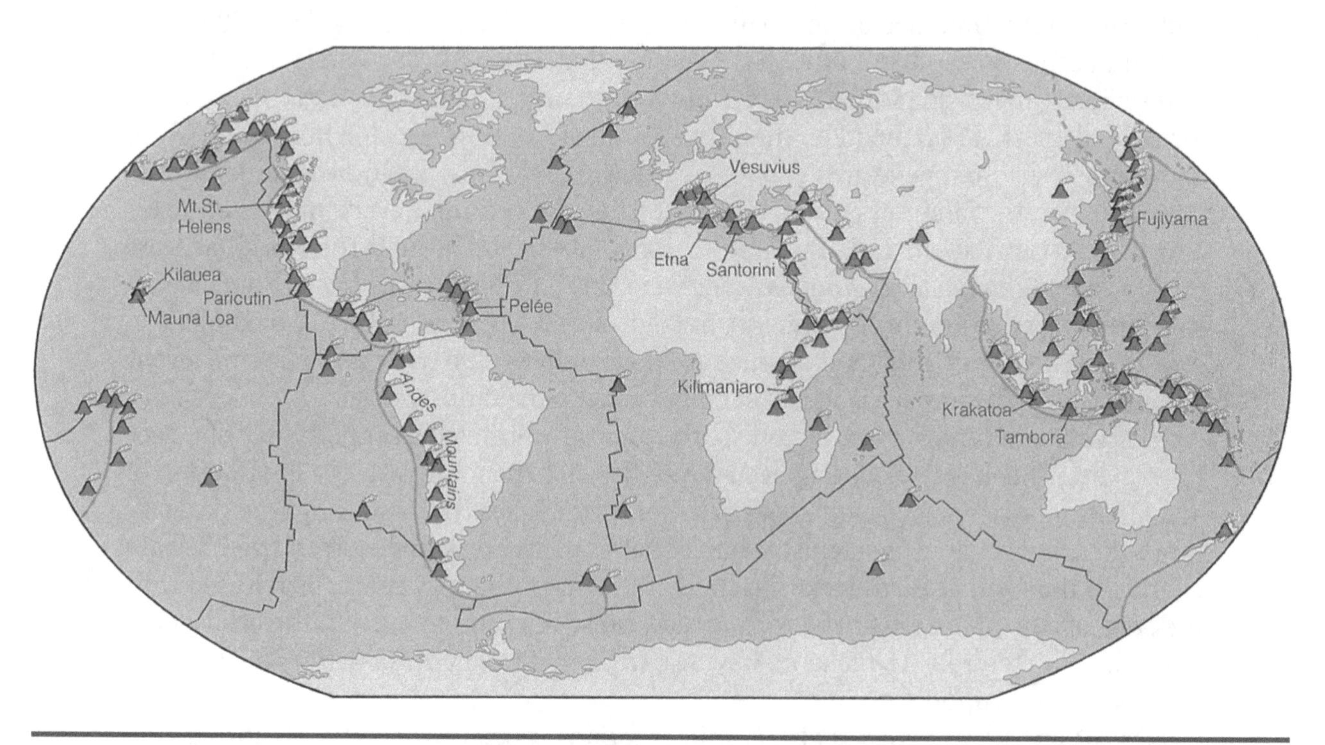

FIGURE 4.1a *Most of Earth's active and dormant volcanoes are associated with convergent plate margins with a few located within plate boundaries. The greatest concentration of volcanoes is found around the Pacific Ocean basin, giving rise to the name "Ring of Fire" for the Pacific rim.*

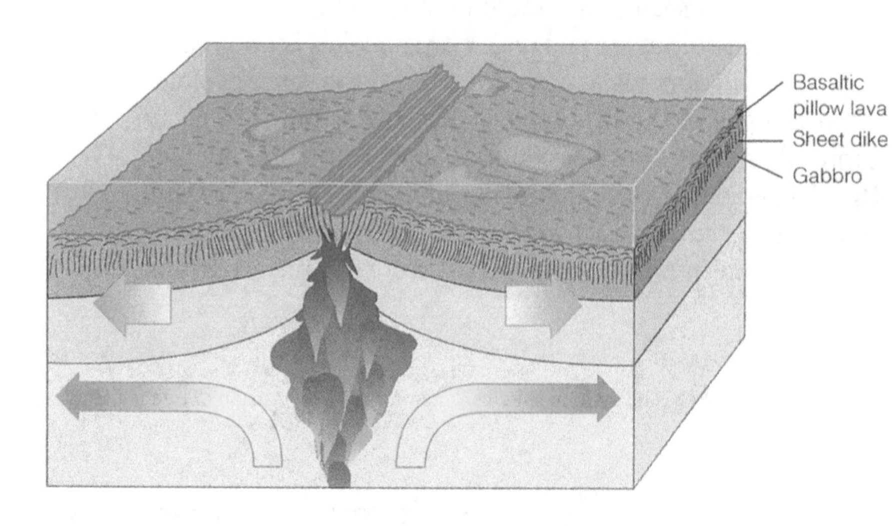

FIGURE 4.1b *Most of Earth's volcanic activity is associated with the oceanic ridges where molten rock constantly wells to the surface to create new oceanic lithosphere.*

DISTRIBUTION OF VOLCANOES

When the locations of Earth's active volcanoes are plotted on a map, a distinctive pattern emerges. Volcanoes are found along the margins of the lithospheric plates with some intraplate volcanism, especially beneath the Pacific plate (Figure 4.1a).

Relationship to Plate Boundaries

Most volcanic activity is associated with **divergent margins** where basaltic magmas well up along the oceanic ridges to continuously form new oceanic crust (Figure 4.1b.) Because it is largely submarine, divergent boundary volcanism generally goes unnoticed. The island of Iceland where a small portion of the North Atlantic Oceanic Ridge builds above sea level provides a rare opportunity to observe oceanic ridge volcanism (Figure 1.8). Several examples of divergent boundary volcanism are located within continents, the best known being the East African Rift Valley (Figure 4.2). A possible example of an incipient divergent boundary is the Rio Grande Rift of the southwestern United States (Figure 4.3).

Most of the observable and familiar examples of active volcanoes are associated with **convergent margins.** In instances where two oceanic plates converge, chains of volcanic islands called **island arcs** form on the overriding plate (Figure 4.4). The Aleutian Islands are an example.

Where an oceanic lithospheric plate and a continental lithospheric plate converge at a plate boundary, the volcanism builds a chain of volcanoes called a **continental arc** on the edge of the continental plate (Figure 4.5). Examples include the Cascade Mountains of the northwestern United States and the Andes Mountains along the western margin of South America.

Most of Earth's 600 or so active or dormant volcanoes are associated with the subduction zones that encircle the Pacific Ocean basin, a band known poetically as the **"Ring of Fire."** In contrast to the relatively quiet eruption of basaltic magmas along rift zones or oceanic ridges, subduction zone volcanism is characterized by the explosive eruption of more silica-rich magmas.

Intraplate Volcanism

A third source of volcanism is associated with mantle plumes generated by **hot spots** located deep within the mantle (Figure 4.6). Mantle plumes deliver heat to the lithosphere in much the same way as a Bunsen burner heats a beaker (Figure 1.9). The location of the hot spot and the mantle plume remains relatively fixed over geologic time while the overlying lithosphere moves laterally with the movement of the plate. Many of Earth's hot spots are located beneath the Pacific plate. As the mantle rocks rise within the plume and the pressure drops, the rocks at the top

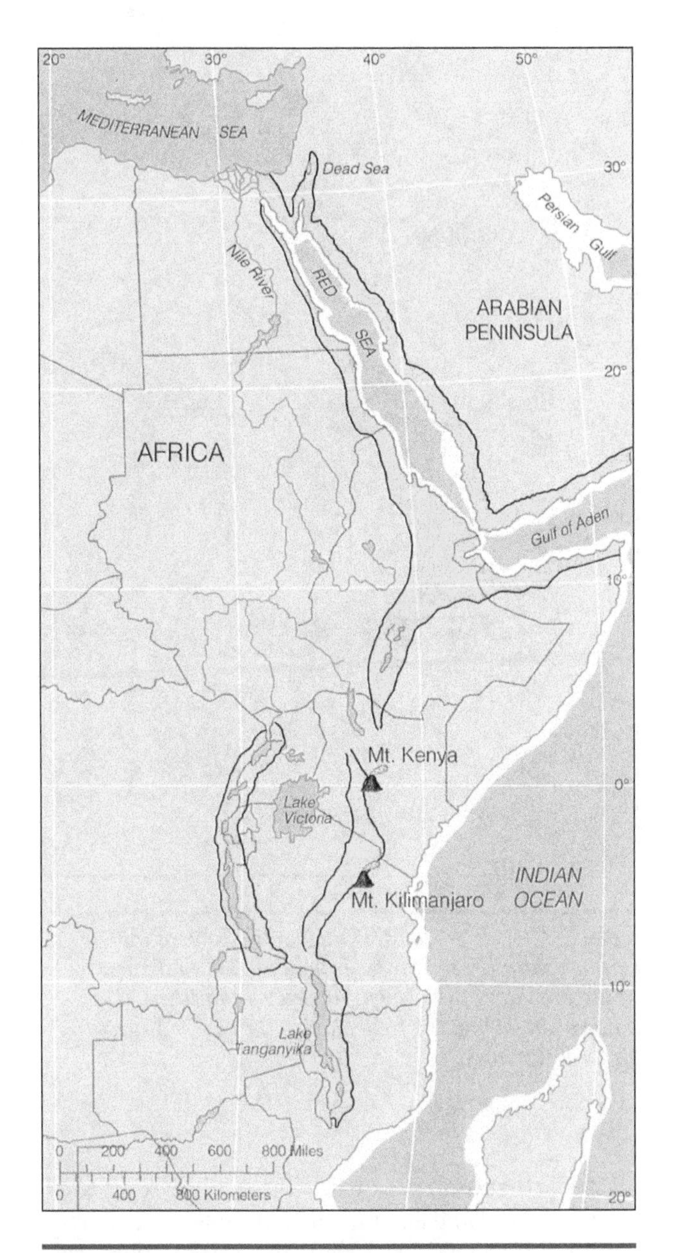

FIGURE 4.2 *The best-known example of a rift valley is the* East African Rift Valley. *At the present time, seawater is beginning to encroach into the northern end of the valley. Soon, the rift valley will begin to flood and may be converted into Earth's newest* linear ocean.

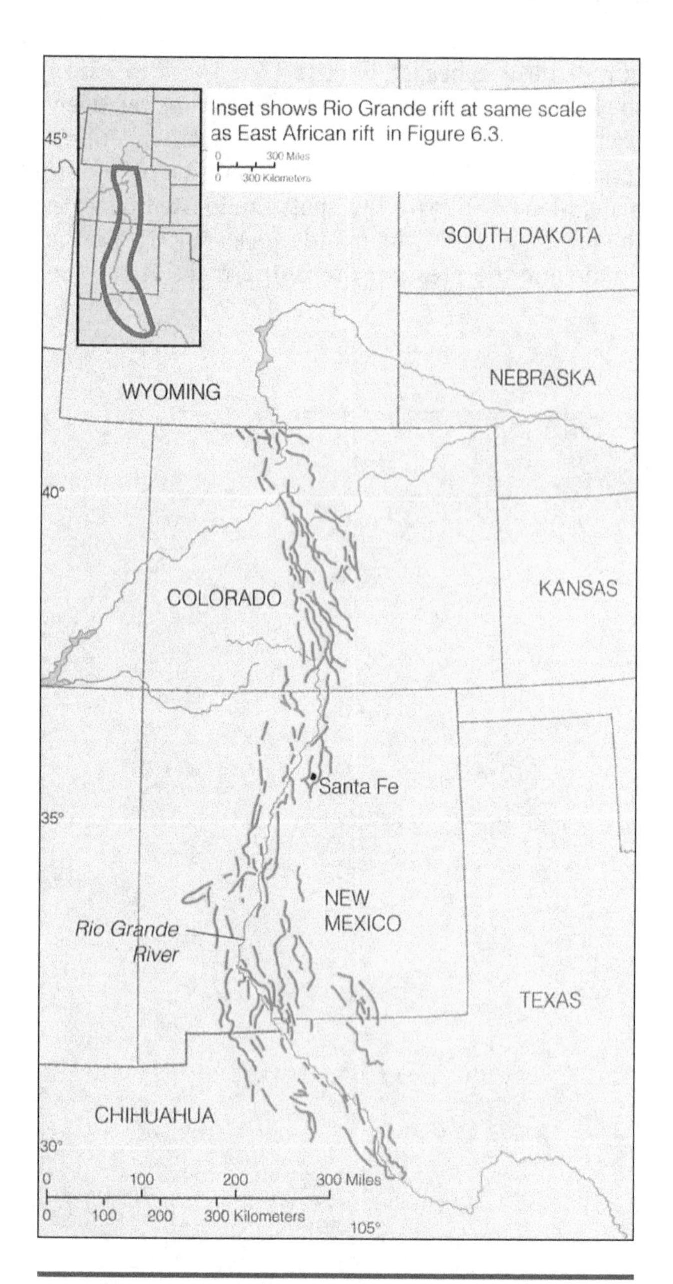

FIGURE 4.3 *The* Rio Grande Rift Zone *of the southwestern United States is possibly the first stage in the breakup of the North American continent. (Data made available by the New Mexico Bureau of Mines/ Mineral Resources)*

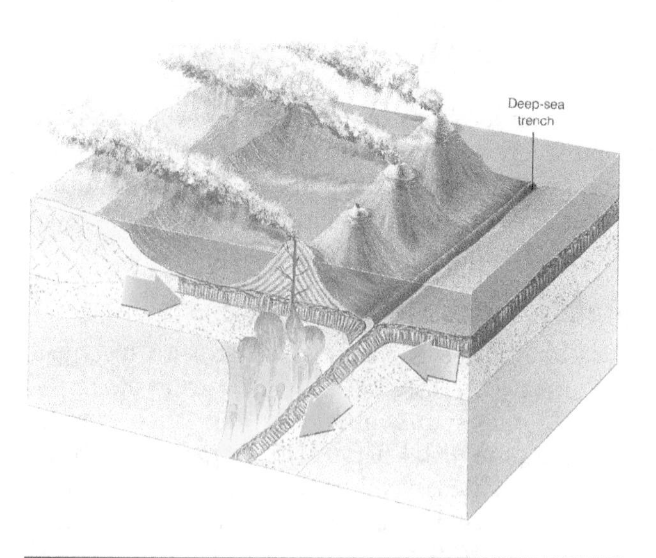

FIGURE 4.4 Island arcs, *such as the* Aleutian Islands, *form as chains of volcanoes build along the edge of the overriding plate at an ocean-ocean convergence.*

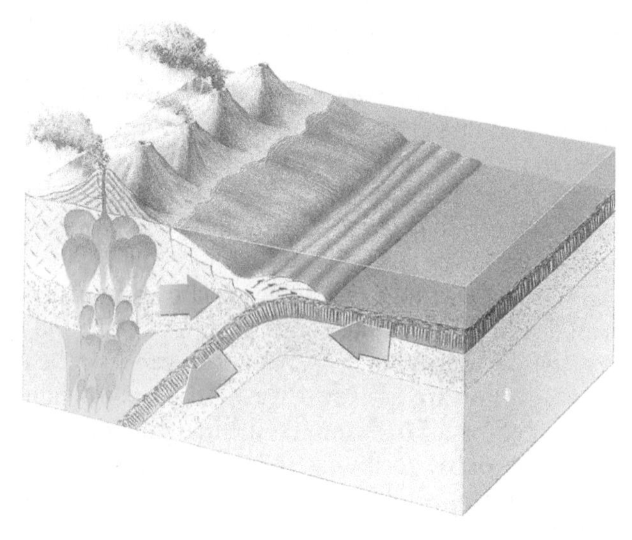

FIGURE 4.5 Continental arcs, *such as the* Cascade Mountains *of Oregon and Washington and the* Andes Mountains *of South America, form along the edge of the overriding plate at an ocean-continent convergence.*

of the asthenosphere melt. The molten rock then moves upward and accumulates in a magma chamber within the lithosphere. From the magma chamber, the molten rock erupts to the surface where it gives rise to a large number of submarine volcanoes called *seamounts.*

The movement of the Pacific oceanic plate over three separate hot spots has created three parallel strings of volcanic islands and seamounts in the Pacific Ocean: (1) the Hawaiian Islands–Emperor Seamount chain, (2) the Tuamotu Archipelago–Line Island chain, and (3) the Austral, Gilbert, and Marshall Islands (Figure 4.7). Like divergent margin volcanism, hot spot volcanism associated with oceanic plates usually involves the relatively quiescent eruption of basaltic magmas. Hot spot

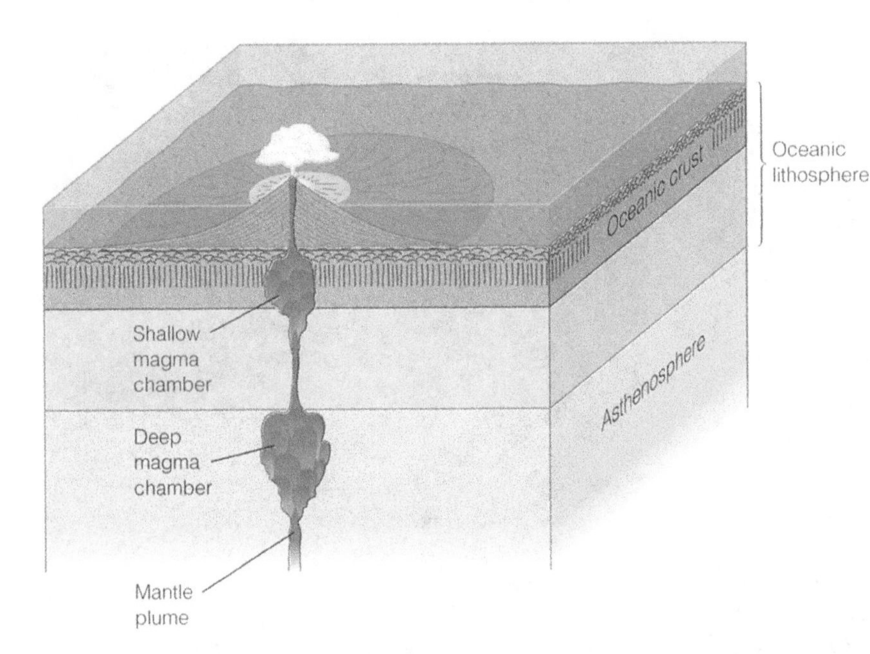

FIGURE 4.6 *Volcanism within the lithospheric plates is located over* mantle plumes *that produce* hot spots *in the overlying lithosphere. For reasons that are not fully understood, most hot spots are located under the Pacific Ocean lithospheric plate.*

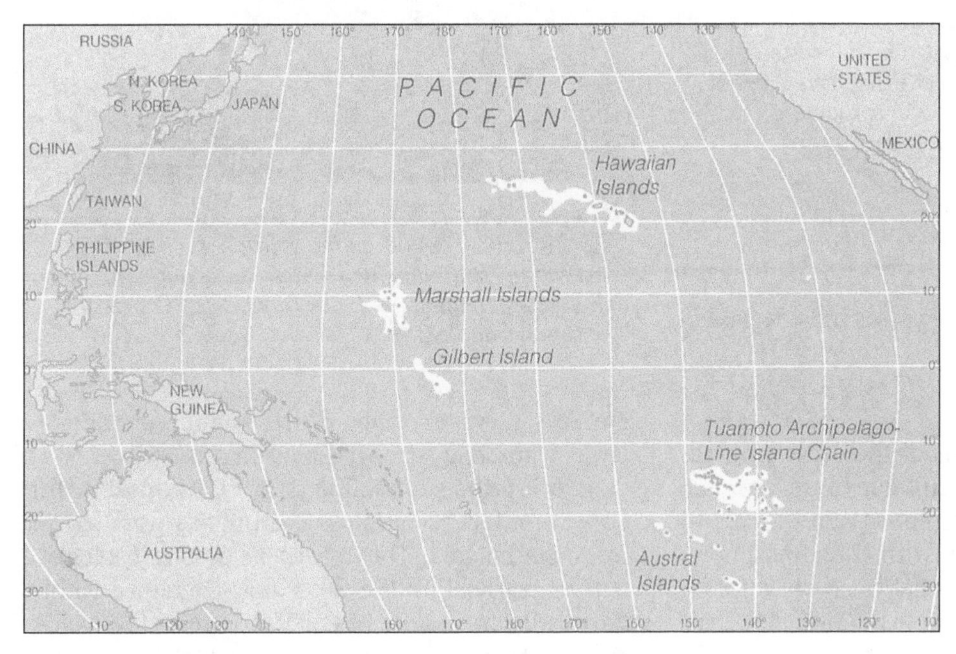

FIGURE 4.7 *Three chains of volcanic islands that formed over stationary mantle plumes exist in the Pacific Ocean. Hawaii is a part of one of these chains.*

eruptions involving continental crustal plates are generally quite different. For example, the hot spot located beneath Yellowstone Park resulted in the highly explosive eruption of viscous rhyolitic magma (Figure 4.8). The Yellowstone volcanism will be discussed later in the chapter.

SPOT REVIEW

1. Describe the geologic settings under which most of Earth's volcanoes form.
2. How do the geologic settings responsible for island arc and continental arc volcanism differ? In what ways are they similar?

GURE 4.8 *The rocks exposed in* Yellowstone k, *Wyoming, are pyroclastic rocks that were created* *he violent eruption of highly viscous rhyolitic* *ma. (Courtesy of W. B. Hamilton/USGS)*

FIGURE 4.9 *An erupting volcano, such as Mount Pinatubo in the Philippine Islands, spews out a mixture of hot gases, magma, and rock fragments. (Courtesy of T. J. Casadevall/USGS)*

EJ CTED MATERIALS

The aterials emitted from volcanoes include gases, liqui , and solids.

Gas

Obtai g an accurate analysis of the gases emitted by an ipting volcano is difficult due to the obvious proble involved in direct sampling and the rapidity wit which the gases mix with the atmosphere (Figure 9). Data have shown that the evolved gases not onl ffer from one volcano to the next, but that the com ition of gas evolved from a particular volcano wil ange during an eruption.

Wate apor is the most abundant gas emitted fro upting volcano. Much of the billowing cloud observed above an erupting volcano is condensed water vapor or steam. Some of the water is derived from the magma; however, most of the water is derived from groundwater, the water contained in the rocks beneath the surface of the Earth. As the magma rises toward the surface and encounters groundwater, the water is heated to boiling and emerges as water vapor. If the volume of water is large, a phreatic (steam) explosion can occur.

During the primeval days of Earth, most of the water emitted from volcanic eruptions was derived from the magma. The consensus among geologists is that nearly all of the water now contained within Earth's hydrosphere has derived from the volcanic eruptions that dominated the surface of Earth during its early evolution. Scientists once thought that the primeval Earth was completely molten at some time in its early history. Although this idea is no longer accepted, many geologists are convinced that "magma oceans" may have existed during the very early history of Earth. During these early stages, Earth's solid surface was very hot and dominated by widespread volcanism, which expelled enormous volumes of water from Earth's interior. With the surface too hot to allow the rainwater to accumulate, the water collected in a thick cloud cover. Eventually, Earth cooled to the temperature at which rainwater was not vaporized. The water then accumulated and accelerated the cooling of the surface rocks; torrential rains washed the high areas and flowed into the low areas. In time, the low areas filled with water to form the oceans, and the high areas became land surfaces with lakes and streams. As Earth's hydrosphere formed, the basic geologic processes of erosion were initiated.

In addition to water, volcanic eruptions release other gases in significant volumes. Of these, the most common are *sulfur dioxide* and *carbon dioxide*. Sulfur dioxide, the acrid smell of "fire and brimstone," has an unpleasant odor that is characteristic of areas of active volcanism. Carbon dioxide is the atmospheric component that is an important contributor to the "greenhouse effect," in which heat reflected from the Earth's surface is trapped and retained by the atmosphere. Throughout geologic time, the two major natural sources of carbon dioxide have been volcanic eruptions and the decay of organic matter. The rate at which carbon dioxide was added to the atmosphere eventually became balanced with the rate at which it was removed by dissolution in water, the formation of the shells of many marine organisms, incorporation into the body tissues of plants and animals, and storage in limestone and in the fossil fuels, oil, gas, and coal. With this balance, Earth achieved a mean atmospheric temperature of about 58°F (14.5°C) that has been maintained over most of geologic time. The current increase in the rate at which carbon dioxide is being introduced into the atmosphere due to the burning of fossil fuels has the potential to significantly unbalance the carbon dioxide cycle, enhance the greenhouse effect, and subsequently increase Earth's atmospheric temperature.

Several gases emitted during eruptions, including chlorine and sulfur dioxide, react with water to produce strong acids. If inhaled, the fumes can cause acute discomfort. For this reason, individuals who suffer from respiratory problems are usually advised to avoid areas of active volcanism.

Liquids

Most lavas are basaltic in composition. Because of their relatively low viscosity, basaltic magmas rise to the surface through fractures with relative ease (Figure 4.10). From our discussions of plate tectonics, you will remember that basaltic lavas are associated with oceanic ridges, oceanic hot spots, fissure eruptions, and continental hot spots.

FIGURE 4.10 *Typical of Hawaiian eruptions, low viscosity basaltic magma is extruded quietly onto the surface and flows downslope as a river of molten rock. (Courtesy of USGS/HVO)*

Although lavas erupting in subduction zones vary widely in composition, the great majority are andesitic. Basaltic lavas are associated with some subduction zone volcanism, for example, in the Cascades, and late-stage subduction zone eruptions may evolve highly viscous rhyolitic lavas that form thick, semisolid plugs within the volcanoes' craters (Figure 4.11).

Depending on whether the lava solidifies exposed to the atmosphere (subaerial) or under water (subaqueous), basaltic lavas take on one of three forms: (1) *pahoehoe*, (2) *aa*, or (3) *pillow lava*. Subaerial solidification of hot, highly fluid lavas produces a flow with a smooth surface often with the appearance of twisted ropes (Figure 4.12).

FIGURE 4.11 *Following the typically violent eruption of volcanoes associated with zones of subduction, such as Mount St. Helens, highly viscous rhyolitic magmas often move into the crater and solidify to form a plug. (Courtesy of USGS/CVO-F)*

FIGURE 4.12 *Low-viscosity basaltic magmas typically solidify to form a smooth or "ropy" surface that volcanologists refer to as* pahoehoe lava.

Although this type of lava is sometimes called "ropy," volcanologists refer to it by the Hawaiian name, **pahoehoe.**

As the basaltic lavas cool subaerially and lose dissolved gas, they become more viscous and flow more slowly (Figure 4.13). The surface of the flow solidifies, but because the interior of the flow is still molten and moving, the solidified layer breaks into angular blocks that are carried along on the surface of the molten flow. Eventually, when the entire mass is solidified, the lava is characterized by a rough, angular surface (Figure 4.14). Because of its appearance, this type of lava is sometimes referred to as "blocky." Volcanologists, however, almost universally use the Hawaiian name, **aa,** to describe this type of lava.

Pillow lava solidifies under water. It derives its name from the flow's resemblance to a pile of pillows. Pillow lavas form at oceanic hot spots and at oceanic ridges where the lavas spread laterally to make up much of the oceanic crust. Because the oceanic crust represents 71% of Earth's surface, basaltic pillow lavas can be considered to be the most common rock type on or near Earth's surface. In addition to the lavas that are extruded under water, pillow lavas may also form from lavas that originate on land and subsequently flow into the water (Figure 4.15).

Solids (Tephra)

The solid fragments ejected from an erupting volcano are collectively called **tephra.** Such an eruption and the rocks formed by it are considered to be

FIGURE 4.13 *As basaltic magma cools and loses its gas content, it becomes more viscous and moves more slowly. (Courtesy of USGS/HVO)*

FIGURE 4.14 *When the more viscous basaltic magma solidifies, it forms a lava with a sharp, angular surface called* aa. *(Courtesy of USGS/HVO)*

pyroclastic. The prefix *pyro-* comes from the Greek word meaning "fire," and *clastic* refers to broken bits and pieces of minerals and rocks. Pyroclastic material is derived from a number of sources including the old volcano walls, shattered fragments of solidified magma, and pieces of lava that were thrown into the air and solidified in flight or on landing. Pyroclastic materials are of two kinds: (1) *tephra falls* and (2) *ash flows*.

Tephra-fall materials are classified based on shape and size into five types: (1) *blocks*, (2) *bombs*, (3) *cinders*, (4) *ash*, and (5) *dust*. **Blocks** and **bombs** are particles with mean diameters in excess of 2 1/2 inches (64 mm), about the size of a baseball (Figure 4.16). Blocks are angular fragments that are generated when portions of the crater walls or volcanic cone are blown apart by explosions. Bombs are blobs of molten rock that are thrown out of the crater, take on a streamlined shape as they pass through the air, and solidify only partly before returning to Earth's surface. **Cinders** are particles that range in diameter from 2 1/2 inches (64 mm) down to about 1/16 inch (2 mm), about the size of a pinhead. Cinders form primarily molten spatter than cools either in midair or upon striking the ground (Figure 4.17). **Ash** and **dust** are particles with diameters smaller than 1/16 inch (2 mm), with dust being the finer.

Most tephra-fall materials larger than ash accumulate in the immediate vicinity of the volcano. The fine-sized materials, however, especially dust, can be carried long distances by the prevailing winds (Figure 4.18). Dust-sized materials from major volcanic eruptions may be carried around Earth in the upper atmosphere where they are responsible for many brilliant sunsets before finally settling back to the surface.

Volcanically derived dust in the upper atmosphere has also been responsible for significant reductions in mean atmospheric temperatures. One of the first scientists to suggest that volcanic eruptions could affect weather was Benjamin Franklin who, in 1784, suggested that the extreme winter of 1783–1784 was the result of eruptions in Japan and Iceland. We have already commented on the climatic effects of the 1815 eruption of Tambora.

The rocks that form from the accumulated tephra-fall materials are called **tuff** in the case of the

FIGURE 4.15 *Pillow lava forms where molten lavas pour into the ocean, as is occurring along the coast of Big Island in Hawaii.*

FIGURE 4.16 *When relatively large blobs of molten lava are thrown into the air, they may solidify after taking on an aerodynamic shape and land as* bombs. *(Courtesy of USGS/HVO)*

fine-grained ash and dust (Figure 4.19) and **volcanic breccia** when made up of the larger particles.

Pyroclastic Flows

If the concentration of pyroclastic materials ejected during the eruption is very high, the eruption column may collapse, the ash will mix with superheated gases, and the combined mass, called an **ash flow**, will move down the mountainside like an avalanche. Because they ride on a cushion of hot air, pyroclastic flows move with little frictional resistance and, depending upon the angle of the slope, can attain

FIGURE 4.17 *Cinders are commonly generated as basaltic magmas fountain, break into small pieces, and solidify in midair. Should the eruption continue for a period of time, the cinders will accumulate around the vent to form a* cinder cone. *(Courtesy of USGS/HVO)*

FIGURE 4.18 *Ash- and dust-sized materials thrown into the atmosphere by an erupting volcano can be carried by the prevailing wind for long distances. (Courtesy of USGS/MSH, July 22, 1980)*

FIGURE 4.19 *When the ash- and dust-sized materials blown into the atmosphere return to Earth, they commonly fall over a wide area and form extensive layers of a pyroclastic rock called* tuff. *(Courtesy of W. B. Hamilton/USGS)*

speeds of over one hundred miles per hour. In some cases, an exceptionally hot cloud of gas called a **nuée ardente** (French for "glowing cloud"), often heated to glowing, may override the ash flow. Needless to say, such a mass of fast-moving, hot materials can be enormously destructive.

The solidified debris from ash flows or nuées ardentes are called **ash-flow tuffs** (Figure 4.19). If the fragments are partially fused by compaction after movement stops, the deposit is called a **welded tuff.** The term **ignimbrite** is used to describe a deposit consisting of avariety of tuffs.

Mudflows or Lahars

Even in equatorial regions, volcanoes such as Mount Kilimanjaro commonly build to such elevations that the summits are covered with snow and capped with ice. During an eruption, some snow and ice may melt, producing large volumes of water. In addition, volcanic eruptions are commonly accompanied by torrential rains triggered by the atmospheric disturbances above the erupting volcano. These waters mix with the loose tephra-fall materials and materials already accumulated on the slopes of the volcano and create a flow called a **lahar** that moves downslope, following an existing stream valley. Lahars usually move more slowly than ash flows, but they have still been clocked at 50 to 60 miles per hour (80–95 KPH). Lahars accounted for much of the destruction resulting from the 1980 eruption of Mount St. Helens (Figure 4.20).

SPOT REVIEW

1. Explain how three different kinds of lava can form from molten rock of the same composition.
2. In what ways have the gases emitted from volcanoes affected the atmosphere since the formation of Earth?
3. What do ash flows, nuées ardentes, and lahars have in common?

FIGURE 4.20 *Most of the damage caused by the eruption of Mount St. Helens was the result of* mudflows *or* lahars. *(Courtesy of MSH-Banks/USGS)*

(a)

(b)

FIGURE 4.21 *(a) Short-lived eruptions of basaltic lavas commonly result in the formation of* cinder cones *around the vent. (b) Cinder cones in Arizona (photo by Daria Nikitina)*

KINDS OF VOLCANOES

Volcanologists define a volcano as the conical structure that builds around a vent or the vent itself. As in the case of intrusive igneous rock bodies, the definition of a volcano does not include a size requirement, another example of geologists being more concerned with the process by which a feature forms than with how large or small it may be. We will discuss four kinds of volcanoes: (1) *cinder cones,* (2) *spatter cones,* (3) *shield volcanoes* and (4) *strato-volcanoes* or *composite volcanoes*.

Cinder Cones

As the name implies, a **cinder cone** is an accumulation of largely cinder-sized pyroclastic material around a vent. The cinders commonly form by the midair solidification of molten rock spatter expelled by a lava fountain during the eruption (Figure 4.21). Most of the materials within the cone are unconsol-

(a)

(b)

FIGURE 4.22 *(a) When insufficient gas exists to create a fountain of lava, blobs of lava may lap out of the vent and solidify as a* spatter cone. *(Courtesy of USGS/HVO) (b) Spatter Cones in the Craters of the Moon National Monument and Preserve (Courtesy of Ksenia Svertilova)*

idated, although some welding of semisolid particles may take place as the materials accumulate. Most cinder cones are relatively small, usually not growing to elevations of more than about 1,000 feet (300 m) above the surrounding land surface. The slope angle of cinder cones, usually about 40°, is the *angle of repose*, or the highest angle at which unconsolidated, relatively large, irregularly shaped particles can reside without sliding downslope. The summit of a cinder cone is commonly occupied by a crater that may be enlarged following the eruption by the collapse of the inner walls.

Spatter Cones

Vents erupting basaltic magma under low gas pressure typically produce large blobs of molten rock or spatter fragments that fall to the outer edge of the vent where they congeal and solidify (Figure 4.22). Repeated expulsion of spatter builds a conical structure called a **spatter cone.** Like cinder cones, spatter cones do not grow to great height, usually not more than a few tens of feet (meters).

Shield Volcanoes

Shield volcanoes form as massive volumes of basaltic lavas are extruded from central vents and flank eruptions; they are usually associated with oceanic hot spots (Figure 4.23). Because of the low viscosity of the basaltic lavas, the molten rock readily flows long distances from the vent and solidifies. Repeated eruptions build the conical structure of the volcano. The volcanic cone is broad at the base with average slopes on the submarine portion of about 17°. In some cases, the cone may build above the surface of the ocean and become a volcanic island. Because the lavas cool more slowly in the atmosphere than under water and are therefore more fluid, the subaerial (exposed) portion of the cone typically has slopes of from 5° to 10°. The structures are called "shields" because they resemble the defensive armor carried by warriors. Shield volcanoes that build over oceanic hot spots remain submarine are called *seamounts*. The best-known examples of shield volcanoes that have grown to become islands, the Hawaiian Islands, will be discussed in detail in the next chapter.

Stratovolcanoes or Composite Volcanoes

Stratovolcanoes or **composite volcanoes** are commonly associated with zones of subduction. The lavas involved in the construction of the cone are primarily andesitic. Because of the higher viscosity and inherently higher gas content of subduction zone magmas, andesitic magmas tend to contain a greater volume of dissolved gases when they erupt to the surface. As a result, the eruptions are more violent than those associated with shield volcanoes.

The explosive eruptions of subduction zone volcanoes generate vast quantities of pyroclastic material. The coarse pyroclastic materials fall in the immediate vicinity of the vent to create a steep-sided cone that is coated and sealed by the lavas of subsequent eruptions. Because the cone can accumulate with a relatively high angle of repose, the combined pyroclastic-lava layering develops a cone with a relatively steep slope. As successive eruptions repeat the pyroclastic-lava layering, the cone builds rapidly to

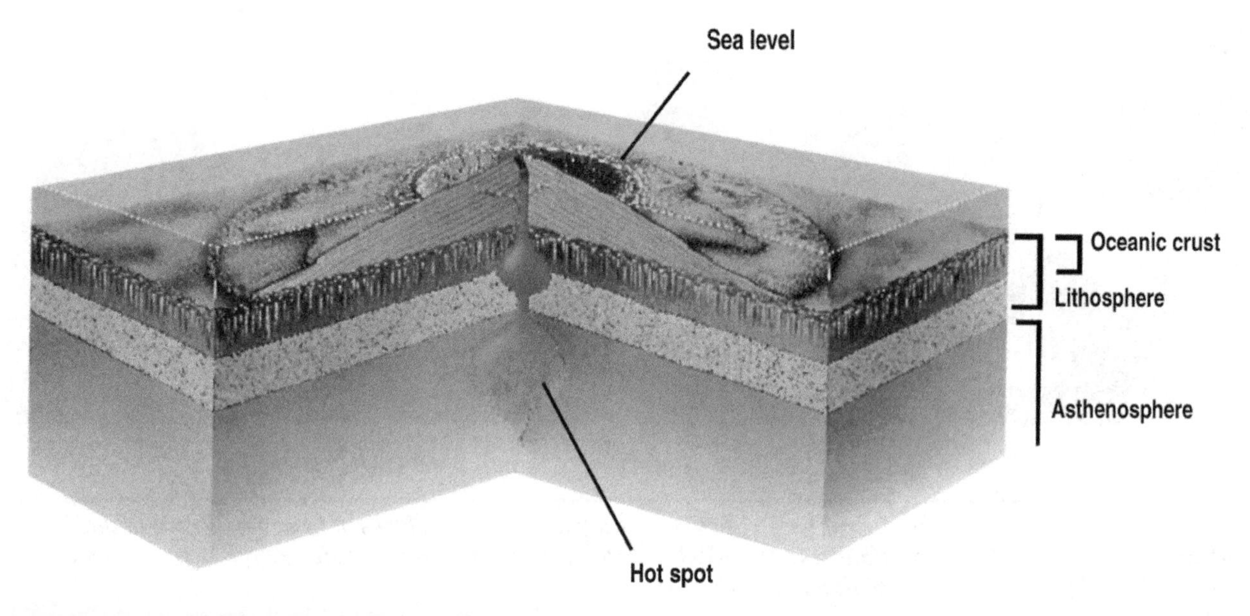

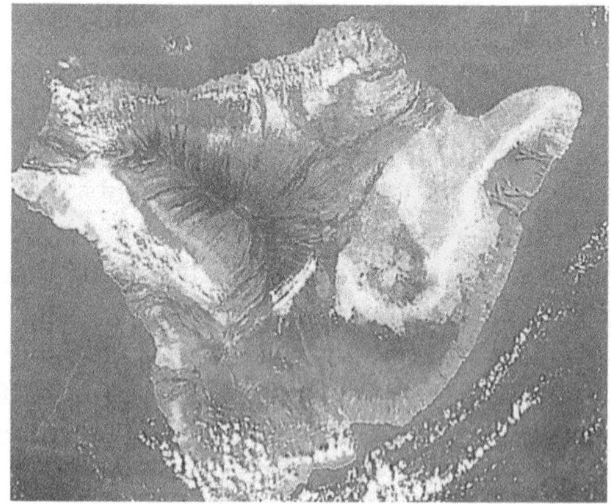

FIGURE 4.23 Shield volcanoes *are constructed of repeated flows of* basaltic lavas, *usually from vents associated with* hot spots. *Most of Earth's shield volcanoes are in the Pacific Ocean basin. The best-known examples of shield volcanoes are the* Hawaiian Islands. *(Photo courtesy of USGS/HVO)*

relatively great heights compared to the diameter of the base (Figure 4.24). The "composite" or "strato" designations refer to the interlayering of the pyroclastic rocks and lava. Lateral penetrations of magma through fractures in the cone produce radiating dikes and result in flank eruptions that add to the complexity of the cone structure. Many of the best-known volcanoes are stratovolcanoes or composite structures, including the famous Fujiyama near Tokyo, Japan.

MAGMA PROPERTIES AND ERUPTIVE PATTERN

The differences in the eruptive intensities of volcanoes are graphically portrayed in Figure 4.25, which compares the 1980 eruption of Mount St. Helens with a typical eruption of Kilauea volcano on the island of Hawaii. The question that arises is, how can two volcanic eruptions differ so much in style? Why, for example, does one volcano erupt with the potential to destroy everything within miles while the other's eruption can be watched with relative safety from a few hundred feet? The answer lies in the fundamental difference between basaltic and andesitic-to-rhyolitic magmas and the solubility of gases in solutions.

The eruptive style of a volcano is primarily determined by two magma parameters: (1) **gas content** and (2) **viscosity**. The eruption of a volcano has been likened to the uncorking of a bottle of warm champagne. Indeed, the chemistry and physics involved are quite similar. Simply stated, explosions are the result of the violent expulsion of gas from the magma as it reaches the surface. Magmas that contain little or no dissolved gas when they reach Earth's surface will erupt quietly. If the magma has high viscosity because of low temperature and high silica content, the potential for violent eruptions increases with increasing gas content.

FIGURE 4.24 *(a) Constructed of alternating layers of andesitic lavas and pyroclastic materials* stratovolcanoes *or* composite volcanoes *form the island and continental arcs associated with zones of subduction. Stratovolcano © daulon, 2014. Used under license from Shutterstock, Inc. Because of the higher viscosity and gas content of andesitic magmas, eruptions of stratovolcanoes are usually violent (b) Peak Sarychev (2008) is one of the most active stratovolcano in the Kuril Island chain, Russia, and is located on the northwestern end of Matua Island. Peak Sarychev, photo by Daria Nikitina. (c) An early stage of eruption on June 12, 2009 captured by International Space Station. Prior to the June 2012 eruption http://www.nasa.gov/multimedia/imagegallery/image_feature_1397.html. The last explosive eruption occurred in 1989 with eruptions in 1986, 1976, 1954 and 1946 also producing lava flows. Commercial airline flights were diverted from the region to minimize the danger of engine failures from ash intake (Courtesy of NASA).*

In a solvent, the solubility of any gas increases with increasing pressure. Dissolved gases can escape from solution when the pressure decreases. Consider, for example, your favorite carbonated beverage. The effervescence is due to the release of carbon dioxide the manufacturer forced into solution during the bottling or canning process. As long as the container is closed, the gas remains in solution, but when the container is opened and the pressure is relieved, the gas comes out of solution and bubbles toward the surface. In addition, the warmer the solution, the faster the gas will escape; this explains why keeping the beverage cold prolongs the effervescence and why the drink becomes "flat" when the container is opened and allowed to warm.

Magma viscosity controls not only the rate at which the gas is released from the magma but also

(a)

(b)

FIGURE 4.25 *The difference in eruptive styles is best illustrated by comparing the eruption of a stratovolcano such as* Mount St. Helens *with that of a shield volcano such as* Kilauea. *(A: Courtesy of MSH-Swanson/USGS; B: Courtesy of USGS/HVO)*

the ease with which the gas bubbles can exit the magma mass. As low-viscosity basaltic magmas begin their ascent, gas begins to escape at greater depth, at a higher rate, and more easily than from more viscous andesitic or rhyolitic magmas. By the time basaltic magmas reach the surface, most of the dissolved gas has already been released and vented. As the magmas emerge from vents or fractures, enough gas may still remain to drive the molten rock a few thousand feet into the air, creating an impressive lava fountain, but the gas content is not sufficient to cause truly violent explosions. The result is a relatively quiet eruption. Volcanism associated with oceanic ridges, oceanic hot spots, fissure eruptions, and most continental hot spots involves basaltic lavas and is therefore rather quiescent.

The eruptive intensity of volcanism associated with subduction zones is quite different. Most of the magmas associated with subduction zone volcanism are andesitic to rhyolitic. Because of their higher viscosities, the release of the gas from solution as these magmas rise toward the surface is retarded. Large volumes of gas may be maintained in solution until the magma breaks through to the surface and the pressures are totally relieved. At this time, the gas is violently released, triggering an explosion such as the 1980 eruption of Mount St. Helens.

The explosions associated with eruptions of rhyolitic magma are often even more violent than those associated with andesitic eruptions and produce huge volumes of pyroclastic materials. Examples include the eruptions that formed most of the rocks in Yellowstone National Park.

Classification of Eruptive Intensity

Volcanologists use five classifications to describe individual volcanic eruptions (1) *Hawaiian*, (2) *Strombolian*, (3) *Vulcanian*, (4) *Pelean*, and (5) *Plinian*.

Hawaiian Type Eruptions

The **Hawaiian** type of volcanic activity has been described as "the quiet evolution of lava" and is typical of basaltic eruptions (Figure 4.26). As the name suggests, this eruptive style is typical of the volcanoes on Hawaii. The few explosive eruptions reported on the Hawaiian Islands have been ascribed to groundwater entering the magma conduit. The ensuing explosions were steam or phreatic explosions and were not the result of either the magma composition or the original gas content of the magma.

Strombolian-Type Eruptions

The second type of volcanic activity is characterized by frequent but mild explosions that discharge incandescent pyroclastic materials and commonly build cinder cones around the eruption vents. The name, **Strombolian,** is derived from the volcano Stromboli, an island just north of Sicily. Stromboli is one of several volcanoes associated with the zone of subduction between Africa and Europe. Historically, Stromboli has erupted with a small explosion about every 15 minutes. Since gas is released in each of these frequent small eruptions, the gas content of the magma

FIGURE 4.26 *A typical eruption on Hawaii usually involves little more than the quiet evolution of basaltic magmas. The few violent eruptions that have occurred on Hawaii have been attributed to the rising magma having come into contact with groundwater, resulting in a steam explosion. (Courtesy of USGS/HVO)*

never builds to the point where a violent explosion can occur. The frequent eruptions constantly stir the magma in the crater, and a cloud of ash, dust, and steam is maintained above the volcano. The cloud can be seen for miles during the day; at night it glows due to the reflection from the red-hot magma below. As a result, Stromboli has been known for centuries as the "Lighthouse of the Mediterranean."

Vulcanian-Type Eruptions The third type of volcanic activity is characterized by "infrequent but severe explosions." **Vulcanian** eruptions explosively eject fragments of incandescent lava, which cool and solidify in midair into bombs, along with large volumes of ash. Vulcanian activity is more typical of the eruptions associated with the zones of subduction. The name is derived from the island of Volcano located about 35 miles (56 km) southwest of Stromboli. The distinctly different eruptive styles of the two volcanoes demonstrate that eruptive style is one of the unpredictable characteristics of volcanoes.

Pelean-Type Eruptions An explosive type of volcanic activity, the **Pelean** eruptions is named after Mont Pelée on the island of Martinique in the West Indies (Figure 4.27). Mont Pelée is one of the active volcanoes associated with the Caribbean subduction zone. The 1902 eruption of Mont Pelée allowed volcanologists to observe for the first time the most violent of explosive phenomena, the nuée ardente.

Mont Pelée had been active for centuries before the 1902 eruption, but had exhibited only minimal signs of activity such as the release of gas and steam. The relative peace of the island was broken in April 1902, when the volcano began to erupt with increasing intensity. The eruptions continued into May, culminating on May 8.

Many of the residents of the island had moved into the port of St. Pierre to vote in a local election. Overnight, the population of the town had grown from a few thousand to an estimated 29,000 people. Unfortunately, St. Pierre was located at the base of Mont Pelée. Frightened by the growing violence of the eruptions, many attempted to leave the island.

The events of the day were reported by a ship's captain who was able to set sail and escape the climactic eruption, but nevertheless lost more than half of his crew. He reported the volcano was erupting violently, sounding like hundreds of cannons firing simultaneously. Finally, with one gigantic explosion, the mountain was enveloped in a dense, glowing cloud.

As we now interpret what must have happened, from the safety of hindsight, the magma chamber literally disemboweled, erupting an enormous mixture of superheated gases, molten rock, and pyroclastic material that formed a super ash flow. This nuée ardente, or "glowing cloud" descended on the town of St. Pierre, destroyed everything in its path, and continued to the harbor where it capsized and set fire to the anchored ships. Within moments, many ships were destroyed, the town was gone, and the entire population was dead, except for two. The first, Leon Compere-Leandre, has been described as one of those individuals who would survive any disaster. Fortunately for him, he was in the basement of his home at the time of the disaster. The other survivor, Auguste Ciparis, was the town reprobate. Having been convicted of murder, Ciparis was incarcerated in the dungeon beneath the town hall, waiting to be hanged. Ironically, the eruption of Mont Pelée

FIGURE 4.27 *The city of St. Pierre on the island of Martinique was destroyed along with 29,000 inhabitants during the 1902 eruption of Mont Pelée. Only two individuals survived the eruption, which allowed geologists to observe a nuée ardente for the first time. (Courtesy of I. C. Russell/USGS)*

occurred on the day set for his execution. He was found by search parties after the ruins of St. Pierre had cooled. Apparently feeling that he had suffered enough, the authorities suspended his sentence. Stories differ as to his fate, but the most popular tale is that he joined Barnum and Bailey's Circus and toured as the "Man Who Defied the Wrath of Pelée, the Goddess of Fire."

Plinian-Type Eruptions **Plinian** eruptions may be even more violent than Pelean. In the typical Plinian eruption, a highly turbulent, high-velocity stream of intermixed fragmented magma and superheated gas is released from a vent, creating eruption columns of great height (Figure 4.28). The name derives from Pliny the Younger who described the eruption of Mount Vesuvius in A.D. 79. Modern examples of Plinian eruptions include the eruption of Tambora in 1815 and Krakatoa in 1883. We will discuss the eruptions of Mount Vesuvius and Krakatoa later in the chapter.

Phreatic Eruptions Phreatic eruption at the summit of Mount St. Helens, Washington. Hundreds of these steam-driven explosive eruptions occurred as magma steadily rose into the cone and boiled groundwater (Figure 4.29). These phreatic erup-

FIGURE 4.28 *The* Plinian-type *eruption is characterized by a column of pyroclastic material driven to great heights by a high-velocity jet of superheated gas. Phreatic eruption at the summit of Mount St. Helens, Washington. Hundreds of these steam-driven explosive eruptions occurred as magma steadily rose into the cone and boiled groundwater. These phreatic eruptions preceded the volcano's plinian eruption on 18 May 1980 (Courtesy of USGS/Photograph by D.A. Swanson on 4 April 1980).*

FIGURE 4.29 *Phreatic Eruption*

tions preceded the volcano's plinian eruption on 18 May 1980.

FISSURE ERUPTIONS

Fissure eruptions occur along fractures in Earth's surface. Most fissure eruptions involve highly fluid basaltic lava, but a few produce felsic lavas.

Oceanic Ridges

The best examples of basaltic fissure eruptions are those along the **oceanic ridges**, where fractures are constantly being formed along the margins of the divergent plates. As the plates drift apart, fissures open and become the conduits of the basaltic magmas being generated at the top of the asthenosphere. The lavas well out from the fractures along the crest of the oceanic ridge and flow laterally down the shallow ridge slopes where they solidify to create new oceanic crust. Beneath the ridge, magma solidifies within the fractures to create gabbroic **sheet dikes** (Figure 4.30). New fissures then open along the ridge to perpetuate the process as the oceanic plates continue to move away from each other.

Historically, the only active fissure eruption ever observed was a fissure 20 miles (32 km) long that opened on the island of Iceland in 1783. Before the eruption ended, an estimated 3 cubic miles (12.5 km^3) of basaltic lava had poured out onto the island. Because Iceland is, in part, an exposed portion of the Atlantic Oceanic Ridge, the Icelandic eruption is thought to be comparable to those occurring along the oceanic ridge system except that the ridge eruptions are submarine and produce pillow lavas.

Flood Basalts

No continental fissure eruption has been recorded in historic time, although there are several geologic examples. The **flood basalts** of the Columbia Plateau of Washington, Oregon, and Idaho are an excellent example (Figure 4.31). Beginning about 20 million years ago, enormous volumes of basaltic lavas flowed from fractures that had opened throughout the area. These eruptions may have been associated with the movement of the North American plate over the Yellowstone hot spot. Over a period of several million years, repeated eruptions resulted in hundreds of layers of basaltic lavas. The accumulated flows buried the existing landscape under thousands of feet of lavas; individual lava flows ranged in thickness from less than 10 feet (3m) to more than 300 feet (90 m). It is estimated that over a period of 10 million years, a total of about 25,000 cubic miles (100,000 km^3) of lava spewed from the fissures, covering an area of about 75,000 square miles (200,000 km^2). The low viscosity of the lava allowed some of the flows to travel up to 100 miles (160 km) from the vent.

Felsic Eruptions

Although most fissure eruptions involve basaltic lavas, some produce more felsic lavas. An excellent example is the area around Yellowstone National Park in northwestern Wyoming. Beginning about 2 million years ago, and then repeating 1 million and 600,000 years ago, the area was the site of enormous eruptions. Volcanologists theorize that

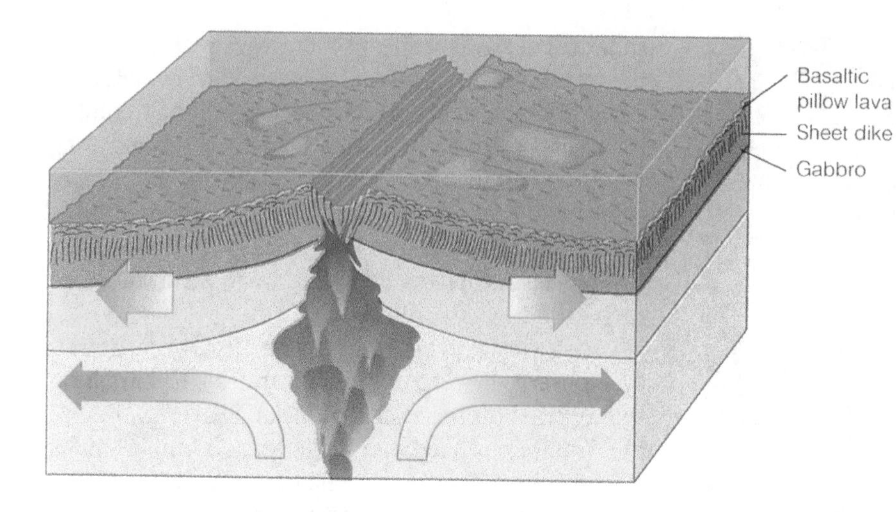

FIGURE 4.30 *Oceanic crust consists of three distinct layers. The uppermost layer is made up of basaltic* pillow lava *and* flows; *the intermediate layer is made of* sheet dikes *that form as basaltic magma solidifies in the fractures that are constantly being formed as the plates diverge; and the lowest layer is made up of coarser-grained* gabbro.

FIGURE 4.31 *Among the best examples of* flood basalts *are those that make up the* Columbia Plateau *of Washington, Oregon, and Idaho. (Courtesy of P. Weis/USGS)*

basaltic magmas formed over a hot spot at depths of perhaps 60 miles (100 km) and were injected into the granitic continental crust. There rhyolitic magmas were generated as heat from the rising basaltic magmas partially melted the continental rocks. Magma chambers containing the highly viscous, gas-charged rhyolitic magmas broke violently through to the surface. Hundreds of cubic miles of pyroclastic materials were erupted, generating ash flows and ash falls that covered hundreds of square miles with deposits hundreds of feet thick. The remains of these eruptive materials can be seen in the area today as extensive tuff and breccia deposits. Although the hot spot is still volcanically active, present activity is restricted to hot springs, fumaroles, and geysers.

CRATERS AND CALDERAS

Craters and **calderas** are circular to elliptical depressions characteristic of volcanic areas. Some workers distinguish them by size, with craters being the smaller. Most craters have diameters of less than a half to three quarters of a mile, while calderas may have diameters of several tens of miles.

Although craters and calderas do differ in size, most geologists would argue that the important distinction between them is not size but rather the mode of origin. Craters are primarily *constructional* structures, being sculpted in the summit area of the volcano by the products of eruption as they emerge from the vent. Calderas, on the other hand, are

FIGURE 4.32 *Once thought to be the result of enormous eruptions that blew away the tops of volcanoes, we now know that* calderas *are* collapse structures *and can form anywhere large volumes of magma are erupted from a magma chamber.*

destructional structures formed by the collapse of the crater walls after an eruption has voided a significant portion of the underlying magma chamber. Although most craters are explosive structures, they may be enlarged by a collapse of the inner crater walls following the eruption. In this sense, the origin of some craters includes a certain element of collapse.

Calderas are the result of exceptionally massive volcanic eruptions that significantly reduce the volume of magma within the underlying magma chamber. With the loss of volume, the overlying rocks no longer have sufficient support, and the roof of the magma chamber collapses into the void (Figure 4.32). Calderas can be very large. The disappearance of the island of Krakatoa following the 1883 eruption is attributed to the collapse of the cone into the voided magma chamber, forming a caldera on the ocean floor that extended 1,000 feet (300 m) below sea level (Figure 4.41).

Calderas are common summit features in the Hawaiian Islands. Unlike most calderas, which form by collapse after a major explosive eruption, the summit calderas of the Hawaiian volcanoes form when magma moves laterally from the summit magma chamber into adjacent, radiating rift zones to feed new eruption sites (Figure 4.33). Collapse is initiated when supporting magma is removed from under the rocks of the crater floor. Thus, like other calderas, the Hawaiian calderas form when the volume of magma is reduced, leaving the rocks overlying the magma chamber without support. In the Hawaiian case, however, the collapse is not preceded by a massive eruption from the crater of the volcano itself, but rather by a flank eruption remote from the summit.

Kilauea Caldera is the best known of the Hawaiian calderas, largely as a result of the much-visited Hawaiian Volcano Observatory perched on its rim (Figure 4.34). Most of Kilauea Caldera has formed since the 1790 eruption although detailed mapping by volcanologists from the observatory shows that an earlier collapse followed a massive eruption about 1,500 years ago. Within the caldera, Halemaumau Crater has been the site of almost continuous activity for more than a century, building the lava surface that now fills the bottom of the caldera.

In the continental United States, the best-known caldera is Crater Lake, Oregon (Figure 4.35). The name illustrates the difficulty of distinguishing between craters and calderas. Originally, the depression occupied by the lake was thought to be a massive explosive structure formed when a huge eruption literally blew the top off a volcano named Mount Mazama. Detailed mapping in the vicinity of the volcano, however, showed that a very small fraction of the total volume of rock fragments accumulated around the mountain had been derived from the original cone. Volcanologists then recognized that the depression was not primarily the direct result of a gigantic explosion; rather, it had formed when the summit collapsed following a massive eruption. Wizard Island, located within the lake, is a cinder cone formed long after the collapse of the structure.

Calderas are not always associated with the collapse of the summits of volcanoes. Calderas may form anywhere magmas come close enough to Earth's surface to create surface fractures around the margins of the chamber (Figure 4.36). The fractures then become the conduits of massive ash-flow eruptions that cover the surrounding area with tuff and breccia without constructing any conical structures that

FIGURE 4.33 *The* Kilauea caldera *did not form after a major eruption of the volcano. Instead the caldera is the result of large volumes of magma being withdrawn from the magma chamber underlying the crater floor and transferred to the rift zone that has become the site of the recent eruptions. (Courtesy of USGS/HVO)*

FIGURE 4.34 *Much of what we know about volcanism is the result of work carried out by the scientists of the* Hawaiian Volcano Observatory *for nearly a century. (Courtesy of J. D. Griggs/USGS)*

FIGURE 4.35 *One of the best-known, and misnamed, calderas in the United States is* Crater Lake, *Oregon. Our understanding of how calderas form came when scientists realized that the amount of pryoclastic material surrounding the present mountain could not account for the volume of the cone that had apparently been lost during the eruption of ancient Mount Mazama. (Courtesy of USGS/CVO-F)*

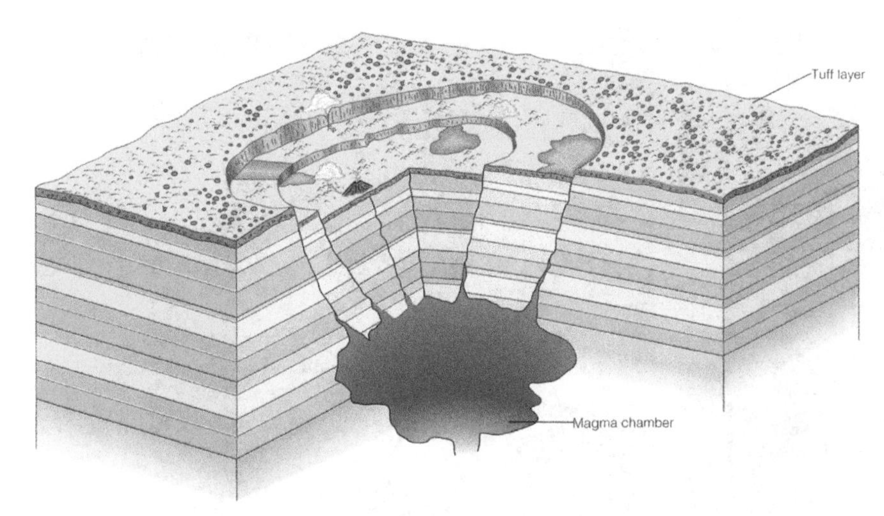

FIGURE 4.36 *Although most calderas form by the collapse of a volcano following a massive eruption, calderas can form anywhere a magma chamber erupts a large portion of its contents.*

would be identified as volcanoes. As before, with the removal of the chamber contents, the surface rocks collapse into the voided chamber. Each of the three eruptions in the area of Yellowstone Park described previously resulted in the formation of a caldera, one of which is the site of Yellowstone Lake.

SPOT REVIEW

1. Compare the geologic settings under which shield volcanoes and stratovolcanoes form.
2. What factors control the intensity of volcanic eruptions?
3. Eruptions of felsic magmas are not as common as eruptions of basaltic magmas. Using the Yellowstone eruption as an example of a felsic eruption, explain how felsic eruptions are thought to occur.
4. Compare the modes of formation of craters and calderas.

OTHER VOLCANIC FEATURES

A number of other geologic features are also of volcanic origin. For the most part, these features are the result of the interaction between the groundwater and the heat derived from the magma, although some of the water involved may be of direct magmatic origin.

Hot Springs

As the name indicates, **hot springs** are emanations of heated waters from vents. Hot springs are very common in volcanic areas. Water temperatures in some springs are barely above body temperature. Other springs, such as those in Lassen National Park in northern California and in Yellowstone National Park in northwestern Wyoming, produce boiling water (Figure 4.37).

Geysers

Some hot springs boil so violently that at regular or irregular intervals, the conduit leading to the vent is explosively cleared by a burst of steam. The hot spring then becomes a **geyser**. Following the eruption of steam and hot water, also known as phreatic eruption, the conduit requires a period of time to refill with groundwater before the sequence of events can be repeated.

Geysers are not common volcanic features, being found only in New Zealand, Iceland, Kamchatka in Russia, and Yellowstone Park. The best-known Yellowstone geyser is "Old Faithful," which erupts "faithfully" for expectant tourists about every 70 minutes. During each eruption, Old Faithful (Figure 4.38) blows about 12,000 gallons (45,000 liters) of water to a height of about 150 feet (45 m). Old Faithful is only one of about 200 geysers in Yellowstone Park. Most, however, are smaller and neither as dependable nor as spectacular as Old Faithful. Yellowstone Park is quite unusual in that it contains about 3,000 thermal vents of one kind or another (Figure 4.37).

Fumaroles

Fumaroles are pipelike vents that emit heated gases, usually water vapor. Fumaroles are common features found in all volcanic areas and may persist long after

(a)

(b)

FIGURE 4.37 *(a) Grand Prismatic Spring; (b) Midway Geyser Basin, Yellowstone National Park. (Photos by Daria Nikitina)*

FIGURE 4.38 *Although* Old Faithful *is neither the only geyser in Yellowstone Park nor the largest, it is the most faithful in that it erupts on schedule once every 70 minutes. (Courtesy of J. R. Stacy/USGS)*

the active emission of molten rock has subsided. The gases emitted are very hot, not uncommonly reaching temperatures of 1,100°F to 1,300 F (600°C to 700°C). Fumaroles that emit hot sulfurous gases are called *solfateras*.

Geothermal Power

In many areas where fumaroles are common, superheated steam at sufficiently high temperatures and pressures is tapped by drilling wells and diverted to drive turbines, which in turn generate electricity (the standard steam-powered turbine requires steam at 950°F (500°C) and 150 atmospheres pressure). Natural steam is extensively used in Iceland, not only for power generation, but also to provide hot water and to heat homes and commercial buildings. The use of natural steam in this way is referred to as **geothermal power.** In the United States, the largest commercial natural steam driven power plant is located near the town of Geysers, California.

PREDICTION OF VOLCANIC ERUPTIONS

Volcanoes do not erupt without warning. A number of events precede an eruption, most notably increased **seismic activity.** As magma rises and gas pressure builds below the mountain, rocks are broken and earthquakes are produced, increasing in both frequency and intensity as the magma continues to rise. Networks of seismic sensors allow seismologists to construct three-dimensional computer drawings of the earthquake sources that help to outline the rising magma body. In addition to the seismic activity, the rising magma and increasing gas pressure cause the surface to **bulge** and **tilt**. These movements can be monitored with delicate instruments called **tiltmeters.** All of these events were recorded in detail before each of the Hawaiian eruptions and before the eruption of Mount St. Helens, but except for announcing that an eruption was "imminent," scientists were unable to pinpoint either the time or the severity of the final eruption.

At the present time, scientists are monitoring volcanic areas that, based on experience, seem ripe for an eruption. In the United States, for example, Mammoth Lakes, California, is considered by some to be the most likely candidate for the next major volcanic eruption (Figure 4.39). Mammoth Lakes is located in the Long Valley Caldera, which was formed by a massive eruption 730,000 years ago. The Long Valley eruption generated an estimated 146 cubic miles (600 km^3) of tephra, covering an area of 570 square miles (1,500 km^2). Some of the fine ash has been detected in soils as far away as Kansas and Nebraska.

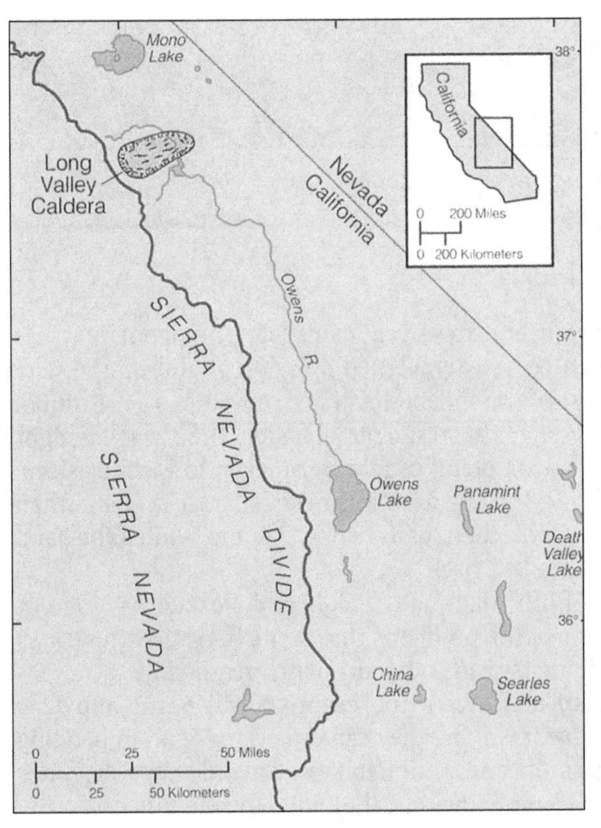

(a)

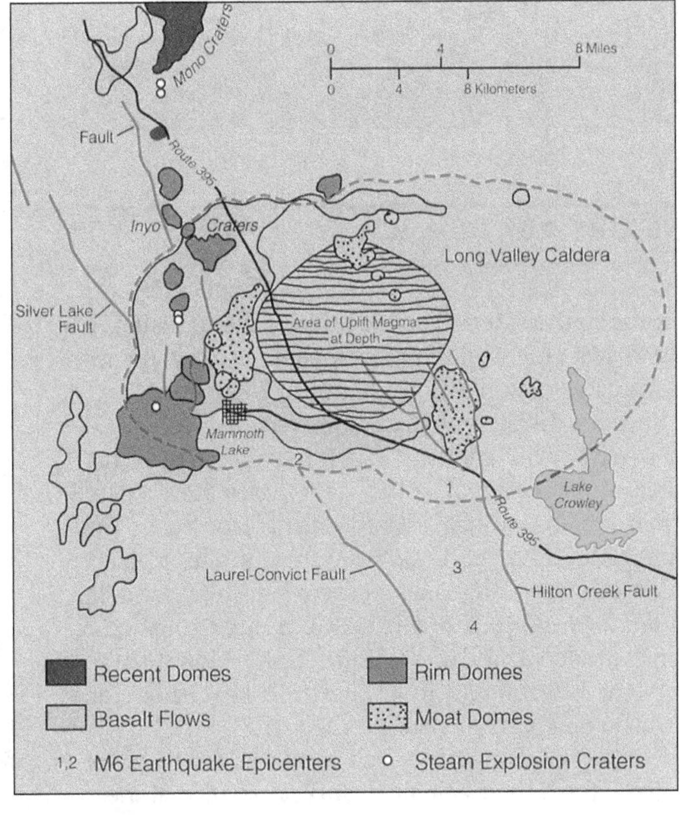

(b)

FIGURE 4.39 Mammoth Lakes, *California, may be the site of the next supereruption in North America. Located near the California-Nevada border, the geology of the area shows that it has been volcanically active for quite some time. The area of recent uplift indicates quite clearly that magma is rising beneath the site with significant new intrusions within the past decade. If the distribution of ash from the last major eruption at Mammoth Lakes is any indication, an eruption of comparable magnitude will be destructive beyond comprehension. (Parts* b *and* c *from R. A. Bailey, 1983. Mammoth Lakes Earthquakes and Ground Uplift: Precursor to Possible Volcanic Activity? In* U.S. Geological Survey Yearbook, Fiscal Year 1982; *Part* d *from C. D. Miller, D. R. Mullineaux, D. R. Crandell, and R. A. Bailey, 1982. Potential Hazards from Future Volcanic Eruptions in the Long Valley-Mono Lake Area, East-Central California and Southwest Nevada—A Preliminary Assessment. U.S. Geological Survey Circular 877.)* (c) and (d) continued on next page

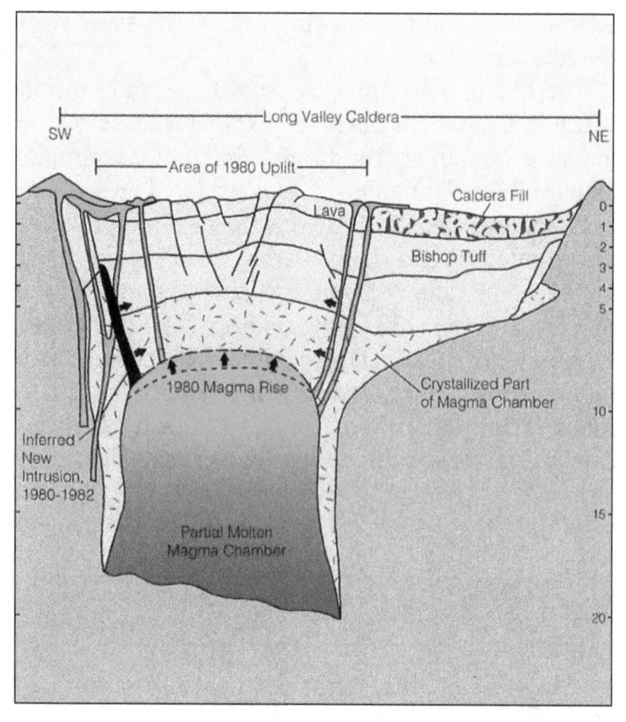

(c)

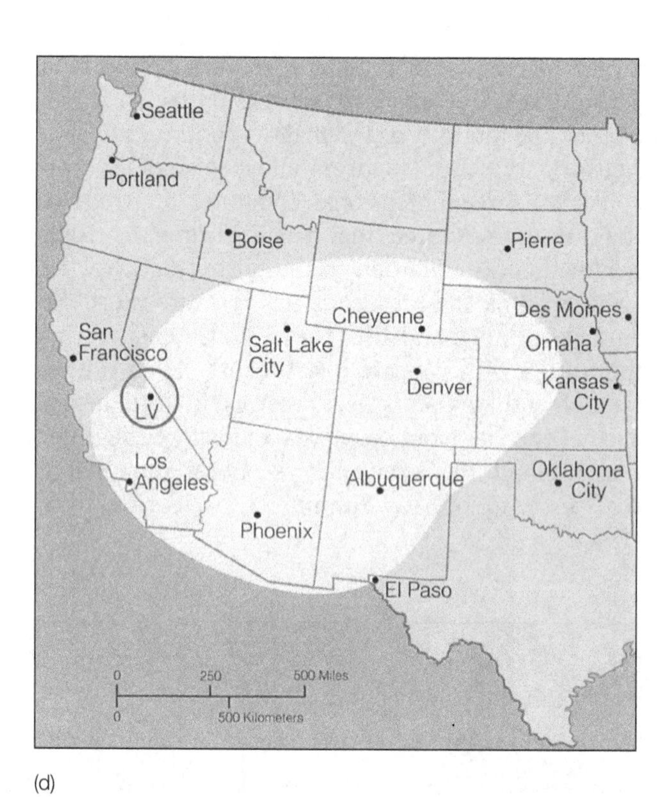

(d)

FIGURE 4.39 *Continued*

In the central portion of the caldera, the tuff is up to 4,920 feet (1,500 m) thick. Tephra falls up to 40 inches thick (1 m) accumulated 120 miles (75 km) away.

Since 1975, geologists have recorded increased seismic activity and have measured a bulging in the area of nearly a foot (25–39 cm). Since 1980, more than 3,000 earthquakes have been recorded in the area. Although most were low magnitude, at least four strong earthquakes occurred in 1980, and in 1986, a magnitude 6 earthquake occurred just east of the caldera. All of the data clearly indicate that magma is forming 3 to 4 miles (5–6 km) below the caldera and is rising beneath the area.

By 1982, the activity had increased to the point where the U.S. Geological Survey deemed it necessary to issue a formal warning of a potential eruption. Many local residents, fearing the loss of tourists, objected to the warning. No eruption has yet occurred.

In another part of the world, the fishing community of Pozzuoli near Naples, Italy, and Mt. Vesuvius, has been experiencing both seismic activity and uplift for more than 10 years (Figure 4.40). During the last 10 years, the area has experienced more than 4,000 earthquakes, and the land has been uplifted nearly 10 feet (3 m). Many of the homes and buildings have been extensively damaged. The population has been repeatedly warned of a potential eruption, and most of the inhabitants have moved to safer ground. Although the residents abandon the town at night for fear of being caught by an eruption while asleep, during the day it is "business as usual" as fishermen go about their daily work. All the while, the land quakes and rises.

Both Mammoth Lakes and Pozzuoli point out some of the problems that prediction still faces. Until scientists can pinpoint with reasonable accuracy *when* and *where* the eruption will occur and *how severe* it will be, they can only warn of an impending eruption, and the inhabitants can only hope that once the eruption begins, they will have time to escape.

EXAMPLES OF OTHER NOTEWORTHY VOLCANOES

We cannot leave the topic of volcanoes without discussing a few of the most famous eruptions.

Mount Vesuvius

One of the best-known volcanoes, Mount Vesuvius, is the only active volcano on the European mainland.

FIGURE 4.40 *The increasing frequency of seismic activity combined with changes in the relative elevations of the land surrounding Mount Vesuvius indicate that the mountain may be building another major eruption.*

Located just outside Naples, Italy, Mount Vesuvius has had a long, eventful, and closely watched history. Before Mount Vesuvius came into being, another volcano called *Mount Somma* stood on the exact same spot. Mount Somma was created about 100,000 B.C. After an eruption ended its activity about 17,000 years ago, Mount Vesuvius grew from the remnants. The remains of Mount Somma can still be seen around the northern flank of Mount Vesuvius.

Since the beginning of historic time, the Romans had considered Mount Vesuvius to be extinct. With vineyards covering its slopes and crater floor, the mountain had never shown even the slightest indication that it was active. In 150 B.C., some minor internal rumblings and newly opened vents showed that Mount Vesuvius had not been extinct but simply dormant. Over the next two centuries, an occasional tremor or minor vent eruption let everyone know that the mountain was still alive. Beginning in 63 B.C. and continuing for 142 years, the vicinity was subjected to repeated and relatively intense earthquakes that were signaling the rise of molten rock below the mountain.

On August 24, in A.D. 79, Mount Vesuvius erupted in a violent Plinian-type eruption (Figure 4.40). At the time of the eruption, Naples did not yet exist, but two small coastal towns were located at the base of the mountain. *Herculaneum*, a name few would recognize today, lay to the west of the mountain, while to the south was the famous town of *Pompeii.*

The eruption on the morning of August 24 blew away the western flank of Mount Vesuvius and created a combined ash and mud flow that overwhelmed the town of Herculaneum within a matter of minutes. The inhabitants, although aware that an eruption was going on, had little warning of the impending avalanche and thus had virtually no time to escape before the town was completely buried.

The demise of Pompeii was quite different. Its fate was recorded by a young Roman historian, *Pliny the Younger,* who described the town being pelted with hot falling ash. He recalled frightened citizens desperately rushing to the harbor in the hope that they would be safe in the water only to find that the hot ash falls had heated the water in the harbor nearly to boiling. Pliny's uncle, *Pliny the Elder,* an admiral in the Roman navy, sailed his ship into the harbor. In an attempt to calm the people and demonstrate that they were in no real danger, he ignored all warnings and went ashore where he died of a heart attack.

As the ash fall slowly inundated the town, noxious gases asphyxiated the inhabitants as they tried to escape. Within a few hours, both Herculaneum and Pompeii were buried. Although most of the inhabitants of the two towns escaped, about 2,000 individuals were trapped and buried. Both towns lay beneath the ash until they were discovered in 1748. Since that time, both Herculaneum and Pompeii have been largely uncovered and restored, adding enormously to our understanding of the life of the times.

The remains of many individuals were preserved in the form of cavities or molds as the ash packed tightly around their bodies. After they were found, the molds were filled with plaster of paris to make a cast or statue of the original body (Figure 4.41). In many cases, the degree of preservation was so good that details of facial and body features could be distinguished as could impressions of hairstyles, jewelry, and clothing.

Since A.D. 79, Mount Vesuvius has experienced two major eruptions, one in 1631 and the other in 1906. Between these two dates, the mountain was more or less continuously active, although the individual eruptions were rather mild. The 1906 eruption, a Plinian type, lasted 18 days and resulted in a crater nearly 2,000 feet (600 m) wide and 2,000 feet deep. The crater was primarily the result of an enormous gas explosion that erupted continuously for an entire day and blew ash to an altitude of 8 miles (13 k). Another eruption of Mount Vesuvius occurred on March 12, 1944, just as the Allied troops liberated Naples during World War II. The mountain is still active today. The people of Naples and the surrounding areas live under the potential wrath of Mount Vesuvius.

FIGURE 4.41 *Much of what we know about life during the Roman era has been obtained from the remains of individuals and artifacts buried during the eruption of Mount Vesuvius in A.D. 79. (© Leonard von Matt/Photo Researchers)*

Santorini

North of Crete in the Aegean Sea is a group of islands called Santoria (Figure 4.42a). The largest of the islands, Thera, surrounds a picturesque bay adjoined by steep cliffs exposing layers of volcanic ash (Figure 4.42c). A more detailed study shows that the islands are the remnants of a volcano historians call *Santorini.* Until 1500 B.C., Santorini was the home of a colony founded by one of the most advanced civilizations of the time, the *Minoans.* Based on the island of Crete and deriving their name from the legendary King Minos, the Minoans were among the first to study the night sky. The names of the constellations, first recorded in Homer's *Iliad* and *Odyssey* about 900–800 B.C., are thought to be based on legends originating with the Minoans. About 1500 B.C., the Minoan civilization on Thera came to an abrupt end with the eruption of Santorini. Earthquakes and eruptions preceding the main event convinced the Minoans to abandon their capitol of Akrotira and seek safety elsewhere. But such was not to be. An eruption rivaling that of Mount Somma destroyed most of the island, buried what remained in ash, and generated a seawave that devastated the surrounding islands (Figure 4.42b). Although the eruption damaged the Minoan centers on Crete, the main Minoan civilization survived for another half century until it was conquered by the Mycenaeans from mainland Greece.

The eruption created a caldera covering 30 square miles (80 km^2), and most of what was left of Santorini sank to the sea bottom (Figure 4.42g). Some historians have suggested that the disappearance of the flourishing island was the origin of the legend of the lost continent of Atlantis. The islands of Santoria now stand as a monument to the cataclysmic event and to the civilization that perished (Figure 4.42e).

Krakatoa

One of the most violent volcanic eruptions ever witnessed was the eruption of Krakatoa on August 27, 1883. Krakatoa was a volcanic island located between the islands of Java and Sumatra, all part of the Java Trench subduction zone that exists between the Australia-India plate and the Asian mainland

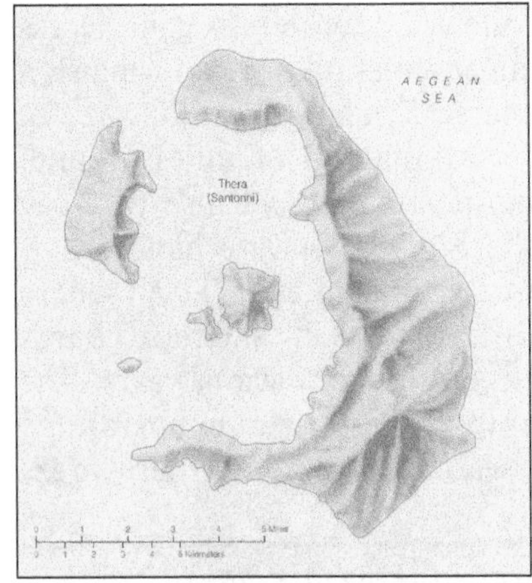

(a)

FIGURE 4.42 *(a) The formation of a caldera and the subsequent collapse of the sea bottom following the eruption of* Santorini *is probably the basis for the tale of the* Lost Continent of Atlantis. *(b) The shaded area on the map shows the zone of significant ash fall from the eruption. (c) Layers of pumice, scoria and volcanic ash exposed in the wall of Santorini caldera (photo by Daria Nikitina). (d, e) Volcano is active, as there have been several eruptions in the 19th and 20th centuries, with the most recent occurring in 1950. Recent lava flows on the central Nea Kameni Islands (photo by Daria Nikitina). (f, g) Present Day Landsat Image of Aegean Sea and Thera (Santorini), Greece (Courtesy of NASA)*

(b) (c) (d) (e) (f) (g)

(Figure 4.43). The eruption was seen from the British ship *Charles Bal* as it sailed about 10 miles (16 km) to the south. The crew reported seeing lightning flashing around the summit of the mountain and being pelted with tremendous volumes of hot ash. The gale force winds that arose allowed the ship to set sail and escape what would have been certain destruction. The next day, the entire island was destroyed in a massive, culminating explosion.

Volcanologists theorize that during one of the mountain's final convulsions, a fracture opened on the sea floor that allowed seawater to gain access to the magma chamber, creating a steam explosion that threw an estimated 5 cubic miles (20 km^3) of rock into the air. For hundreds of miles around, the ash in the atmosphere was so dense that day became night as the Sun's rays were blotted out. Ash fell over an area estimated at 300,000 square miles (780,000 km^2). Fine dust encircled Earth and caused brilliant sunsets for several years. The worldwide decline in atmospheric temperatures shortened growing seasons in more northerly areas

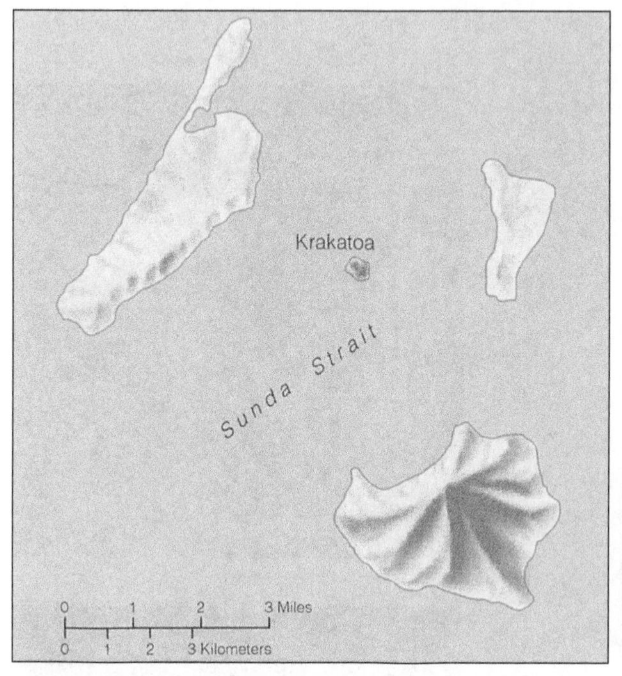

FIGURE 4.43 *All that now remains of the island of* Krakatoa *following one of the most violent volcanic eruptions ever witnessed is a tiny cone called* Anak Karkatau, or Child of Krakatoa, *appeared above the sea level in August 1930 (photo taken in 2010, a courtesy of A. Kushlin).*

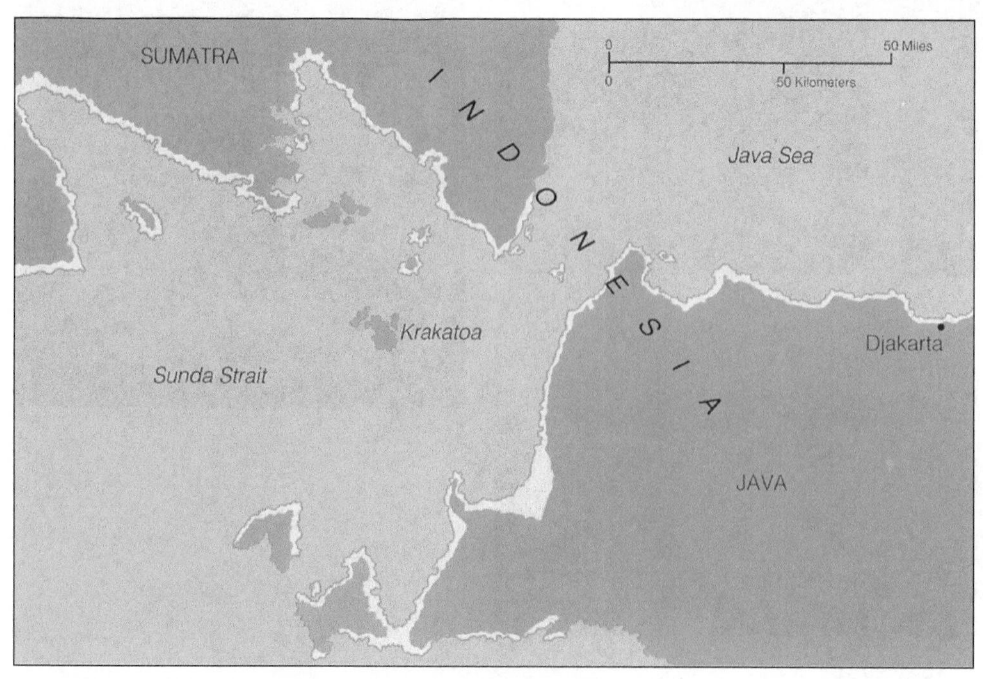

of Europe and Asia. The explosion generated a giant seawave that overwhelmed surrounding islands and killed nearly 40,000 people, while smaller seawaves were measured at seashores halfway around the world. The sound of the blast was heard in Australia 1,500 miles (2,400 km) away. After about 50 years, a new cone began to build from the remains and now stands 3,200 feet (980 m) above sea level; it is called Anak Krakatoa (Child of Krakatoa).

Mount St. Helens

Few volcanoes have been more studied than Mount St. Helens in the Cascade Mountains of the Pacific Northwest. Mount St. Helens is located about 45 miles north-northeast of Portland, Oregon, in southwestern Washington (Figure 4.44). The mountain is the youngest of the 15 major stratovolcanoes or composite volcanoes that make up the Cascade Mountains. The volcanic history of the mountain goes back nearly 40,000 years.

Studies of eruptive deposits indicate that Mount St. Helens has been active throughout its 40,000-year history. Following a 123-year period of dormancy since 1857, a few low-magnitude earthquakes on March 20, 1980, signaled its return to life. Within a week, fissures had opened in the summit area, and gases and ash began to be ejected. The summit area began to bulge outward like an inflating balloon as magma made its way upward beneath the mountain. By April, the bulge was expanding horizontally on the northern flank at a rate of about 6 feet (2 m) per day. Although the mountain was obviously still in a state of eruption, the signs of activity began to taper off. On the morning of the eruption, May 18, 1980, the mountain was quiet.

The main eruption was triggered by two closely spaced earthquakes that produced a landslide in the vicinity of the bulge. The removal of rock by the landslide took the roof off the magma chamber, released the pressure on the magma below, and resulted in the violent eruption of gases and magma. The event was observed by a young geologist, David Johnston, in an observation station nearly 5 miles (8 km) away. Johnston was to become the first known victim of the eruption. Within moments of radioing his last communication to the base station, "Vancouver, Vancouver, this is it," he was overwhelmed by the blast that roared out of the breached flank of the mountain and devastated an area nearly 20 miles

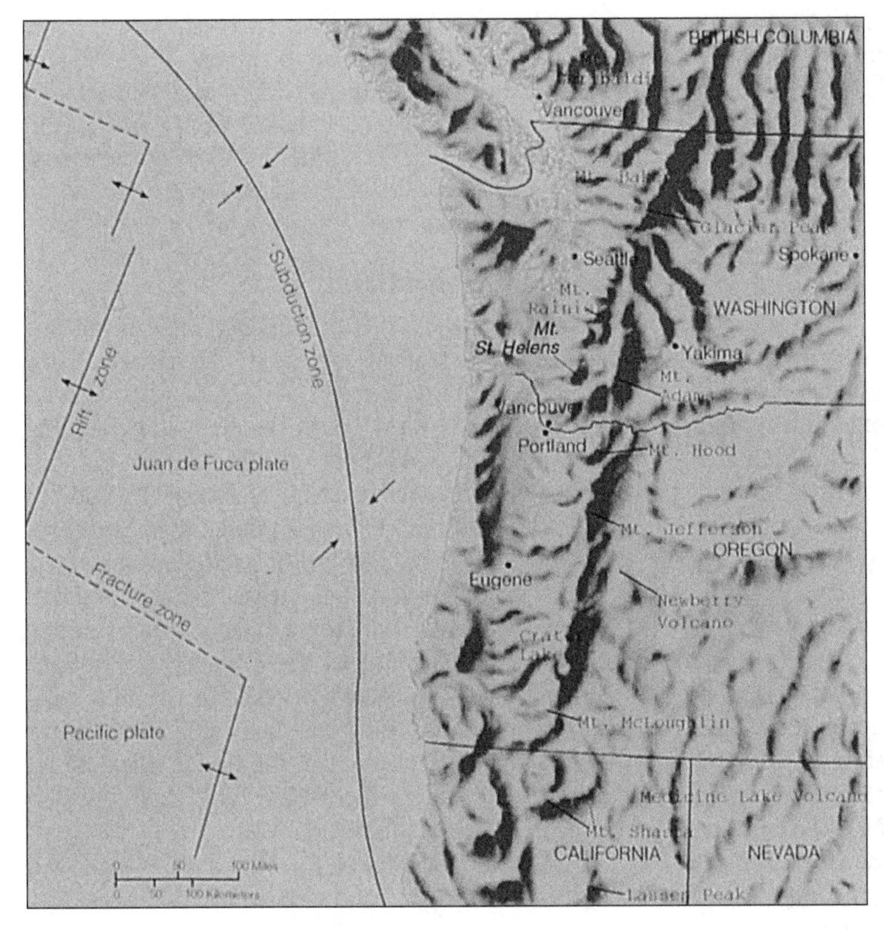

FIGURE 4.44 Mount St. Helens *is one of 15 volcanoes that make up the continental arc associated with the subducting Juan de Fuca oceanic plate.*

FIGURE 4.45 *The major eruption of Mount St. Helens was triggered by a landslide that removed rock from the roof of the magma chamber.*

FIGURE 4.46 *The damage resulting from the eruption of Mount St. Helens made many people, who possibly had become complacent, more aware of the potential for destruction of volcanic eruptions. It should be kept in mind, however, that the eruption of Mount St. Helens was only a minor event when compared to the eruption of another member of the Cascades that formed Crater Lake.*

(32 km) wide extending out 12 miles (19 km) from the mountain (Figure 4.45).

The plume from the eruption rose to an elevation of 15 miles (25 km). The destruction was truly impressive. Out to a distance of 6 miles (10 km), the forest was covered with a blanket of searing hot ash. At a distance of 10 miles (16 km), trees were leveled, stripped of branches, and aligned like so many matchsticks (Figure 4.46). Water from the melting snowcap generated lahars that filled Spirit Lake and moved down the valley of the Toutle River, depositing debris 200 feet (61 m) deep on the valley floor.

Within an hour, ash began to fall in Yakima, Washington, about 80 miles (129 km) to the east. The ash became so thick that residents had difficulty breathing and were told to stay indoors. Auto engines stalled as ash clogged the air filters. Ash accumulated throughout eastern Washington and parts of Idaho and Montana. Within a few days, the dust cloud had crossed the continental United States.

As destructive as the eruptions was, the total amount of pyroclastic material produced was only about 20% of that generated by the Krakatoa eruption and only 7% of that produced by the Crater Lake eruption (Figure 4.35). Imagine what would happen today if an eruption of the magnitude that created Crater Lake or the Long Valley Caldera were to occur anywhere within the Cascade Mountains. Not only *could* such an eruption occur, it is almost certain that it *will* occur. The Cascade Mountains are still a young mountain range with a long and potentially violent future. Indications are that the present eruptive episode of Mount St. Helens is waning. Declining seismic activity indicates there has been no magma movement since 1986. As long as the North American plate continues its relentless westward movement, however, the Cascades will remain an active volcanic mountain range (Figure 4.44).

The Hawaiian Islands

The state of Hawaii has the distinction of being the only state constructed almost in its entirety from the products of volcanism. The Hawaiian Islands are a series of shield volcanoes that stretch west-northwestward 1,146 miles (1,844 km) across the Pacific Ocean from the island of Hawaii to Midway Island. The Hawaiian chain of volcanoes then continues for another 491 miles (790 km) as a series of submerged volcanic peaks or seamounts. At that point, the chain takes an abrupt turn northward and continues another 1,296 miles (2,085 km) as the *Emperor Seamounts* (Figure 4.47). Some geologists believe that the redirection occurred as India and Asia collided, changing the direction of movement of the Pacific plate. At its northern end, the Emperor Seamount chain reaches the Aleutian Island subduction zone.

The volcanic peaks of the Hawaiian Islands, the Hawaiian seamounts, and the Emperor Seamounts

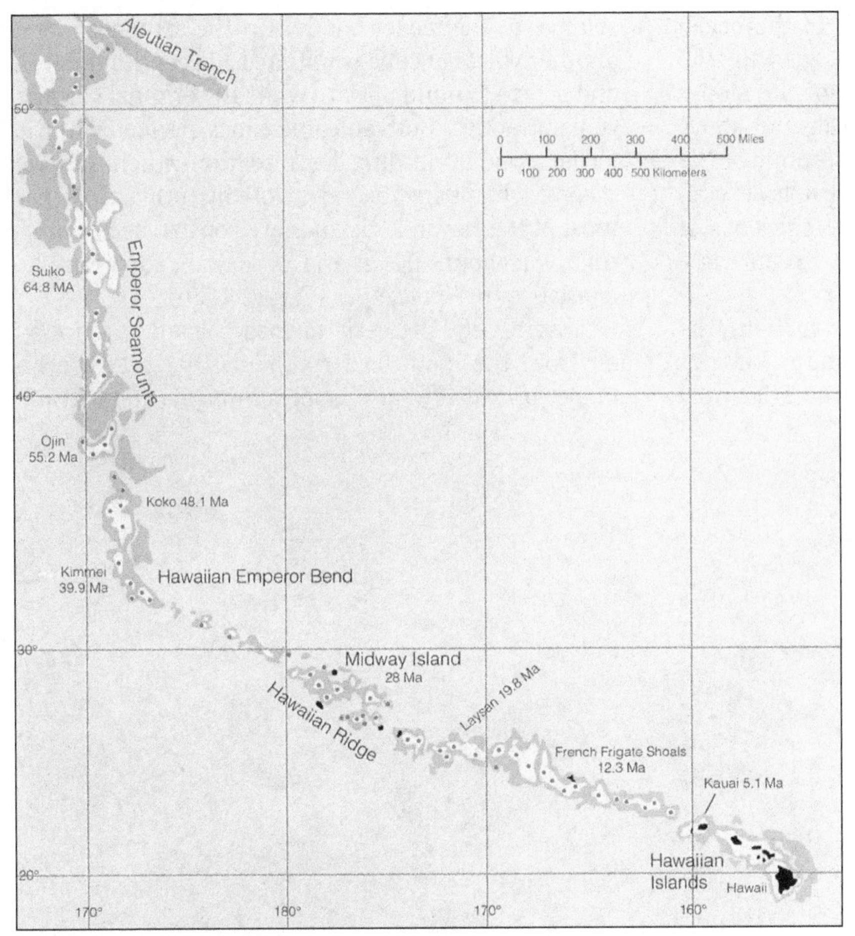

FIGURE 4.47 *Over the past 80 million years, a string of shield volcanoes has been produced by the Hawaiian hot spot that extends nearly 2,500 miles from Hawaii to the deep-sea trench off the southern coast of Aleutian Islands.*

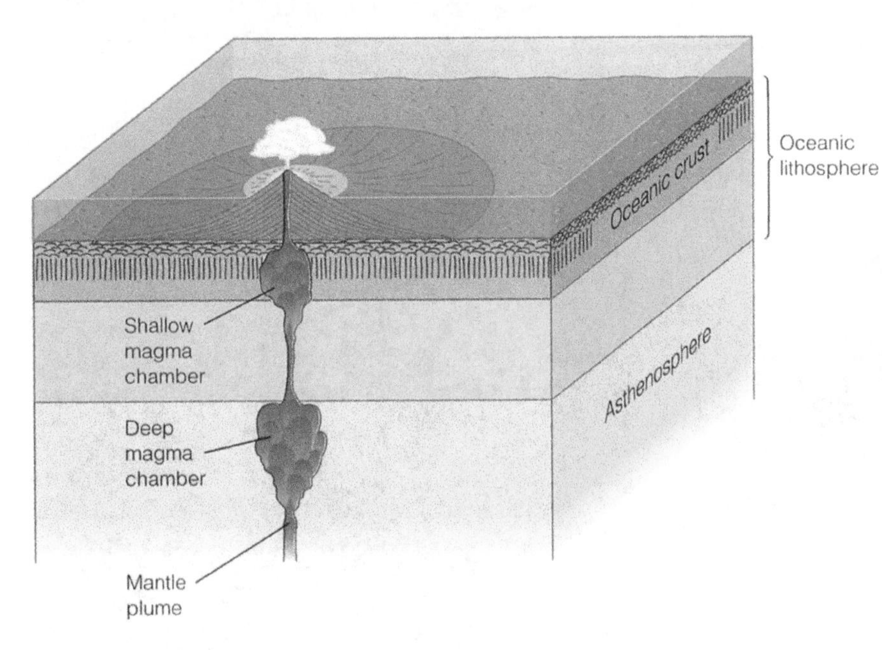

FIGURE 4.48 *Volcanologists believe that two magma chambers are associated with hot spot volcanism, one at the top of the asthenosphere, which serves as a source of the magma, and a second within the lithosphere, which stores the magma below the cone.*

formed over a mantle plume located at the southeastern end of the chain (155° west longitude and 20° north latitude). Radiometric dating of the rocks from the most northerly seamount of the Emperor chain indicates that the plume has been active for at least 80 million years.

According to volcanologists, a basaltic magma body exists at the top of the asthenosphere and feeds a second magma chamber within the overlying lithosphere (Figure 4.48). The volcanoes are constructed as the magma rises from the second magma chamber and breaks through to the ocean floor. During the submarine phase of growth, the basaltic lavas construct a shield volcano with slopes of about 17°. Eventually, the cone breaks out above sea level and continues to build the summit of the volcano as a volcanic island. Because the lavas cool more slowly when extruded subaerially, the slope of the exposed cone becomes less steep with an angle of about 5° or 10° (Figure 4.49).

While the island undergoes construction, the Pacific plate moves northwestward at about 3 inches (8 cm) per year. As the lithosphere moves laterally relative to the feeder conduit in the asthenosphere, another volcanic cone is initiated adjacent to the first and begins to build upward while the original volcano remains active. Both volcanic cones are fed from the feeder conduit leading back to the asthenospheric magma chamber. The result of this process is that most of the Hawaiian Islands are constructed of multiple volcanoes. The island of Hawaii, for example, consists of five volcanoes (Figure 4.50).

Eventually, the early formed volcanoes move so far from the main feeder conduit that they are no longer provided with a supply of magma and become

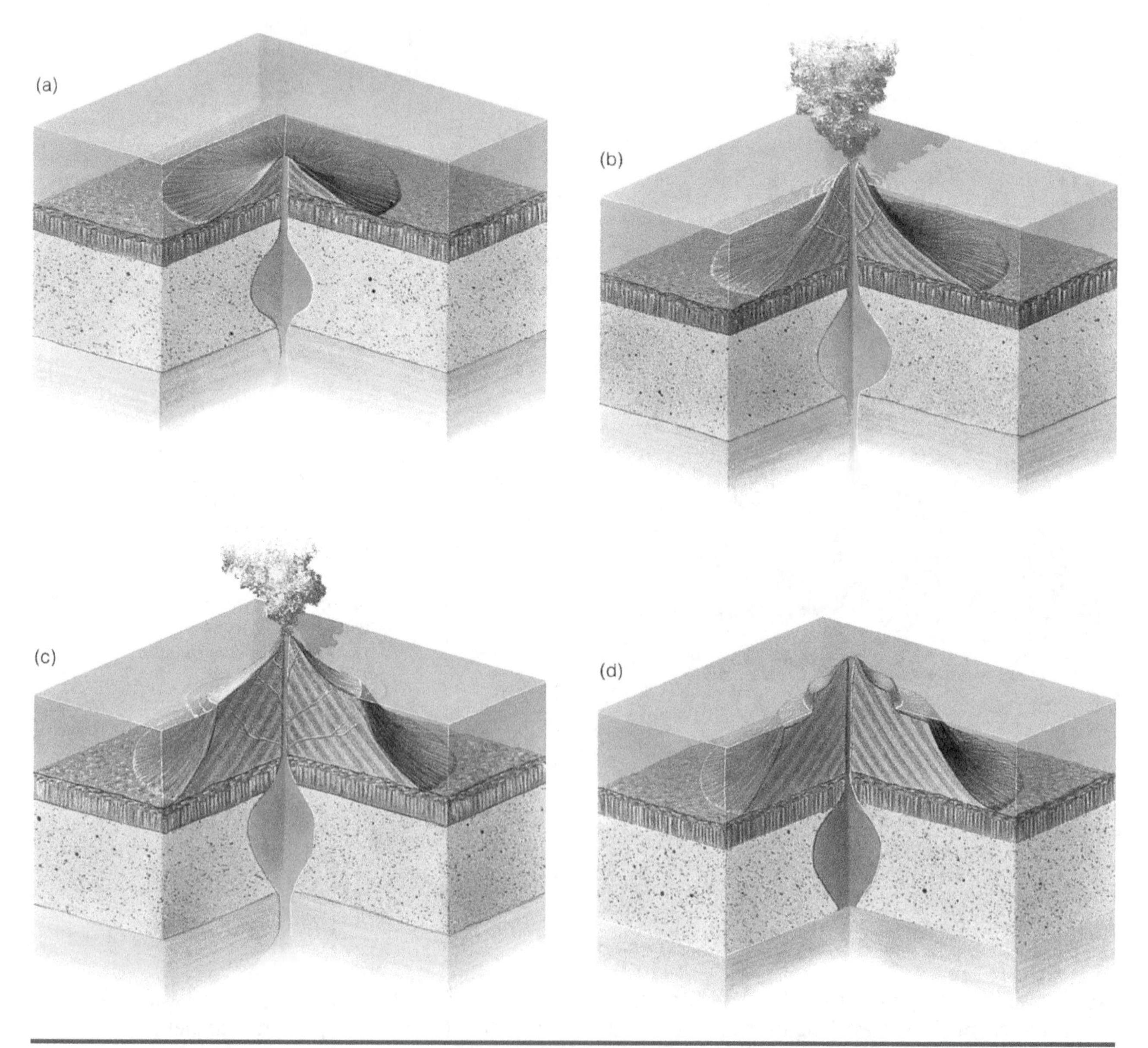

FIGURE 4.49 *The formation of the string of seamounts and volcanic islands associated with a mantle plume is believed to be due to cyclic heating and cooling of the mantle plume combined with the lateral movement of the oceanic lithosphere.*

extinct. Of the five volcanoes on the island of Hawaii, the two oldest, *Kohala* and *Mauna Kea,* are now considered to be extinct. The next oldest, *Hualalai,* although still active, is thought to be nearing extinction. The two active volcanoes on the island are *Mauna Loa* and *Kilauea*. At the time of this writing, Kilauea is in active eruption. A new volcano, *Loihi,* is now building offshore to the south of Hawaii. It is estimated that Loihi will become part of the island in about 50,000 years.

After the construction of an island, which takes about 500,000 to 1 million years, the volcanic activity apparently shuts down. Why and how this happens is one of the most controversial topics in the science of volcanology today. Although the upward flow of magma ceases and the construction of the volcanoes stops, the oceanic plate is still moving. During the period of volcanic inactivity, the first-built island moves completely off the location of the mantle plume. The magma then begins to be generated once again and is delivered upward. Upon breaking through to the ocean floor, the erupting lavas initiate the construction of a new shield volcano that will eventually rise above sea level to become the second island of the chain. The chain of volcanic islands lengthens as successive islands are created.

As the lithosphere moves laterally, the individual islands begin to sink as the ocean floor is depressed under their weight. The subsidence, combined with erosion and slumping, results in the islands becoming progressively smaller. Eventually, each island will disappear beneath the sea (Figure 4.51).

Geologists agree that the Hawaiian Islands formed as the oceanic plate moved continuously over a mantle plume. Except for Hawaii, none of the islands currently has any direct connection to the magma chamber at the top of the asthenosphere, explaining why only Hawaii, the youngest island,

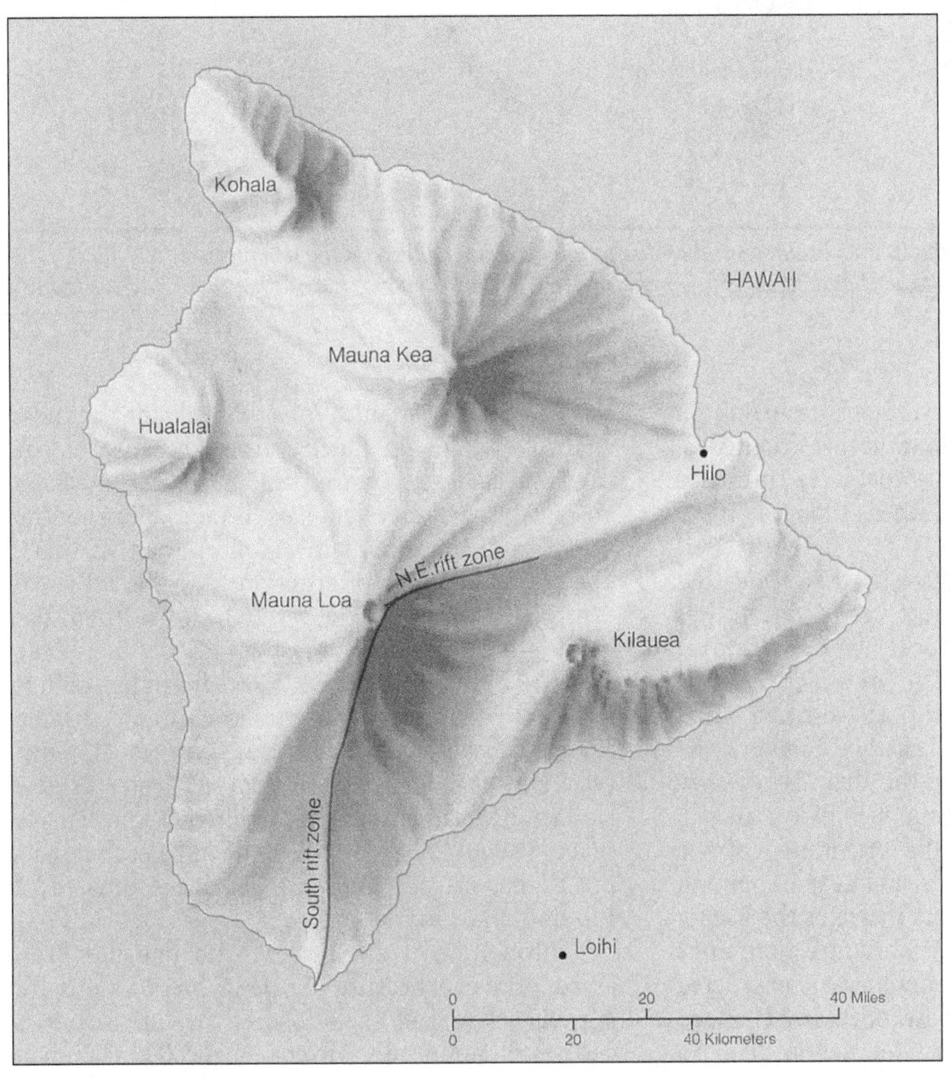

FIGURE 4.50 *The island of Hawaii is made of five volcanoes, only two of which are active. Just offshore, a new volcano named Loihi is building from the ocean floor and will become part of the island in about 50,000 years.*

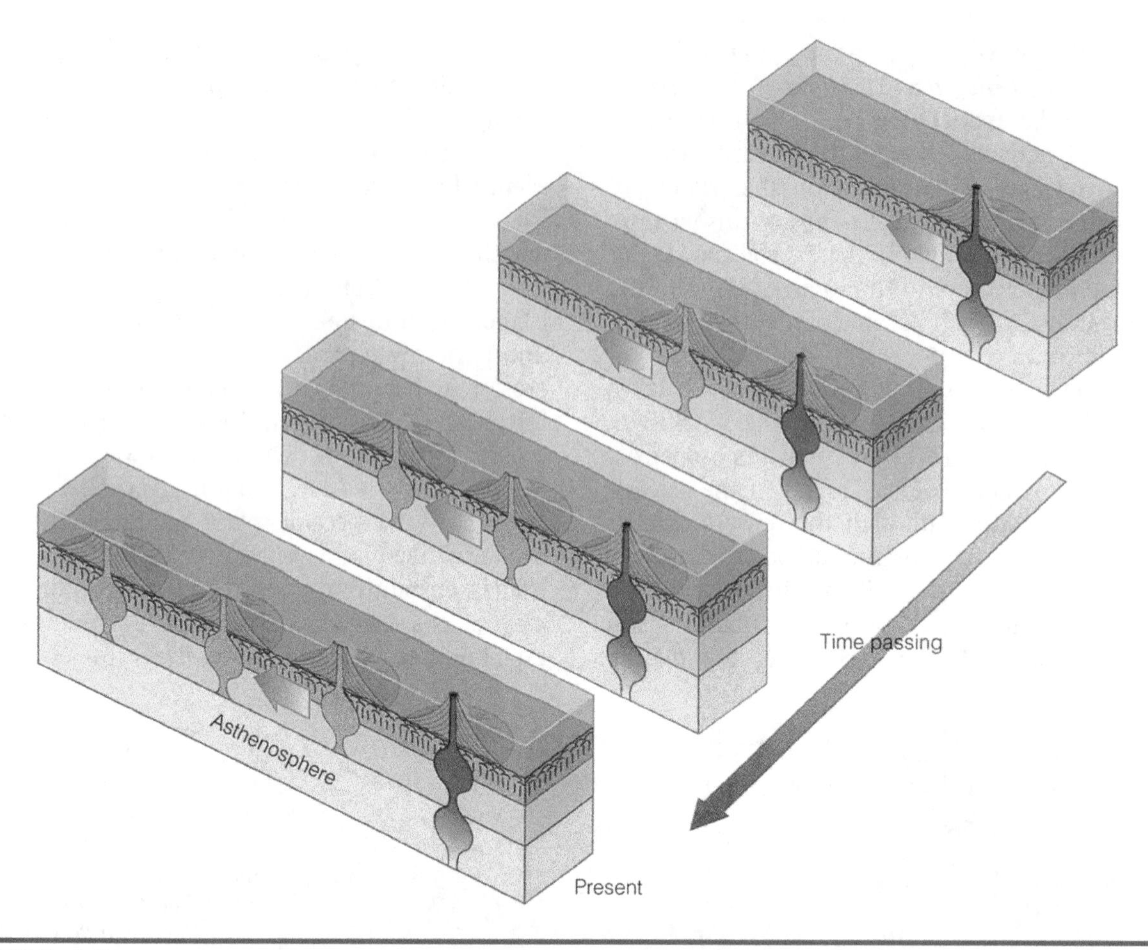

FIGURE 4.51 *With time, the islands become smaller due to the combined effects of the downwarping of the oceanic crust under the weight of the island, weathering and erosion of the exposed portion of the island, and slumping of the subaqueous flanks of the cone.*

has active volcanism. Small eruptions occurring on the other Hawaiian Islands have been attributed to the expulsion of magma masses that were entrapped within the rocks of the island when it moved off the mantle plume.

The scenario also explains why the islands are aligned and why they become older as one travels northwestward. The oldest rocks on the island of Hawaii are just under 1 million years old while the oldest rocks of Kauai, the most northwestward island usually visited by tourists, are nearly 6 million years old. The most distant island in the chain before they disappear beneath the sea is about 30 million years old.

The marked change in the orientation of the trend of the Hawaiian Islands and seamounts and that of the Emperor Seamounts indicates the Pacific plate initially moved in a northward direction. After the last of the Emperor Seamounts was constructed as an island, the direction of plate movement changed to the present west-northwesterly direction.

The volcano Kilauea is without doubt the most studied volcano on Earth, largely because of the work and dedication of the personnel of the Hawaiian Volcano Observatory who have been monitoring the activities of the mountain for more than 75 years. The present episode of eruption from the *Pu'u O'o* vent began in 1983 (Figure 4.52). Since then, the volcano has produced more than 27.5 billion cubic feet (850 million m^3) of lava. Flows from the summit and from rift eruptions fed from the summit magma chamber have reached the sea 7 miles (11 km) away and have added more than 200 acres of new land to the island. At the same time, the flows have covered about 25 square miles (65 km^2) of the existing island, causing property damage estimated at $10 million or more.

Throughout it all, however, no one has been killed, primarily because the Hawaiian eruptions are normally devoid of explosions of the intensity that destroyed Mount St. Helens. Typically, Hawaiian

FIGURE 4.52 *The present episode of activity on Hawaii began in 1983 with the eruption of the* Pu'u O'o *vent along the southeast fracture zone. (Courtesy of USGS/HVO)*

FIGURE 4.53 *New land is being created along the south shore of Hawaii as the lava reaches the sea and builds the edge of the land seaward. (Courtesy of USGS/HVO)*

eruptions are characterized by lava fountains that tower from hundreds to nearly 2,000 feet (600 m) into the air and rivers of molten rock that flow down the mountain slopes, cascade over cliffs, and sometimes pour through clouds of billowing steam into the sea (Figure 4.53). The relatively quiescent nature of Kilauean eruptions allows geologists, and tourists, to stand within feet of a flowing tongue of lava, to venture out onto the surface of a newly solidified lava flow while it is still too hot to touch, and to watch the slow movement of the oncoming front of an aa flow while peering at the molten rock inside the solidified blocks. Few human experiences can be more awe inspiring than these.

Sakurajima, Japan

One last volcano deserves mention, if only because it has the highest frequency of eruption. The island of Sakurajima, Japan, is dominated by a volcano that has been erupting ash nearly 200 times per year. The 9,000 inhabitants of the island have adjusted to this ever-present threat to their lives. They live in homes built to exclude dust. Special channels line their streets so that the ash can be washed away. Parents send their children off to school with the reminder to "wear your hard hat." Five miles across the bay, Kagoshima, a city of half a million people, also thrives under the constant outpourings of Sakurajuma volcano. Here too, citizens are prepared to shovel away daily ash falls.

SPOT REVIEW

1. How do hot springs, fumaroles, and geysers differ?
2. What kinds of information do geologists use to predict an impending volcanic eruption?
3. Using examples from noteworthy volcanic eruptions, describe how eruptions can affect human lives far beyond the immediate site of the eruption.

CONCEPTS AND TERMS TO REMEMBER

- Active, dormant, and extinct volcanoes
- Distribution of volcanoes
  - divergent margins
  - convergent margins
  - island arc
  - continental arc
  - Ring of Fire
  - hot spots
- Ejected materials
  - gases
  - liquids
    - magma
    - lava
      - pahoehoe
      - aa
      - pillow lava
  - solids
    - tephra
    - tephra falls
      - blocks
      - bombs
      - cinders
      - ash
      - dust
      - tuff
      - volcanic breccia
    - ash flows
      - nuée ardente
      - ash-flow tuff
      - welded tuff
        - ignimbrite
    - mudflow or lahar
- Kinds of volcanoes
  - cinder cone
  - spatter cone
  - shield volcano
  - stratovolcano
  - composite volcano
- Eruptive intensity
  - gas content
  - viscosity of magma
- Eruptive styles
  - Hawaiian
  - Strombolian
  - Vulcanian
  - Pelean
  - Plinian
- Fissure eruptions
  - oceanic ridges
  - sheet dikes
  - flood basalts
  - felsic eruptions
- Craters and calderas
  - crater
  - caldera
- Other volcanic features
  - hot springs
  - geyser
  - fumarole
    - geothermal power
- Prediction of volcanic eruptions
  - seismic activity
  - bulging and tilting
    - tiltmeter

REVIEW QUESTIONS

1. Most of the Earth's active volcanoes are associated with
   a. hot spots beneath oceanic plates.
   b. zones of subduction.
   c. transform faults.
   d. oceanic spreading centers.
2. The type of lava associated with stratovolcanoes or composite volcanoes is
   a. andesitic. c. basaltic.
   b. rhyolitic. d. felsic.
3. Stratovolcanoes or composite volcanoes are associated with
   a. mid-oceanic ridges.
   b. oceanic hot spots.
   c. continental hot spots.
   d. zones of subduction.
4. The eruptive feature known as a nuée ardente is a super
   a. ash fall. c. fissure eruption.
   b. mudflow. d. ash flow.
5. The most violent volcanic eruptive style is the
   a. Strombolian. c. Hawaiian.
   b. Plinian. d. Vulcanian.
6. The type of lava that forms by the underwater solidification of molten lava is called
   a. pahoehoe. c. pillow.
   b. aa. d. ropy.
7. The most abundant lava composition is
   a. rhyolitic. c. basaltic.
   b. andesitic. d. felsic.
8. Subduction zone volcanism is more violent than hot spot volcanism because the magmas
   a. come from greater depths.
   b. are hotter.
   c. are more viscous.
   d. are less viscous.
9. Calderas differ from craters in that calderas
   a. are smaller.
   b. form by collapse.
   c. form during the eruptive phase.
   d. are always associated with volcanic cones.
10. The 1980 eruption of Mount St. Helens was an example of a ____ style eruption.
    a. Strombolian c. Pelean
    b. Hawaiian d. Vulcanian
11. Why are island arc or continental arc volcanic eruptions more violent than those associated with most hot spots?
12. What is the basic difference between a shield volcano and a stratovolcano or composite volcano? Why are stratovolcanoes associated with zones of subduction while shield volcanoes are not?
13. Why are chains of volcanic islands such as the Hawaiian Islands aligned, and why is only one island volcanically active?
14. Name the different kinds of lava and describe the characteristics of each.
15. Explain how felsic and mafic magmas differ. How will the differences affect the style of an associated eruption?
16. What are the difficulties involved in predicting the future eruption of a volcano?
17. How does an increase in silicon content result in an increase in the viscosity of a magma?

THOUGHT PROBLEMS

1. Why are the Cascade Mountains limited in linear extent from northern California to southern British Columbia, Canada?
2. Even after plate movements have served the connection between an oceanic shield volcano and the underlying hot spot, volcanic activity can continue for a short period of time. Explain how this can happen.
3. What steps can the population of a town like Mammoth Lakes, California, take to protect themselves from a possible volcanic eruption?

FOR YOUR NOTEBOOK

Most of us do not live in areas of active volcanism or in the shadows of dormant volcanoes. Many of us, however, do live in areas where rocks recording ancient episodes of volcanic activity can be found. The Palisades along the west bank of the Hudson River, for example, is a sill. Perhaps the best source of information concerning the presence of volcanic rocks and the volcanic history of your area is the state geological survey or an office of the U.S. Geological Survey. They will be able to provide information about the locations of any volcanic rocks that may be in your area.

Hardly a year goes by without a volcanic eruption occurring somewhere in the world. Newspapers and weekly news magazines will provide much information that you can cut out and place in your notebook. Information about past eruptions can be acquired by a visit to the cataloged newspapers in your local library.

The activity on the island of Hawaii has certainly been well documented as has the 1980 eruption of Mount St. Helen. A number of videotapes of the Hawaiian eruptions are available from the Hawaiian Volcano Observatory, Hilo, Hawaii. The library at your college or university is also a possible source of video materials.

An excellent book written for a general audience is *Volcanoes,* by Robert and Barbara Decker, published by W. H. Freeman. Should you want to research the subject of volcanism further, the book includes a rather extensive bibliography.

CHAPTER 5

Types of Eruptions

KATMAI NATIONAL PARK AND PRESERVE SOUTHWEST ALASKA	
Area: 4,090,000 acres; 8809 square miles	**Phone:** 907–246–3305
Proclaimed a National Monument: September 24, 1918	**E-mail:** KATM_Visitor_Information@nps.gov
Established as a National Park and Preserve: December 2, 1980	**Website:** www.nps.gov/katm/index.htm
Address: P.O. Box 7 #1 King Salmon Hall, King Salmon, AK 99613	

Katmai's active volcanoes are in the Alaskan-Aleutian volcanic belt. The violent eruption of Novarupta in 1912, created the ash-filled Valley of Ten Thousand Smokes.

Powerful waves attack an emergent coast of fiords, cliffs, and waterfalls.

Wildlife abounds in an ecosystem of mountains, glaciers, lakes, streams, forests, tundra, and marshes. The Alagnak Wild River, rising in a northern lake in the preserve, teems with salmon.

FIGURE 5.1 *Mount Griggs was named after the botanist Robert F. Griggs from Ohio who made several trips to Katmai to study the plant regeneration in the Valley of Ten Thousand Smokes. National Park Service photo.*

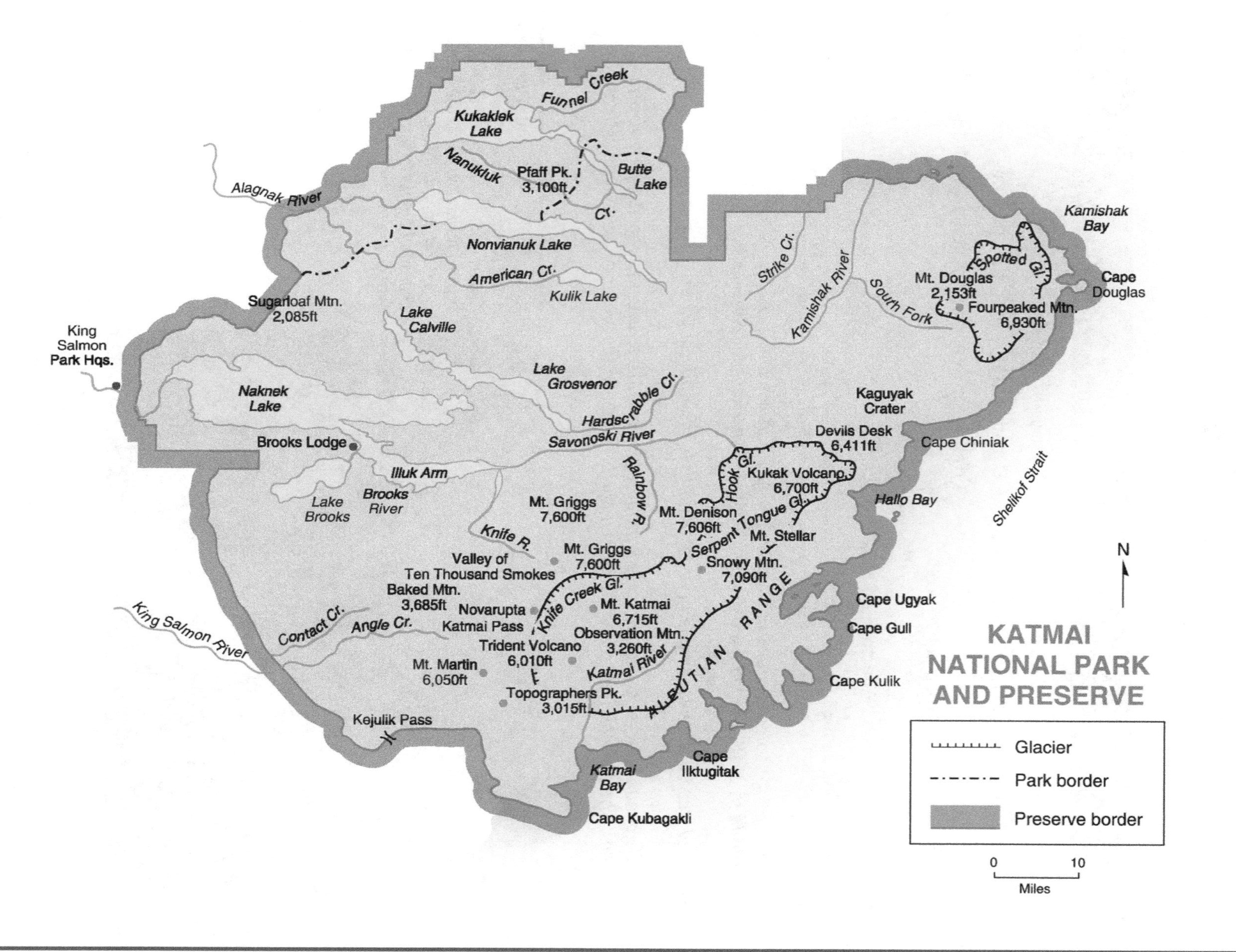

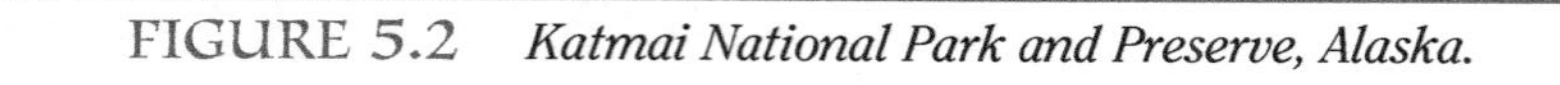
FIGURE 5.2 *Katmai National Park and Preserve, Alaska.*

GEOGRAPHIC SETTING

Located in the northern part of the Alaska Peninsula, overlooking the Shelikof Strait, Katmai National Park and Preserve is 290 air miles southwest of Anchorage. The park/preserve extends along the coast from Kamishak Bay south to the Becharof National Wildlife Refuge and westward nearly to the community of King Salmon. To the north, at the head of the peninsula, is Lake Clark National Park and Preserve. The active volcanoes in the two parks are part of the Alaskan-Aleutian volcanic belt (Figure 5.3).

The Katmai volcanoes, like those in the Lake Clark region, trend from northeast to southwest, parallel to the eastern coast of the Alaska Peninsula. In this part of the peninsula, the volcanoes rise over 7000 feet from the shores of Shelikof Strait. Inland from the coast, the mountains of the Aleutian Range have elevations of 3000 to 5000 feet; and the whole range, including the volcanoes, is about 100 miles wide from east to west. Several hundred miles down the peninsula, the range breaks up into the Aleutian Island chain, which stretches in an arcuate pattern westward for another thousand miles out into the North Pacific. Throughout its length, the Aleutian Range has 60 volcanoes, of which 47 are known to have been active since 1760. In the spring and summer of 1986, Augustine Volcano, which rises from the floor of Kamishak Bay (between Katmai National Park and Preserve and Lake Clark National Park and Preserve), erupted large quantities of ash.

The Katmai climate is wet and stormy. From whatever direction the wind blows, moisture-laden winds are forced to rise over the mountainous Alaska Peninsula. Light rains can last for days, and strong winds and gusts ("williwaws") frequently sweep the area. In summer, skies are clear perhaps 20 percent of the time. However, summer temperatures (averaging 50° to 60°F) are mild by Alaska standards.

The world's largest predatory land mammals, the Alaska brown bears, thrive in the wild, remote

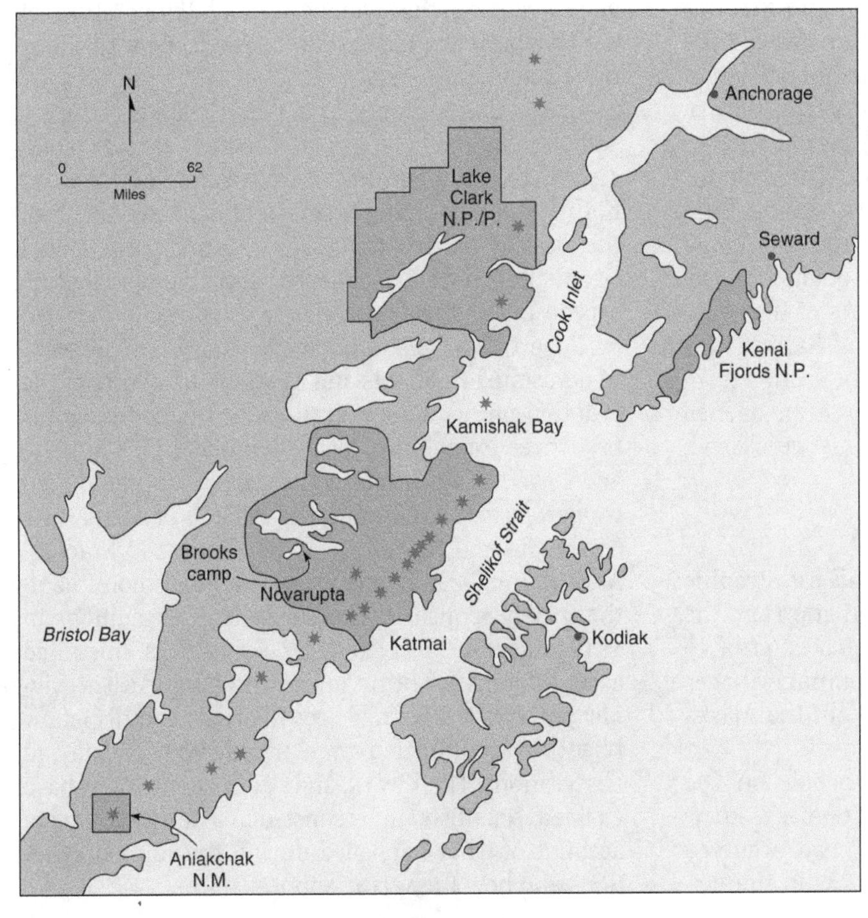

FIGURE 5.3 *Active volcanoes of the upper Alaska Peninsula. The line of volcanoes marks the inner edge of an ongoing subduction zone. In this part of the Alaska-Aleutian volcanic arc, the magma that builds the volcanoes rises through continental crust, then erupts as lavas of variable composition. Modified after Fierstein 1984.*

Katmai area. The huge size of these Ice Age survivors, weighing from 600 to 1500 pounds, is attributed to fish oils in their rich diet. The bears feed on the red salmon that spawn in the lakes and streams of the Alaska Peninsula.

Facilities in the park/preserve are open only from June to September, although charter flights over Katmai can be made during the winter. In summer, access to Katmai is by air from Anchorage to King Salmon, where the park headquarters is located, outside the western edge of the park/preserve. Float planes ferry visitors from there to the Brooks River Camp on the shore of Lake Naknek. A four-wheel drive bus makes daily trips to an overlook on the rim of the Valley of Ten Thousand Smokes. This is the only road in the park/preserve.

The Aleutian Range was such a difficult barrier to cross that separate native cultures developed on the Gulf of Alaska and Bristol Bay sides of the Alaska Peninsula. Russian fur traders, based on Kodiak Island across Shelikof Strait from Katmai Bay, were the first Europeans to live in the area. They were succeeded by American traders after the Alaska purchase in 1867. The establishment of salmon canneries on the Bristol Bay side of the Alaska Peninsula and the Nome gold rush brought more people into the area in the late nineteenth century. In order to avoid the long, stormy passage around the Aleutian Islands, traders and prospectors followed the old Katmai Trail, used since prehistoric time as a shortcut across the mountains of the Alaska Peninsula. This arduous route went from Katmai Bay through Katmai Pass and down to Bristol Bay, where prospectors could board sailing ships to Nome and the goldfields. The 1912 eruption destroyed the trail. Parts of the route have been restored on the west side of Katmai Pass and can be used by backpackers, but the section from Katmai Pass to Katmai Bay has not been opened because of heavy brush, braided streams, quicksand, quickmud, and frequent slides.

The 1912 Eruption

Apparently no lives were lost during this catastrophic event, probably due to the sparse population and the fact that a week-long series of earthquakes, prior to the explosion, prompted natives in mountain villages to flee to coastal areas on either side of the Alaska Peninsula.

Around noon of June 6, 1912, people on the northern and eastern sides of the Gulf of Alaska knew that an extraordinary volcanic event had occurred because they heard loud booms and saw in the distance great clouds of ash. Around midnight of that day, a severe earthquake occurred, estimated to be at least magnitude 6.4 on the Richter scale.

The ashfall lasted for three days in the immediate vicinity of the volcano, disrupting wireless communications. No word reached the outside world until June 9. More than a week passed before a rescue party, paddling through chunks of floating pumice, could reach Katmai Bay from Kodiak Island. They found the native village deserted and ash from 3 to 15 feet deep over the whole area. Kodiak Island was covered by a foot of ash and many homes were destroyed (Figure 5.4).

A U.S. revenue cutter happened to be in Kodiak Harbor (100 miles east of the volcano) on June 6. The crew took local residents on board in complete darkness during the ashfall. The ship lay at anchor until the skies cleared and it was safe to go ashore on the morning of the third day after the first explosions.

Although much of the ash and pumice fell in the Gulf of Alaska, over 3000 square miles of land were covered by about a foot of pyroclastic debris. Ash fell in Juneau on the Alaska Panhandle, in British Columbia, and as far north as Fairbanks. Ash reaching the upper atmosphere was carried around the globe, causing the average annual temperature of the Northern Hemisphere to be reduced by 1.8°F for more than two years.

Later that summer, the U.S. Geological Survey and the National Geographic Society sent a geologist, George C. Martin, from Washington, D.C. to Katmai to make the first scientific assessment of the disaster. During the next few years, the National Geographic Society sponsored four more scientific expeditions to the eruption area. All but one were led by Robert F. Griggs, a botanist, who originally went to devastated Kodiak Island to study the effects of the eruption on plant regeneration. In the course of his first three expeditions to the mainland (1915, 1916, and 1917), Griggs explored, under difficult and often dangerous field conditions, the calderas of Katmai and Novarupta and the Valley of Ten Thousand Smokes. After Katmai was proclaimed a national monument, the society sponsored two subsequent expeditions in 1919 and 1930, both led by Dr. Griggs. His illustrated account of his scientific adventures (Griggs 1922) vividly portrays the Katmai scene. Griggs and his fellow scientists accumulated a remarkable body of data and observations. Their work, and studies done since, have enabled scientists to reconstruct a fairly complete account of what happened during the Katmai eruptions and how they came about.

FIGURE 5.4 *A succession of photographs showing the aftereffects of the 1912 eruption. National Park Service photos by C. G. Martin.*

FIGURE 5.4 *Continued*

GEOLOGIC FEATURES

The Katmai-Novarupta Connection

Katmai Volcano, which was probably about 7500 feet in elevation before its summit collapsed, and its lower neighbor, Novarupta (elevation, 2757 feet), were connected in the eruptive sequence in a rather unusual way. The first visitors to the area thought that the Katmai vent was the main source of the approximately 4 to 6 cubic miles of volcanic debris ejected during the eruptions. The entire top of Katmai had collapsed, leaving in place a great caldera. Not until 1953 was it determined that all the ash flows and pumice had been ejected from Novarupta's vent rather than from the Katmai outlet. Why, then, did Katmai collapse in on itself?

The earthquakes and fault movements preceding the eruptions may have allowed magma from Katmai to flow laterally through fissures and connect with conduits containing magmas of differing chemical composition rising beneath Novarupta. As the magma masses mingled, they frothed and exploded, ejecting from Novarupta great clouds of fine, white rhyolitic pumice, followed by dacitic ash, and finally andesitic scoria. Fissures opened in the smaller volcano, releasing colossal pyroclastic flows that roared northwestward down the valley of the Ukak River. The incandescent mass devastated everything in its path, buried the valley floor, and covered 40 square miles with ash as much as 700 feet deep. When Griggs named this area the Valley of Ten Thousand Smokes in 1916, countless fumaroles of steam, some a thousand feet high, still issued from the plain of ash. At the present time, only a few active vents remain.

As the magma chamber emptied, the unsupported summit of Katmai collapsed, forming a caldera three miles long, two miles wide, and 3700 feet deep from the rim down. Only minor explosive eruptions have occurred in these volcanoes since 1912, the most recent activity on Katmai being in 1968. The post-eruption elevation of the rim of Katmai is 6715 feet (Figure 5.5).

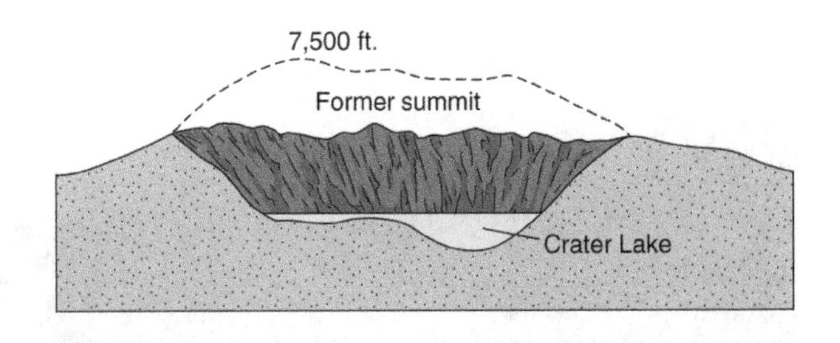

FIGURE 5.5 *Cross section of Katmai after the summit collapsed in 1912. The dashed line shows the volcano's projected height before the eruption occurred (after Griggs 1922).*

Crater Lake formed in the Katmai caldera not long after the initial cooling. In recent years the lake level has risen, but the lake probably won't run over the caldera rim because water seeps through the porous pyroclastics making up the caldera walls.

The Valley of Ten Thousand Smokes

On his second expedition in 1916, Robert F. Griggs went up the Katmai River valley. On their way to Katmai Pass, he and his companions made the first ascent of Katmai to the caldera rim and looked down at the lake, more than 1000 feet below. They saw hundreds of steam jets shooting into the air, roaring "like a great locomotive when the safety valve let go," as Griggs described it.

Wanting to press on to the Ukak River valley, they continued toward Katmai Pass. But when Griggs crossed the divide, he saw a most awesome sight. Thousands of columns of steam rose 500 feet and more from the ash-filled valley below. Escaping steam hissed and roared, and the sulfurous odor was overwhelming. Brightly colored encrustations sparkled around the spouting vents. The river was no more, but hot water bubbled up and ran off in innumerable rivulets. He knew that in this desolate place, which he named the Valley of Ten Thousand Smokes, no living thing could have survived the eruptions.

Scientists who later studied the Valley of Ten Thousand Smokes found that the heat is "rootless"; that is, the fumaroles, escaping gases, steam jets, and hot water were generated by the gradually cooling ash-flow deposits rather than by magmatic heat at some depth below the surface. Fumarolic activity in the valley had died out by the 1930s. Near the Novarupta caldera, however, a few fumaroles still give off steam as groundwater percolates down to the hot rocks below the volcanic vent (Figure 5.6).

The Valley of Ten Thousand Smokes is still barren of vegetation; but new growth now covers much of the ash-fall layers outside the valley. Clumps of grass and low plants struggle to survive in sheltered crannies in the ash-flows.

Tephra in the valley consists of fine ash, *lapilli* (small fragments), and *volcanic bombs* (larger fragments). Massive pyroclastic flow deposits fill the valley. Air-fall deposits cover the ash flow deposits and are spread over the region. The chemistry of

FIGURE 5.6 *A few fumaroles near Novarupta caldera still give off steam as groundwater percolates down to the hot rocks below the volcanic vent. The photograph was taken in July 1953. Note the helicopter for scale. National Park Service photo.*

the ejecta shows that the magma was of mixed composition—40 percent rhyolite, 50 percent dacite, and 10 percent andesite.

Novarupta

In 1917, Griggs discovered what he recognized as a new caldera west of Katmai. He named the feature Novarupta, meaning "newly erupted." Novarupta's pre-1912 configuration is unknown, but Griggs hypothesized that a crater did not exist at that location before 1912. The explosion of Novarupta produced a shallower, smaller caldera, less impressive than Katmai's broad, steep-walled caldera. It is not surprising that the early investigators were convinced that most of the ejected pumice came from Katmai.

In later years, detailed mapping and study of ash deposits by Garniss Curtis, Howel Williams, and other geologists show that the ash layers thicken away from Katmai and toward Novarupta, despite changes in wind direction that occurred during the eruptions. Clearly, Novarupta was the true outlet for the tephra thrown out in the 1912 explosions and eruptions.

The eruption probably began with deafening booms and roars as a great high plume of gas and ash shot up into the atmosphere, like a jet stream from the nozzle of a giant fire hose. It "played" for perhaps 60 hours. Curtis (1968) has estimated the rate of expulsion of ejecta during this phase as 35 billion cubic feet per hour. As the plume began to subside, pyroclastic flows, or glowing avalanches, swept down the slopes, and falling ash blanketed all the surrounding terrain. Some of the ash and pumice welded into dense rock due to great heat and pressure from overburden. The mixing of the magmas also produced blocks of "banded pumice" with striking black and white swirls.

In a late phase of the 1912 eruption, a rhyolitic dome was pushed up from the floor of Novarupta's collapsed caldera, and possibly a dacite dome emerged on the floor of Katmai's caldera (Figure 5.7).

Related Volcanoes and Eruptions

Kaguyak, about 40 miles northeast of Katmai, has a caldera and a lake, but it has been inactive in historic time. Fourpeaked Mountain, farther to the northeast, has had no volcanic activity in historic time and is almost completely covered by an ice field and glaciers. Neighboring Mt. Douglas, also ice-covered except for its 7063-foot peak, steams intermittently. Between Kaguyak and Katmai are four inactive volcanoes—Devils Desk, Stellar, Denison, and Snowy—and one, Kukak, that is fumarolic. Mount Griggs, which rises above the Valley of Ten Thousand Smokes, across from Novarupta, has been inactive in historic time but has two sulfurous fumarole fields *(solfataras)*. Trident, Mageik, and Mount Martin, which trend southwest from Katmai, have been active in historic time. Trident has had a number of small eruptions. The most recent was in 1974 when a dome formed. Between 1953 and 1968, several flows of blocky

FIGURE 5.7 *During the Katmai-Novarupta eruption of 1912, ash and pumice ejected from the Novarupta vent buried a 40 square mile area with up to 700 feet of the welded tuff that formed from the hot debris. Streams are gradually cutting through this material. National Park Service photo.*

andesitic lava moved down Trident's flanks, covering 1912 ash. Trident's neighbor, Mageik, had its most recent ash eruption in 1953, simultaneously with Trident. Another feature of Mageik is the "boiling lake" in its crater. The warm, acidic water in the crater bubbles vigorously due to escaping gases. Mount Martin steams intermittently and had a small ash eruption in 1951.

GLACIAL FEATURES

The active glaciers in the park/preserve, all of which are on the flanks of the coastal volcanoes, are nourished by moisture-bearing winds and protected from melting by the nearly continual cloud cover. Serpent Tongue glacier, between Mt. Denison and Snowy Mountain, is the largest glacier in the park/preserve. The Knife Creek glaciers, which flow down Katmai's western slopes, are so well insulated by their cover of ash that they have changed little since the 1912 ejecta blanketed them. They look more like mud slides than glaciers.

In the foothills on the western side of the Aleutian Range, a half-dozen large, elongated lakes and many small lakes occupy glacial troughs, basins and cirques excavated by the mightier glaciers of the Pleistocene Epoch. Moraines dam most of the lakes. Naknek Lake, which is the largest, was for a time connected to Bristol Bay. Kukaklek Lake, in the preserve (northwest side of the park), is the source of the Alagnak Wild River, noted for whitewater rafting and sport fishing.

GEOLOGIC HISTORY

1. Paleozoic rocks.

North of Puale Bay, about five miles south of the park/preserve's southern boundary, limestones bearing mid-Permian fossils have been found, interbedded with volcanic flows and breccias. The limestones are part of a 4000-foot sequence of sedimentary and volcanic rocks.

2. Mesozoic rocks.

Thick sequences of Jurassic and Lower Cretaceous rocks, consisting of clastic and volcanic sediments, make up the main mass of this part of the Aleutian Range. In early to middle Jurassic time, the accumulation of more than 15,000 feet of such material suggests the long-continued presence of an area of subsidence, probably an oceanic trench. Uplifting and rapid erosion of coastal areas to the northwest brought arkosic sediments, which are interbedded with a considerable volume of volcanic debris, presumably from an island arc. These rocks make up the Shelikof Formation and the overlying Naknek Formation that are widespread throughout the region. The Cretaceous Kaguyak

Formation is frequently found overlying the Shelikof and the Naknek, sometimes conformably and sometimes unconformably. Fossils are abundant in these Jurassic and Cretaceous formations.

Granitic plutons, part of the Aleutian batholith, began to intrude the sedimentary rocks early in Jurassic time. Plutonic activity continued in a series of pulses through most of the period. Late in Cretaceous and early in Tertiary time, tectonostratigraphic terranes, rafted by a subducting oceanic plate, docked in the area, causing intermittent folding, uplift, and volcanism.

3. **Cenozoic rocks.**

In the Katmai area, mainly in the southern coastal section of the park/preserve, nonmarine Eocene beds, including coal deposits, are exposed. The coal was probably laid down in coastal swamps. Lava flows are interbedded with the sedimentary rocks. Small intrusions of quartz diorite and quartz monzonite, with radiometric ages ranging from early to middle Tertiary time, have cut the Eocene rocks. Younger Tertiary rocks are absent in the park/preserve and were probably stripped off by erosion.

Convergence of plates and subduction of oceanic crust continues. A major fault system, the Bruin Bay fault, which trends northeast-southwest down the Aleutian Range, probably represents a suture, or "seam," between the Peninsular terrane on the southeast side and older rocks on the northwest side. Strike-slip displacement has moved rocks on the southeastern side of the fault to the southwest. Vertical movement has also occurred.

4. **Quaternary volcanic activity.**

Fifteen young volcanoes rise from a volcanic pile on top of Tertiary rocks in Katmai National Park and Preserve. Some cones are presently inactive, but six

TABLE 5.1

GENERALIZED GEOLOGIC COLUMN, KATMAI NATIONAL PARK AND PRESERVE

Time Units			Rock Units	Geologic Events
Era	Period	Epoch		
Cenozoic	Quaternary	Holocene	Volcanic rocks: basalt, andesite, rhyolite, dacite, pumice, welded tuff, etc.	1912 eruption, with ash falls, ash flows, explosions
		Pleistocene		Volcanic activity preceding and following climactic eruption, ranging from steam eruptions, minor ash falls, lava flows to violent eruptions
			Glacial drift	Glacial advances and retreats from late Tertiary to Holocene
	Tertiary		Volcanic rocks Small plutons	Buildup of cones Subduction, deformation, uplift Lava flows; intrusive activity Docking of terranes
		Eocene	Nonmarine sediments; coal beds; volcanics	Erosion of highlands Compaction of coal in swamps and coastal marshes
Mesozoic	Cretaceous		Kaguyak Formation	Intermittent volcanism Fossiliferous marine sediments
	Jurassic		Naknek Formation Shelikof Formation Granitic plutons	Ongoing subduction Accumulation of marine sediments in subsiding basin Emplacement of Aleutian batholith Erosion
Paleozoic	Permian		Lava flows, breccias, fossiliferous limestones (adjacent to south edge of park/preserve)	

Sources: Keller and Reiser, 1959; Buck, 1965.

have erupted since 1912. Studies of Holocene ash layers have revealed at least 12 major volcanic events that preceded the 1912 eruption. The oldest ash layer that has been identified and dated fell in 5400 B.C. Other ash layers in the marine sediments of the Gulf of Alaska carry the record back 30,000 years.

With more than 6 cubic miles of ash and pumice discharged, the 1912 Novarupta eruption was the most voluminous of this century in the United States. The volume of ejecta thrown out by Katmai was about 12 times that of Mount St. Helens in 1980. Steam plumes, heat, and occasional eruptions from the more active volcanoes in the park/preserve, plus frequent earthquakes, are indications that we can expect more volcanic events to occur in Katmai National Park and Preserve in the future.

5. **Quaternary glaciation.**

Wisconsinan glaciation in Katmai National Park and Preserve was extensive and severe. Evidences of earlier glaciations were destroyed. In some places, glacial deposits and volcanic deposits are interbedded, indicating that glaciers were melted by eruptions and then reformed. Holocene glaciers advanced and retreated as patterns of climate changed. The most recent major advance ended about 4000 B.C. However, the fact that two small glaciers reformed on the inner walls of Katmai's caldera since 1912 provides evidence that at high elevations, given sufficient precipitation, glaciers continue to thrive in Katmai National Park and Preserve.

SOURCES

Curtis, G. H. 1968. The Stratigraphy of the Ejecta from the 1912 Eruption of Mt. Katmai and Novarupta, Alaska. In *Studies in Volcanology: A Memoir in Honor of Howel Williams.* Geological Society of America Memoir 116, p. 153–210.

Fierstein, J. 1984. *The Valley of Ten Thousand Smokes, Katmai National Park and Preserve.* Anchorage, Alaska: Alaska Natural History Association. 16 p.

Griggs, R. F. 1922. *The Valley of Ten Thousand Smokes.* National Geographic Society. 340 p.

Hildreth, W. 1983. The Compositionally Zoned Eruption of 1912 in the Valley of Ten Thousand Smokes, Katmai National Park, Alaska. In S. Aramaki and S. Kuchivo (editors). Arc Volcanism, *Journal of Volcanology and Geothermal Research* 18:1–56.

Hildreth, W., and Fierstein, J. 1987. Valley of Ten Thousand Smokes, Katmai National Park, Alaska. Centennial Field Guide, Cordilleran Section, Geological Society of America, v. 1, p. 425–432.

Hildreth, W., Fierstein, J. E., Grunder, A., and Jager, L. 1981. The 1912 Eruption in the Valley of Ten Thousand Smokes, Katmai National Park: A Summary of the Stratigraphy and Petrology of the Ejecta. U.S. Geological Survey Circular 868, p. 37–39.

Montague, R. W. et al., editors. 1975. *Exploring Katmai National Monument and the Valley of Ten Thousand Smokes.* Anchorage, Alaska: Alaska Travel Publications. 276 p.

Snyder, G. L. 1954. Eruption of Trident Volcano, Katmai National Monument, Alaska, February-June 1953. U.S. Geological Survey Circular 318. p. 1–7.

Wallman, P. C., Pollard, D. D., Hildreth, W., and Eichelberger, J. C. 1990. New Structural Limits on Magma Chamber Locations at the Valley of Ten Thousand Smokes, Katmai National Park. *Geology* 18:1200–1244.

HAWAII VOLCANOES NATIONAL PARK ISLAND OF HAWAII, HAWAII	
Area: 299,177 acres; 358 square miles	**Address:** P.O. Box 52, Hawaii National Park, HI, 96718
Established as a National Park: August 1, 1916	**Phone:** 808-985-6000
Redesignated a National Park: September 22, 1961	**E-mail:** HAVO_Interpretation@nps.gov
Designated a Biosphere Reserve: 1980	**Website:** www.nps.gov/havo/index.htm
Designated a World Heritage Site: 1987	

Magma, melted by a "hot spot," rises from reservoirs beneath the ocean floor and builds Hawaiian volcanoes. Cones attain elevations above sea level of more than 10,000 feet. As oceanic crust passes slowly over the hot spot, old volcanoes become dormant; new ones form and grow from the ocean depths.

Frequent, "gentle" eruptions are characteristic of volcanic activity on Hawaii Island; but lava may squirt in high fountains from volcanic vents, or cascade down slopes in rivers of incandescent rock. Observers have recorded frequent eruptions of Mauna Loa and Kilauea since the early 1800s.

A new volcano, Loihi, is being born in the ocean south of Hawaii.

Earthquakes and precise measurements on the volcanoes yield data that

allow forecasts of eruptive behavior.

FIGURE 5.8 *Lava from Pu'u' O'o covered most of the highway, but a few sections were left to show where it had been. Photo by Ann G. Harris.*

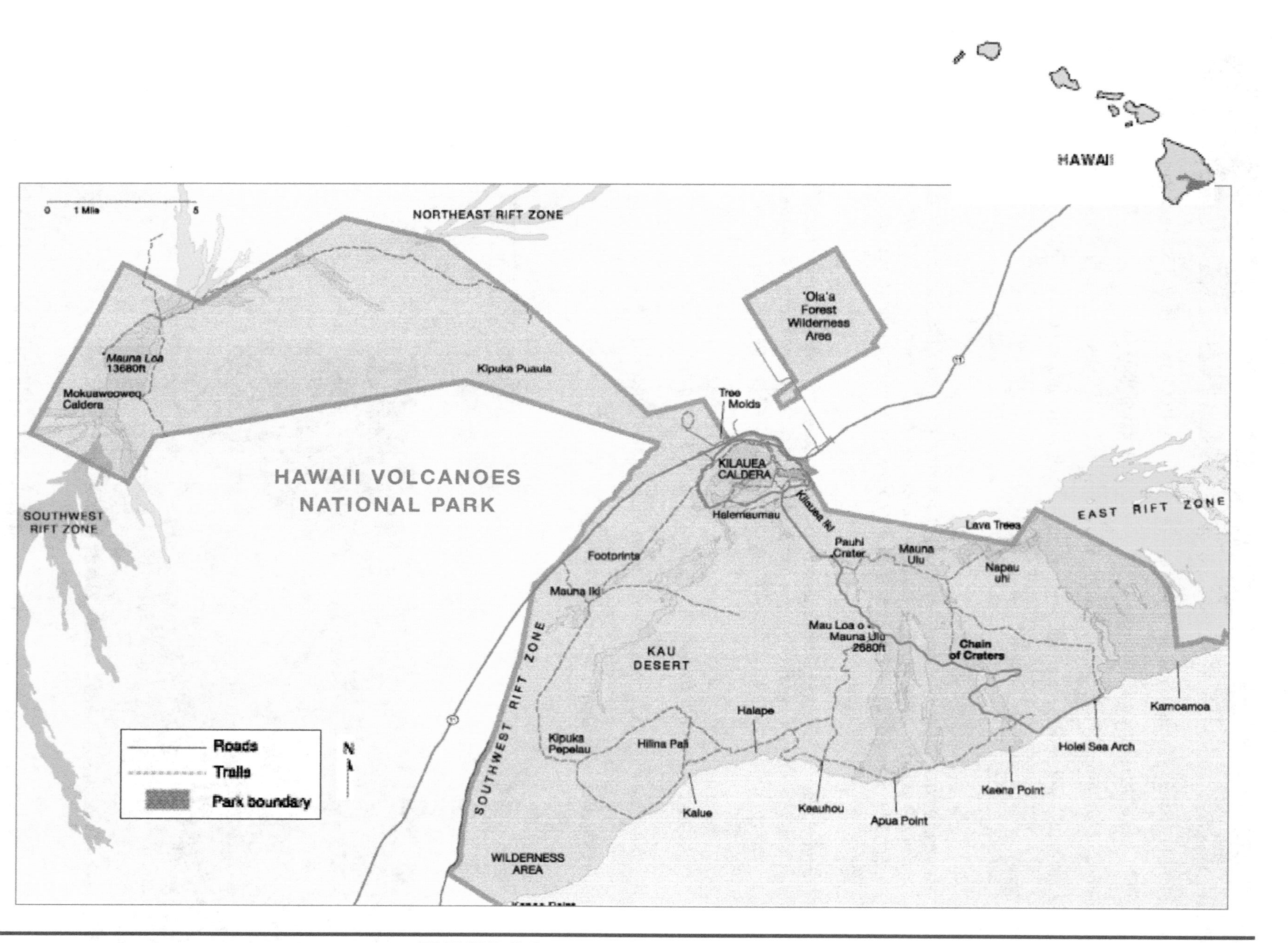

FIGURE 5.9 *Hawaii Volcanoes National Park.*

Polynesian voyagers from the South Pacific settled the Hawaiian Islands. These courageous, seafaring people began arriving at the islands about 2000 years ago, their final migration being around 750 A.D. Volcanoes, believed to be the abodes of deities, were central to Polynesian culture and religion. Pele, the goddess of all volcanoes, lived in Halema'uma'u, the fire pit in Kilauea caldera on Hawaii.

Captain James Cook discovered the islands in 1778 on a voyage of exploration for the British Admiralty. From Captain Cook on, visitors to the islands have been fascinated by the Hawaiian volcanoes. Travelers from all over the world came to see the amazing spectacle of Halema'uma'u, the boiling lake of lava; among them was Mark Twain, who said that, by comparison, Vesuvius was a "soup kettle"!

James Dwight Dana, a young geologist attached to a U.S. Navy expedition commanded by Lieutenant Wilkes, conducted the first scientific investigation of the Hawaii volcanoes in 1840-41. Dana recognized the trend of decreasing age of the volcanoes from northwest to southeast because he observed differences in the degree of erosion that had taken place on the islands since eruptions had occurred. He theorized that a great fissure zone, from which lava erupted, might have opened on the ocean floor.

The movement to protect and preserve the land around the volcanoes on Maui and Hawaii Islands was led by L.A. Thurston, editor of the *Honolulu Advertiser.* As founder of the Hawaiian Volcano Research Association, Thurston was also active in helping to establish the Hawaii Volcano Observatory. It was formally organized in 1911 under the auspices of the Massachusetts Institute of Technology and the Carnegie Institute of Washington. Professor T.A. Jagger, a geologist and volcanologist at Massachusetts Institute of Technology, chose Kilauea as the site for the proposed observatory because of the constancy of the volcano's activity and because investigators could get close to active lava without exposing themselves to too much danger. Continuous recording of seismic and volcanic activity has been carried on at the Hawaii Volcano Observatory since 1912 under the successive sponsorship of the Massachusetts Institute of Technology, the U.S. Weather Bureau, the National Park Service, and (since 1956) the U.S. Geological Survey. The Observatory buildings are located in the park on the rim of Kilauea Crater, overlooking Halema'uma'u.

Included in the park is the summit area of Mauna Loa, a strip of land between Mauna Loa and Kilauea, and most of the southern flank of Kilauea, from the caldera to the coast (Figure 5.9). Mauna Loa and Kilauea make ideal natural laboratories for studying volcanic activity. They erupt frequently but seldom explosively; they emit enormous amounts of lava; and they have large, relatively simple structures.

THE GEOGRAPHY AND GEOLOGY OF THE HAWAIIAN ARCHIPELAGO

The Hawaiian Archipelago in the Pacific Ocean is a group of shoals, atolls, and islands, stretching 1700 miles on a northwest-southeast linear trend, and a little over 2400 miles west of Los Angeles. The southeast part of the chain, made up of eight main islands and numerous islets, includes: Hawaii, at the southeast end, largest island and location of Hawaii Volcanoes National Park; Maui, the second largest island and site of Haleakala National Park (Chapter 41); Kaho'olawe; Lana'i; Moloka'i; Oahu, where Honolulu, the state capital, is located; Kaua'i; and Ni'ihau.

Extending toward the northwest from Ni'ihau are the small, mostly uninhabited atolls and shoals of the Hawaiian Islands National Wildlife Refuge. The Midway Islands and Kure, at the far end of the chain, are some 1100 miles northwest of Honolulu. The state of Hawaii includes all the islands and reefs in the chain from Hawaii to Kure, except the Midway Islands, which are administered by the U.S. Navy. Ka Lae (South Cape) at the southern tip of Hawaii is the southernmost point in the United States.

Climate. The subtropical climate is pleasant and mild. Air temperatures range from below freezing at high elevations to the high 90s on the lee coasts. Because the Hawaiian Islands lie in the path of the northeast trade winds, rainfall is heavy on the windward sides of the islands. Occasionally in winter, winds from the south or southwest ("kona winds") bring heavy storms and may dump as

THE GEOGRAPHY AND GEOLOGY *continued*

much as 24 inches of rain in four hours. Maximum precipitation on the islands occurs between 2000 and 6000 feet above sea level on windward slopes (i.e., facing northeast), but rain shadow zones reduce precipitation on some islands. Annual rainfall ranges from about 10 inches on lee slopes to about 450 inches in the wettest belts. Stream densities are high and stream erosion is intense on the windward sides of the islands.

Geologic Development. The islands in this long chain consist mostly of inactive volcanoes built of basaltic lavas that punched through the crust beneath the ocean floor. Repeated eruptions of thin, fluid lava flows (2 to 3 feet thick) brought the volcanoes up to the surface from depths of some 2.5 to 4 miles below sea level in less than a million years. Above water, their evolution as volcanic islands has followed a consistent pattern. Some reached greater heights than others; but regardless of how high above sea level the volcanoes grew, each one evolved into an "alkalic cap stage," which means that the basaltic lavas became poorer in silica and richer in sodium and potassium. Eruptions were less frequent and more explosive during this stage and lava flows became more viscous.

As eruptive activity waned, erosion cut deep valleys into the soft rock of the volcano slopes. On the older islands, Ni'ihau, Kaua'i, Oahu, Moloka'i, Maui, and Kaho'olawe, erosion was interrupted by renewed volcanism of even more alkalic and explosive lavas. Diamond Head on Oahu, for example, is the result of a period of renewed volcanism. As each volcanic island aged and was eroded, it sank as oceanic crust beneath adjusted to the rapid growth of newer volcanoes in the chain. Coral reefs, like those on Midway, encircled the older volcanoes. With continued subsidence, waves eroded the tops of extinct volcanoes, and fringing reefs became atolls surrounding lagoons.

Radiogenic dates derived from rock samples taken at various places along the Hawaiian chain confirm that the volcanoes get progressively younger toward the southeast. Ages of principal lava flows, in millions of years, beginning with Kaua'i (the oldest) are 5.8 to 3.4; Oahu, 3.5 to 2.3; Moloka'i, 1.5 to 1.3; Maui, 1.3 to 0.8; and Hawaii, less than 0.7 and still growing (Clague 1989).

A seamount named Loihi, about 20 miles south of Hawaii Island, is expected to be the next island in the chain. Loihi has risen 12,000 feet from the ocean floor and grown to within 3000 feet of water level. Surveys show its shape as similar to Mauna Loa and Kilauea, with a flat summit area, a caldera, and two probable rift zones.

What caused the Hawaiian chain to form this linear pattern of older to younger volcanoes across the ocean floor? Since the volcanoes are erupting near the middle of the Pacific plate, more than 2000 miles from the nearest plate boundary, the volcanism cannot be associated with tectonic activity occurring at plate margins. Moreover, the oceanic crust in that part of the Pacific is much older (Cretaceous) than the lava flows that constructed the Hawaiian volcanoes. Most of them are younger in age than Miocene (5.3–23.7 millions of years ago).

The most likely hypothesis suggests the existence of an immobile hot spot, a thermal high at depth, perhaps as much as 200 miles across, that melts rock material below the overriding Pacific plate. The magma accumulates in pockets that bulge up and eventually break through a stretched or weakened place in the crust. Hot lava then rises rapidly through conduits and fissures to the ocean floor. The release of great quantities of basaltic lava builds up shield volcanoes that rise above water level in about 300,000 years. As the Pacific plate drifts slowly northwestward, successive volcanoes form over the hot spot, are rafted away on the moving plate, and become dormant as they are cut off from the source of fresh magma. The hot spot theory thus explains the pattern of a volcanic chain with active volcanoes at the southeast end and older volcanoes becoming inactive or extinct toward the northwest.

GEOLOGIC FEATURES

Five shield volcanoes make up the island of Hawaii, called "Big Island." Its area is 4030 square miles. Mauna Loa and Kilauea are both active volcanoes; Kohala, at the northern end of the island, and Mauna Kea, on Kohala's south side, have been inactive in historic time. Haulalai, on the west coast of Big Island, erupted last in 1801. Mauna Kea, the highest cone (elevation 13,784 feet), is also the world's highest peak on an island. At 13,680 feet, Mauna Loa is a close second in elevation.

Approaching the island of Hawaii, whether on board ship or in a plane, travelers see the summits of Mauna Loa and Mauna Kea from great distances if the weather is clear. The closer one gets, the more impressive their bulks appear as they rise from the ocean. Yet much more of their great size is hidden beneath the water; the ocean floor on which the volcanoes stand is 18,000 feet below. Thus Mauna Loa's true height is 32,000 feet, or more than 6 miles! Kilauea, whose highest point is 4090 feet above sea level, is nearly 10,000 feet lower than Mauna Loa and Mauna Kea. All of the rock above sea level on Big Island has been erupted during the past million years, which gives us an idea of the rapidity with which this tremendous volume of lava has been extruded.

Because Mauna Loa and Kilauea have been active during historic time, many changes in their features have been documented. During the past century and a half, Mauna Loa has produced a lava flow on the average of every seven years and has exhibited minor eruptive activity in the years between. Its shield-shaped dome is about 70 miles long and 30 miles wide. Even more impressive is its volume of about 10,000 cubic miles, which makes it the world's largest active volcano. At the summit, the Moku'aweoweo caldera coalesces with adjacent pit craters. Formed by collapse rather than explosion, the caldera has changed its size, shape, and depth frequently during historic time as lava flows have spread out over the caldera floor or breached the rim.

Rising from the southeastern flank of Mauna Loa is Kilauea, which has a shield dome about 50 miles long and 14 miles wide. Its caldera is three miles long and two miles wide (Figure 5.10). Within the caldera is the fire pit (active crater) of Halema'uma'u that from time to time contains a lake of liquid lava. A *lava lake* forms when magma rises within the volcano and floods the pit in the floor of the caldera. Sometimes a lava lake lasts for years, bubbling and glowing (Figure 5.11).*

*In *Roughing It,* Mark Twain describes in awed terms seeing the lava lake after dark: "The greater part of the vast floor of the . . . [caldera] under us was as black as ink and apparently smooth and level; but over a mile square of it was ringed and streaked and striped with a thousand branching streams of liquid fire! . . . Imagine a coal-black sky shivered into a tangled network of angry fire." Mark Twain visited Halema'uma'u in 1866. The fiery "branching streams" he described were lava fountains.

If gas pressure raises the level of the molten rock high enough, lava spills out over the rim of the fire pit and spreads over the caldera floor. More often, lava breaks out through the flanks of the volcano above or below sea level. As lava pours out of the flanks, the level of the lake falls and the lake may disappear, perhaps for years at a time. If the whole plumbing system of the volcano is drained, then the caldera floor collapses. After a time, more magma rises into the reservoir, pushes up through the conduit, and the whole process—with modifications—may be repeated.

FIGURE 5.10 *Kilauea Caldera with Halemaumau Pit Crater in the background. Steam is coming from the molten lava deep in the volcanic neck. Photo by Ann G. Harris.*

FIGURE 5.11 *Inside of Pu'u 'O'o crater showing the lava lake and its crust. Photo by Ann G. Harris.*

Volcanic Rocks That Form from the Lavas

The Hawaiian basalts are of two general types, or *suites—tholeiitic* and *alkalic,* both believed to be derived from the upper mantle, but perhaps from different levels. The tholeiitic basalts make up the bulk of the shields (99%). The alkalic basalts (1%) are found mainly on the caps of the shields. The predominant phenocrysts in both types of Hawaiian basalts are olivine. Tholeiitic basalts commonly contain low- and high-calcium pyroxenes, both poor in aluminum. Pyroxene phenocrysts are rare in these rocks. Alkalic basalts contain only one pyroxene, a high-calcium and moderately high-aluminum variety that occurs commonly as phenocrysts as well as groundmass grains.

Types of Lava and Lava Features

Since most of the Hawaiian lavas are fluid, they flow from vents relatively quietly and congeal with a smooth, billowy or ropy surface called *pahoehoe* (pah-hoay-hoay) (Figure 5.12). Typically, lava erupts in a series of pulses that build up layers of pahoehoe, one on top of the other. Sometimes a flow hardening as pahoehoe changes to another common type of congealing termed *aa* (ah-ah), which has a rough, jagged, or clinkery surface. The basalt in the interior of aa flows is dense and thick with many stretched vesicles left by escaping gas. The front of the flow moves along as a red-hot, clinkery wall. Massive aa flows may show columnar jointing. The change from a pahoehoe flow to aa involves the initial gas content of the lava and changes in lava viscosity. An aa flow never reverts to pahoehoe, but a fresh pahoehoe flow may be erupted on top of an earlier aa flow.

When a lava lake is active, or when a flank vent is first erupting, gases bubbling up from depth produce *lava fountains* that may spout from a few inches to 30 feet in height, rising and falling on the hot lava surface (Figure 5.13). Occasionally lava fountains shoot up much higher, as happened at Kilauea Iki in 1959 when a lava fountain reached a height of 1900 feet, the highest ever observed. Pu'u 'O'o, which began erupting in 1983, delighted volcano-watchers with its exceptionally high and frequent fountaining.

During high fountaining, *thread-lace scoria,* or *reticulate,* which is a light, feathery type of pumice, may form and be carried many miles downwind. Tiny strands and droplets of fountaining lava, also caught by the wind, harden in the air and fall. These glassy pellets and strings, called *Pele's tears* and *Pele's hair,* are fragile and short-lived.

Lava tubes and *lava tunnels* beneath the congealing surface of a pahoehoe flow convey fresh, hot lava out to the moving front (Figure 5.14). Later, when the whole mass has cooled, the tunnels and tubes remain. A lava tube serves as a large conduit for the movement of lava down a slope far from the vent. The tube's outer crust acts as an insulator, keeping lava from cooling. The Thurston lava tube on the east side of Kilauea, which is 20 feet high and 22 feet wide in places, was formed by the crusting over of a pahoehoe flow from one of the Twin Craters. Some lava tubes

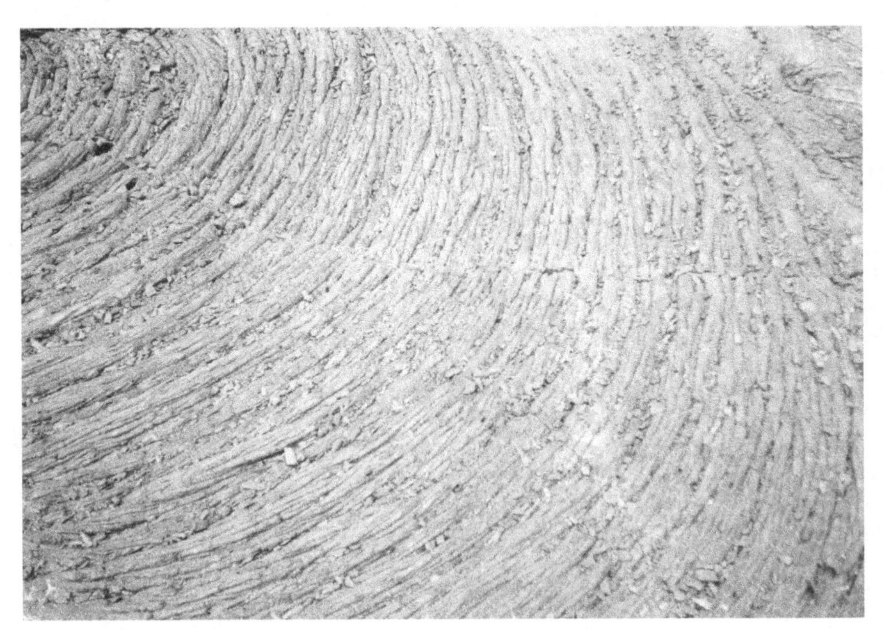

FIGURE 5.12 *A pahoehoe lava flow is a basaltic flow having a ropy or billowy surface. It forms when a crust forms over the flow; then the flow moves again, wrinkling the crust. Photo by Ann G. Harris.*

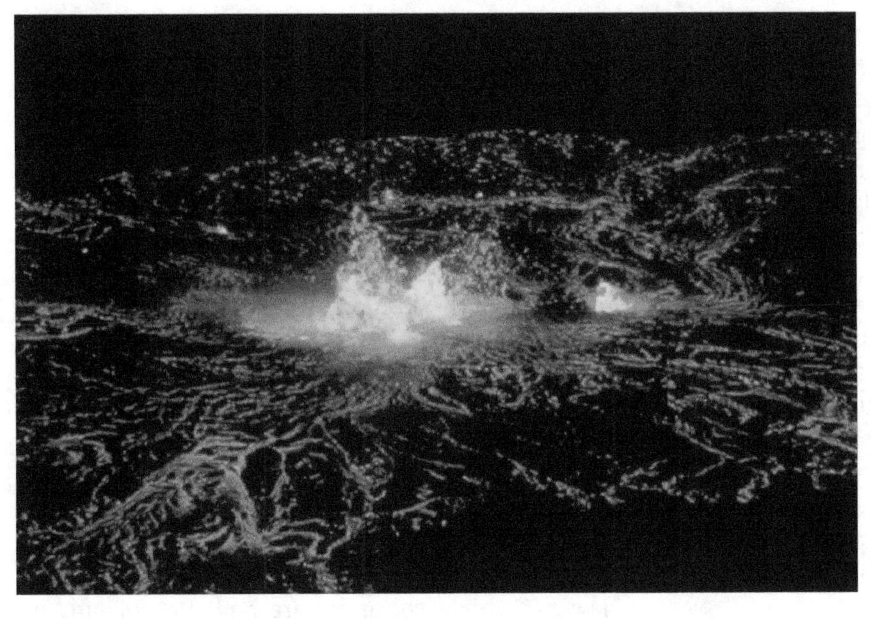

FIGURE 5.13 *The fire pit Kilauea Iki had a lava lake in it in the 1950s. In 1959 there was a spectacular eruption, with numerous fountains erupting in the lava lake. National Park Service photo.*

contain *lava stalactites,* produced as liquid lava hardened while dripping from the tube roof.

Tree molds and *lava trees* may form in pahoehoe lava when a flow inundates a forest. Hot lava, chilled by its initial contact with a tree, either sticks up against the bark, hardening quickly, or flows around the trunk. After the charred tree rots away, the lava "trees" are left standing. Tree molds, resembling empty pails in the flow, show where tree trunks stood. Tree molds are numerous in Hawaii Volcanoes National Park and in Lava Tree State Park.

Mounds or hillocks on the surface of pahoehoe lava are called *tumuli* (singular, *tumulus*). They tend to form parallel to the flow front in places where lava is confined, especially in craters. Solid crust at the top of the flow is dragged along and domed up by the faster-moving fluid lava underneath. Sometimes fresh lava dribbles or spatters out, hardening into *driblet spires* of fantastic shapes.

Lava balls are accretionary features characteristic of aa flows and range in size from a few inches to 10 feet or more. They form when a fragment of

FIGURE 5.14 *Lava tubes and lava tunnels beneath the congealing surface of a pahoehoe flow convey fresh, hot lava out to the moving front. Occasionally a section of the roof will fall in, creating a "skylight" through which the flowing lava can be seen. Photo by Ann G. Harris.*

solidified lava rolls along, enlarging as it picks up sticky lava. (Children in snowy regions make snowballs large in much the same way by rolling them in sticky snow until they're too big to push.)

A Hawaiian term, *kipuka,* meaning "opening," refers to island-like areas of older land surrounded by more recently erupted lava. Kipukas, ranging in size from a few square feet to several square miles, are usually the result of irregularities in the land surface. A kipuka surface may be higher or lower than the level of the surrounding lava, the latter case being more common. New lava is sterile and barren of vegetation, but some kipukas have weathered long enough to have soil that can support plant life, making them islands of greenery in a landscape of bare volcanic rock. Several kipukas are in the park. The best known is Kipuka Puaulu, or Bird Park, northwest of Hawaii Volcano Observatory.

Ash Explosions and Pyroclastics

Kilauea has been an ash producer, probably because more moisture is trapped on its windward slopes. Ash explosions are commonly associated with the sudden mixing of ground (phreatic) water and hot lava or magma. Violent explosions have occurred when water drained down into newly emptied vents or conduits and reached magma. The result has been a *phreatomagmatic explosion.* In 1790 such an explosion killed a band of Hawaiian warriors who were marching across the Ka'u desert (a rain shadow zone on the southwest slope of Kilauea). Despite years of weathering, the warriors' footprints are still faintly visible on an ash surface close to the park's western boundary.

Phreatic explosions occur any time that hot lava encounters ground water and turns it to steam. The most explosive steam explosions in historic time occurred during the 1924 eruptions from Kilauea when large blocks of old lava (that can be seen scattered around the summit area) were ejected from the volcano's main vent. More typically, phreatic explosions produce ash, cinder cones, and pyroclastics of various sizes, especially at or near the shore where moisture is plentiful. Most cones, such as those along Kilauea's east rift zone, are a combination of cinders and spatters.

Rift Zones on the Volcanoes

Hawaiian shield volcanoes are built up not only by summit eruptions but also by eruptions from *rift zones* that radiate from the summit calderas and extend down into the ocean. Their orientation is influenced by the buttressing effects of neighboring volcanoes and by gravitational stresses. The rift zones are scored at the surface by many cracks, some of which have served as vents for erupting lava. Below the surface, dikes of congealed lava fill fissures. Each upwelling of magma from depth causes swelling of the summit and flanks, which creates new stresses on the rift zones and opens new cracks and fissures or reopens old ones. An eruption relieves pressure in the magma reservoir, and the volcano then deflates.

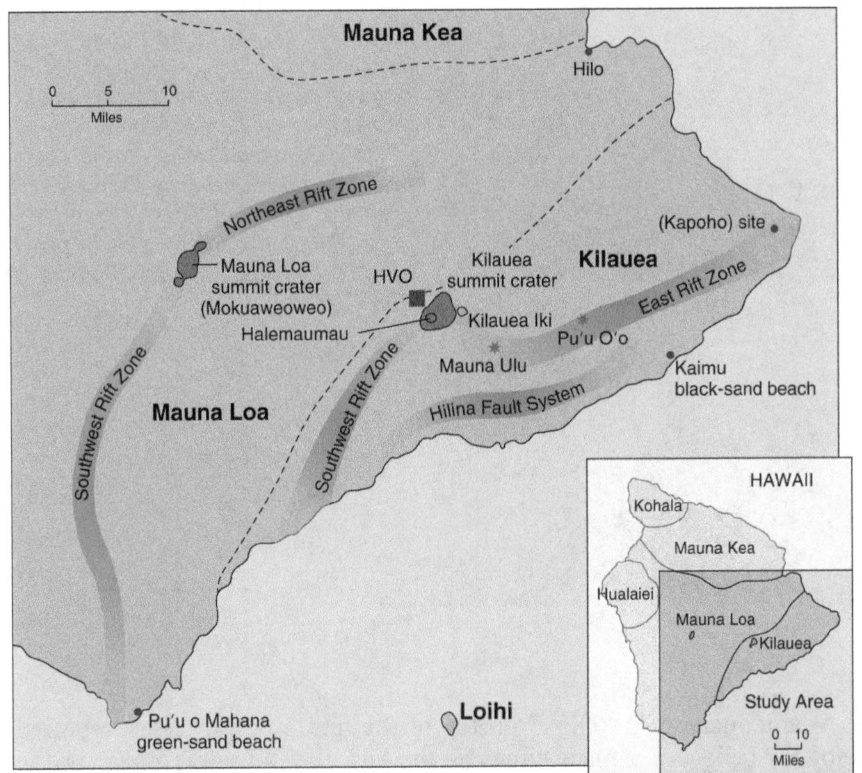

FIGURE 5.15 *Major rift zones associated with Mauna Loa and Kilauea and the Hilina fault system. Note underwater location of Loihi offshore. From Tilling, Heliker, and Wright 1987.*

Mauna Loa and Kilauea each have two principal rift zones (Figure 5.15). Mauna Loa's main growth has been along the two principal rift zones that extend from the caldera to the southwest and to the northeast. Kilauea's south flank is pushing into the sea, growing a few inches per year due to repeated intrusions of magma into the southwest and east rift zones. Since it began erupting in 1983 from the east rift zone, Pu'u 'O'o has built its cone to a height of more than 830 feet, making it the highest volcanic landform that has grown and been monitored by scientists in historic time. In 1986, after the Pu'u 'O'o conduit ruptured, lava erupted to the east of the cone from a new vent and began to build up a shield. Flows from the vent, named Kupaianaha, formed a system of lava tubes that carried fresh lava over a fault scarp and down to the ocean (Figure 5.16).

Fault Systems and Landslides

Rift zone eruptions tend to make the volcano flanks out of balance or unstable. Earthquakes accompanied by sudden movements along fault lines release the accumulated strain. A fault system between Mauna Loa and Kilauea appears to act as an "expansion joint" between the two volcanoes. A jolting 6.6 earthquake on this fault system in November 1983 damaged nearby roads, trails, and some buildings. Along Mauna Loa's southeastern base, step faulting, by which the mountain side has moved downward in response to gravity, has produced a series of fault scarps (called "palis") facing southeast.

Similar step faulting takes place from time to time in the Hilina fault system along Kilauea's southern slope. Here the fault blocks form a series of steps down to the coast, with the trend being parallel to the shore. Measurements after a magnitude 7.2 earthquake in November 1975 revealed that some rocks along the Hilina fault had dropped 11 feet and shifted southward as much as 24 feet (Tilling, 1987). The fault scarps of the Hilina fault system show up on Kilauea's south flank as steep cliffs overlooking the sea. Lava flows have cascaded down over the fault scarps into the ocean.

The U.S. Geological Survey recently completed imaging of the sea floor around the Hawaiian Islands by means of a sidescan sonar device (Geologic Long Range Inclined Asdic, or "GLORIA"). The sonar monitor revealed many undersea landslides consisting of giant slumps that had slid gradually off the volcanoes into the sea. However, in some places, the slides were debris avalanches that had given way explosively, leaving on the volcano flanks headwalls of great

FIGURE 5.16 *A lava flow derived from Pu'u 'O'o entering the ocean. Note how the cliff is gradually breaking away, forming terraces. Much of the lava entering the ocean is chilled instantly, forming obsidian, a natural black volcanic glass. This is the composition of the sand of the black beaches. Photo by Ann G. Harris.*

"amphitheaters," as huge sections of volcanic material dropped to the sea bottom. Geologists believe that the major fault systems on the Big Island are related to the upper parts of the giant landslides hidden below sea level. At present, the south side of the Big Island is slumping toward the sea while the East Rift Zone is migrating southward. Scientists at the Hawaiian Volcano Observatory continue to carefully monitor the area.

SUBMARINE AND SHORE FEATURES

Lava flows that reach the ocean surface produce cinders and ash when the hot lava mixes with the sea water. Cinder cones and benches are formed when lava enters the ocean. Waves erode and transport volcanic sediment, depositing "black sands" on beaches. Since the source of supply is limited and some sand is washed out to sea each year, the black sands may last only two or three centuries.

Older sea cliffs and sea arches undergo wave erosion and are retreating, but in some places along the shore, recent lava flows protect the cliffs and have enlarged them.

When fluid lava erupts under water or flows from a tube that opens in the ocean, *pillow structures* may form. With water removing heat rapidly, the lava quickly crusts over and solidifies as a pillowlike or tonguelike protrusion. More lava squeezes out (like toothpaste from a tube) and stacks another "pillow" on top of the first. The process is repeated many times as the lava flow advances under water. Because the lava is under pressure from the weight of the overlying water, explosive interactions seldom occur between the cold ocean water and the hot lava. In recent years scuba-diving scientists have observed and filmed the forming of pillow structures during underwater eruptions of lava near Big Island.

GEOLOGIC HISTORY

1. **Initial eruptions on the sea floor, probably early in Pleistocene time.**

The age of the rock at the base of the volcanic pile has not been determined. Putting down drill holes in order to get core samples for age determination is a difficult and costly operation. Moreover, in some locations, drill bits might hit magma that would melt them.

2. **Construction above sea level of Hawaii Island.**

Five shield volcanoes, each with its own magma reservoir but all having the same magma source in the upper mantle, built up the island during Pleistocene time. Kilauea, the youngest volcano, is still enlarging at the southeastern end of the island. South of Hawaii, Loihi seamount is rising. It is regarded as the newest expression of the hot spot beneath the Pacific plate. Using a submersible, serviced by a research vessel, scientists monitor Loihi's development and collect samples of vent fluids, gases, and lavas. Loihi is expected to reach the ocean surface in 60,000 years, more or less.

3. The development of Mauna Loa.

Rock units. Mauna Loa's rocks are mainly olivine basalts. The oldest that are exposed are the lavas of the Ninole Formation that make up the core of the mountain. Ninole rocks crop out in low hills on Mauna Loa's southeast slope. Deep canyons were cut in these lavas during an erosional period, producing an unconformity between Ninole beds and the overlying lavas of the Kahuku Formation. The Kahuku lavas partially filled the canyons and built up the slopes of Mauna Loa. Pahala ash, 5 to 50 feet thick, mantles the Kahuku rocks. Most of the Pahala ash in this area came from Kilauea and was probably the result of violent phreatomagmatic explosions that accompanied the collapse of Kilauea's caldera. Much of the Pahala ash has been covered by younger lavas of the Ka'u Formation. The fairly fresh and relatively unweathered rocks of the Ka'u include lavas from eruptions during historic time. Mauna Loa had probably attained its present height by the end of the Pleistocene Epoch.

Eruptions in historic time. Mauna Loa's eruptions since early in the nineteenth century have been mainly of two types: (1) summit eruptions without flank flows and (2) the more typical flows from the rift zones that may be accompanied by brief summit eruptions. Eruptions generally begin with harmonic tremors (swarms of little earthquake tremors) and the release of clouds of steam and gas. Lava fountains may shoot up many feet in Moku'aweoweo, the summit caldera. Strands of Pele's hair are blown out and away by uprushing winds. During a summit eruption, lava pours from cracks in the caldera wall and spreads over the floor. Pit craters may develop or enlarge in and around the caldera.

When eruptions begin in the rift zones, lava and hot rising gases shoot up, making a "curtain of fire" along the length of a fissure opening. Clots, blobs of pumice, and threads of volcanic glass harden in the air and then fall glistening to the ground. During the first few days, very fluid pahoehoe pours down over the slopes like a slow-moving river (10 to 25 miles per hour). Later flows may change to aa.

The 1859 eruption, which lasted for 10 months, produced a 33-mile flow, the longest recorded in historic time. Several eruptions from the northeast rift zone have endangered the city of Hilo and its harbor on the eastern coast of the island. The closest call was a pahoehoe flow 29 miles long in 1881 that stopped on Hilo's outskirts.

A major eruption of Mauna Loa took place in June 1950. Lava first began erupting near the caldera at the top of the southwest rift zone. In minutes the erupting fissures had opened up the southwest rift zone for several miles down the slope. Within 24 hours, lava fountains blazed in a curtain of fire down the mountainside for 8 miles. On both sides of the rift ridge, rivers of lava poured down the slopes. Three flows reached the ocean. The volume of lava (19.37 cubic miles) that was extruded during the three weeks of the 1950 eruption equaled the amount of lava that erupted in the 10-month eruption of 1859. In terms of volume, these two eruptions are the largest that have been recorded on the island of Hawaii. Although Mauna Loa has been quiet most of the time since 1950, more eruptions are expected to occur.

4. The development of Kilauea.

Rock units. Eruptions of Hilina Formation lavas from the southeast flank of Mauna Loa began building Kilauea in about the middle of Pleistocene time. The Hilina lavas are presumed to be roughly equivalent in age to the Kahuku Formation on Mauna Loa. Hilina basalts are exposed in the fault scarps along the southern coast. Flows of both pahoehoe and aa can be distinguished.

Pahala ash overlies the Hilina Formation. Above the Pahala ash are rocks of the younger Puna Formation. Puna lavas were erupting at about the same time as the Ka'u lavas on Mauna Loa, and the flows interfinger in the area between the two volcanoes. Puna flows were erupted largely from vents in Kilauea's caldera and from the rift zones. Spatter and cinder cones formed on both the southwestern and eastern rift zones. Those on the eastern rift zone were more numerous.

Eruptions in historic time. The types of activity have been (1) flank flows, more often than not preceded by summit activity or eruption; (2) shield construction; (3) rising and falling of lava lakes; (4) brief summit eruptions; and (5) violent explosive eruptions.

The lava lake in the fire pit on Kilauea's summit apparently filled not long after the violent explosion of 1790. Throughout the 1800s and the first quarter of this century, the lava lake was almost continuously

FORECASTING ERUPTIONS

Crisscrossing the island of Hawaii is a network of scientific instruments for monitoring the volcanoes. Collecting and analyzing data and making direct observations are the responsibility of volcanologists at the Hawaii Volcano Observatory and the University of Hawaii. Tied in with this network are seismic stations and investigators in other parts of Hawaii, Japan, New Zealand, and many locations throughout the world.

What are changes or signs that scientists look at? What combinations of factors indicate an impending eruption?

1. Periodicity. What is the eruptive history of the volcano? How long are its quiet periods? Such information is useful in a general way, although eruptive patterns of volcanoes may change unexpectedly.
2. Changes in the behavior of fumaroles and in the output, composition, and temperature of gas. Steam and gas emissions are affected by atmospheric temperatures and humidity, but observations in this category are useful. On Kilauea, for example, an increase in the hydrogen content of gas emissions usually signals the onset of an eruption.
3. Shifts in the orientation of the volcano's magnetic field and reduction of magnetic attraction of rocks due to increasing heat. Changes in magnetism are difficult to interpret, but they appear to indicate movement of magma within the interior of a volcano. Often an eruption does follow such movement, but sometimes a volcano quiets down without an extrusive event occurring.
4. Infrared photographs and remote sensing. Increases in heat and changes in the locations of hot areas show up when these techniques are used. Any eruptive activity is invariably preceded by an increase in heat on the volcano's surface.
5. Geoelectrical changes. Variations in electrical currents in a volcano have been observed just before explosions or eruptions, suggesting that geoelectricity may be useful as an indicator.
6. Tumescence and tilting. Kilauea and Mauna Loa have histories of swelling and tilting before eruptions, but occasionally swelling goes down without an eruption occurring. Land deformation on any active volcano is considered an ominous sign. The cause of bulging, rising of the summit, and tilting is assumed to be magma filling the volcano reservoir and rising in the conduits. Tiltmeters on Kilauea and Mauna Loa continuously monitor land surface deformation as well as migration of the center of a tumescent area.
7. Changes in seismicity. From earliest times, people who lived near volcanoes associated earthquakes with eruptions. On Hawaii Island, certain kinds of earth tremors and rumblings were called "Pele's thunder." These were regarded as warning signals of her anger.

Earthquakes are frequent on Big Island; many of them are so slight that they are imperceptible to residents. Most of the earthquakes are shallow, local, and indicate a shifting around of rocks in response to stresses in the volcanic structures. The Big Island's most severe earthquake in historic time, estimated to have had a magnitude of 8, happened in 1868 after days of strong shocks. The series of quakes triggered mud avalanches and a 30-foot drop of the island's south flank. The southern coastline subsided as much as seven feet and a destructive tsunami washed in over the shore. Both Mauna Loa and Kilauea erupted shortly after the earthquakes (Parks 1993). On Maui and the other Hawaiian Islands, earthquakes are less common and are probably caused by isostatic adjustments of the oceanic crust to the load of overlying volcanoes.

Seismic data collected on Hawaii have been studied for many years by scientists seeking clues as to the imminence of eruptions. Several significant patterns have shown up, but they must be combined with other observations to be meaningful. Earthquake "swarms," consisting of thousands of little earthquakes, are recorded on seismographs

FORECASTING ERUPTIONS *continued*

before eruptions occur on Mauna Loa and Kilauea. These quakes, called *harmonic tremors,* are probably the result of minor faulting that accompanies tumescence or the inflation and pushing up of the volcano top. Fissures and troughs may develop along the rift zones. After the magma has drained and the summit starts to shrink, quakes begin again, sometimes stronger than those that accompany the swelling. Evidence from tiltmeters may confirm whether the volcano is shrinking or possibly swelling again. This is important in Hawaii where eruptive periods last for weeks, months, or years.

Forecasts regarding eruptions are necessarily based on several lines of evidence and are expressed in probabilistic terms. Nevertheless, even though anticipating volcano behavior remains an inexact science, forecasts can be made that can help to prevent loss of lives when dangerous areas need to be evacuated.

active; although some sinking, followed by refilling, occurred from time to time after lava had been drained by flank eruptions below sea level. The lava lake was at a high level at the beginning of 1924 when a series of earthquakes began. They first occurred at the summit and then shifted outward along the eastern rift zone, suggesting that magma was moving in that direction. Meanwhile, the level of the lake dropped several hundred feet. The explosions began in May and lasted for about three weeks. Ash-filled clouds of steam and gas shot up. Sulfur fumes killed vegetation over a wide area. Blocks of caldera rock were thrown a mile into the air. Local earthquakes, lightning, and heavy downpours of rain accompanied the explosions. Afterwards Halema'uma'u had increased in diameter from 500 feet to over 3000 feet, and the floor of the caldera had sunk 11 feet. Even the caldera rim was lowered a few feet. What caused the explosions and collapse was ground water pouring into the hot vents as lava drained away. Instantaneous boiling resulted, and then the violent escape of the steam. The explosions were not magmatic since no fresh lava was ejected with volcanic debris.

Since 1924, the lava lake that had bubbled and flamed for so many years has refilled partially, but only for very brief intervals. The fire pit is quiet except for wisps of steam. Summit eruptions have been short and infrequent. Flank eruptions, on the other hand, have been numerous and sustained. One summit eruption, however, was spectacular. In 1959, Kilauea Iki, a large pit crater close to the caldera, had a major eruption. The lava fountains that shot up were the highest on record, and cinders and pumice devastated a nearby forest.

Before the lava lake disappeared, the southwest rift zone was more active than the eastern rift zone. Mauna Iki is a secondary lava cone southwest of the caldera that was built by an eruption in 1920. Fresh lava flowed out near Mauna Iki in 1971.

The east rift zone has been exceedingly active for many years. From the caldera, the east rift zone goes southeastward and then takes a sharp bend almost due east to Cape Kumukahi and on out into the ocean. The Chain of Craters, on the upper part of the eastern rift zone, used to be a row of 13 pit craters coming down from the caldera. A highway went through the area from the Crater Rim road southeastward to the coast. The road was buried by lava in 1969 when an entirely new vent opened up and Mauna Ulu, a satellite cone, began to grow. Mauna Ulu continued to erupt until 1974, and many of its flows reached the ocean. Next to Mauna Ulu is the Alae shield, which was built up over an old pit crater that was 440 feet deep. Other nearby craters were partly filled or buried, and some reopened and began pouring out lava or filling up and forming lava lakes.

The following year, in 1960, lava erupted 28 miles away from the summit near Cape Kumukahi at the extreme eastern tip of the island. Lava flowing from the eastern rift zone destroyed the village of Kapoho and created some 500 acres of new land where the flow went into the ocean. (The question of who owns new land has been in litigation.)

The Pu'u 'O'o eruptive series, which has been continuously active since 1983, has become the

TABLE 5.2

GEOLOGIC COLUMN, HAWAII VOLCANOES NATIONAL PART

Time Units			Rock Units and Geologic Events	
Era	**Period**	**Epoch**	**Mauna Loa**	**Kilauea**
Cenozoic	Quaternary	Holocene — "Historic"	*Younger Ka'u Formation* Summit eruptions, 1975, 1984 Numerous summit and rift zone eruptions, 1832–1950 Devastating earthquake and landslide, 1868 Continuing shield construction by flows of aa and pahoehoe Enlargement of caldera by pit craters	*Younger Puna Formation* Construction of satellite cones Mauna Ulu, Kilauea Iki, Pu'u 'O'o, Kupaianaha since 1968 Numerous summit and east and west rift zone eruptions, 1790 to present Continuing shield construction by flows of aa and pahoehoe Explosive eruptions; 1790, 1924 Appearance, disappearance, reappearance of lava lakes; enlargement of caldera
		Holocene — "Prehistoric" —?—	*Older Ka'u Formation* Shield construction by aa and pahoehoe; cinder cones, spatter cones, etc. Development and collapse of Moku'aweoweo caldera; pit craters *Kahuku Formation,* capped by *Pahala Ash* Building up of shield by thick accumulations of aa and pahoehoe	*Older Puna Formation* Shield construction by aa and pahoehoe flows; ash falls, cinder cones, spatter cones Development and collapse of Halema'uma'u in summit caldera; pit craters *Hilina Formation,* capped by Pahala ash Beginning of shield construction by many flows, both aa and pahoehoe, building up thick accumulations; some ash layers, indicating past explosive eruptions
		Pleistocene	Great erosional unconformity Fluctuations at sea level and submergence; intense erosion of shield; canyon cutting	
			Ninole Formation Construction of Ninole volcano Volcanic "basement" rocks (not exposed at surface); age unknown	Volcanic "basement" rocks (not exposed at surface); age unknown

Sources: Modified after Macdonald et al., 1983; Tilling et al., 1987.

FIGURE 5.17 *Pu'u 'O'o is a satellite cone on the flank of Kilauea Volcano; the most recent lava flows have come from this vent. Photo by Ann G. Harris.*

longest and largest eruption in Kilauea's history (Figure 5.17). Flows from Pu'u 'O'o and its associated vent Kupaianaha have constructed a new shield and added new land to the island where lava has entered the ocean. The coastal community of Kalapana, which lay in the path of the flows, was wiped out by repeated lava flows. More than 189 homes and businesses were destroyed, as well as a National Park Visitor Center, archeological sites, roads, trails, and other structures.

5. **Glaciation.**

A small icecap existed on Mauna Kea, Hawaii's highest volcano, during the Pleistocene Epoch. Glaciers were also present on Mauna Loa at the same time, but all evidence has been destroyed or buried by recent lava flows. In today's climate, snow falls every winter on Mauna Kea and Mauna Loa, but does not last long.

6. **Changes in sea level.**

Old shorelines, some higher than the ocean's present stand and some lower, testify to the fluctuating ocean level around the Hawaiian Islands during their comparatively brief geologic history. The most significant long-term trend is land submergence. As the volcanic pile has gotten higher and heavier, the sea floor has tended to sink. However, older sea levels are not consistent from place to place because local tilting and warping have occurred. Thus tectonic changes have affected the relative position of land and ocean. The islands' shorelines were also affected by worldwide changes in sea level that occurred during Pleistocene time.

SOURCES

Abbott, A. T. and Peterson, F. L. 1983, 2nd edition. *Volcanoes in the Sea: The Geology of Hawaii.* Honolulu: University of Hawaii Press.

Clague, D. A., Dalrymple, G. B., Wright, T. L., Klein, F. W., Kayanagi, R. Y., Decker, R. W., and Thomas, D. M. 1989. The Hawaiian-Emperor Chain (chapter 12) in Winterer, E. L., Hussong, D. M., and Decker, R. W. eds. *The Eastern Pacific Ocean and Hawaii.* Boulder, Colorado: The Geology of North America, v. N. p. 187–287.

Decker, B., and Decker, R. W. 1987, 2nd edition. *Road Guide to Hawaii Volcanoes National Park.* Mariposa, California: Double Decker Press.

Decker, R. W., Wright, T. L., and Stauffer, P.H. (editors). 1987. *Volcanism in Hawaii.* U.S. Geological Survey Professional Paper 1350, 2 vol. 1667 p.

Griggs, J. D., Takahashi, J. T., and Wright, T. L. 1986. Volcano Monitoring at the U.S. Geological Survey's Hawaiian Volcano Observatory, *Earthquakes and Volcanoes* 18 (1):3–71. U.S. Geological Survey.

Hazlett, R. W. 1987. Kilauea Caldera and Adjoining Rift Zones. Geological Society of America Centennial Field Guide, Cordilleran section, v. 1, p. 15–20.

_____ 1990. *Geological Field Guide, Kilauea Volcano.* Hawaii Natural History Association.

Heliker, C. 1990. *Volcanic and Seismic Hazards on the Island of Hawaii.* U.S. Geological Survey General Interest Publication.

Herbert, D.; and Bardossi, F. 1968. *Kilauea: Case History of a Volcano,* New York: Harper & Row Publishers.

Macdonald, G. A. 1972. *Volcanoes.* Englewood Cliffs, New Jersey: Prentice-Hall.

_____ and Hubbard, D. H. 1989, 9th edition (revised by Jon W. Erickson). *Volcanoes of the National Parks in Hawaii.* Hawaii Natural History Association and National Park Service.

Parks, N. 1993. The Fragile Volcano. *Earth* 2(6): 42–49.

SOURCES *continued*

Stearns, H. T. 1985. *Geology of the State of Hawaii.* Palo Alto, California: Pacific Books.

—— 1978, 2nd edition. *Road Guide to Points of Geologic Interest in the Hawaiian Islands.* Palo Alto, California: Pacific Press.

Macdonald, G. A. 1946. Geology and Groundwater Resources of the Island of Hawaii. Hawaii Division of Hydrography Bulletin 9.

Swanson, D. A., and Peterson, D. W. 1972. Partial Draining and Crustal Subsidence of Alae Lava Lake, Kilauea Volcano, Hawaii. U.S. Geological Survey Professional Paper 800-C.

Tilling, R. I, Heliker, C., and Wright, T. L. 1987. *Eruptions of Hawaiian Volcanoes: Past, Present, and Future.* U.S. Geological Survey, General Interest Publication. 55 p.

ESS 101

Lab 5: Measuring the Rate of Plate Motion

Student Name: ____

In this Lab you will investigate volcanic structures in the Pacific Ocean and along the Pacific coast of North America.

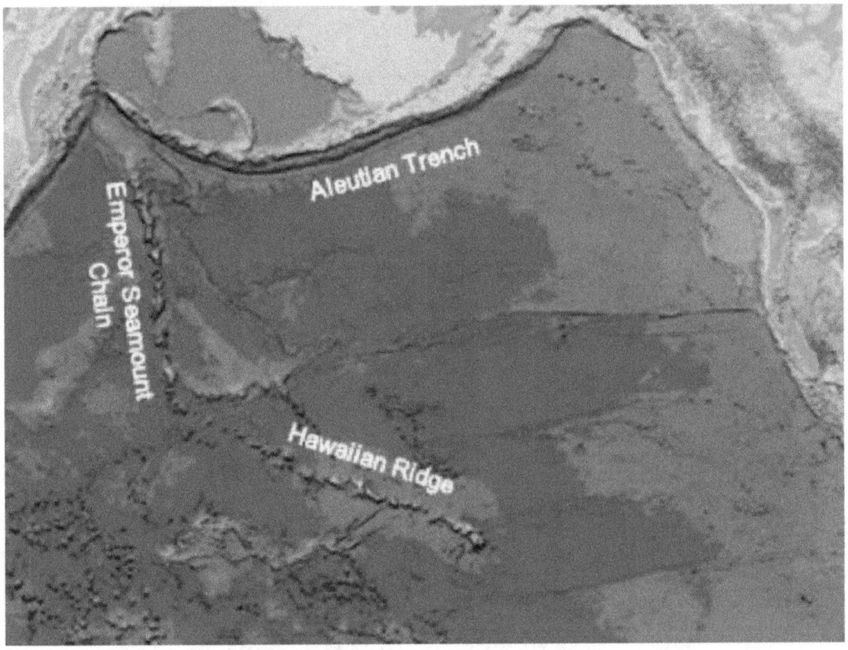

FIGURE 5.18 *Map of Pacific Ocean showing the location of Hawaiian Ridge, Emperor Seamount Chain, and the Aleutian trench locations. (Base map from "This Dynamic Earth," USGS).*

1. Hawaiian volcanic islands are located in the middle of the Pacific Plate and are known as 'hot spot' volcanoes. Refer to pages 105–108 and 124–129 to answer the following questions:
   a. What is a hot spot?

   ____

   ____

   ____

   ____

   b. What kinds of volcanoes are in Hawaii?

   ____

   c. What is the source (sources) of magma that sustains a hot spot?

   ____

   ____

d. What type of geologic features make up Emperor Seamounts?

____

____

____

e. How did Emperor Seamounts form?

____

____

____

____

2. Study the map in Fig. 4.47, page 125. Notice that the Hawaiian Islands are just the eastern-most segment of the chain of volcanic structures that extends to Kimmei (at the Hawaiian Emperor Bend). All of these volcanic structures are extinct (no longer active) except for Hawaii. Age of the rocks sampled on these islands were estimated using radiometric dating. Ages are shown on the map in millions of years (Ma).
   a. In what direction has the Pacific Plate been moving during the past 39.9 Ma?

   b. What is the distance from Hawaii to Kimmei in km (kilometers)? ________ km

   Convert this distance into mm (millimeters). ________ mm
   (Hint: There are 1000 mm/meter and 1000 meters/km.)

   c. What has been the average rate of Pacific Plate motion, in mm/yr, for the past 39.9 Ma? (Show your work below, and draw a line of that length.)

   d. How long is the Emperor Seamount chain (use the map to measure the distance)? ________ km.

   e. In what direction did the Pacific Plate move from 64.8 to 39.9 Ma? ________________ .

   f. What was the average rate of Pacific Plate motion, in mm/yr, over that time? (Show your work.)

   g. Based on the map (p. 125), how long has the Hawaiian Hot Spot been active? ________________ Ma.

h. Do you think geologists know exactly when the Hawaii hot spot began to operate? Use argument to support your answer.

____

____

____

____

i. What did cause the change in the direction of the plate movement?

____

____

____

____

3. Study the map in Figure 4.44 on page 123. Notice that there is a "rift zone" between the Pacific Plate and Juan de Fuca Plate.

a. What kind of plate boundary is this? ______________________

b. What kind of feature would you expect to see at that boundary if you could look beneath the ocean surface and see it?

____

____

____

____

c. Would you expect to see any volcanoes at that boundary? If so, then what kind (basaltic or andesitic)?

____

____

____

____

d. If there were erupting volcanoes at the boundary, then would you expect to see a'a, pahoehoe, or pillow lavas? Why?

e. Notice the red line on Figure 4.44.
What kind of seafloor feature would you expect to see there?

f. Small black arrows indicate plate motions on Figure 4.44.

The North American Plate is moving in what direction? ______________________

The Juan de Fuca Plate is moving in what direction? ______________________

g. Notice the continental volcanic arc (line of active volcanoes), the Cascade Range of Washington and Oregon. How does magma form beneath these volcanoes, and what kind of volcanoes are they?

4. The Hawaiian lava eruptions are mild compared to the violent eruptions of the Cascade region. Which lava is more viscous, and how does that affect the style of eruption?

CHAPTER 6

Weathering

INTRODUCTION To most people, rocks are permanent. Euphemisms such as "rock of ages" and "solid as the Rock of Gibraltar" reflect the common impression that rocks are extremely durable. Certainly, in terms of human lifetimes, some rocks do indeed seem to last forever. Rock of Ages, for example, is the trade name for a commercial brand of granite monuments. Granite was the rock used by the Egyptians to construct the cores of the pyramids, which many view as the epitome of antiquity (Figure 6.1). Today granite is still frequently used in the construction of buildings and monuments, especially those designed to last a long time. Granite is indeed a fairly durable rock for reasons that you will learn in this chapter.

The longevity of granite is, however, simply a question of perspective. Using a human lifetime as a measuring stick, granite does seem to be the "rock of ages," but in terms of geologic time, even granite soon succumbs. The process that dooms granite and all rocks is **weathering.** Weathering is the first of a series of processes collectively called **erosion.** Erosion itself is defined as any process by which Earth's surface is loosened, dissolved, and worn away, and simultaneously moved from one place to another, by natural agents. No matter how durable a rock may appear to be, no rock will survive these processes for very long.

Weathering is of critical importance to life on Earth. A major end product of the process of weathering is *soil.* Without soil, nothing but the lowest forms of plant life could exist on land. Without soil, the continents of Earth would be as lifeless as the Moon.

FIGURE 6.1 *Much of the longevity of structures such as the* pyramids *is due to the limited availability of water in their desert location. (Courtesy of E. D. McKee/USGS)*

WEATHERING DEFINED

Several definitions of weathering can be found, each of which concentrates on a specific aspect of the overall process. For example, weathering has been defined as any reaction between a rock surface and the agents of the atmosphere. This definition reflects the fact that the process can go on anywhere the atmosphere can penetrate, be it a hairline crack in a rock, a cave, or a mine.

Another definition specifies that rocks will undergo weathering wherever they are exposed to circulating surface water. In addition to emphasizing the role of water in the weathering process, this definition indicates that weathering can go on at depth and indirectly implies that the "agents of the atmosphere" responsible for the breakdown of rocks are transported dissolved in water.

From the standpoint of describing its mechanisms, weathering is best defined as any process whereby rocks either *disintegrate* or *decompose*. **Disintegration** is any process that breaks rocks into smaller pieces without changing the composition. **Decomposition** includes any process that results in either a partial or complete change in the mineral and elemental composition of the original rock.

Disintegration and decomposition are the basis for the two modes of weathering, **mechanical weathering** and **chemical weathering**, respectively. Although these two processes can and do occur independently, as one might expect, they usually go on simultaneously. Depending largely on the climate, however, one process usually dominates over the other.

MECHANICAL WEATHERING

Mechanical weathering is accompanied by a variety of physical processes, all of which break the rock materials into smaller pieces that retain the chemical composition of the parent rock.

Frost Action

During the spring, throughout the more temperate, humid parts of the world, roadways below road cuts and the bases of cliffs are commonly littered with boulders and rock fragments, the end products of one of the most common mechanical weathering processes, **frost action**. Frost action is the result of a special property of water. Water is one of the few materials that expands as it converts from the liquid to the solid state. A volume of water will expand about 9% as it freezes. This may not sound like much, but water freezing within a completely filled, enclosed container could theoretically generate a pressure of about 1,500 pounds per square inch (680 km/cm^2). If the container happens to be the water pipes in your home, the expansion of the ice will cause them to burst long before this pressure is attained. Expanding ice will also break the strongest rock. In fact, very few solid materials can long withstand the internal stresses generated by freezing water. When water penetrates even the narrowest crack or crevice within a rock and freezes, the crack is enlarged. When the temperature rises and the ice melts, more water enters the expanded crack and penetrates deeper into the rock. The next freeze opens the crack even further. Each successive **freeze-thaw cycle** continues to enlarge the crack, eventually separating the block of rock from the main rock body. An end product of frost action is the loose rock debris called *talus* that accumulates along the base of a cliff or a road cut.

Frost action is responsible for other examples of mechanical rock disruption. Most of the rock litter that frequently covers the ground in high mountainous areas is produced by frost action. This same process is a major contributor to the formation of the chuckholes or potholes that make driving such an exciting experience in areas where freeze-thaw is common.

It is important to note that the process of frost action requires numerous freeze-thaw cycles. Many cycles are needed to enable water to enter the enlarged cracks, refreeze, and ultimately cause the physical disruption of the rock. Consequently, regions subjected to prolonged freezing conditions such as the polar regions, do not exhibit extensive frost action.

Exfoliation

Exfoliation is a common process of weathering in which concentric layers are removed from the rock mass. The layers, ranging in thickness from fractions of an inch (a few millimeters) to several feet or meters, are formed and removed by pressures exerted on the outer layers of rock by a number of processes operating below the rock surface (Figure 6.2). An

FIGURE 6.2 *A number of processes result in the surface of the rock being removed layer by layer, a process called* exfoliation.

example of such a process involves the growth of water-soluble crystals. Water moving through layers of rock carries minerals in solution. At the face of a cliff or road cut where rock layers are exposed, the water may seep out and flow down the rock face. Often during times of low flow or on especially warm days, the water may evaporate and deposit the minerals, producing the white or reddish stains commonly observed on cliff faces (Figure 6.3). Consider, however, a special situation where the rock exposure faces toward the Sun and is heated by its radiant energy. Should the flow of water through the rock layers be sufficiently low, heat generated by the Sun's rays can cause the water to evaporate down to a depth of a few millimeters below the rock surface. As the water evaporates, the minerals crystallize in a layer and constantly grow in mass as they are "fed" by fresh solutions from behind (Figure 6.4). In time, the pressures of the growing crystal mass will cause the rock to split parallel to the rock surface and be removed, or **spall**, in sheets or flakes. As fresh rock surfaces are exposed, evaporation of water begins anew just beneath the surface and the process continues. Slowly, the cliff face retreats.

FIGURE 6.3 *The white to reddish stains commonly seen on rocks exposed in cliffs and road cuts are the result of minerals deposited as seeping groundwater evaporates on the surface of the rock. (Courtesy of W. B. Hamilton/USGS)*

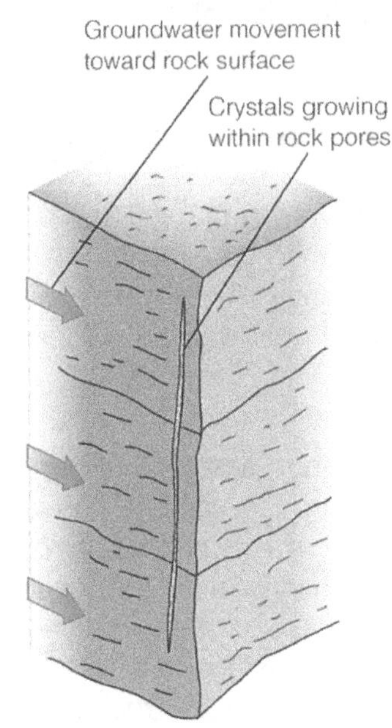

(a) Deposition and growth of salt crystals 1–2 mm below rock surface.

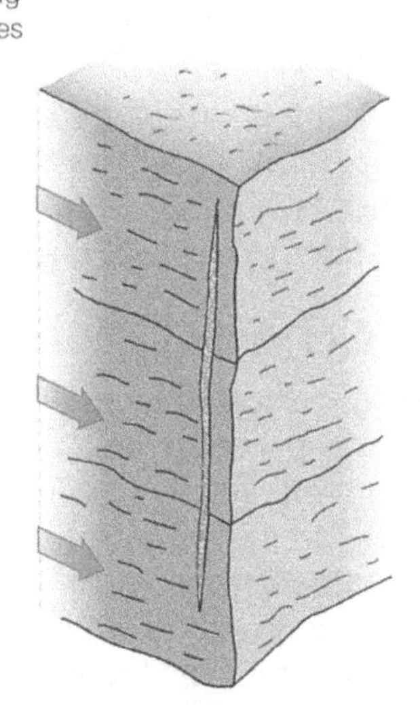

(b) Continued growth of crystals causes development of fracture parallel to rock surface.

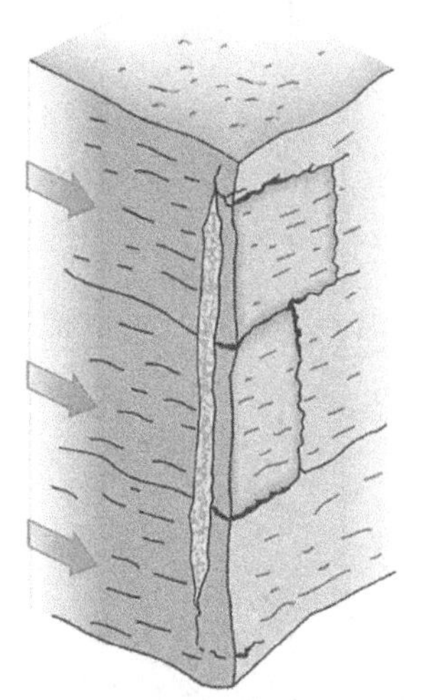

(c) Detached rock slabs begin to be displaced by growing crystals.

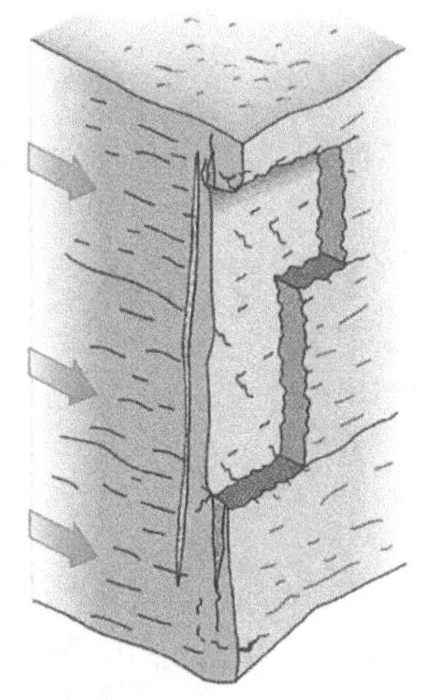

(d) Growing crystals, perhaps aided by other physical processes, cause rock slabs to be completely displaced. New layer of crystals begins to deposit below rock surface.

FIGURE 6.4 *Crystals that form and grow as groundwater evaporates just beneath the surface of sunlit outcrops are responsible for the spalling of the surface layers of rock.*

Oftentimes, local variations in rock composition result in "passageways" within the rock through which water is conducted to the surface more readily. At sites where these passageways reach the rock surface, the process of mineral growth is accelerated and causes the rock surface to retreat at a faster rate than the adjoining rock. The result is a weathering phenomenon called **honeycomb weathering** (Figure 6.5).

The effect of crystal growth can also be seen operating on the sunny side of stone walls or buildings. Water leaking from gutters may dissolve some of the mortar between the stone blocks. Driven by gravity, the water seeps out toward the face of the blocks and evaporates just beneath the surface, causing the rock surfaces to exfoliate in thin layers. If you observe the blocks closely, you can see the rock flakes being dislodged and just below, a white powder, most likely *gypsum* ($CaSO_4 \cdot 2H_2O$).

A related phenomenon can often be seen attacking the rocks exposed along the coast. Wind blowing onshore carries sea spray containing dissolved salts. As this moisture penetrates into the rocks along more porous zones or through cracks and evaporates, the pressure of growing salt crystals can loosen grains from the rocks. If you live near the ocean, you may try to find rocks that show evidence of physical deterioration due to the growth of sea salts.

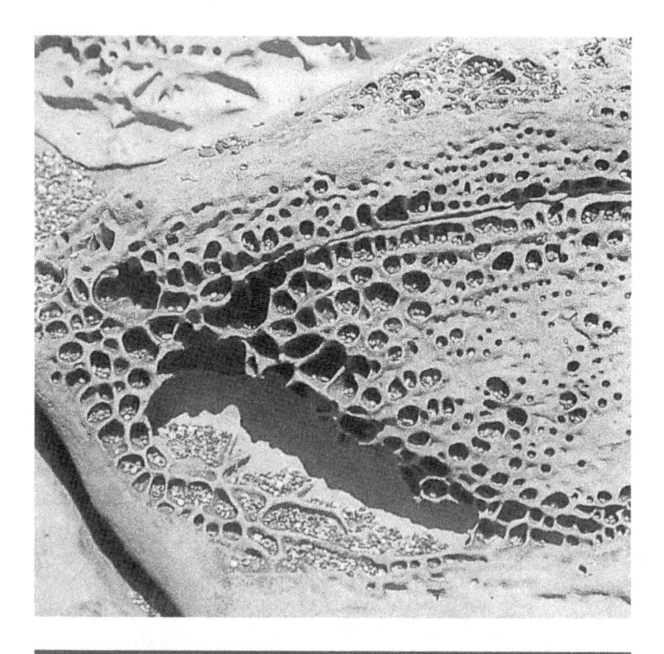

FIGURE 6.5 Honeycomb weathering *is a special type of weathering resulting from the subsurface growth of crystals in outcrops, usually sandstone, where variations in porosity control the amount of groundwater that is conducted to the surface.*

Sheeting Another kind of exfoliation that is especially well developed at the surface of exposed intrusive igneous rock bodies is called **sheeting.** As erosion removes the overlying rock and the stresses that accumulated within the igneous rock body during its initial emplacement and solidification within Earth's crust are released, cracks develop parallel to the rock surface. The rock surface may then fall prey to processes such as frost action, which removes the rock in layers or sheets. A well-known site that shows the process of sheeting is seen on a mountain in Yosemite National Park (Figure 6.6).

Spheroidal Weathering

The end result of weathering is the rounding of rocks. When first exposed by processes such as frost action, rock fragments of all sizes are invariably blocky with sharp edges and corners. Weathering processes preferentially attack the edges and corners, smoothing and rounding them (Figure 6.7). The same effect can often be seen where exposed rock masses are broken by intersecting sets of fractures called joints. As water penetrates the rock and weathering attacks the rock adjacent to the fracture, the rock at the intersection of the fractures is particularly vulnerable. With time, as the weathered portions of the rock are removed grain by grain, or by exfoliation (Figure 6.8), the shape of the rock mass tends to become increasingly spherical. Because of this tendency toward sphericity, the process is called **spheroidal weathering.**

Effect of Temperature

Daily temperature cycles in hot desert areas are commonly presented as being especially effective at promoting the grain-by-grain disintegration of rock. Every solid compound expands and contracts at a specific rate with changes in temperature. Theoretically, as a result of this expansion-contraction process, the minerals within the rock may break apart. However, in experiments subjecting a variety of rocks to ranges of temperature equal to those experienced on Earth, no disintegration was observed. Under normal ranges of daily temperature fluctuation, the differences in the thermal responses of minerals are insignificant. Rock disintegration due to daily temperature fluctuations therefore may be questioned as a mechanism of physical weathering.

FIGURE 6.6 Sheeting *is a type of exfoliation that is commonly associated with exposed granitic plutons. (Courtesy of N. K. Huber/ USGS)*

If a rock is subjected to an extreme temperature change, however, such as being heated by a forest fire or a bolt of lightning and then quickly cooled by a sudden downpour, the rapid contraction may cause it to break.

Plant Wedging

We should not forget to include the role of plants in our discussion of mechanical weathering. As plants grow and send their roots down in search of nutrients and water, they may enter cracks in rocks. As the plant roots extend and grow in diameter, they force the rock apart, a process called **plant wedging** (Figure 6.9). Plant roots are extremely efficient agents of physical weathering. Because their roots exude acids, they are also major contributors to certain processes of chemical weathering.

Before leaving the topic of mechanical weathering, take particular note of how water and changes in temperature are involved, either directly or indirectly, in nearly all mechanical weathering processes.

FIGURE 6.7 *Regardless of the size of the exposure, the combined efforts of mechanical and chemical weathering result in the* rounding *of rocks as sharp edges and corners are preferentially attacked.*

SPOT REVIEW

1. How do mechanical weathering and chemical weathering differ?
2. Why is the freezing of water such an effective agent of mechanical weathering? What are some common examples of this process?
3. Why are daily changes in temperature not thought to have any significant direct effect in mechanical weathering?

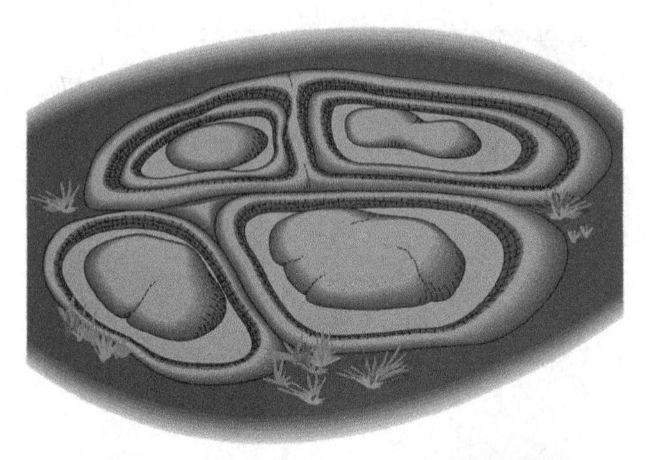

FIGURE 6.8 Spheroidal weathering *is the common result of the tendency of all weathering processes to eliminate sharp edges and corners from exposed rocks. (VU/© Paul Bierman)*

FIGURE 6.9 Plant roots *growing within cracks are extremely effective agents of mechanical weathering. Over time the roots result in the splitting of rocks by a process called* plant wedging. *(VU/© Nada Pecnik)*

CHEMICAL WEATHERING

Water and temperature also play major roles in all processes of chemical weathering. For example, very few chemical reactions go on without the intervention of water. If you have had a course in chemistry, think of all the laboratory experiments you have been asked to conduct. Very often, water was one of the reactants called for in the experimental menu. Similarly, few chemical experiments are conducted at room temperature; most are conducted at elevated temperatures. The importance of water and temperature will be borne out in the following discussions of the individual chemical weathering process.

Oxidation

A chemist might define the process of oxidation in several different ways. For our purposes, **oxidation** is any reaction combining an element and oxygen. An important point, however, is that, except when combustible materials are rapidly oxidized by fire, oxygen must be dissolved in water to be an effective oxidizing agent. A shiny iron nail kept in a closed container filled with oxygen will not rust. If the nail is thrown outside, however, and comes in contact with the moisture in the ground, rust will be visible within hours.

Of all the common rock-forming elements, iron (Fe) most readily reacts with oxygen. For this reason, oxidation is especially important in the chemical decomposition of the ferromagnesian minerals and mafic rocks. The process can be illustrated by the reaction of oxygenated water with the iron olivine, fayalite:

$$2Fe_2SiO_4 + O_2 + 4H_2O \rightarrow 2Fe_2O_3 + 2H_4SiO_4$$

fayalite + dissolved oxygen → hematite + silica in solution

The silica is shown as being removed in solution as silicic acid although chemists are not sure precisely how silica exists in solution.

A common product of oxidation is the material known as *limonite,* which is a variable mixture of iron oxides, usually *goethite* ($FeOOH$ or $HFeO_2$). Goethite is unstable and, in time, dehydrates to form hematite by the following reaction:

$$2FeOOH \text{ or } 2HFeO_2 \rightarrow Fe_2O_3 + H_2O$$

goethite → hematite + water

Once formed, the oxides of iron are extremely stable and accumulate with other products of weathering. Their bright colors are conspicuous, ranging from the red of hematite to the yellows and browns of the more hydrated forms. The iron oxides produced by chemical weathering are largely responsible for the colors of most sedimentary rocks.

Another reaction of environmental significance is the oxidation of the sulfide minerals, which always results in the formation of acid. The process can be illustrated with the reaction of oxygenated water with pyrite (FeS_2):

$$2FeS_2 + 7\,{}^{1}/{}_{2}\,O_2 + 4H_2O \rightarrow Fe_2O_3 + 4SO_4^{2-} + 8H^{1+}$$

pyrite + dissolved oxygen → hematite + sulfuric acid

The acid generated by the oxidation of the sulfide minerals is the source of the acid mine drainage that plagues the mining industry from the sulfide metals–mining regions of the western United States to the coal-mining areas of the east.

Dissolution

We usually do not think of rocks dissolving in the common sense of the word, but some rocks will dissolve almost instantly upon contact with water. For example, the salt with which we season our food may have existed as *rock salt* composed of the mineral *halite* ($NaCl$). Rock salt quickly dissolves in water. Combined with the fact that rock salt is not an abundant rock, its high solubility in water explains why we rarely see it exposed at Earth's surface except in desert areas where it may occur in and around seasonal lakes, salt licks, and salt domes (Figure 6.10).

Water dissolves more materials than any other known solvent. But before we discuss the process of dissolution further, we must consider another important characteristic of water that helps to explain its power as a solvent.

The formula for water is H_2O, which might seem to imply that a glass of water contains nothing except hydrogen and oxygen, that is, *pure* water. Pure water does not exist. Even the distilled water used in the chemistry laboratory is not pure water. As long as water is exposed to the atmosphere, it will contain all the atmospheric gases in solution, the "agents of the atmosphere" mentioned in one of the definitions

FIGURE 6.10 *Water-soluble salts can only occur in abundance in arid regions such as Death Valley, California.*

of weathering. We have already seen that oxygen dissolved in (reacted with) water is an effective oxidizing agent. The other important gas that dissolves in water is carbon dioxide, which reacts to form **carbonic acid:**

$$H_2O + CO_2 \rightarrow H_2CO_3$$

water + carbon dioxide → carbonic acid

Carbonic acid is a very weak acid—so weak, in fact, that you can bathe in it or drink it. But regardless of how dilute or weak it may be, carbonic acid is nevertheless an acid and dissociates to provide hydrogen ions and bicarbonate ions according to the following equation:

$$H_2CO_3 \rightarrow H^{1+} + HCO_3^{1-}$$

carbonic acid → hydrogen ion + bicarbonate ion

The reaction between a mineral and dissociated carbonic acid is called **carbonation.** An important example of rock dissolution by carbonation is the chemical weathering of limestone. Limestone is largely made up of the carbonate mineral, calcite ($CaCO_3$). All carbonate minerals are acid soluble. When limestone is subjected to a dilute solution of carbonic acid, it dissolves according to the following equation, which summarizes several intermediate reactions:

$$CaCO_3 + H^{1+} + HCO_3^{1-} \rightarrow Ca^{2+} + 2HCO_3^{1-}$$

calcite + dissociated carbonic acid → in solution

Although calcite dissolves at a much slower rate than rock salt, the rate is sufficiently fast for limestone, newly exposed in a quarry or as a tombstone, to begin showing evidence of water dissolution within 10 or 20 years. Another impressive example of the dissolution of calcite is the formation of caves and caverns (Figure 6.11).

A very spectacular, but unfortunate example of acid dissolution is the deterioration of antique statuary where limestone and marble (a metamorphic rock also composed of calcite) are being attacked by rain acidified with strong acids such as sulfuric and nitric acid. These acids are created by the reaction between rainwater and the sulfurous and nitrous oxide gases produced by the burning of fossil fuels by industry and motor vehicles. Priceless artwork that would have survived for thousands of years is now being destroyed within a few decades by the as yet uncontrolled atmospheric pollution of modern civilization.

Although it is important to note that human activity has substantially increased the gases in the atmosphere and that dissolve in water to form "acid rain," it is equally important to appreciate that for billions of years the rainwater falling to Earth's surface has contained carbonic acid.

Carbonation/Hydrolysis

As we saw previously, *carbonation* refers to chemical processes involving the reaction between a mineral and dissociated carbonic acid. **Hydrolysis** is a decomposition reaction involving water. Of all the chemical weathering processes, the combination

FIGURE 6.11 *Limestone* caves and caverns *are among the most spectacular products of limestone dissolution. (Courtesy of W. B. Hamilton/USGS)*

of carbonation and hydrolysis is perhaps the most important because it is the chemical process by which the major rock-forming minerals, the silicates, decompose. The combined process of carbonation/hydrolysis may be represented by the following chemical equation:

$$2KAlSi_3O_8 + 2H^{1+} + 2HCO_3^{1-} + HOH \rightarrow$$

potassium feldspar + dissociated carbonic acid + water

$$Al_2Si_2O_5(OH)_4 + 2K^{1+} + 2HCO_3^{1-} + 4H^{1+} + SiO_4^{4-}$$

clay minerals + metal ions and silica in solution

In this equation, potassium feldspar is the representative silicate mineral, but it could be replaced with almost any common silicate mineral except olivine and quartz to illustrate the process. Olivine, $(Fe,Mg)_2SiO_4$, which decomposes by oxidation and carbonation, will not weather to produce aluminum silicate (the clay minerals) because it does not contain aluminum. The other ferromagnesian minerals may contain aluminum and therefore have the potential to produce aluminum silicates by carbonation/hydrolysis as well as iron oxides by oxidation. Quartz is chemically stable because it does not react with dissolved carbon dioxide.

In the carbonation portion of the process, the bicarbonate ion reacts with whatever major metal ion is contained within the mineral to form a soluble bicarbonate of that metal ion. In this case, the metal ion is K^{1+}. If the mineral had been one of the plagioclase feldspars, the metal ions would have been calcium or sodium ions (Ca^{2+} or Na^{1+}). Once the metal ion, the "glue" with which the crystal structure is being held together, is removed, the structure begins to disintegrate.

Hydrolysis then takes over as water reacts with the remaining silicate framework to form an important group of silicate minerals, the *clay minerals.* Generally defined as hydrous (water-containing) aluminum silicates, the clay minerals are represented by the formula for the clay mineral *kaolinite.* Later, we will see that the clay minerals are major constituents of soil. As was shown in the equations describing the oxidation of olivine, any remaining silica is usually removed in solution, probably as silicic acid, H_4SiO_4. In summary, we can say that the carbonation/hydrolysis of almost any common silicate mineral except olivine and quartz will reduce the original mineral to clay minerals and a variety of soluble materials that are carried off in solution.

SPOT REVIEW

1. Why are the ferromagnesian minerals so susceptible to the chemical weathering process of oxidation?
2. Describe the chemical weathering process by which most silicate minerals decompose.

RATES OF CHEMICAL WEATHERING

Up to this point we have not considered the rate at which the chemical reactions may proceed. Common experience tells us that some reactions are slow while others are fast. Gold, for example, reacts with the atmosphere so slowly that it will retain its luster for centuries. Silver, on the other hand, will tarnish so quickly that silver table settings usually need to be polished between uses. Rocks are much the same, in that they decompose at widely differing rates depending upon their own characteristics and those of the environment to which they are exposed. We will now consider the main controls of chemical weathering.

Temperature

Certain instructions included in the menu of a chemistry experiment are designed to affect the rate of the chemical reaction. In many experiments, for example, you are told to assemble the ingredients in a beaker or a test tube, which you then heat over a Bunsen burner. The reason for heating the ingredients is simply to increase reaction rates and shorten the experiment. In most cases, the reaction would have occurred without the application of heat, but the *rate* of the reaction would have been so slow that it might have taken years to complete.

The rates of most chemical reactions increase with an increase in temperature. With one notable exception, which we will discuss later, this is also true for the chemical weathering of rocks and minerals. Due to this dependence of chemical reactions on temperature, chemical weathering proceeds at the fastest rates in the tropics, decreases through the temperate areas, and occurs at the slowest rates in the polar regions.

Availability of Water

Our discussions of both mechanical and chemical weathering have emphasized the importance of water. Because of this dependence on the availability of water, areas of high rainfall show the highest rates of chemical weathering while areas of low water availability have correspondingly low weathering rates. Once again, the highest rates of chemical weathering are found in the tropics where yearly rainfalls are measured in hundreds of inches or centimeters. Combined with the high tropical temperatures, it is no wonder that rock outcrops in the tropics are conspicuous by their absence.

Conversely, chemical weathering is relatively slow in desert areas because of the scarcity of water. This explains why the remains of iron tools of the early pioneers are found throughout the deserts of the southwestern United States; although rusted, they are reasonably intact after lying on the desert floor for a hundred years or more. An equally impressive, though unfortunate, example of the ability of the desert to preserve metals is the story of the "Lucky Lady," a World War II bomber that ran out of fuel and crash-landed in the Sahara. Twenty years later, a geologic mapping crew found the aircraft in almost new condition with virtually no indication of weathering; it had been protected from oxidation by the lack of water.

Because temperatures are low and the available water is confined in ice, rocks exposed in the polar regions show the least effect of either mechanical or chemical weathering of any place on Earth. For this reason, geologists who are interested in observing rocks with minimal alternations due to weathering have gone to the ice-free regions of the Antarctic.

Particle Size

Another common instruction in chemistry experiments is to reduce all solid materials to a powder by grinding. For any mass of material, as the particle size decreases, the surface area increases (Figure 6.12). The rates of all chemical reactions involving solids are directly proportional to the total surface area of the solid components. For this reason, all processes of mechanical weathering serve to increase the rates of chemical weathering by reducing the particle size of rocks.

The effect of particle size on the rate of chemical weathering also means that, all other things being equal, fine-grained rocks should weather faster than coarse-grained rocks. Actual field observations, however, usually indicate that this is not always the case. Granite, for example, is often observed to disintegrate and decompose to a loose assemblage of sand grains faster than its fine-grained equivalent, rhyolite. The reason for this seemingly anomalous situation is that the weathering of the coarser-grained granite removes relatively large crystals of feldspar and ferromagnesian minerals, thereby creating larger holes and passageways than will be formed by the weathering of the finer-grained rhyolite. As a result, water penetrates deeper into granite and enhances both chemical and physical weathering processes.

Composition

If one parameter had to be chosen as the prime controller of chemical reaction rates, it would be composition. Because of fundamental differences in composition, when exposed to the atmosphere, a shiny new iron nail will rust quickly while silver will develop a coating of tarnish much more slowly, and gold will remain bright and shiny for centuries. The same is true of rocks and minerals. To understand the effect of mineral composition on chemical reaction rate, let's consider the minerals that make up the great volume of the rocks of Earth's crust, the silicates.

In the discussion of igneous rocks in Chapter 3, the major rock-forming silicate minerals were divided into two groups, ferromagnesian and nonferromagnesian, and listed in order of their crystallization

(a) Number of cubes = 1

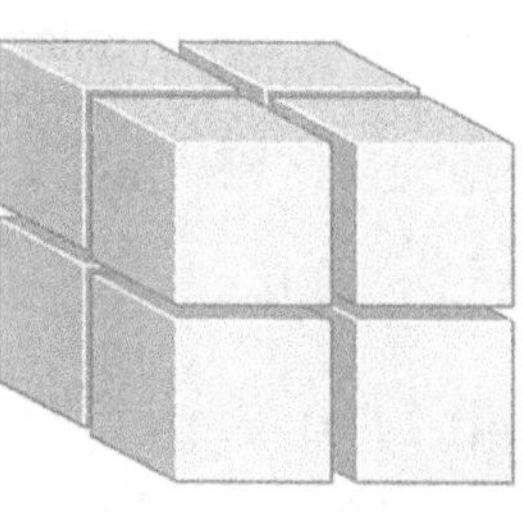

(b) Number of cubes = 8
Increase in surface area = 2×

(c) Number of cubes = 64
Increase in surface area = 4×

FIGURE 6.12 *As the particle size of solids is reduced by either mechanical grinding or mechanical weathering, the material experiences an increase in the total surface area and, thereby, in the rate of chemical reactivity.*

temperatures. Let us now consider these same minerals in terms of their relative rates of decomposition when affected by a combination of oxidation and carbonation/hydrolysis.

If one were to experimentally determine the relative reaction rates at which the major silicate minerals succumb to oxidation and carbonation/hydrolysis by reacting them with air-saturated water and observing the rate at which soluble ions form, the ranking of the minerals based upon the observed chemical reactivity would be generally the same as the order of crystallization from a melt (Table 6.1). The minerals that crystallize *first* at the *highest* temperatures will show the *highest rates* of chemical weathering. Those minerals that crytallize *last* at the *lowest* temperatures will show the *slowest rates* of chemical weathering. Most of the major silicate minerals in igneous rocks, except for quartz, decompose by a combination of oxidation and carbonation/hydrolysis to form iron oxides and/or clay minerals, indicating that clay minerals and the oxides of iron are stable minerals at atmospheric temperature and pressure.

Of the common rock-forming silicate minerals, quartz is the most resistant to chemical weathering. Probably the main reason quartz exhibits such low rates of chemical reactivity is that its crystal lattice does not contain metal ions such as sodium, potassium, calcium, magnesium, and iron that tend to react readily with either dissolved oxygen or the bicarbonate ion produced by the dissolution of carbonic acid. Another reason for the chemical stability of quartz is that it can crystallize at surface temperatures. Although most of the silica precipitated from solution at atmospheric temperatures consists of amorphous opaline materials ($SiO_2 \cdot nH_2O$ where n is the number of water molecules), quartz crystalizes in modern marshes and swamps from silica placed into solution by the carbonation/hydrolysis of other silicate minerals and from the decomposition of plants. Crystals grow and become embedded in the edge of "saw grass" where they may cut the skin of some unwary intruder. Crystals also precipitate within empty stem cells of certain aquatic plants to produce what are commonly called "scouring rushes." In fact, the name *scouring rush* comes from the early settlers who mashed rush stalks and used them as homemade "Brillo pads."

In summary, the major solid products of weathering include (1) the *clay minerals,* (2) residual *quartz,* (3) *oxides of iron* from the chemical weathering of the silicate minerals, and (4) *rock fragments* remaining from mechanical weathering.

Metastable Materials

Before leaving the topic of chemical weathering, a comment is in order concerning minerals such as graphite, diamond, and gold that form at high temperatures and pressures, yet can exist indefinitely at Earth's surface. These minerals survive the attack of weathering because, like quartz, they do not react with either oxygenated or carbonated water at any significant rate; as a result, they remain inert over geologic periods of time. A chemist would say that these materials have extremely high **activation energies,** meaning that unless they exist in an environment of very high energy, they will be reluctant to react with any other element. An example of an environment with levels of energy sufficiently high to result in such reactions is the interior of a mass of molten rock. The amount of energy available at Earth's surface is nowhere near sufficient.

TABLE 6.1
GOLDICH WEATHERABILITY SERIES

High Temperature	Olivine		First to crystalize
	Augite	Anorthite	
	Hornblende		
	Biotite	Albite	
	Orthoclase		
	Muscovite		
Low Temperature	Quartz		Last to crystalize

SPOT REVIEW

1. Why is water essential for nearly all chemical weathering processes?
2. Although chemical reaction rates increase as particle size decreases, why are some fine-grained rocks more resistant to chemical weathering than their coarse-grained equivalents?
3. Why are certain silicate minerals relatively resistant to the chemical weathering process of carbonation/hydrolysis?
4. Why is quartz, one of the most common silicate minerals, resistant to all processes of chemical weathering?

REGOLITH

Geology is the study of rocks, yet students often comment that when they look about, they usually don't see any rocks. They are quite correct. In most places, the rocks are covered with the products of weathering, commonly referred to as "dirt," a material more scientifically termed **regolith** (*rego* = blanket; *lith* = stone). Regolith is the loose material atop the bedrock; it combines the products of both mechanical and chemical weathering and consists primarily of rock fragments, clay minerals, and quartz. In addition, certain other minerals such as the iron oxides (Fe_2O_3) and oxyhydroxides ($FeOOH$), calcite ($CaCO_3$), and gypsum ($CaSO_4 \cdot 2H_2O$) may precipitate from groundwater, coat individual mineral grains and rock fragments, and fill spaces between the solid materials. These precipitated minerals may not only determine the color of the materials, but may also significantly influence their chemical characteristics.

The thickness of regolith varies. In areas where weathering rates are extremely slow or where the products of weathering are being rapidly removed, there may be little or no regolith and bedrock will be exposed. Because all weathered material, once formed, begins a downhill journey that eventually ends in the sea, regolith in general will be thinnest over hill tops and thickest over valley floors (Figure 6.13). The processes by which the regolith is moved will be considered later in this text.

In most areas, if the regolith is allowed to accumulate undisturbed by natural events such as landslides or human activities such as plowing, a **soil** will develop. Soil has been defined as "that part of the regolith that supports plant life" or "that part of the regolith down to the deepest penetration of plant roots." Soils will be discussed in the next chapter.

ENVIRONMENTAL CONCERNS

When we think of elements entering the environment, especially metals, pollution arising from human activities come to mind. No doubt, many human activities are responsible for the contamination of the environment. However, what we fail to consider is that elements are constantly being released into the environment as chemical weathering decomposes the mineral components of rocks.

With the exception of iron, the elements released by weathering in the greatest abundance *are light elements,* that is, elements with atomic numbers of 20 or less. These elements are among the elements that make up more than 90% of most living tissues; the list includes sodium, magnesium, potassium, calcium, carbon, hydrogen, nitrogen, oxygen, phosphorous, sulfur, and chlorine. The only heavy element associated with life that is released in abundance by chemical weathering is iron, a major ingredient in blood hemoglobin.

Trace Elements

Unless concentrated by mineralization, most elements with atomic numbers higher than 20, the *heavy elements,* are contained in rocks as *trace elements* with concentrations of less (often, much less) than 1 weight percent of the original rock. Once released into the environment by chemical weathering, however, trace elements can become concentrated by various geologic and biologic processes. For example, it is not uncommon for the concentration

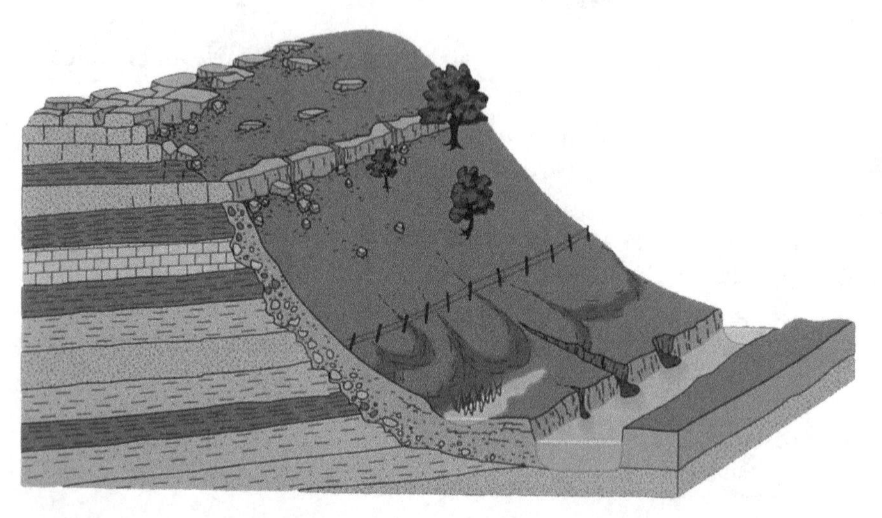

FIGURE 6.13 *Because gravity causes weathered material to move downslope, the regolith generally thickens from the hilltops to the valley floor.*

of an element in a sedimentary rock to be significantly greater than its concentration in the rock from which the sediment was derived.

Once released into the groundwater, trace elements may become concentrated on the clay minerals by the process of cation absorption. We will discuss cation absorption in the next chapter. As the soil is subsequently leached by acidic rainwater, these elements may be remobilized, become part of the groundwater, and pass up through the food chain, where they experience further concentration in plant and animal tissues. Eventually, these elements at their elevated levels may be ingested by humans as they eat the plants and animals.

Fluorine Many trace elements are beneficial to our well-being. A well-known example, flourine, increases the strength of bones and tooth enamel by increasing the crystallinity of apatite, a major constituent of both.

Most of us acquire our daily requirement of flourine by consuming water supplied by municipal water treatment plants where flourine is maintained at concentrations of about 1 part per million (10,000 ppm=1%). Over the past several decades, the daily consumption of flourine-treated drinking water has been responsible for significant reductions in the incidence of tooth decay in the United States and for a reduction in the number of bone fractures suffered by the elderly.

Selenium Some elements that are beneficial to life at trace abundances become toxic when ingested by animals at higher levels. An example is the metal selenium. The products of volcanic eruptions are the main source of selenium. Widely dispersed by the prevailing winds, volcanic ash and dust may become incorporated into soils where they undergo chemical weathering.

Once selenium is released from the volcanic materials, its mobility in the environment depends on the pH of the soil. In acid soils, selenium is highly insoluble and is therefore not available to either plants or animals. In alkaline soils such as those found throughout the Great Plains, however, selenium oxidizes to a highly soluble form and is taken up by the plants that serve as food for grazing animals. At concentrations above 4ppm, selenium becomes toxic to most animals.

Selenium is also concentrated in organic-rich rocks. This explains why elevated concentrations of selenium are commonly found in the acid mine drainage associated with the weathering of coal and coal-related rocks, especially when they are exposed to the atmosphere by mining.

A much-publicized example of pollution by high levels of selenium occurred in the San Joaquin Valley of California. The San Joaquin Valley is one of the most productive agricultural areas in the world. With an annual rainfall of less than 10 inches (25 cm), however, its productivity is dependent upon extensive irrigation.

One of the problems encountered in irrigated areas, especially where the downward percolation of the water is inhibited, is that the water levels may rise to the root zone of the plants and literally drown them. Such a problem exists in the western side of the valley where clay layers at shallow depths within the soil interrupt the percolation of the water. The solution to the problem is to install drains to carry the excess water off.

Another problem that occurs in the areas of low rainfall is that soluble salts accumulate in the soil. To keep the concentration of salts from rising to toxic levels, the soils are periodically flushed with large volumes of fresh water. The salt-laden water is then removed by the drainage system.

Between 1968 and 1975, a concrete canal was constructed to carry off the agricultural water from the valley. Originally, the canal was to empty into San Francisco Bay, but lack of funding prevented the canal from being completed, and the agricultural waters drained into Kesterson Reservoir. In the first years after the reservoir was built, when most of the water entering it was fresh, the Kesterson Wildlife Refuge purchased the water to establish a protected area for waterfowl. By 1981, however, the water entering the reservoir was the salt-laden agricultural drainage from the valley.

Among the elements that the irrigation water leached from the agricultural soils was selenium derived from the weathering of the rocks in the Coastal Range that forms the western margin of the valley. High levels of selenium were detected in fish from the reservoir in 1982, and by 1983, dead and deformed waterfowl chicks were being reported in the wildlife refuge. Suspecting selenium poisoning, scientists analyzed the waters in both the refuge and the drainage canal. Concentrations of selenium as high as 4,000 parts per billion (ppb) were reported in the drainage waters and up to 400 ppb in the ponds of the refuge. The Environmental Protection Agency limit for selenium in drinking water is 10 ppb. The Kesterson drainage water was declared a hazardous waste in 1985. The problem still remains with conflicting views as to how it should be solved. Even if a plan of action is eventually implemented that eliminates the selenium contamination of the agricultural waters draining into Kesterson Reservoir, the fact remains that selenium is still being leached from the rocks of the Coastal Range by natural weathering processes and may become a problem in other areas.

A problem faced by those who deal with the potential impact of specific trace elements such as selenium on human health is that insufficient data exist to determine the concentration at which a particular element stops being beneficial and becomes toxic. For example, while the concentration at which selenium becomes toxic to cattle has been well documented, the comparable concentration for humans has not been established, and the overall effects of selenium on human health are still not well understood.

Iodine In many cases, health problems develop in areas where there is a *deficiency* of a particular trace element. It is known, for example, that the soil and water in areas of the United States affected by the most recent Pleistocene glaciation are deficient in iodine. Deficiencies of iodine are directly related to the incidence of goiter, the swelling of the thyroid gland located at the base of the neck. It has also been shown that babies born to mothers suffering from iodine deficiency are more prone to exhibit stunted growth and mental retardation. Another medical problem that has been linked to the iodine deficiency of the same area is breast cancer, the leading cause of death from all types of cancer among women between 35 and 55 years of age.

Although there is little agreement as to why the soils in these areas are deficient in iodine, some geochemists believe that the deficiency is due to the fact that insufficient time has lapsed since the retreat of the ice and deposition of the moraines for adequate iodine to accumulate in the newly developed soils.

Zinc Another example of health problems related to elemental deficiency involves zinc. Zinc is considered an essential element for both plants and animals. Zinc deficiencies in plants cause a variety of plant diseases that result in a range of problems from low crop yields to total crop failure. In animals, zinc deficiency is linked to skin disorders, the slow healing of wounds, and disorders of bones and joints that have the potential to affect growth, especially among the young.

Water Quality

Unless strongly influenced by pollution generated by human activities, the chemical character of surface and ground water in any region is primarily due to the chemical weathering of the exposed rocks. For example, in regions immediately underlain by limestones, water is generally *hard* due to the high concentrations of calcium and magnesium while waters in areas deficient in limestones are generally *soft* because of the lack of these two elements. The original definition of water hardness is "the ability of the water to precipitate soap." If either calcium or magnesium is present in the water when soap dissolves, an insoluble precipitate forms and produces the familiar ring that collects around the washbasin or bathtub. Water softeners remove calcium and magnesium from the water by having them displace either sodium or potassium ions that are held on the cation exchange positions of resins specifically designed to attract and exchange the alkali and alkaline earth elements.

Water Quality and Heart Disease Studies have shown a higher number of deaths resulting from heart disease in areas characterized by soft water. The real significance of this correlation is debated, however. Some argue that the increased heart disease is not due to the drinking of soft water per se, but rather to heavy metals that the acidic soft water leaches out of the supply pipes.

Another example of a health problem associated with water quality has been documented in Ohio where a correlation has been shown between increased incidence of deaths due to heart attacks and elevated levels of sulfate ion (SO_4^{2-}) in the drinking water. The affected area is the coal-producing counties of southeastern Ohio where sulfate ion is generated by the weathering of the pyrite contained in the coal and coal-associated rocks. In comparison, the incidence of deaths due to heart attacks and the sulfate content of drinking water in the glaciated portions of northern Ohio are both lower.

As was the case with the correlation between heart disease and soft drinking water, some argue that the increase in the number of deaths due to heart attacks is not caused by the higher concentration of sulfate ion in the drinking water, but rather is the result of another component, or components, that are present or absent in the sulfate-rich water.

Other studies have shown lower incidences of death due to heart disease in areas where there are trace levels of certain elements. For example, fewer cases of heart disease are found in areas where the drinking water contains trace levels of manganese, chromium, vanadium, and copper. Presumably, deficiencies in these metals would result in an increase in the heart disease that promotes heart attacks. As these few examples clearly demonstrate, the relationship between heart disease and water quality is poorly understood at best.

Water Quality and Stroke A study in Japan reported a correlation between water chemistry and health. The study related the ratio of sulfate and bicarbonate ions (SO_4^{2-}/HCO_3^{1-}) in the water to the incidence of stroke. Stroke is a loss of body functions resulting from either a rupture or a blockage of blood vessels in the brain. In northern Japan, where the ratio was greater than 0.6, the death rate from stroke was more than 120 per 100,000 of population. In southern Japan, where the ratio was less than 0.6, the death rate was below 120 per 100,000 with many areas experiencing fewer than 80 deaths per 100,000 population. The increased sulfate content of the waters in northern Japan was attributed to the chemical weathering of abundant sulfur-rich volcanic rocks compared to southern Japan where more of the exposed rocks are sulfur-poor sedimentary rocks.

The Effect of Mining on Water Quality

Human activities commonly result in an increase in the concentration of metals in surface and ground waters. Notwithstanding both air and water pollution associated with a wide variety of industrial operations such as processing and manufacturing plants, the generation of large volumes of rock fragments by mining operations and the disposal of ore processing refuse contribute directly to increased production of metal ions by increasing the surface area of the rocks exposed to the processes of chemical weathering. This is especially true in areas where metallic sulfide minerals, such as those of lead, zinc, copper, and iron, are mined. The chemical weathering of sulfide minerals, especially pyrite, FeS_2, produces sulfuric acid that both increases the dissolution rate of metals and provides a chemical environment in which most metals remain in solution.

CONCEPTS AND TERMS TO REMEMBER

- Weathering
- Erosion
- Definition of weathering
  - disintegration
  - decomposition
  - mechanical weathering
  - chemical weathering
- Processes of mechanical weathering
  - frost action
  - freeze-thaw cycle
  - exfoliation
    - spalling
    - honeycomb weathering
    - sheeting
  - spheroidal weathering
  - plant wedging
- Processes of chemical weathering
  - oxidation
  - carbonation
    - carbonic acid
  - hydrolysis
- Metastable materials
  - activation energy
- Products of weathering
  - regolith
  - soil

REVIEW QUESTIONS

1. The major agent of physical/mechanical weathering is
   a. plant roots.
   b. daily temperature changes.
   c. freezing water.
   d. growing salt crystals.
2. Oxidation primarily involves the element
   a. silicon c. calcium.
   b. iron. d. sodium.
3. Water is such an effective agent of chemical weathering because it
   a. is present everywhere on Earth.
   b. contains dissolved oxygen and carbon dioxide.
   c. is itself a strong acid.
   d. can penetrate into small fractures.

REVIEW QUESTIONS *continued*

4. Which of the following igneous rocks would you expect to have the fastest rate of chemical weathering?
   a. granite c. basalt
   b. rhyolite d. granodiorite
5. Which of the following minerals would you expect to weather at the fastest rate?
   a. olivine c. quartz
   b. albite d. orthoclase
6. The major mineral that forms by the chemical weathering of most silicate minerals is
   a. quartz. c. clay.
   b. regolith. d. feldspar.
7. The term "regolith" refers to
   a. the accumulated products of chemical weathering.
   b. the unweathered rock that is actively undergoing weathering.
   c. the total accumulated products of weathering.
   d. the layers of rock that result from the process of spalling.
8. In which of the following areas would you expect the process of frost wedging to be most effective?
   a. the polar regions
   b. regions of temperate, humid climate
   c. deserts
   d. subtropical areas
9. What are the special properties of water that make its presence so vital to both physical and chemical weathering processes?
10. If one of the major controllers of chemical weathering processes is temperature, why are chemical weathering processes not nearly as important as physical weathering processes in deserts?
11. What makes granite deserve the accolade "rock of ages"?
12. Why is the order in which the silicate minerals crystallize from molten rock identical to the order of the rates at which they chemically weather?
13. Why is the mineral quartz excluded from the process of carbonation/hydrolysis that chemically decomposes the great majority of the silicate minerals?
14. Why is it significant that the clay minerals are the end product of the chemical weathering of most of the silicate minerals?

THOUGHT PROBLEMS

1. In many areas of northwestern North America and throughout much of Europe, the rates of chemical weathering are being accelerated by the presence of acid rain. What are the strong acids in acid rain and what are their sources? Short of totally eliminating the sources for these acids, what can be done to reduce the impact of acid rain? Why is the statuary throughout Europe so susceptible to the ravages of acid rain and what can be done to protect it from further damage?
2. Until the evolution of green plants, the composition of Earth's primeval atmosphere was quite different from that of the modern atmosphere. Consider what possible effects these compositional differences would have on both the kinds and rates of weathering.

   The composition of the present atmosphere is being affected by pollution. Assuming that we continue to modify the composition of the atmosphere, discuss possible effects the changes might have on the processes of weathering in the future.

FOR YOUR NOTEBOOK

Another field trip is in order to seek examples of the chemical and physical attack on rocks. Because the processes of weathering attack not only natural rock outcrops but also rocks exposed at the surfaces of buildings and other structures, a trip through your town should reveal numerous examples of both chemical and physical weathering. The same processes that attack rocks are effective at reducing construction materials such as concrete to rubble. All-too-familiar examples are the potholes and the slow destruction of concrete driveways. You might want to consider what steps can be taken to minimize these kinds of problems.

C H A P T E R 7

Soils

INTRODUCTION Few natural materials are more important to the existence of every terrestrial species than soil. Soils support the growth of the green plants that make up the basic components of the land-based food chain. Without these species, the remaining members of the terrestrial food chain, including humans, could not exist. The study of soils, a science called **pedology** (Greek: *pedon* = earth, soil), deals with the composition, formation, and distribution of soils and with their use and management. Soils often provide geologists with a means of determining the relative age of rocks, sedimentary deposits, and special landforms as well as providing information about past climate.

SOIL DEFINED

Soil scientists define **soil** as *"that part of the regolith that contains living matter and supports, or is capable of supporting, plant life out of doors."* You will remember that regolith is the layer of weathered material that accumulates above bedrock. A geologist might define soil as *"the outermost layer of the terrestrial crust composed of organic material, gases, water, and regolith"* (Figure 7.1). In this sense, soil is the overlap of Earth's biosphere, atmosphere, hydrosphere, and lithosphere; the combination is called the *ecosphere* (Figure 7.2).

An important point made by the definition is that soil is a *part* of the regolith. While regolith may exist without either the development of soil or the presence of plants, both regolith and plants must be present for soil to develop.

In warm, humid areas where the rates of chemical weathering are high, the regolith is generally thick, may be totally converted to soil, and supports a continuous plant cover. In drier climates, either warm or cold, both plant cover and soil development may be minimal.

FACTORS CONTROLLING SOIL FORMATION

Soil is the end product of a complex interaction between the minerals contained within the regolith and a variety of chemical, physical, and biochemical processes. Its development is affected by factors such as the parent material, the topographic relief of the area, the orientation of the slopes with respect to the Sun, and the climate, especially temperature and the availability of water.

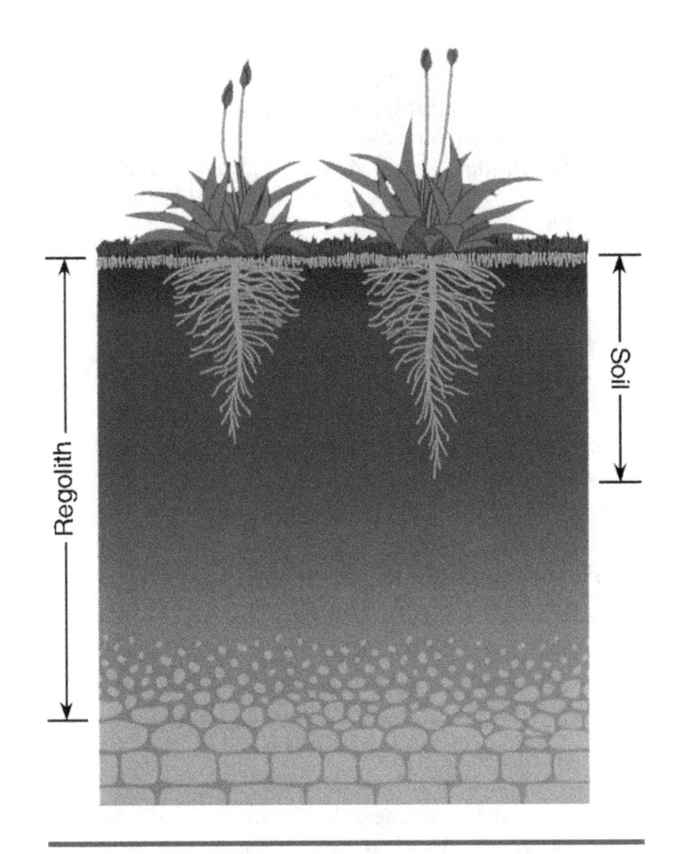

FIGURE 7.1 *Any accumulation of weathered material above bedrock is* regolith; soil *is that part of the regolith that supports plant life. When plant roots extend to bedrock, the entire regolith becomes soil.*

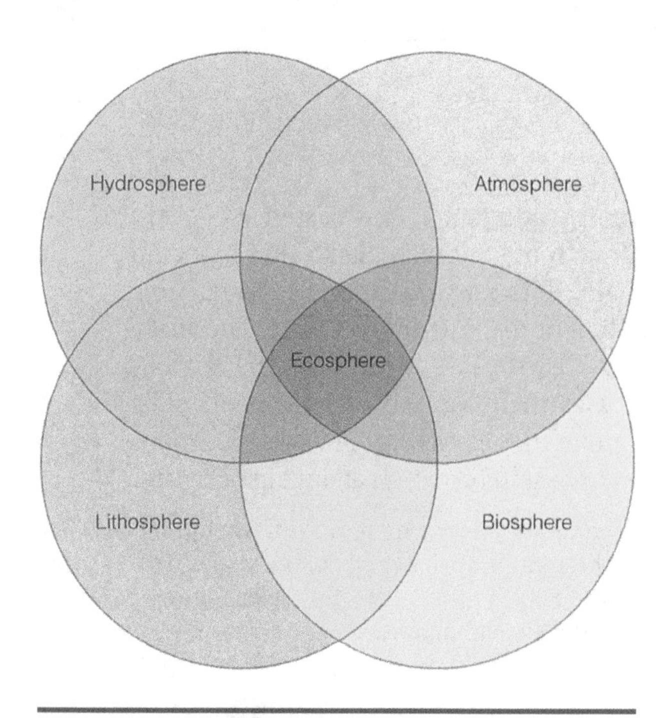

FIGURE 7.2 *The* ecosphere *is created by the interaction of the* atmosphere, *the gases that surround Earth; the* hydrosphere, *the water that exists on and within Earth; the* lithosphere, *the rocks of which Earth is made; and the* biosphere, *those regions of Earth that support life.*

Biological Factors

The biological processes are especially important in the formation of soil. Although the primary role of the soil is to support plant life, certain organisms, in particular, microorganisms such as bacteria, play an essential role in both the initiation and the continued development of the soil. As microscopic organisms decompose the organic debris accumulated at the surface of the soil, essential plant nutrients are released to be used by the existing plant population. Various chemicals, including organic acids, are generated by the biochemical activity and attack the particles of regolith as agents of chemical weathering. As larger plants become established and send their roots into the regolith, acids exuded from the root surfaces promote chemical weathering of the regolith; at the same time, the roots penetrate into fractures within the larger fragments of regolith where they serve as an agent of mechanical weathering, further preparing the regolith for soil formation. Thus, plants, once established, play a key role in the development of soil.

In addition to plants, certain animals play an important part in soil formation. Foremost among these are earthworms, which ingest the clay-rich component of soil in search of nutrients and chemically alter the material as it passes through their digestive track. The end product is an organic-rich material that possesses all the characteristics of a good fertilizer. In addition to modifying the chemical makeup of the soil, the activity of earthworms tends to loosen and stir the soil and in general makes it more porous and conductive for the penetration of plant roots, gases, and nutrient-bearing water.

To complete our discussion of organisms, we should mention the contribution of burrowing animals that overturn the soil within the area of their activity. Moles, voles, groundhogs, prairie does, and a host of other burrowers all contribute to the formation of this most important product.

Physical Factors

In addition to these biological factors, several physical factors affect soil formation including topography and climate. Topography affects the formation of soil in two ways: (1) by controlling the rate of downhill movement of the regolith and (2) by affecting the temperature and moisture content of the regolith.

On steep slopes, the downhill movement of materials under the force of gravity, a process called *mass wasting,* and the movement of a thin surface layer of water called *slope wash* combine to remove the smaller particles, leaving behind a layer largely composed of rock fragments. This coarse material requires additional weathering in order to provide both the finer particles required for soil development and the nutrients needed for initial plant growth. As we discussed in Chapter 6, the thickness of regolith generally increases with decreasing angle of slope. The thickest regolith is usually found over valley floors where the fine-grained materials removed from the slopes by mass wasting combine with those deposited on the valley floors by floods. Because these deposits commonly contain relatively high concentrations of clay minerals and because ample supplies of water are available, floodplains have been extensively used for crop production. In some of the more poorly drained areas within the floodplain, however, certain soil development processes may be inhibited, and as a result, agricultural soils may not be well developed.

The orientation of the slopes is extremely important in soil formation. Because slopes that face the Sun are warmer, the rates of all chemical processes tend to be higher. These include chemical weathering and the processes involved in the formation of soil, in particular, microbial activity that promotes the decomposition of organic matter and subsequent soil development. On the other hand, sunlit slopes are also drier, a condition that tends to retard chemical reactions including those that form soils. The effect of slope orientation on soil temperature and moisture is also apparent in the kinds of abundances of plants that populate the slopes. North- and south-facing slopes commonly have different plant populations whereas the orientation of east- and west-facing slopes has an intermediate impact upon soil development.

Climate

Of all the parameters affecting the development of soil, *climate* is the most important. Climate controls both the temperature and the availability of water, which, in turn, control not only the rates of weathering and erosion and the kinds of plants that will dominate, but also the rates of the various chemical soil-forming reactions. In general, the thickest soils form in the moist tropics with soil thickness decreasing toward the polar regions as both temperature and the availability of water decline.

Time

Time is a soil-forming parameter that is often overlooked. By now, you have likely been impressed with the importance of time in all geologic processes and appreciate that a slow process acting over a long period can perform incredible feats. In general, the formation of soil is a slow process usually requiring hundreds of years. In regions where the temperatures are high and water is abundantly available, less time is required for soil to form on newly generated regolith than in areas where temperatures are low and water availability is limited. For this reason, soils develop most rapidly in the tropics, decreasing in rate toward the poles. Note, however, that in many areas, such as the midwestern United States, soils that have taken as long as 10 million years to develop are now being removed by erosion faster than they are being formed.

SPOT REVIEW

1. What is the difference between regolith and soil?
2. In what ways do organisms, topography, and climate affect the formation of soil?

THE ROLE OF CLAY MINERALS

Because of the singular importance of the **clay minerals** in supplying plant nutrients, let us pause in our discussion of soils to consider the chemistry of the clay minerals in some detail. The clay minerals are silicates with sheet structures similar to those of the micas. The clay minerals are the only minerals whose crystal lattices do not achieve the status of internal *electric neutrality* that is required for most other crystalline compounds. Most clay minerals have a deficiency of positive ions within the crystal lattice. As a result, the individual clay mineral particles are negatively charged. A small percentage of clay minerals develop with an internal surplus of positive ions and are therefore positively charged. Because the negatively charged clay minerals are by far the more abundant, we will discuss them in detail. Once the chemistry of the negatively charged clay particles is understood, the comparable mechanism involving the positively charged clays should be evident.

Cation Adsorption

Even though the clay minerals crystallize without fulfilling the usual requirement of electric neutrality, they are not exempt from the rule that crystal lattices must be electrically neutral. If electric neutrality cannot be attained internally, it must be attained *externally*. Exposed to the dissolved components of soil water and groundwater, the *negatively charged* clay particles attract *positively charged* **cations**, which are *adsorbed on* (adhere to) the surfaces of the clay particles (Figure 7.3). The number and concentration of cations adsorbed to the surfaces of the clay particles depend upon the magnitude of the negative charge within the particles and the charge of the particular cations. The process by which cations are attracted and adhere to the surface of the clay minerals is called **cation adsorption**.

Cation Exchange Capacity

Cations will remain adsorbed to the surface of a clay particle until other cations appear in the groundwater with a higher concentration or a higher positive charge than those occupying the clay mineral adsorption sites. The adsorbed cations will then be *displaced* and *replaced* by the second cations (Figure 7.4). The process by which a cation on an adsorption site is replaced by another cation from solution is called **cation exchange**. Depending on the magnitude of the internal charge deficiency, individual clay minerals have a greater or lesser capacity to engage in the

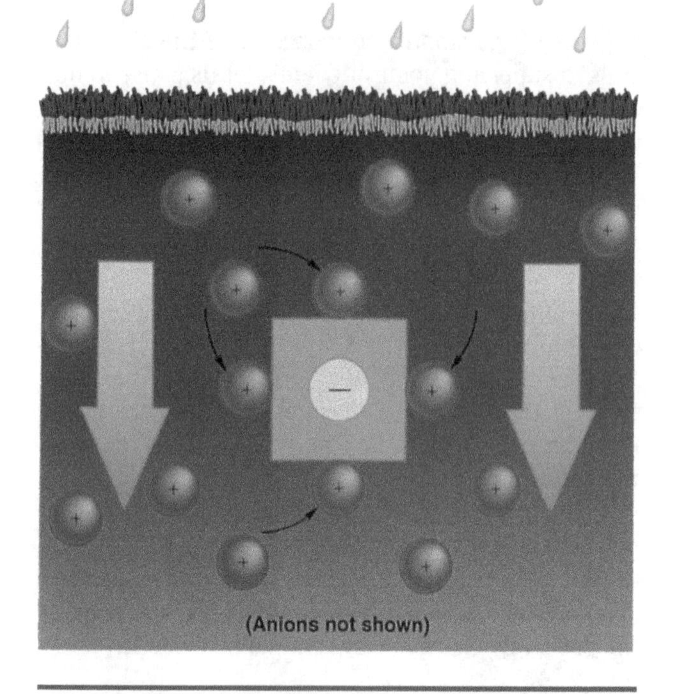

FIGURE 7.3 *During the process of* cation adsorption, *cations are removed from the groundwater and adsorbed to the surface of the clay particles to make up for the cation deficiencies that exist within the crystal lattices of most clay minerals.*

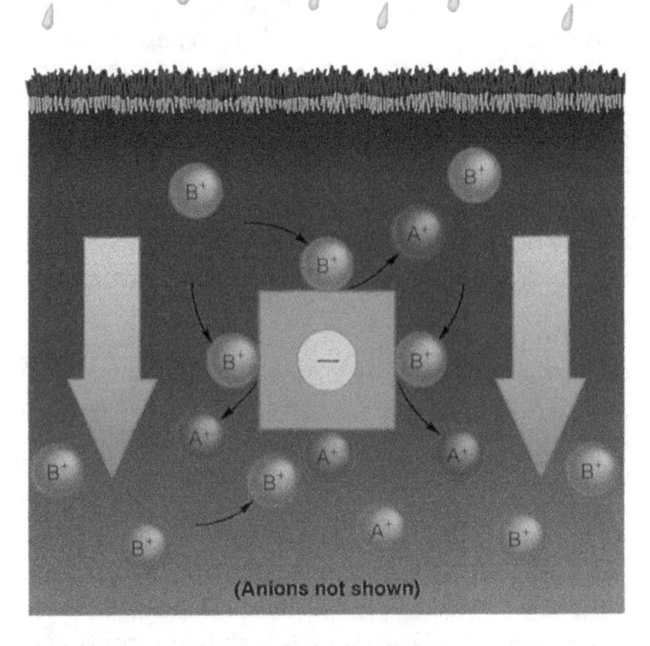

FIGURE 7.4 *Depending on their relative concentrations, cations carried by the groundwater may displace and replace those previously adsorbed to the surface of the clay minerals, a process called* cation exchange.

process of cation exchange. Through a laboratory procedure, soil scientists can determine the **cation exchange capacity (CEC)**, which measures the ability of a particular clay mineral to exchange one cation for another. The cation exchange capacity of the clay minerals is important because it provides a mechanism by which plant nutrients can not only be stored within the soil but also can be made available to the plants upon demand.

SPOT REVIEW

1. Explain what is meant by "cation exchange." What characteristics of the clay minerals favor its formation?
2. How is cation exchange involved in providing plant nutrients?
3. What role could cation exchange play in chemical weathering?

SOIL HORIZONS

If a soil is allowed to develop undisturbed, that is, without being overturned by a natural process such as a landslide or by human activities such as deep plowing, it will develop distinct layers called **horizons** (Figure 7.5). Horizons usually develop parallel to the surface of the soil and are characterized by distinctive compositions and textures.

The O Horizon

The uppermost horizon is called the **O horizon** (the O stands for *organic*). The O horizon consists of plant debris that accumulates on the surface, including the leaves that are shed each fall, dead grass, broken twigs, limbs, and fallen trees. Once accumulated, these materials are immediately attacked and decomposed by bacteria and fungi to gain nutrients for their own growth and energy needs. As part of the decomposition process, the microorganisms begin to convert the plant material, especially cellulose

THE INTERACTION BETWEEN SOIL AND PLANTS

To illustrate how plants and clay minerals interact, let us consider a situation that is the topic of numerous advertisements for soil additives. Soils formed on noncarbonate rocks in temperate areas that receive more than 20 inches (50 cm) or rainfall per year are said to be *"acid."* In the United States, this includes most of the country east of the Mississippi River and a small portion of the extreme Pacific Northwest (Figure 7.B1). But, what actually is an acid soil?

With more than 20 inches (50 cm) of rainfall per year, the dominant direction of water movement through the soil is downward. From our previous discussion of chemical weathering, you will remember that rainwater is really a dilute solution of carbonic acid. Therefore, the dominant cation being carried through the soil is the hydrogen ion. As the acidified rainwater moves downward through the soil, the hydrogen ions displace whatever cations were being held on the adsorption sites of the clay particles. Through this process, the clay particles become *hydrogenated.*

Once the cation exchange sites are occupied by hydrogen ions, each clay particle could be presented by the chemical formula H^{1+} $(clay)^{1-}$. If this clay particle were placed into a solution containing other ions, the clay particle would *dissociate* and give up the hydrogen ion into solution by *exchanging* it for a different ion. To a chemist, one of the basic definitions of an acid is "any compound that will provide a hydrogen ion into solution." Because the hydrogen-clay compound may contribute a hydrogen ion into solution as a result of cation exchange, it is by definition an acid.

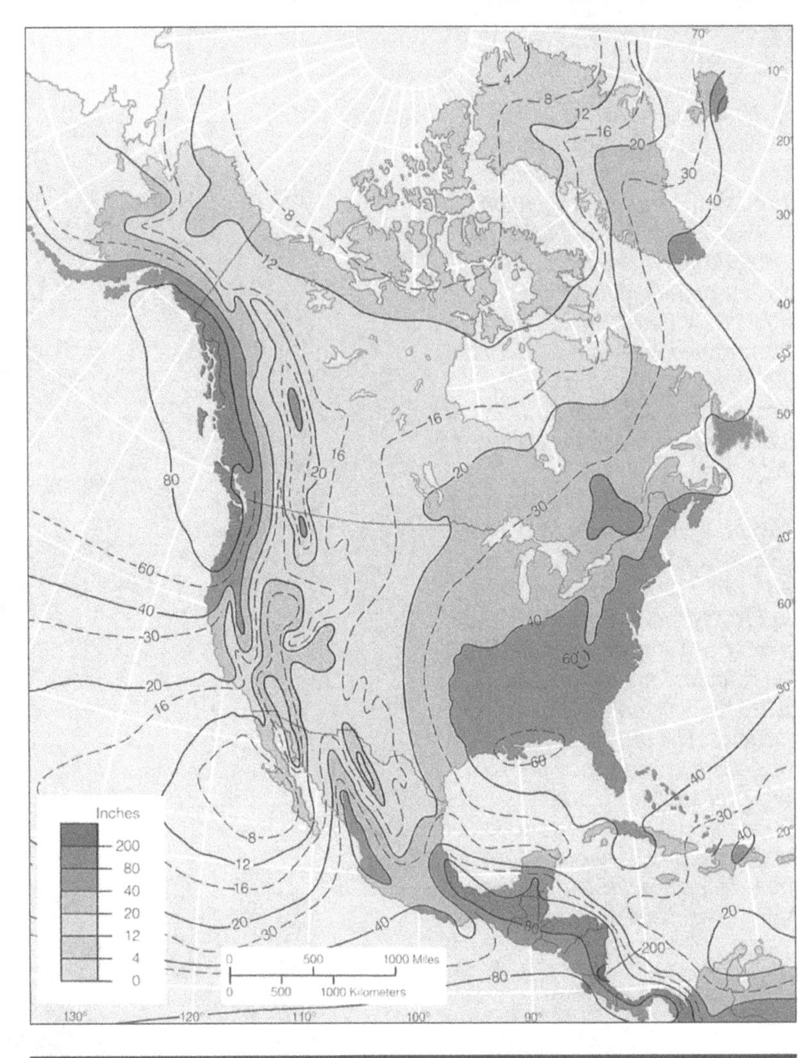

FIGURE 7.B1 *The distribution of rainfall throughout North America is the result of basic patterns of worldwide air movements.*

Advertisements for various kinds of agricultural products stress that acid soils are not necessarily the best for growing the kinds of plants we often like to see in our gardens or lawns. Grass, in particular, will not grow well in an acid soil. Consequently, advertisements for lawn products tell us that before grass will grow with any real success, the acid soil must be *neutralized,* and they suggest that this task can be accomplished by utilizing the *lime* that the manufacturer is trying to sell.

(continues on next page)

THE INTERACTION BETWEEN SOIL AND PLANTS *continued*

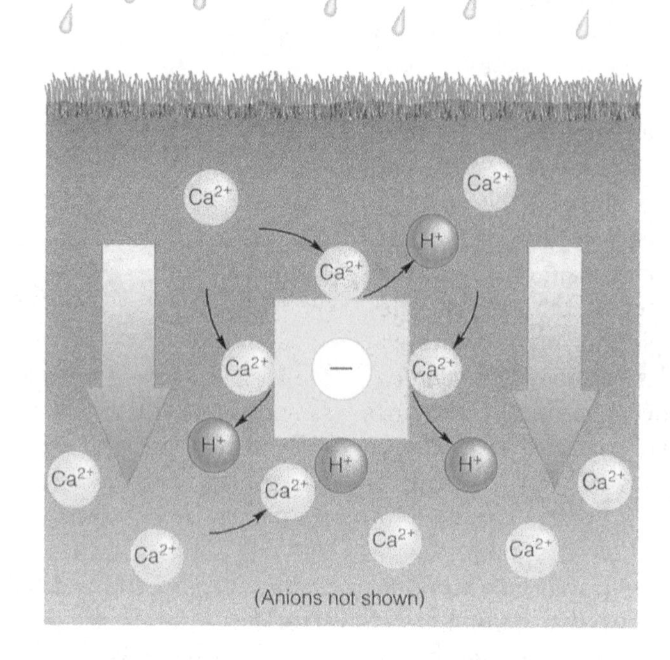

FIGURE 7.B2 *Rainwater reacts with carbon dioxide (CO_2) in the atmosphere and in soil humus, to produce carbonic acid. As the rainwater moves down through the soil, hydrogen ions produced by the dissociation of carbonic acid are adsorbed to the surface of individual clay mineral particles and neutralize the internal negative charge. Because the hydrogen ions can be released into solution, the clay minerals are sources of hydrogen ions. Thus the soil is acid.*

The surface of the lawn is treated with agricultural lime (powdered $CaCO_3$), which reacts with rainwater to produce Ca^{2+} and HCO_3^{1-} ions. Because of their greater concentration and charge, the calcium ions replace the hydrogen ions on the clay mineral particles. The displaced hydrogen ions combine with the hydroxyl ions (not shown) to produce water. Since the clay particles are no longer sources of hydrogen, they are no longer acid. The soil has been neutralized.

Agricultural lime is calcium carbonate ($CaCO_3$) and is manufactured by pulverizing limestone. The agricultural lime that the advertiser suggests that you spread on the ground dissolves with the next rain or in water provided by a lawn sprinkler to produce a solution concentrated in Ca^{2+} ions. As the solution moves down through the soil, the H^{1+} ions originally present on the surfaces of the *acid* clay particles are displaced and replaced by the Ca^{2+} ions (Figure 7.B2). Once the Ca^{2+} ions are adsorbed, the new clay compound can be represented by the formula Ca^{2+} $(clay)^{2-}$. Note that once the H^{1+} ions are replaced by the Ca^{2+} ions, the clay particles are no longer sources of H^{1+} ions and are therefore no longer acid. They have been neutralized.

Not only have the clay particles been neutralized, but the H^{1+} ions displaced from the clay particles combine with the $(OH)^{-1}$ ions formed by the dissolution of the lime to form water. At this point, the entire system has been neutralized.

The final question is, how does the neutralized soil provide the nutrients necessary for the growth of grass? Once again, the mechanism is an example of cation exchange. Calcium ions are important nutrients for grasses of all kinds. Once grass seeds germinate, the roots penetrate the soil seeking the Ca^{2+} ions adsorbed on the surfaces of the clay particles. In order to obtain the Ca^{2+} ions, the plant roots secrete carbonic acid. Hydrogen ions generated as the acid dissociates are exchanged for the Ca^{2+} ions located on the surfaces of the clay particles. The Ca^{2+} ions are then taken up by the plant for its continued growth. Note, however, that as the Ca^{2+} ions are progressively replaced by H^{1+} ions, the clay particles once again become *hydrogenated*. In other words, as the grass acquires the Ca^{2+} ions needed for growth, the soil becomes progressively more acid. Eventually, the grass will begin to suffer from a lack of nutrients until the situation is alleviated by applying more agricultural lime to the surface.

In general, the successful growth of any plant, other than those that grow naturally, requires that the chemistry of the soil be modified by the addition of the necessary nutrient ions. The negatively charged clay minerals are the recipients of the added positive ions such as Ca^{2+} and K^{1+} while the positively charged clay minerals are the storehouses for the negative functional groups such as phosphates and nitrates. Because the average soil contains significantly fewer positively charged clay minerals than the negatively charged varieties and thus has a low anion exchange capacity, negative ions must be added to the soil more frequently. As the plants consume the added nutrients, they must be periodically replaced with further treatments.

and lignin, into a variety of organic compounds that combine to make up a dark brown to black substance called *humus,* which accumulates and colors the upper portion of the soil. Humus is an important source of nutrients for many plants.

The A Horizon

Below the O horizon is the **A horizon,** which is a mixture of humus and chemically decomposed regolith, mainly clay minerals intermixed with quartz. The A horizon is usually black to dark brown in color due to its humic content. The combination of the O and A horizons is often referred to as *topsoil.*

The E Horizon

Below the A horizon is the **E horizon,** commonly referred to as the *zone of leaching.* Acids generated by the bacterial decomposition of the organic debris in the overlying O and A horizons move down through the E horizon layers and remove most materials except quartz. Consisting almost entirely of quartz, the E horizon is easy to recognize by its coarse texture and white to gray color. Because of its low clay mineral content, the E horizon has little nutrient-holding capacity.

The B Horizon

Below the O, A, and E horizons is the heart of the soil, the **B horizon.** The B horizon is commonly referred to as the *zone of deposition* because many of the materials leached from the overlying horizons accumulate here. In most soils, the B horizon contains the greatest concentration of the clay minerals, making it the major storehouse of plant nutrients. Other components of the B horizon are relatively small concentrations of iron oxides and oxyhydroxides, which are responsible for the brown to red to yellow colors of most soils.

Depending upon the type of soil, the B horizon may contain water-soluble minerals, most commonly, calcite. The presence or absence of calcite is important,

FIGURE 7.5 *When allowed sufficient time to develop, layers called* horizons *develop within the soil. Within any soil, the thickness, compositions, and colors of the individual horizons depend in great part on the* climate *of the region, in particular, the amount of rainfall.* (a *VU/© Doug Sokell;* b *VU/© Albert J. Copley)*

not only in determining the soil type, but also in establishing the chemical characteristics of the soil.

The C Horizon

The bottommost soil horizon is the C **horizon**, which is usually referred to as the *parent material.* The C horizon is the part of the regolith where the soil-forming process is initiated.

HORIZON DEVELOPMENT

The thickness and composition of the different soil horizons vary substantially from region to region. Here we examine not only the factors that contribute to these differences but also the types of soil horizons characteristic of regions ranging from the humid tropics to arid deserts.

The Influence of Rainfall

One of the major factors controlling the development of the individual soil horizons and the type of soil that subsequently develops is the *amount* of rainfall. Water moves both upward and downward within the regolith. The relative volumes of water moving in the two directions depend upon the amount of rainfall. In moist tropical areas where it rains nearly every day and the rainfall is measured in hundreds of inches or centimeters per year, the movement of water is totally downward. In humid, temperate areas receiving more than 20 inches (50 cm) of rainfall per year, the dominant movement is downward as rainwater makes its way to the water table. During the low-rainfall periods of the year, however, upward movement of groundwater by *capillary action* becomes increasingly dominant. Capillary action is generated when the attraction of a liquid to a solid surface is greater than the internal cohesion within the liquid itself. The force causes the liquid to rise along vertical surfaces. Drawn up by capillary action, the groundwater evaporates in the upper layers of the soil. In arid regions where the rainfall is less than 10 inches (25 cm) per year, the upward movement of groundwater followed by evaporation dominates over the downward movement of rainwater.

The Effect of Water Volume and Chemistry

The marked differences in zonation and composition that evolve within a soil profile depend upon both the relative volumes and the compositional differences of the rainwater and groundwater moving within the regolith. Rainwater, as you will recall, is a dilute solution of carbonic acid. In many areas, the natural acidity of the rain has been significantly increased by the introduction of strong acids formed from the reaction of rainwater and the oxides of sulfur and nitrogen introduced into the atmosphere by industry, power-generating stations, and motor vehicles. These acidic solutions, the so-called *acid rain,* leach most soluble salts from the soil profiles and may render the soil exceptionally **acid.** While acid rain may be causing some soils to turn more acid, it is important to point out that not all soils are becoming acid because of acid rain. Neutral or calcite-rich soils may be only marginally affected by acid rain, if at all, while some soils turn acid naturally regardless of the rain quality.

In contrast to the acidic downward-moving rainwater, the upward-moving groundwater contains the soluble products of chemical weathering and is mostly neutral to alkaline. As groundwater evaporates in the soil, salts of the **alkali** (Na^{1+} and K^{1+}) and **alkaline earth** (Ca^{2+} and Mg^{2+}) elements are deposited and become part of the soil composition. Within any region, the volumes of these two sources of water moving within the soil establish a *balance* that, combined with their respective chemical components, profoundly affects the character of the resultant soil.

In the subtropical and humid, temperate areas where the annual rainfall exceeds 20 inches (50 cm), the downward movement of rainwater dominates over the upward movement of groundwater. The rainfall is sufficient to support a continuous plant cover, consisting of a wide range of plant types including trees. With abundant plant cover, both the O horizon and the humic A horizon are well developed. The downward-moving rainwater also produces a well-developed E horizon.

The B horizon is also well developed with abundant clay minerals to serve as potential storehouses of plant nutrients. Because of the dominant downward movement of rainwater, however, the clay particles are hydrogenated and the soil is acid. Consequently, these soils have to be neutralized by the application of lime before they can be used for most agricultural purposes, except for growing acid-tolerant tuber plants such as potatoes or carrots.

In semiarid regions that receive 10 to 20 inches (25 to 50 cm) of rainfall per year, the amount of rainwater moving downward through the soil during the rainy season is approximately offset by the amount of groundwater moving upward during the dry season. The upward-moving groundwater brings various ions, in particular, calcium ions that displace cations held in the exchange positions of the clay particles. As a result, these soils are self-neutralized and normally do not require any kind of pretreatment for agricultural use. These are some of the most fertile soils on Earth.

With such limited rainfall, large plants such as trees cannot be sustained. However, the amount of rainfall and the available nutrients, particularly the availability of the calcium ions, are perfect for grasses. These soils are the great **grassland soils.** The Great Plains of North America stretching from Oklahoma into south-central Canada, the pampas of South America, the plains of Ukraine, the veldt of Africa, and the great grassland of Australian interior all are based on these soils. The original grasses that dominated these soils were native grasses. With the coming of settlers, the natural grasses were plowed under and replaced with other grasses such as wheat, oats, and rye. With the exception of rice, most of the world's grains are produced by these soils, and they are therefore often referred to as the *breadbasket soils.*

Another characteristic of these soils is their black color. Because the dense grasses die from one season to the next, plant parts both above and below the surface decompose and generate humus, which is distributed throughout the entire soil profile. To fly over the Great Plains during plowing season and see the endless stretches of coal-black soils awaiting the next growing season is always impressive. Under the old classification of soils, these soils were called *chernozems,* a Russian word meaning "black soil."

In arid regions where the rainfall is less than 10 inches (25 cm) per year, the development of soil is severely restricted if, in fact, a soil develops at all. Without seasonal accumulations of plant debris from a continuous plant cover, humus will be nearly nonexistent. Without the downward movement of rainwater, horizonation does not develop. In many arid regions, soil development is minimal, and the bedrock is either exposed or overlain by unaltered regolith.

In areas that have a covering of regolith, the extremely low humidity of the atmosphere initiates the dominant upward movement of groundwater. Subsequently, the water evaporates, resulting in the deposition of large volumes of salts within the regolith. In some areas, the accumulation of salts is so extensive that it forms a rock-hard layer up to several inches thick called a **hardpan.** Consisting mostly of the salts of the alkali and alkaline earth elements, these deposits render the regolith totally inhospitable for the growth of most plants. Only the desert plants are, by design, able to survive under these normally toxic conditions. Cellular adaptions within their roots allow these plants to filter out the excessive concentrations of dissolved ions.

In some desert areas where an ancient soil profile may exist from a more humid time, the concentration of toxic salts within the paleo-soil can be reduced by extensive irrigation that reverses the relative dominance of the upward movement of groundwater within the soil profile by providing artificial "rainwater." As the irrigation waters dissolve the salts and carry them off, the soil becomes agriculturally productive. In many desert areas around the world including the southwestern United States and the Middle East, irrigation has converted deserts into highly productive agricultural regions.

High levels of desert irrigation are not without potential problems, however. As the irrigation waters dissolve and flush the toxic salts from the soil, the salts may enter both the groundwater and streams, rendering them too saline for use. In other areas such as the Imperial Valley of California, soils have become saline due to the intense evaporation of water containing trace quantities of salt added to the soil by irrigation.

Tropical Regions

In tropical areas where the rainfall is very high, only downward movement of rainwater occurs. The superdominant downward movement of rainwater leaches the soils and, combined with the high temperatures, develops a unique soil profile. Little O horizon exists due both to extreme chemical and biochemical attacks upon the plant debris and to the fact that tropical plants acquire their nutrients directly from the developing humus. Clay minerals, which are dominant ingredients in most soils, are nearly nonexistent in tropical soils. Although generally considered to be chemically stable at surface conditions, under the extreme weathering conditions that exist in the tropics, the clay minerals decompose into the hydrated oxides of aluminum (*alumina*) and silicon dioxide (*silica*). In some areas, the accumulations of hydrated alumina are so concentrated that they constitute a commercial aluminum ore called *bauxite.* Silica precipitates as quartz and adds to the general low fertility of the tropical soil. Oxygenated water precipitates iron, if present in the soil, as hematite, which is responsible for the characteristic red color of the tropical soils. When iron is not present, tropical soils are usually yellow. As is the case with alumina, the concentrations of hematite in some tropical areas may become sufficiently high to be considered a commercial iron ore.

Although the casual observer might think the lush tropical rain forests reflect high soil fertility, such is not the case. The fertility of tropical forest soils is actually very low. Most plant nutrients are leached from the soil by the high downward flows of acidic rainwaters. Tropical plants acquire their nutrients either directly from the decomposing humus or from mineral dust that falls on the surfaces of their leaves.

In many tropical areas, rain forests are being cut down in order to obtain agricultural and grazing lands. Unfortunately, the soil is generally not suited to growing the high-protein-yielding plants that are subsequently planted. In only a few growing seasons, the limited concentration of plant nutrients in the original soil in the form of minerals or humus is exhausted, necessitating the cutting down of more rain forest. It is estimated that rain forests are being destroyed at the rate of an acre per second. Unfortunately, these lush forests not only serve as a source of atmospheric oxygen but also play a role in maintaining the carbon dioxide content of the atmosphere. Carbon dioxide, you will remember, is one of the major components controlling the atmospheric temperature via the greenhouse effect. The reduction in the areas covered by rain forests will certainly add to the already increasing greenhouse effect.

SPOT REVIEW

1. Compare the various soil horizons.
2. Under what conditions do soils become "acid?"
3. What characteristics of semiarid soils make them especially good for the production of grains such as wheat and corn?
4. Why are tropical soils ill-suited for agriculture and the grazing of cattle?

TYPES OF SOIL

Although no scheme for soil classification is universally accepted, the system that has been widely used since 1975 is **soil taxonomy**. This system is based upon the physical, chemical, mineralogical, and morphological properties of the soil, which are in turn related to the processes under which the soil formed. The classification scheme utilizes a hierarchy beginning with orders subdivided into suborders, great groups, subgroups, families, and series. As in most hierarchal classification systems, the number of components increases exponentially with each subdivision. At the series level, for example, there are about 14,000 different kinds of soils. We will only consider the 11 basic orders. The suffix, *-sol,* in each of the orders is derived from the Latin word *solum* meaning soil. A simplified soil map of the United States showing the regional distribution of the basic soil orders is shown in Figure 7.6. The basic soil orders are summarized in Table 7.1.

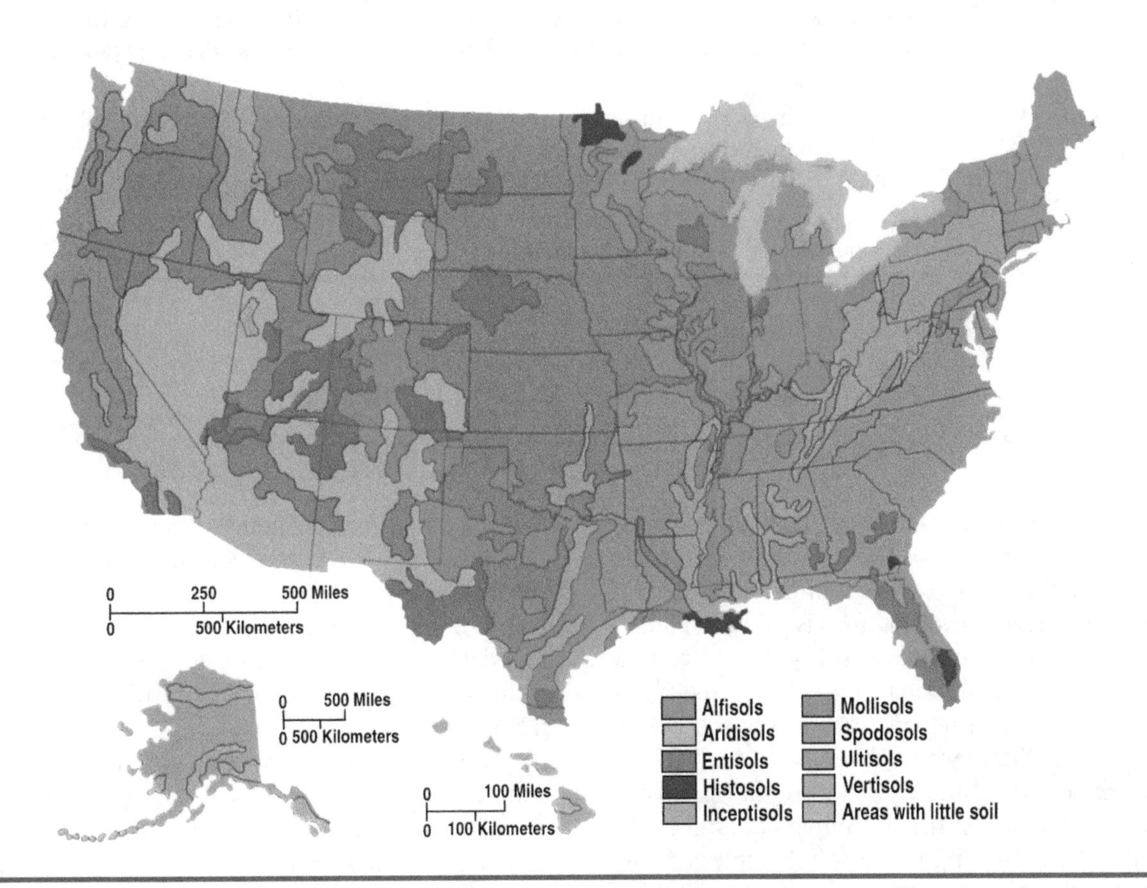

FIGURE 7.6 *The distribution of the basic soil orders throughout the United States reflects the regional variation in climate.*

TABLE 7.1

THE BASIC SOIL ORDERS	
ENTISOLS	Entisols are regolith that is being subjected to the very beginnings of soil fermentation.
INCEPTISOLS	Inceptisols show the beginnings of horizonation and include materials accumulated on steep slopes, floodplains, and the surface of slowly weathering bedrock.
ARIDISOLS	Aridisols form in areas of low rainfall and are characteristically composed of rock fragments with little or no humus. High levels of salts may be present in some aridisols, rendering them highly alkaline.
MOLLISOLS	Mollisols generally develop in areas receiving 10–20 inches (25–50 cm) of rainfall per year. Vegetation is continuous and dominated by grass. Mollisols are usually neutral to slightly alkaline and are among the most fertile agricultural soils on earth.
SPODOSOLS	Spodosols develop under conifer forest cover in cool, humid areas and are characteristically acid. Spodosols can be converted to marginally fertile agricultural soils with proper lime treatment and the addition of nutrients.
ULTISOLS	Ultisols are similar to spodosols except that they develop under broadleaf forests in areas of higher mean temperatures. Ultisols may be converted to highly productive agricultural soils with appropriate lime neutralization and the addition of nutrients.
ALFISOLS	Alfisols are transitional soils between the mollisols of the grasslands and the spodosols and ultisols of the forested regions. Alfisols are widespread and are highly productive food producers.
OXISOLS	Oxisols develop in moist, tropical areas. They are the most highly weathered of all soils and commonly lack both clay minerals and humus. In general, they are quite infertile. In many cases, they consist of little more than quartz and the hydrated oxides of aluminum and iron.
VERTISOLS	Vertisols are unusual soils that develop in semiarid areas underlain by expandable clay minerals. The name refers to vertical cracks that develop in the soil during the dry season. As the soils wet and dry, they experience a natural mixing. When irrigated, vertisols are highly productive agricultural soils.
HISTOSOLS	Histosols develop in poorly drained areas such as swamps and marshes. They commonly contain peat, which when drained and dried can be used as a fuel or as a humic soil additive. When buried and subjected to millions of years of elevated temperatures, the peat may convert to coal.
ANDISOLS	Andisols develop on various kinds of volcanic materials such as lava or tuff. They are the newest soil order, previously being classified as inceptisols.

ENVIRONMENTAL CONCERNS

Although most environmental problems are perpetrated directly or indirectly by humans, some problems are the result of natural processes. The most serious problem involving soils is a case in point.

As Table 7.1 indicated, the vertisols are soils that contain appreciable contents of expandable clay minerals. In regions such as the southwestern United States where the soils are subjected to seasonal wetting and drying, the cyclic expansion and contraction of the clay minerals result in the natural overturning of the soil. Samples of pure expandable clays can expand up to 10 or 15 times their original volume as water is absorbed into the crystal lattices. Although vertisols may contain less than 5% expandable clay minerals by volume, the pressures that develop within the soil are often sufficient to cause damage to foundations of all kinds. In the United States, an estimated $2 billion of damage results each year from soil expansion.

The problems of expandable soils can be controlled by a combination of soil testing and proper construction methods. Because the expansion and contraction of the soil are critically affected by soil moisture, drainage of the water immediately adjacent to the foundation is essential.

FIGURE 7.7 *Poor agricultural practices have accelerated the loss of soil, especially during times of drought. (Bettmann Archive)*

In many cases, the foundation can be protected from the pressures generated by soil expansion by employing buffers such as crushed rock that both absorb the forces generated by expansion and prevent the soil from coming in direct contact with the surface of the foundation.

Although the problem of expandable soils stems from natural causes, other serious problems involving soils are attributable in varying degrees to human actions. The loss of soil by erosion, for example, is often the result of poor agricultural practices. The removal of a protective layer of vegetation combined with the loss of soil strength as a result of plowing and tilling accelerates the rates of erosion by both water and wind. The loss of soil from agricultural lands is aggravated by long periods of drought when the wind becomes increasingly effective. A case in point was the erosion that took place in the *Dust Bowl* of the central plains states during the early 1930s (Figure 7.7). The human impact was poignantly portrayed in John Steinbeck's *The Grapes of Wrath,* which describes the plight of a family of farmers driven from the land. During a single dust storm in 1935, an estimated 5 million tons of dust were in suspension over an area of about 30 square miles (78 km^2). Reports of nearly a foot of soil being removed within a single day were not uncommon. Although accelerated soil loss due to cultivation cannot be eliminated altogether, practices such as contour plowing and planting combined with the rotation of row and total coverage crops can alleviate the problem.

In many cases, overgrazing of pasture land has led to similar soil loss by erosion. The impact of too many animals, especially those that tend to close-crop the grass, during times of low rainfall can initiate the erosion of soil by destroying the protective layer of grass.

More recent sources of environmental damage are the estimated 12 million **off-the-road vehicles (ORVs)** that are driven through for-

ENVIRONMENTAL CONCERNS *continued*

ested areas, deserts, and dunes throughout the country. Aside from the direct damage caused by the passage of the vehicle, the disruption of the soil surface exposes the disturbed materials to increased erosion by water and wind. During periods of heavy rain, increased sediment loads clog streams and contribute to the infilling of ponds and lakes. As the popularity of ORVs increases, their encroachment into new areas poses a constant problem to those who manage public lands and dramatizes the clash between those dedicated to the preservation of wildlands and those who claim equal access.

Urbanization creates a host of environmental problems. Construction exposes soils to increased erosion. The near-continuous cover of pavement drastically changes the rates of runoff into streams and subsequent stream erosion. Growing populations increase the potential of soil pollution from improperly installed septic systems, leaking storage tanks, effluents from solid waste disposal sites, and the chemical treatments utilized by some homeowners to maintain perfect lawns. The adsorptive ability of the clay minerals renders the soil ready receptors and storage sites for a host of pollutants.

One of the most notorious cases of soil contamination is Love Canal. Located in Niagra Falls, New York, Love Canal was originally excavated in 1892 as a portion of a transportation system that was supposed to connect various industrial sites in the area. After the venture failed, the canal became the dumping ground for waste of all kinds. In particular, from the 1920s, the canal became the repository for a wide variety of chemical wastes, many of which were later determined to be carcinogenic. In 1953, the land was sold to the school board for the sum of one dollar, and homes were built on the old dump site. Unusually heavy rainfalls and snow melts in 1976 caused the toxic materials to surface. Vegetation in the area began to die. Rubber products such as the soles of shoes and bicycle tires began to disintegrate after coming in contact with the contaminated soil. Chemicals of unknown composition began to seep to the surface and collect in puddles.

The federal government bought the homes along Love Canal and relocated the residents. Subsequently, $20 million was made available to more than a thousand current and former residents to compensate them for medical expenses, property damage, and the costs of relocation. At the present time, nearly $100 million has been spent to clean up the area, and the task is not yet complete. How much the ultimate cost will be is anyone's guess. The real question, however, is what is being done to ensure that there will be no more Love Canals.

CONCEPTS AND TERMS TO REMEMBER

- Soil science
  - pedology
  - soil
- Clay mineral chemistry
  - clay particles and cations
  - cation adsorption
  - cation exchange
  - cation exchange capacity (CEC)
- Soil horizonation
  - kinds of horizons
    - O horizon
    - A horizon
    - E horizon
    - B horizon
    - C horizon
- Soil formation
  - rainwater versus groundwater
    - acid soils
    - alkali and alkaline earth elements
    - alkaline soils
      - grassland soils
      - superalkaline soils
        - hardpan
      - tropical soils
- Soil taxonomy
- Environmental problems
  - soil erosion
  - off-the-road vehicles (ORVs)
  - urbanization

REVIEW QUESTIONS

1. The specific source of the cation exchange capacity of the clay mineral is
   a. the crystal structure.
   b. the particle size.
   c. deficiencies of cations within the crystal lattices.
   d. a surplus of cations within the crystal lattices.
2. Except for tropical soils, plant nutrients are stored within the
   a. O horizon. c. B horizon.
   b. A horizon. d. C horizon.
3. In general, soils that develop in regions receiving more than 20 inches (50 cm) of rainfall annually will
   a. be acid.
   b. be thin.
   c. be saturated with water.
   d. have low cation exchange capacities.
4. Commercial deposits of bauxite, the major source of aluminum, are found in
   a. mollisols. c. entisols.
   b. oxisols. d. aridisols.
5. The soils that support the world's grasslands are the
   a. vertisols. c. inceptisols.
   b. mollisols. d. aridisols.
6. In terms of soil horizons, laterites are characterized by an overdevelopment of the _____ horizon.
   a. E c. B
   b. O d. A
7. Chernozems are important because they
   a. are the most widespread of all the various kinds of soil.
   b. support the world's grasslands.
   c. support the tropical rain forests.
   d. are the beginnings of soil development.
8. Our major source of aluminum, bauxite, develops in _____ climates.
   a. cold, wet c. hot, dry
   b. cold, dry d. hot, wet
9. Most of the soils in the eastern United States are
   a. mollisols. c. oxisols.
   b. ultisols. d. histosols.
10. Some desert soils are converted into productive agricultural soils by irrigation. What characteristics do many arid soils have that preclude their being used for agriculture without irrigation, and what changes are brought about by irrigation that allow these soils to become productive?
11. What is the cation exchange capacity of a soil? Explain what determines the magnitude of the CEC and how it is involved in the supply of plant nutrients.
12. Why is it common practice for homeowners in the eastern United States to spread lime on their lawns each year? Why don't homeowners in the Midwest do the same?
13. What are the major factors involved in the topographic control of soil formation?
14. How does climate affect the formation of soil?

THOUGHT PROBLEMS

1. Debate the statement that the introduction of clay minerals to tropical soils would allow them to be converted to permanently productive agricultural soils.
2. A conflict exists between the owners of ORVs who feel that they have a right to access to public lands and those who feel that ORVs should be banned because of the damage they cause. What steps could be taken to resolve the problem?

FOR YOUR NOTEBOOK

You will want to determine the kinds of soils in your area. Soil maps are available from the Soil Conservation Service (SCS) of the U.S. Department of Agriculture. If there is no local SCS office, copies may be available in your library or from the departments of soil science or agriculture of a university in your state.

Note the kinds of soils in your immediate area. Investigate the relationship between the local geology and topography and the distribution and types of soil.

If the soils in your area are used for agriculture, investigate whether specific soils are dedicated to certain crops and whether the soils require treatments before they are used. The general topic of chemical treatments of agricultural products and the potential that exists for soil and groundwater contamination is very controversial. You may want to find out whether there are any restrictions on the use of chemical additives in your area.

In light of our discussion of Love Canal, are there any local sources of soil contamination that may present health concerns? Not all sources of contamination are industrial. A growing problem in urban areas is the chemical treatment of lawns. In some areas, local governments have restricted and even prohibited the use of pesticides and herbicides. In other areas, treatments can only be applied if written releases have been acquired from neighbors.

CHAPTER 8

Sedimentary Rocks

INTRODUCTION **Sedimentary rocks** are composed of the transported and subsequently redeposited insoluble and soluble products of weathering. The products of weathering, when deposited by the various agents of erosion, are referred to collectively as **sediment.** Sediments include the insoluble products of weathering such as *rock fragments, quartz,* residual *feldspar,* and *clay minerals* as well as minerals precipitated from materials originally in aqueous solution.

Sedimentary rocks are the most abundantly observed rocks at Earth's surface. Common examples of sedimentary rocks—*shales, sandstones,* and to a lesser degree *limestones*—appear in most road cuts, cliffs, and valley walls. This abundant display is due to the fact that sedimentary rocks cover 75% of Earth's exposed land surface. Though they constitute most of the surface rocks, sedimentary rocks represent only 5% of the volume of Earth's crust. The reason for this apparent discrepancy is that sedimentary rocks form a thin veneer covering the crystalline core of the continents in much the same fashion as a thin layer of icing covers the much larger volume of a cake. The thickness of sedimentary rocks over most of the continental surface averages a few thousand feet or meters with the thickest sections being found either associated with foldbelt mountains, where the total thickness of sedimentary rocks may measure tens of thousands of feet, or in deep basins.

The geologic and economic importance of sedimentary rocks is disproportionate to their limited representation within Earth's crust. Sedimentary rocks allow geologists to achieve one of the major goals of the science of geology, namely, to interpret the history of Earth. Sedimentary rocks literally record the geologic (and biologic) history of Earth. The kind of sedimentary rock, the physical features found within the rocks, the occurrence of fossils, and the minerals composing the rocks record many of the events and environmental conditions in existence at the time the sedimentary materials were deposited. Because sedimentary rocks contain the major clues to unraveling Earth's history, a large portion of a geologist's education is spent studying them to obtain the tools needed to interpret the historical record they contain.

Economically, sedimentary rocks contain many of the natural resources needed for modern society, not the least of which are the sources of energy. The primary sources of energy in the world today, oil and gas, are largely contained within the sedimentary rocks. Coal is also found as layers within sequences of sedimentary rocks. Our present supplies of uranium are primarily extracted from deposits contained in sedimentary rocks. In addition to supplies of energy, sedimentary rocks provide us with a wide variety of other valuable and necessary commodities such as iron and copper ore, bauxite for aluminum, basic building materials such as sand, clay, and stone, raw materials for the manufacture of cement, and essentials for life as phosphate, salt, and water.

CLASSIFICATION OF SEDIMENTARY ROCKS

Sedimentary rocks made from the insoluble products of weathering are called **detrital** rocks while those composed of minerals precipitated from solution or formed from biologic accumulations are called **nondetrital.** Nondetrital rocks are further subdivided as *chemical, biochemical,* or *evaporite* rocks depending upon the mechanism by which the minerals were removed from solution.

The name given to a detrital sedimentary rock depends primarily on the *size* and *composition* of the grains comprising the rock. Two other parameters, *sorting* and *shape,* although not specifically required for the classification of sedimentary rocks, provide important information about the origin of the particles that make up detrital sedimentary rocks. Grain size, sorting, and shape are collectively known as *texture.*

Size

Many of the terms used to describe the size of particles are common words such as *cobble, pebble,* and *sand.* To a geologist, these terms refer to specific sizes or, more accurately, to specific ranges of particle sizes. Table 8.1 summarizes the complete list of particle sizes.

TABLE 8.1
PARTICLE SIZES IN MILLIMETERS

	Mean Diameter in Millimeters
Boulder	
	256
Cobble	
	64
Pebble	
	4
Granule	
	2
Sand	
	.06
Silt	
	.004
Clay	

Composition

The composition of the *rock fragments* generated by physical weathering obviously depends upon the kind of rock undergoing weathering and usually includes more than one kind of mineral.

The insoluble products of chemical weathering are primarily particles of *quartz* and *clay minerals* with lesser amounts of *feldspar.* Quartz is present because of its resistance to chemical weathering. The clay minerals are present because they are the stable silicate mineral formed by the chemical weathering of most of the aluminosilicate minerals. Feldspar is a residue of weathering because of its dominance in the rocks of Earth's crust. But since feldspar also has a relatively high susceptibility to chemical weathering, feldspar grains are not preserved in sediments except under conditions that minimize the amount of time the materials are exposed to the atmosphere before being buried. For this reason, significant concentrations of feldspar in a sedimentary rock indicate that the materials of which the rock is made were deposited and buried soon after formation and probably were not transported far from their point of origin.

Quartz and feldspar particles are dominantly sand and silt size while clay mineral grains are almost exclusively clay size. Usually, in fact, the great volume of clay-sized grains are clay minerals, which explains why the term *clay* is used to describe both a particle size and a kind of mineral. An alternative term frequently used for clay-sized material is *mud.*

Sorting

As sediment is transported by the various agents of erosion, the insoluble products of weathering are separated naturally by size. Most rock fragments are granule size and larger. As we have noted, most quartz and feldspar grains are sand size while the clay minerals are predominantly clay size. The process whereby materials are separated by size is called **sorting.** The more uniform in size the particles with sediment or rock become, the more *well sorted* the material is. Conversely, the wider the range of particle sizes, the more *poorly sorted* the materials.

Of the agents of erosion, water is the most efficient at sorting. Examples of the ability of moving water to produce *well-sorted* deposits are the fine-grained muds that cover a stream floodplain after a flood and the sand-sized quartz grains that constitute your favorite beach.

Wind is a unique sorting agent in that it is very selective with respect to the particle sizes it can transport. It is limited to sand-, silt-, and clay-sized materials. Sand-sized materials are moved within a few feet of the ground surface and accumulate in deposits called *dunes.* The finer silt- and clay-sized particles are lifted into the air as dust and may be carried great distances before being deposited as fine-grained deposits called *loess.*

Glacial ice is a relatively ineffective sorting agent, picking up and transporting whatever material it may encounter in its path. When the ice melts, the material is literally *dumped* with little discrimination by size. As a result, most glacial deposits are usually *poorly sorted* and consist of a mixture of particle sizes that may range from huge boulders to clay. Some glacial deposits are well sorted, but these are not deposited directly by the ice but by meltwaters emerging from the terminus of the glacier.

Shape

The two descriptors of particle shape are *roundness* and *sphericity*.

Roundness The **roundness** of a particle is a measure of the sharpness of its corners or edges and is calculated by dividing the average diameter of the circles that can be inscribed into its corners by the diameter of the largest inscribed circle (Figure 8.1). A sphere represents perfect roundness with a roundness factor of 1.0.

Particles newly created by most means of mechanical weathering are characterized by sharp edges and corners and have *low roundness.* Particles with roundness values from zero to 0.15 are commonly referred to as being *angular.* Examples are the particles that accumulate in talus.

The presence of angular particles in sediment or in a sedimentary rock is usually interpreted as indicating that the material was transported only a short way from the point of origin before being deposited or converted into a sedimentary rock. Increasing roundness usually indicates that the particles have been transported greater distances from the point of origin and have undergone progressive rounding by mutual abrasion.

Sphericity The **sphericity** of a partricle is a measure of its three-*dimensional* shape; that is, it is a measure of the relationship of the length, width, and thickness of the particle. Sphericity is calculated as the ratio of the nominal diameter of the particle to the diameter of the largest possible circumscribed sphere (Figure 8.2). The nominal diameter of a particle is the calculated diameter of a sphere having the same volume as the particle.

The difference between roundness and sphericity is somewhat subtle. An equidimensional particle such as a cube has *high* sphericity but *low* roundness. On the other hand, a flat pebble found along a stream or shore may have *high* roundness but *low* sphericity. Figure 8.3 compares sphericity and roundness.

How closely the shape of a particle approaches a sphere is a complex product of crystal structure, mineral hardness, the type of transporting agent, the distance of transport, and the particle size. Most particles become rounded by the attack of chemical weathering and exfoliation and by mutual abrasion as they are being transported either in streams or wave zones.

The process of rounding commonly affects only sand-sized and larger particles. Silt- and clay-sized particles transported by water or wind usually do not

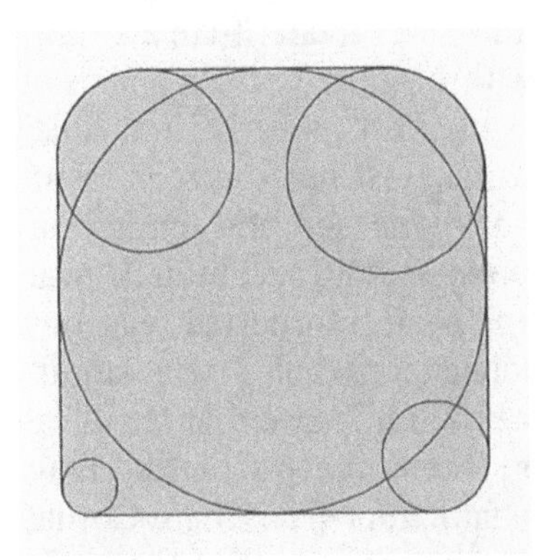

FIGURE 8.1 *Rock fragments are progressively rounded during transport as sharp edges and corners are worn away.* Roundness *is measured by dividing the diameter of the* average *inscribed circle by the diameter of the* largest *inscribed circle. A perfectly round particle would have a circular outline.*

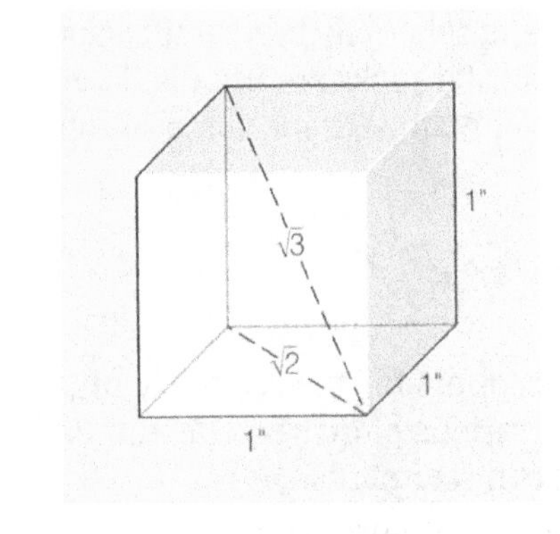

FIGURE 8.2 Sphericity *differs from* roundness *in that the former is a three-dimensional property while the latter is a two-dimensional property. The sphericity of a cubic particle 1 inch on a side is determined as follows: The diameter of the largest circumscribed sphere is equal to $\sqrt{3}$ or 1.73 inch. The diameter of a sphere with a volume of 1 cubic inch is 1.24 inch. Therefore the sphericity of the cube is or 0.72.*

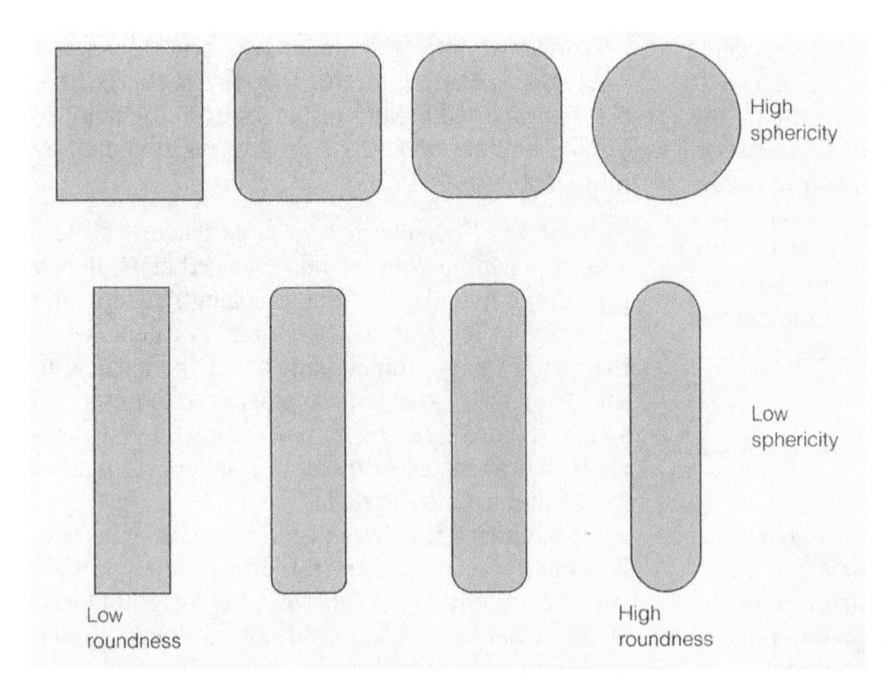

FIGURE 8.3 *Sphericity and roundness are not interrelated; that is, a particle may be highly rounded and yet not be spherical, or it may be highly spherical but not be rounded.*

become rounded because the water or air provides a cushioning effect as the particles are transported in suspension. Consequently, the particles have little opportunity to experience mutual abrasion. The subrounded (almost round) silt-sized particles that have been observed in some glacially transported materials are the one exception to this rule.

Because nondetrital rocks consist of minerals precipitated from solution, the parameters of sorting, roundness, and sphericity have limited application to them. Nondetrital rocks are usually described solely on the basis of grain size and composition.

SPOT REVIEW

1. What accounts for the relatively high concentrations of quartz, clay minerals, and feldspar in sediments?
2. What is meant by "sorting"? How does the agent of erosion determine the sorting of sediments?
3. What is the difference between roundness and sphericity? What kind of information concerning the mode of sediment transport do roundness and sphericity provide?

DETRITAL SEDIMENTARY ROCKS

About 90% of the total volume of sedimentary rocks is made up of detrital sedimentary rocks. These rocks form from the insoluble products of weathering and are named primarily on the basis of the size and composition of their particles.

Breccia

A **breccia** is a sedimentary rock composed primarily of angular grains, granule size (greater than 2.0 mm) and larger, contained within a finer-grained matrix of sand, silt, or clay. For the most part, the individual grains are rock fragments. Most breccias form from rock fragments such as talus and the pyroclastic debris of volcanic eruptions that accumulate near the site of origin. Having been transported relatively short distances, the individual particles have had little chance to experience rounding by either chemical weathering or abrasion. Some breccias, called fault breccias, form within a fault zone from the rock that is broken or crushed by the movement of the rocks on opposite sides of the fault.

Conglomerate

A **conglomerate** is a sedimentary rock with the same size particles as a breccia, differing only in that the

FIGURE 8.4 *The most common sandstone is a* quartz sandstone. *(VU/© A. J. Copley)*

particles are rounded. Although some conglomerates may form from poorly sorted glacial deposits, most form from materials that accumulate in the channels of steep, highly turbulent streams and along rocky coastlines where wave action is intense.

Sandstones

A **sandstone** is a detrital rock consisting primarily of sand-sized (0.062 to 2.0 mm) grains, contained within a finer-grained matrix (Figure 8.4). Sandstones constitute about 20% to 25% of all sedimentary rocks. Because quarts is a major component, sandstones are generally more resistant to weathering and erosion than the enclosing rocks and, as a result, are usually the most conspicuous rock layers in outcrops such as cliffs and road cuts. This often leads beginning students of geology to overestimate the abundance of sandstones relative to other types of sedimentary rocks.

The term *sandstone* is commonly used without qualification to imply a quartz content of at least 85%. The proper term for such a rock is a **quartz sandstone.** Although quartz is the major mineral found in sandstones, some sandstones contain appreciable quantities of feldspar. The term **arkose** is used to describe a sandstone subtype composed of at least 25% angular to subangular sand-sized particles of feldspar. Arkoses commonly represent debris from rapidly disintegrating granite or granitic rocks that has undergone little transport or subsequent chemical weathering before being deposited and buried.

A third kind of sandstone is called a **graywacke.** The term is an old one, dating back to the early 1800s. There is little agreement about the specific definition of a graywacke and, in fact, some geologists have recommended that the term be discarded. The term is still used, however, to describe a dark gray sandstone consisting of a mixture of poorly sorted, angular to subangular, sand-sized particles of quartz, feldspar, rock fragments, and significant concentrations (at least 15%) of clay minerals.

Siltstone

A **siltstone** is a rock composed primarily of silt-sized (0.004 to 0.062 mm) grains. As with sandstones, the composition of siltstones is not restricted to any mineral type. Most siltstones are intermediate in composition between the quartz-rich sandstones and the clay mineral–dominated shales and mudstones.

Shales and Mudstones

Shales and **mudstones** are the most abundant of all sedimentary rocks, constituting about 65% of the total. They are composed primarily of clay minerals, the major solid product of the chemical decomposition of igneous and metamorphic rocks, intermixed with silt- and clay-sized quartz and to a lesser extent feldspar. Shales and mudstones differ in the thickness of their individual layers. Shales characteristically split into thin layers due to the parallel alignment of the clay minerals while mudstones are more massive or thickly layered because the clay minerals are not aligned. Because of their more massive nature, mudstones tend to be more resistant to weathering and erosion. Of the two, shales are the more abundant.

In most exposures of rocks such as cliffs or road cuts, the dominance of shales is not at all obvious because they weather and erode at a relatively rapid rate. However, careful examination of most sites that expose a variety of sedimentary rocks will show that shales are the most abundant single sedimentary rock type.

NONDETRITAL SEDIMENTARY ROCKS

Nondetrital sedimentary rocks form from the soluble products of weathering and are subdivided into three types: (1) **chemical,** (2) **biochemical,** and (3) **evaporite.** The elements taken into solution in the greatest concentration as positive ions are *sodium* (Na^{1+}), *calcium* (Ca^{2+}), and *silicon* (Si^{4+}). The major negative ions resulting from chemical weathering include *sulfate* (SO_4^{2-}), *nitrate* (NO_3^{1-}), *phosphate*

(PO_4^{3-}), *chloride* (Cl^{1-}), and the most abundant and important of all, the *bicarbonate* ion (HCO_3^{1-}). These ions combine to form the minerals that constitute the nondetrital rocks. In contrast to the detrital sedimentary rocks, which usually consist of a mixture of different minerals, nondetrital sedimentary rocks are usually *monomineralic;* that is, they are dominated by a single mineral. This characteristic explains why the classification of nondetrital sedimentary rocks is based primarily on composition.

Chemical Sedimentary Rocks

Limestone, the most abundant of all the nondetrital sedimentary rocks, is the major sedimentary rock type in both the chemical and biochemical categories. Limestones constitute about 10% of all sedimentary rocks by volume and are made primarily of *calcium carbonate* ($CaCO_3$). The calcium carbonate in chemical limestones formed strictly by chemical precipitation from solution. The calcium carbonate in biochemical limestones, on the other hand, was produced by living organisms, plant or animal, although in practice it may be difficult to tell whether the origin was animal or plant.

Chemical Limestones

Calcium carbonate is unusual in that it is more soluble in cold water than in warm water. Because the solubility of carbon dioxide increases with decreasing water temperature, cold water promoted the dissolution of $CaCO_3$ while warm water promotes its precipitation. This explains, at least in part, why shelled animals and the accumulation of shell remains are more abundant in warm waters than in cold. Although some limestones form in freshwater lakes, most limestones form from materials that accumulate in the ocean.

For the most part, the surface waters of Earth's oceans are saturated with $CaCO_3$. The materials for future limestones accumulate in the shallow offshore areas where the waters are warm. In some areas, cold ocean waters saturated with $CaCO_3$ rise from the ocean deep and flood out onto the warm, shallow ocean bottom. The combined effect of the decreasing depth and the warming of the waters results in the precipitation of $CaCO_3$ as a fine-grained carbonate mud. Should this material become a rock, it will be an example of a chemical limestone. In some areas where the carbonate mud is agitated by wave and tidal currents, the $CaCO_3$ precipitates in concentric layers onto the surface of tiny shell and mineral fragments to form spherical grains called *ooids,* which may give rise to *oolitic limestones,* another example of a chemical limestone. Another kind of limestone, *travertine,* forms by the precipitation of $CaCO_3$ from surface and groundwater. Travertine is found in fractures, on rock faces, as vein fillings, and perhaps most spectacularly, in limestone caves and caverns (Figure 8.5). Chemical precipitation, however, accounts for less than 10% of all limestones.

Dolostone

Another common carbonate rock, **dolostone,** is made of the carbonate mineral *dolomite,* $CaMg(CO_3)_2$. Most dolomite forms when magnesium partially replaces the calcium in grains of $CaCO_3$. The carbonate sediments that initially accumulate are mostly composed of the mineral *aragonite.* Aragonite

FIGURE 8.5 Stalactites *and* Stalagmites, *the spectacular structures that adorn limestone caves and caverns, are made from* $CaCO_3$ *precipitated by the evaporation of carbonate-rich groundwater. (VU/© A. J. Copley)*

has the same composition as calcite, $CaCO_3$, differing only in crystal structure. If magnesium-rich solutions come in contact with aragonite, magnesium is emplaced into the carbonate lattice to form dolomite. The process is especially effective in arid regions where the evaporation of seawater preferentially increases the concentration of magnesium by precipitating calcium as gypsum ($CaSO_4{\cdot}2H_2O$). In some cases where leaching is limited to more porous sediments or where carbonate muds have been preferentially dolomitized due to their high surface area, the resultant rock will be a mixed limestone-dolostone. In situations where the leaching is intense or where the magnesium concentration is especially high, the process of dolomitization can convert all the carbonate grains to dolomite, forming a pure dolostone.

Biochemical Sedimentary Rocks

As we have pointed out, biochemical sedimentary rocks form from the remains of plant and animal organisms. Chert and biochemical limestone are among the most important.

Chert **Chert** is a sedimentary rock composed entirely of *silica* (SiO_2). Most chert is described as being *cryptocrystalline* (Greek *kruptos*=hidden), which means that the individual crystals are so small that they are not visible under a light microscope and can only be viewed with the high magnification of an electron microscope.

Some chert forms from silica originally in solution. The dissolved silica may replace $CaCO_3$ to form the *nodules* that are commonly found within beds of limestone. In such a case, the chert would be chemical in origin. Most chert is biochemical in origin, however, and forms from the opaline shells of microscopic animals called *radiolaria,* microscopic plants called *diatoms,* and the siliceous spicules of *sponges* (Figure 8.6) that accumulate on the ocean floor. Eventually, they are transformed into *bedded chert.*

A common variety of chert, showing a more dense texture, black color, and perfect choncoidal fracture, is called *flint.* For hundreds of years, the term has been used to refer to any hard rock. Ever since the evolution of *Homo habilis* 2.5 million years ago, flint has been prized for the manufacture of tools such as knife blades and points for spears and arrows. Other varieties of chert are the red *jasper* and variously colored, banded *agate* where the coloration is due chiefly to impurities of iron oxide.

Biochemical Limestones Most limestones (about 90%) are **biochemical** in origin and form from $CaCO_3$ produced by the biological activity of animals and plants. Many marine animals such as *clams, oysters, snails,* and *coral* as well as many planktonic (floating) organisms secrete shells of $CaCO_3$. When they die, the shells accumulate on the ocean floor. Should this material enter the geological record as a limestone, it forms a biochemical limestone.

Many species of marine plants also secrete microscopic plates made of $CaCO_3$ on their outer surfaces to strengthen and protect their body parts. One common example found extensively on the bottom of

FIGURE 8.6 *A variety of microscopic siliceous shells accumulate as part of the fine* muds or oozes *that coat the deep ocean floor. (Courtesy of the Deep Sea Drilling Project/Scripps Institution of Oceanography)*

Florida Bay is *Penicillus,* commonly called "shaving brush algae" because of their shape. Upon the death and decomposition of the plant, the tiny plates accumulate on the bottom of the bay as a fine-grained mud that may be converted into a biochemical limestone. Studies have shown that much, perhaps most, carbonate mud is derived from the breakdown of the fine skeletal materials of algae and bryozoans and the fragile frameworks of other organisms.

The limestone that forms from carbonate muds composed of microscopic algal materials is quite different in appearance from the limestone that is produced from shell debris. Although both are examples of biochemical limestones, the limestone that forms from the accumulated shell debris is *fossiliferous* with many of the shells and shell fragments identifiable within the rock. An extremely fossiliferous variety of limestone called **coquina** is composed almost entirely of loosely cemented shells and shell fragments (Figure 8.7). On the other hand, the rock formed from muds composed of microscopic algal plates appears as a dense, fine-grained limestone with no visible plant or animal remains.

Evaporitic Sedimentary Rocks

As the name implies, **evaporites** form by the evaporation of water. The two major evaporite minerals are **gypsum** ($CaSO_4 \cdot 2H_2O$) and **rock salt** or **halite** (NaCl). Most evaporites form in hot arid to semiarid regions where water may evaporate in seasonal environments such as desert lakes or along arid coastlines called *sabkhas* where seawaters continuously evaporate on shallow, nearly landlocked embayments. Most of the salt you use to season your food came from such deposits. Because high temperature and low relative humidity are required for their formation, evaporites are important indicators of paleoclimate.

Organic Sedimentary Rocks

An **organic** category is often included in a sedimentary rock classification scheme primarily to classify **coal.** Some textbooks omit coal altogether while others classify it variously as biochemical, nondetrital-biochemical, nondetrital-accumulated plant debris, or chemical-organic.

The problem of coal illustrates the inherent fallibility of all classification schemes; there will always be something that doesn't quite fit. By definition, coal is a rock containing 50% or more organic material by *weight.* Because of the difference in the densities of organic and inorganic materials, the *minimum* volume of organic material in coal is about 60%. Most coal—and all minable coal—forms from plant debris converted into *peat* within freshwater swamps and subsequently converted after burial into *lignite, subbituminous, bituminous,* or *anthracite* coal (Figure 8.8).

FIGURE 8.7 *A* coquina *is an example of a biochemical limestone consisting almost entirely of cemented shell and coral fragments. (VU/© Albert J. Copley)*

FIGURE 8.8 *Coals form largely from the wood of trees growing in swamps dominated by fresh water. (Courtesy of J. F. Elder/USGS)*

Note that the major materials from which coal is made are *not* products of weathering but rather the remains of once-living plants. In the next chapter, we will see that metamorphic rocks form by the solid-state recrystallization of an existing rock through the action of heat, pressure, and/or chemically active fluids. Although the transformation of peat into other types of coal takes place at significantly lower temperatures than are required for most metamorphic processes (less than 400°F or 200°C), and might not be considered metamorphism by a metamorphic petrologist, coal geologists regard this transformation as a metamorphic process. By their definition, coal is a metamorphic rock even though it is found associated with sedimentary rocks. Many geologists who are not involved in coal geochemistry would not classify coal as a metamorphic rock. Instead, because coal is obviously not an igneous rock, they would classify it as a sedimentary rock by default.

SPOT REVIEW

1. Why are shales the most abundant of all the various kinds of sedimentary rocks?
2. What possible differences in the origin of the sediments do the compositions of quartz sandstones, arkoses, and graywackes suggest?
3. In what way do quartz sandstones and cherts differ? In what way are they the same?
4. What different kinds of materials make up biochemical sedimentary rocks?

DEPOSITIONAL ENVIRONMENTS

The products of weathering are carried away by the processes of erosion to be deposited in a number of different depositional sites. Any site where the products of weathering can accumulate is a **depostional environment.** Depending on whether the sediment is deposited on land, at the margins of a continent, or in the ocean, the sites and the sedimentary rocks produced from them are categorized as one of three basic types: (1) *continental,* (2) *transitional,* and (3) *marine.*

Continental Deposits

Continental deposits vary depending upon the particular agent of erosion and the depositional site.

Fluvial **Fluvial** deposits are materials laid down directly from streams in a number of depositional sites including *floodplains, channels, point bars,* and *levees.* Except for deposits of desert streams, most fluvial deposits are well-sorted sands and exhibit well-rounded grains. Materials deposited by desert streams are usually less well sorted and more angular than most other fluvial deposits because they are transported relatively short distances and deposited rapidly.

Glacial **Glacial** deposits include materials that accumulate under ice sheets, at the margins of melting ice, and from glacial meltwaters (Figure 8.9). Except for glacial meltwater deposits, glacial deposits are almost always poorly sorted.

Eolian **Eolian** deposits are materials deposited by the wind. They are found both as sand-sized accumulations in *dunes* (Figure 8.10) and also as silt- and clay-sized materials in *loess.* Some of the spectacular sandstone cliffs in the deserts of the southwestern United States, such as the Navajo and Entrada Sandstones, are examples of ancient dune deposits (Figure 8.11). Because of the extreme degree of grain-to-grain abrasion, wind-transported sand grains commonly become well rounded. The most extensive loess deposits on Earth are found in northern China where fine-grained, wind-transported materials coat the landscape with a blanket that is several hundred feet (meters) thick in places.

Lacustrine **Lacustrine** deposits accumulate in lakes. Geologically, all lakes are short-lived. Sediment introduced by streams will eventually fill the lake basin, converting the lake environment first to a marsh, next to a swamp, and then to a bog. With the complete infilling of the bog, the original lake is eliminated.

Paludal **Paludal** deposits accumulate in *marshes* and *swamps.* Marshes are *wetlands* dominated by grasses and other nonwoody plants. Swamps, on the other hand, are wetlands dominated by wood-rich plants. Because of the slow movement of the water within wetlands and the chemistry of the waters, accumulating plant debris is partially preserved and converted into peat. Depending upon the relative concentrations of preserved organic material and mineral matter, the accumulated material will be the precursor of a range of rock types from organic-rich *black shales* to *coal.*

FIGURE 8.9 *Glacial deposits range from materials deposited directly from the melting ice to those deposited from meltwater streams. (Photos courtesy of D. R. Crandell/USGS)*

FIGURE 8.10 *Accumulations of windblown sand are called* dunes. *(Courtesy of E. D. McKee/USGS)*

FIGURE 8.11 *Sandstones formed from dune deposits, such as those exposed in Zion National Park, commonly exhibit well-developed cross beds. (Courtesy of W. B. Hamilton/USGS)*

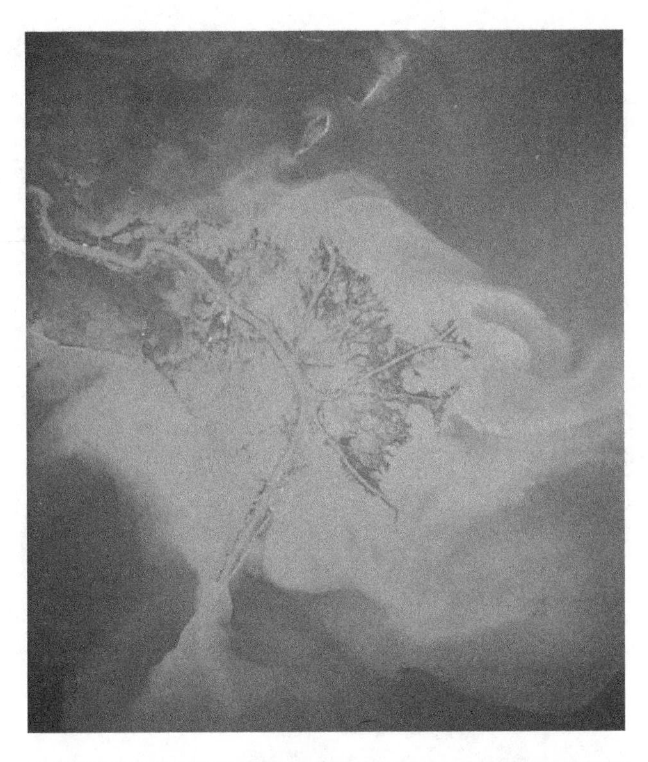

FIGURE 8.12 *Most of the present load of the Mississippi River is being deposited in the Balize delta lobe. Note the plume of suspended load as the lower density fresh river water spreads out over the surface of the denser saline water of the Gulf of Mexico. (Courtesy of NASA)*

Transitional Environments

A number of **transitional** settings between the continental and marine depositional environments provide much of the material for marine sedimentary rocks. These environments include *deltas, beaches, barrier islands, coastal wetlands,* and *mudflats.*

Deltas A coastal **delta** accumulates as a wedge of stream-fed continental detrital sediment, extending the reaches of the land into the marine environment (Figure 8.12). At some point, the seaward development of the delta is terminated by ocean currents that become strong enough to pick up and redistribute the materials on the ocean floor. Deltas may also form where streams enter freshwater lakes.

Beaches **Beaches** are coastal accumulations of unconsolidated materials (Figure 8.13). The portion of the beach nearest the water is usually a relatively flat surface, sloping seaward. Inland from the beach, windblown sand may accumulate as dunes, or the land may rise more abruptly as a cliff. Along shorelines characterized by high-energy waves, the beach may consist of larger rock fragments, up to boulder size (Figure 8.14).

Barrier Islands **Barrier islands** are enlongated sand islands that parallel the shore and are separated from the mainland by a lagoon or a coastal wetland. The islands commonly have dunes at the center and a beach on the seaward side (Figure 8.15). The origin of barrier islands is a subject of debate. Some barrier islands form from land-derived sand transported along the coast by longshore currents. The most widely accepted idea is that most barrier islands originated as beaches and beach ridges during the last glacial episode when the ocean was at a lower level. As the ice age waned and sea level began to rise, the beach ridges

FIGURE 8.13 *The beach sands that accumulate along the continental margin are often the exposed edge of a wider belt of sand that parallels the shore. (Courtesy of S. J. Williams/USGS)*

FIGURE 8.14 *Along high-energy beaches, finer particles are removed by the waves, leaving behind particles that are often cobble size and larger.*

migrated landward. Eventually, the ocean encroached into the low area behind the beach ridges isolating them from the mainland (Figure 8.16).

Coastal Wetlands and Mudflats As the lagoons behind the barrier islands fill with sediment derived from the land and from the ocean by way of the tidal inlets, the water shallows and the bottom of the lagoon become exposed at low tide as a **mudflat**. Once the bottom is exposed, grass begins to grow on the mudflat and traps additional sediments. The sediment surface builds up rapidly and eventually emerges above high tide and develops into a *marsh*. As larger plants and trees begin to encroach into the wetland, the marsh changes into a *swamp*.

Marine Environments

The most important of all the depositional environments is the ocean bottom. Most of the products of

FIGURE 8.15 Barrier islands *can be found in many areas along the coastlines of the Atlantic Ocean and the Gulf of Mexico. (VU/© Frank M. Hanna)*

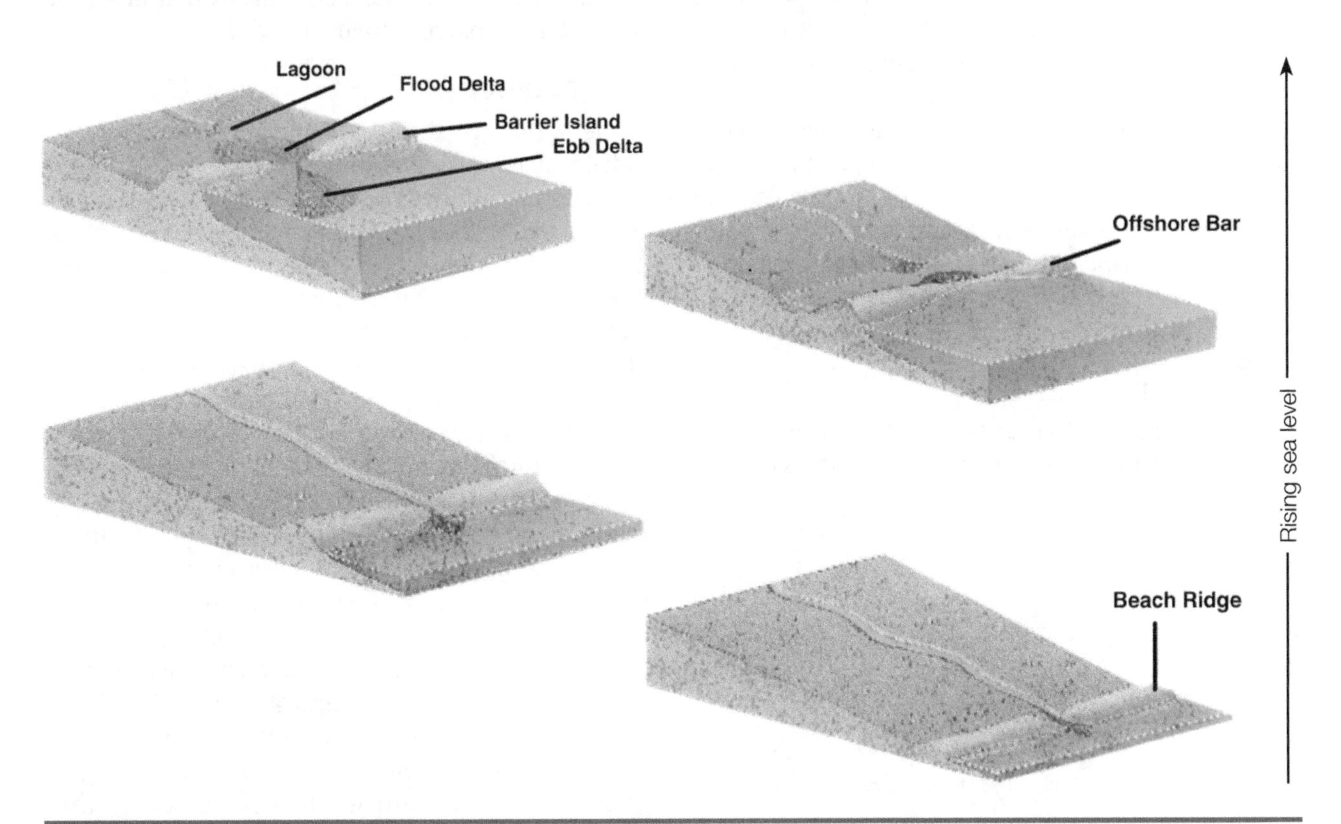

FIGURE 8.16 *Most of the barrier islands that exist along the Atlantic coast and at other places around the world are believed to have formed as sea level rose following the end of the Pleistocene Ice Age.*

weathering and erosion, including materials initially deposited in continental and transitional depositional environments, eventually accumulate on the ocean floor. Once deposited, the sediments are the source materials for *marine* sedimentary rocks, the most voluminous of sedimentary rocks.

The shallow waters of the **continental shelf** are the major depositional site for most of these materials, although some sediments are carried beyond the edge of the continental shelf to accumulate on the **continental slope** and on the **abyssal floor.** Ocean currents distribute the sediments on the surface of the continental shelf in bands parallel to the shoreline (Figure 8.17). The coarse materials, usually sand, accumulate nearest the land with the finer-grained silts and clays being carried farther seaward. Beyond the influence of the detrital sediments, carbonates may accumulate on the ocean bottom, chiefly where the waters are warm enough to promote the chemical precipitation of calcium carbonate and preserve the remains of shells. In some cases, the clay-sized sediments are transported seaward beyond the zone of carbonate accumulation.

LITHIFICATION

The process by which accumulated products of weathering are converted into a rock is called **lithification.** Sedimentary deposits normally do not lithify at Earth's surface although Sun-dried muds and newly deposited evaporite minerals can achieve a certain rock-like character. For the most part, the processes of lithification require the sediments to be buried and subjected to increased temperatures (up to a maximum range of 400°F to 570°F or 200°C to 300°C) and pressures of less than a kilobar (about 14,000 lb/in^2). Once buried, the sediments undergo lithification. The major processes of lithification are *compaction* and *cementation.*

Compaction

The initial change that takes place with burial is **compaction,** which forces the individual grains closer together, decreasing both the volume and the porosity of the sediment layer. When the layers consist primarily of materials that are sand size and coarser, the volume reduction due to compaction is relatively small because of the dominance of relatively rigid, equidimensional grains. If, on the other hand, the material is mud, which consists primarily of clay minerals with silt-sized quartz, the volume reduction can approach 50% as the water is squeezed from the mass of platelike clay mineral particles.

Compaction is the major mechanism by which muds are transformed into *mudstones* and *shales.* As the water is physically expelled from the layer, compacting the clay crystallites closer together, the individual clay particles begin to adhere to each other. In cases where mud is deposited along with larger grains, such as sand-sized quartz, compaction of the clay minerals may create a clay matrix that holds the larger grains together to form a sandstone.

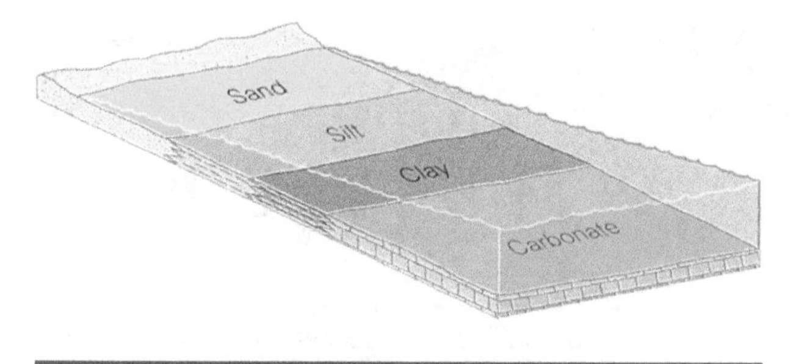

FIGURE 8.17 *The combined effects of the various coastal currents distributes the sediments in strips parallel to the coastline with the largest particles accumulating nearest the shore and decreasing in size seaward. When the water is warm, carbonates accumulate beyond the influence of the clastic sediments.*

Cementation

Cementation is the dominant process by which detrital materials consisting of sand-sized or larger particles are lithified. During cementation, minerals precipitate from solution in the pores between the grains. This precipitated material "cements" the grains together. The most common cements are *calcite* ($CaCO_3$), *quartz* (SiO_2), and *iron oxide* (Fe_2O_3), all provided by the chemical weathering of rocks.

Coarse-grained nondetrital materials such as shells and shell fragments are also lithified by cementation to form biochemical limestones. The cements that form most biochemical limestones are usually carbonate minerals such as *calcite* or *aragonite* (both calcite and aragonite consist of calcium carbonate, $CaCO_3$) and *dolomite,* $CaMg(CO_3)_2$. In some cases, *gypsum* $CaSO_4{\cdot}2H_2O$, is the cementing agent.

Fine-grained, nondetrital sediments such as those that would produce biochemical or chemical limestones and evaporites usually lithify by a combination of compaction and cementation. Lithification begins as the materials are buried and forced closer together with the subsequent expulsion of water and collapse of pore space. The sediments continue to lithify as additional minerals precipitate from solution and the individual mineral crystals grow and interlock. Biochemical rocks characteristically undergo lithification by cementation as calcite precipitates from solution and fills the pore spaces between the shells and shell fragments.

SPOT REVIEW

1. Which of the various sedimentary environments is responsible for the greatest volume of sedimentary rocks? Why?
2. Which depositional environments are considered "fluvial?"
3. What kinds of sediments lithify primarily by cementation? By compaction? By a combination of cementation and compaction?

SEDIMENTARY FEATURES

Many internal features associated with sedimentary rocks are the product of the environment under which the materials accumulated; therefore these features reveal a great deal about the conditions that existed within the depositional environment at the time of accumulation. Geologists are able to establish the history of Earth by studying the kinds and distributions of sedimentary rocks and the various sedimentary features contained within them.

Beds

All sedimentary rocks, regardless of type or mode of formation, are *layered* or *stratified*. The principal layers, which are called **beds** (strata), consist either of detrital rock and mineral fragments deposited by a stream, glacier, or the wind or of mineral grains precipitated from solution. According to the law of original horizontality, the sediments always accumulate in horizontal or near-horizontal layers, regardless of whether the accumulations occur on the bottoms of lakes, in stream floodplains, in desert basins, or, most commonly of all, in marine environments. If sedimentary rocks are observed in road cuts or cliffs with the beds tilted from the horizontal, it means that some force must have acted upon the original layers to change their original horizontality.

Other sedimentary features contained within the individual beds allow geologists to reconstruct certain aspects of the conditions that existed within the depositional basin when the sediments accumulated. Each detail that can be understood adds one more bit of information to the overall goal of establishing the history of Earth.

Ripple Marks

Currents of water or air moving across the surface of a recently deposited layer of sand-sized materials may produce standing waves or dunes on the surface (Figure 8.18). If the currents move in a single direction, as occurs with stream flow, the wind, or a current flowing on the ocean bottom, waves that develop on the surface of the sediment layer will be oriented perpendicular to the direction of the current and will be **asymmetrical** in cross section with the steep side in the direction of current flow

FIGURE 8.18 *The wind forms asymmetrical ripple marks on the surface of sand. (Courtesy of E. D. McKee/USGS)*

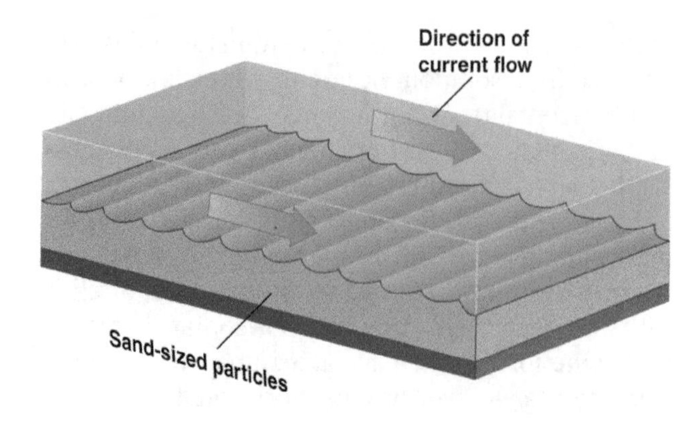

FIGURE 8.19 Asymmetrical ripple marks *are generated as water or wind moves across the surface of layers of sand. The ripples form perpendicular in the direction of flow with the steep side of the ripple in the downflow direction.*

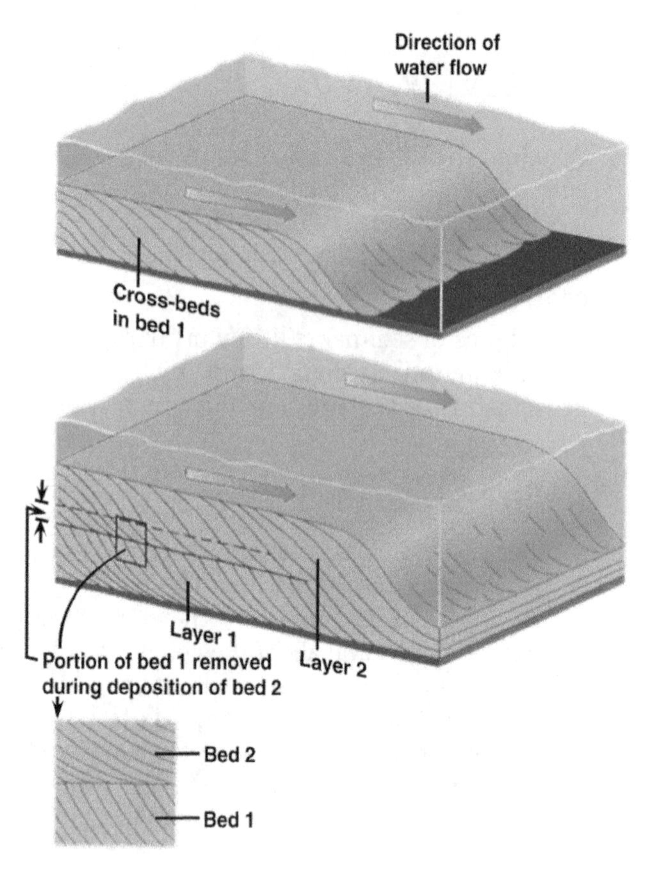

FIGURE 8.21 *The direction of movement of water or wind is recorded by the* cross-beds *that form within layers of sand.*

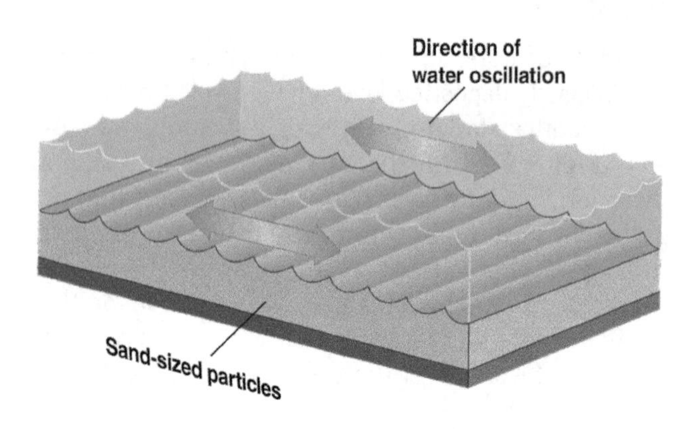

FIGURE 8.20 Symmetrical ripple marks *form where layers of sand are exposed to the oscillating currents in the shallow water near shore.*

(Figure 8.19). If this layer of sediment is quickly buried by the influx of another layer of sediment, the original waveforms may be preserved on the top surface of the bed as **current ripple marks.** Like cross-bedding, current ripple marks are used to interpret the direction of current flow.

When sediments in shallow coastal areas are exposed to the oscillating action of waves, the waveforms that develop on the surface of the sediments are **symmetrical** (Figure 8.20). Again, if an influx of sediment buries the first layer, the waveforms may be preserved as **oscillation ripple marks.** Oscillation ripple marks indicate that the materials accumulated in a shallow, near-shore environment.

Cross-Bedding

As sediments are moved by flowing water along the bed of a stream or are blown across a dune by the wind, grains constantly move along the top surfaces of the bed and cascade down the leading edge of the deposit (Figure 8.21). The result is the progressive development of **cross-bedding** or layers that are oriented at an angle to the major bedding in the downstream direction. Because the cross-beds are contained within the main bedding and are an integral part of it, their formation does not defy the law of original horizontality.

The orientation of the cross-beds records the direction of water or wind movement within the original depositional environment. Spectacular examples of cross-bedding preserved in eolian sandstones record the direction of wind flow that created dune fields in some prehistoric desert (Figure 8.11).

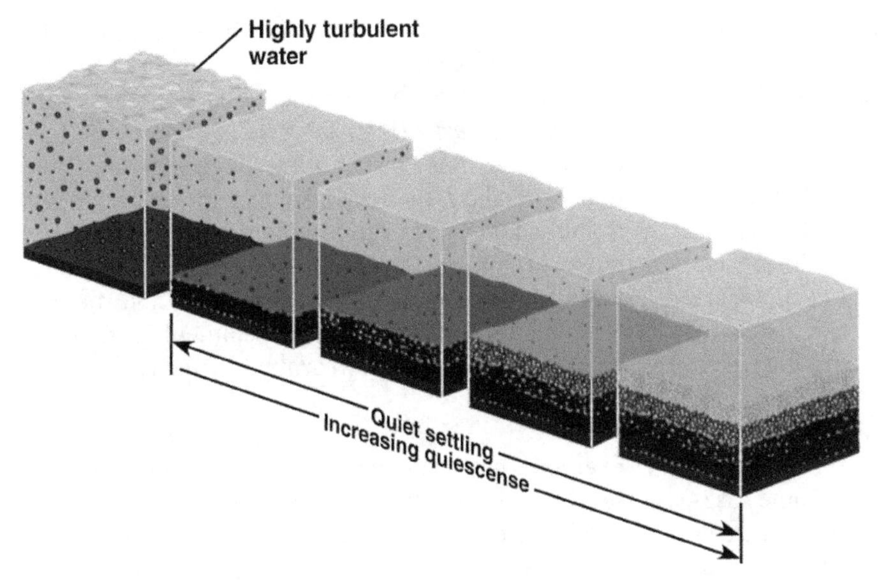

FIGURE 8.22 Graded beds *form as progressively smaller particles settle out of water with time.*

Because the wind commonly changes direction, the patterns of cross-bedding in eolian sandstones can be quite intricate as they record the changing wind directions. In similar fashion, the cross-beds within a fluvial sandstone can be used to determine the direction of stream flow.

Graded Bedding

When clastic sedimentary materials carried in moving currents settle out in standing water or in very slow moving water, the largest particles settle out first, followed by progressively finer-grained materials. The result is a feature called **graded bedding** in which the grain size changes from coarse at the bottom to fine at the top (Figure 8.22). Graded bedding is common in many sedimentary beds, but is especially well developed in certain kinds of deposits. An excellent example of graded bedding is found in the sedimentary rocks called *turbidites.* Sediments carried by ocean currents beyond the edge of the continental shelf are deposited on the continental slope. In many cases, these unconsolidated sediments rest nearly at the angle of repose. This precarious balance may be upset by a shock such as a local earthquake or by the gravitational pull of the deposit's own mass and set into motion downslope as a *turbidity flow.* The turbidity flow carries the sediment into deep water and onto the abyssal ocean floor. As the sediments carried by individual turbidity flow settle out, graded bedding develops in the layer of sediment. Once lithified, these trubidites may eventually be uplifted and exposed on land (Figure 8.23).

FIGURE 8.23 Graded beds, *called* turbidites, *which formed from materials that originally accumulated on the deep-ocean floor, may be uplifted and exposed at the surface following mountain-building episodes.*

Another example of graded bedding commonly results from the seasonal accumulation of sediments within lakes located in temperate climates. During the spring and early summer when the discharge and loads of streams flowing into the lakes are high, the materials that accumulate on the lake bottom will be relatively coarse-grained. As the year progresses and the amount of rainfall lessens, usually during late summer and early fall, stream discharge declines, and the sediments brought into the lake and deposited decrease in both volume and size. During the winter when the streams and lakes may freeze over, the influx of sediments come to a halt. All that remains to settle out are the very finest materials still held in suspension within the lake waters. In some cases, there may be a period during late winter when no sediments remain in suspension and accumulation on the lake bottom ceases. With the spring thaw, the cycle begins again.

The result of the yearly cycle is a single graded bed called a *varve* (Figure 8.24). The age of the lake can be determined by counting the number of varves in a core taken from the lake bottom. The thickness of individual varves can also be used to derive other kinds of information such as variations in yearly rainfall throughout the life of the lake, and the pollen contained within the lake sediments can reveal changes in the vegetation around the lake.

Mud Cracks

Everyone has seen the polygonal, curling plates of semiconsolidated mud in the bottom of a dried mud puddle. If they are preserved by the rapid accumulation of another layer of sediment, they may appear in the rock record as **mud cracks** (Figure 8.25). Mud cracks are evidence that the area was once suf-

FIGURE 8.24 *Graded beds that accumulate in lakes are called* varves. *Because each varve represents one year's accumulation, the varves can be used to calculate the age of the lake and to determine climate changes that occurred during the lifetime of the lake. (Courtesy of P. Carrara/USGS)*

FIGURE 8.25 *Clay mineral-rich sediments commonly shrink and crack upon drying to form* mud cracks, *which can in turn be preserved during lithification. (Photo courtesy of P. Carrara/USGS)*

ficiently shallow to undergo periodic flooding and evaporation, perhaps a flat coastal area or an inland seasonal lake, and usually suggest a dry, warm environment that promoted the drying of the sediment surface.

Animal Trails

Animal marks left in sediment range from the rather nondescript trails of worms crawling across the sediment surface or boring into the upper few inches or feet of the sediment layer to the more spectacular footprints of dinosaurs (Figure 8.26). Although **animal trails** are not the result of the process of sediment accumulation, they provide information not only about the depositional environment at the time of sediment accumulation, but also the kinds of creatures that existed in the area, the climate, and the means of locomotion of these long-extinct creatures.

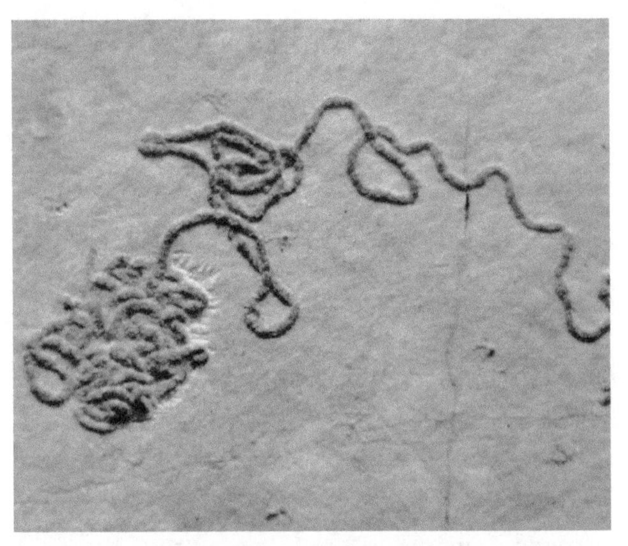

FIGURE 8.26 *Some information about the depositional environment can be obtained from the trails left by animals that lived in the area.* (a *VU/© John D. Cunningham,* b *VU/© Scott Berner)*

SPOT REVIEW

1. Which sedimentary features can be used to determine the direction of sediment transport?
2. How do the conditions for the formation of symmetrical and asymmetrical ripple marks differ?
3. What is a "varve" and how does it form? What kinds of historical information can be obtained from the study of varves?

INTERPRETATION OF ANCIENT DEPOSITIONAL ENVIRONMENTS

A primary goal of geology is to determine the history of Earth. Considering their relatively small crustal volume, sedimentary rocks provide a disproportionately large amount of historical information. As each layer of sediment accumulates, it records information about the depositional environment, such as the kinds and relative abundances of minerals, sedimentary features, bed geometry, and the remains of plants and animals. The following examples illustrate how geologists utilize sedimentary rocks to determine Earth history.

Applying Hutton's concept of uniformitarianism as exemplified by the expression "the present is the key to the past," geologists have achieved considerable success in deciphering the history contained in

FIGURE 8.27 Corals *are among the most prolific rock-forming animals living in the ocean. (Courtesy of S. J. Williams/USGS)*

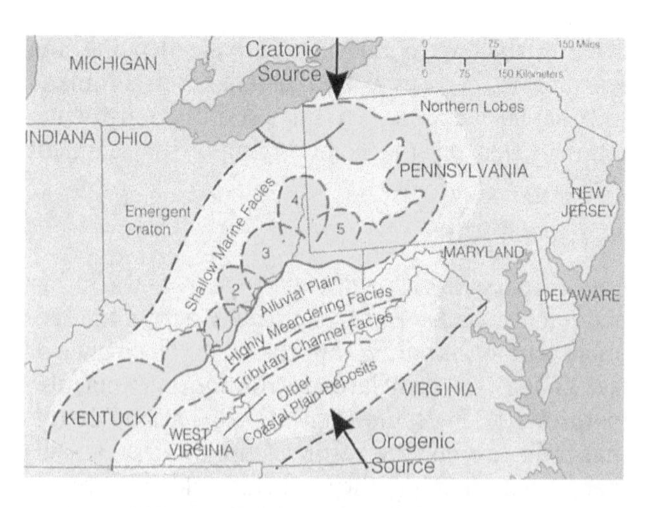

FIGURE 8.28 *Maps prepared by plotting the directions of stream flow based on the cross-beds found in fluvial sandstones have allowed geologists to recreate the patterns of deltaic deposition.*

sedimentary rocks. Fossils are one of the parameters that provide us with a great deal of information. By studying the habitats of modern plants and animals and assuming that comparable ancient organisms has similar affinities, we can use fossils to interpret the ancient depositional setting. For example, in modern marine environments, green plants, plant eaters, and animals that utilize plankton (floating organisms) as a food supply are abundant in water penetrated by sunlight, the *photic zone.* Below the photic zone, scavengers and sediment feeders become dominant. Filter feeders such as the colonial bryozoans that acquire their food by filtering seawater through their bodies can only live in clear water.

The shapes of animal shells reflect the degree to which the water is agitated. The shells of animals living in agitated waters such as clams are usually shorter, rounder, and thicker than those of animals living in quiet waters. Some colonial species modify the shapes of their colonies in response to water depth. For example, branching corals are usually found in shallow water. As the water deepens, the coral colonies become flat and platy while colonies in very deep water are small (Figure 8.27). When fossil remains with these shell or colonial forms are observed, geologists can assign the appropriate depositional setting to the layer of rock in which they are found.

The specific rock lithology can often be used as an indicator of a certain depositional setting. For example, modern studies have shown that although carbonate materials are produced in most aquatic environments, limestones form primarily from materials that accumulate in marine environments where stream-borne detritus is excluded, such as the Florida Keys or the Bahamas. Within carbonate environments, lime muds preferentially accumulate in sheltered environments such as Florida Bay where water movement is restricted. Where waves, tides, and longshore currents agitate the water and winnow out the muds, such as on the seaward side of the Florida Keys, the sediment consists primarily of coarse carbonate particles. The limestones that would form from these two sediments would be quite different in appearance. Lithification of carbonate muds would produce a dense, fine-grained limestone while the materials accumulating in the more agitated waters would form a coarse-grained, highly fossiliferous limestone. Note that these two modern depositional environments (and the rocks that would form from them) are separated only by the narrow width of the Keys.

Sedimentary features provide important information about the depositional environment. In some cases, the same sedimentary feature can provide data at different scales. For example, the attitude of the cross-bedding in a single bed of a fluvial sandstone can be used to determine the direction of stream flow within the stream in which the sand accumulated. By combining the cross-bed orientation data for a large number of fluvial sandstones deposited within a drainage system, we can determine the extent of an ancient delta (Figure 8.28). Perhaps the most impressive examples of cross-bedding are exhibited in eolian sandstones such as those exposed in Zion National Park where the complex patterns of cross-bed orientation record the changing directions of

FIGURE 8.29 *Because the wind can change direction over relatively short periods of time, the orientations of cross-beds in eolian sandstones are more variable than those found in stacked fluvial sandstones.*

sand transport within the dunes of an ancient desert (Figure 8.29).

Certain rocks, such as coal, can provide information about the chemistry of an ancient depositional environment. Coal consists primarily of partially preserved wood that initially accumulated as peat in a swamp. As the wood decomposed, mineral matter originally contained in the wood accumulated within the peat. The relative amounts of preserved wood and mineral matter in the peat are a function of the pH of the swamp matter. At pH values above 5 to 6, the amount of mineral matter in the peat dominates over that of preserved wood. When such a peat lithifies, it forms an organic-rich black shale. As the swamp water pH decreases, the relative portion of preserved wood increases, eventually resulting in the formation of a coal-forming peat. Initially, however, the high mineral matter content results in the formation of a poor-quality coal. As the pH of the swamp water decreases, the relative amount of preserved wood increases, and the quality of the resultant coal improves. High-quality coals with organic contents in excess of 90% by weight form when swamp waters have a pH lower than about 2.5

Walther's Law

The change in rock lithology within a vertical sequence of rocks may record an event such as a change in sea level. A principle called **Walther's law** states that the vertical sequence of rocks observed at any locality reflects the lateral sequence of depositional environments that existed side by side perpendicular to the shoreline at any one time. To illustrate, refer back to Figure 8.17, which is an idealized illustration of the distribution of depositional environments parallel to a coastline. Sand dominates the sediments in the environment immediately adjacent to the coastline. Seaward, the sand-sized materials grade (change) laterally to the silt- and clay-sized sediments that dominate the intermediate environment. These materials, in turn, grade seaward into the carbonates that usually characterize the outermost environment. It is important to note that these three different kinds of sediments are accumulating *simultaneously* in a *single* layer on the ocean floor. It is equally important to note that when these sediments are eventually lithified, the rocks will be in *lateral continuity* within a single layer; that is, they will grade laterally into each other. The sandstone will grade laterally into shale, which will in turn grade into limestone. Rock bodies that differ in lithology from other beds that are deposited at the same time and in lateral continuity are called *facies*.

Consider what would happen if sea level rose due either to a subsidence of the land or to an actual rise in sea level (Figure 8.30). As the coastline moved inland, each of the three depositional environments would follow or *transgress,* with each environment moving *over* the adjacent landward environment. Note that, in time, the carbonate depositional environment could eventually be positioned above the original location of the sand-dominated environment.

The transgression would be recorded by a vertical sequence of rocks with more seaward rocks positioned over more landward rocks. The sequence from bottom to top would be

sandstone → shale → limestone

Note that as Walther's law predicts, the sequence duplicates the distribution of depositional environments

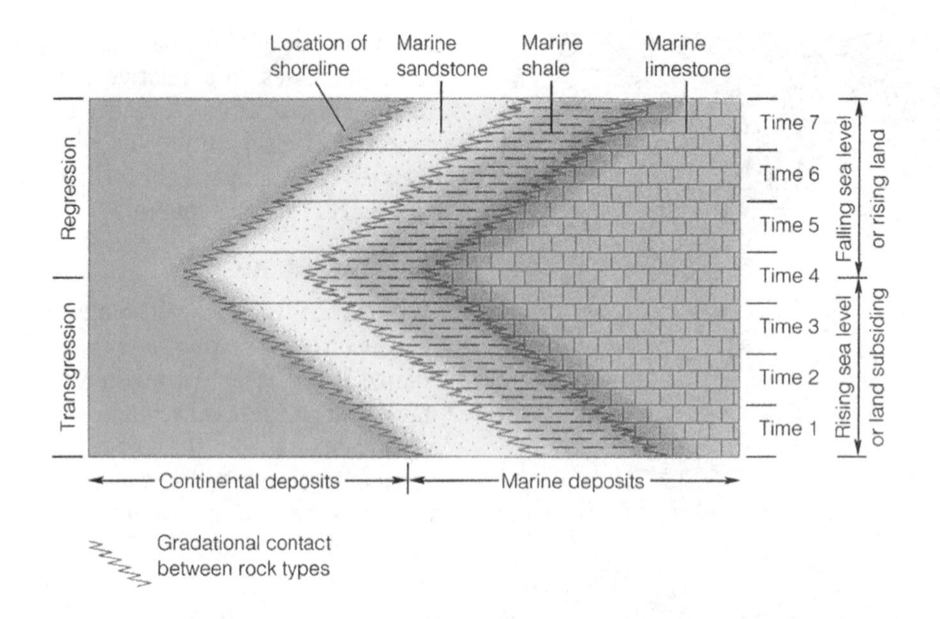

FIGURE 8.30 *The stratigraphic (vertical) sequence of marine rocks exposed in an outcrop provides information about the relationship between the elevation of the land surface and sea level during the period when the sediments were accumulating.*

During a period of rising sea level, the shoreline transgresses or moves inland, and the depositional environments parallel to the shoreline move inland correspondingly. As a result of the transgression, new carbonate deposits that will in time give rise to limestone are positioned over the shales and sandstones of the original depositional environments. During a period of falling sea level, when the sea transgresses or withdraws from the land, the opposite process occurs. Thus, during a transgression, new sediments that will eventually give rise to sandstones are deposited over the shales and limestones of the earlier depositional environments.

from the land seaward. Should the process reverse and the sea *withdraw* from the land, a process called *regression,* the sequence of rocks recording the regression would also reverse and form a regressive sequence with more landward rocks positioned over more seaward rocks (refer to Figure 8.30).

SPOT REVIEW

1. How is the concept that the present is the key to the past used to understand the historical significance of sedimentary rocks?
2. Give an example of the application of Walther's law.
3. What kinds of features within a sedimentary rock can provide information about the climate in the area of the depositional environment at the time of sediment accumulation?
4. What kinds of information contained within sedimentary rocks allow geologists to determine whether a particular sequence of rocks was of continental or marine origin?

ENVIRONMENTAL CONCERNS (Sediments)

Sediment Pollution

Sediments are often the cause of serious pollution problems. Any disruption of the natural land surface has the potential to produce sediments in such volume that they may pollute adjacent land surfaces and water bodies. For example, denuding the land of vegetation during construction, forest clear-cutting, or the conversion of land to agricultural use have all been cited as sources of sediment that pollute streams and lakes. The high suspended loads that are produced as the sediment enters the water body may harm or destroy aquatic life. The useful life span of reservoirs and dams constructed for the generation of hydropower or flood control may significantly shorten as they fill with increased volumes of sediment, while sediment flushed into drainage canals and aqueducts may clog the systems and render them ineffective.

In cases where the generation of sediment cannot be prevented, control measures can be initiated to minimize the potential for pollution. The use of contour plowing significantly reduces the amount of soil lost from agricultural land. The replanting of clear-cut forests and the judicious placement of access roads can reduce the sediment load generated by logging operations. Many urban areas have enacted legislation that requires the establishment of diversion channels and sediment collection ponds in areas undergoing construction. Similar laws have been applied to surface mining operations where large areas are disturbed and large volumes of sediment can be generated when piles of unconsolidated rock materials are exposed to heavy rains and snow melts.

It seems unlikely that the production of sediment by a wide range of human activities can be totally eliminated, and it must therefore be accepted as part of our daily activities. The production of sediment can be controlled, however, and the potential environmental impacts of sediment pollution can be minimized.

Engineering Problems

Sedimentary rocks, especially shales and limestones, may potentially present several engineering problems, including difficulties at construction sites and surface subsidence.

Shales Of the various kinds of sedimentary rocks, shales are, environmentally, the most troublesome. Shales are so prone to failures of various kinds that a design and construction engineer views even the presence of shales in a construction site as a sign of potential trouble. Unfortunately, shales are the most abundant sedimentary rock type.

Most engineering problems arise when poorly cemented shales and mudstones are cyclically wetted and dried, especially if they contain expandable clay minerals. Such rocks can repeatedly absorb and release water, causing the rock to cyclically swell and shrink. Should shale be exposed in the foundation for a building, bridge, or dam, the foundation may be exposed to stresses of sufficient magnitude to result in structural failure.

Road cuts containing expandable shales are prone to slumps and rock slides as the layers of shale absorb water, lose cohesion, and develop slippage planes. The situation is compounded when the bedding dips into the roadway. In fact, good engineering practice usually dictates that, if at all possible, road cuts be oriented such that the bedding does not dip toward the roadbed.

The shales associated with coal often contain appreciable amounts of calcite ($CaCO_3$) and pyrite (FeS_2). When exposed to weathering, products of the chemical decomposition of these two minerals combine to form $CaSO_4$, which subsequently hydrates to gypsum ($CaSO_4 \cdot 2H_2O$). As the gypsum forms, the rock expands. Should the shale layer be located under the foundation of a building or other structure, the swelling of the shales may subject the foundation to stress sufficient to cause serious structural damage.

Limestone Limestones are common sources of environmental problems, especially in humid areas where subsurface dissolution of the beds results in the formation of caves and caverns and the development of karst topography. During times of drought, groundwater that normally fills the cave system and serves to support the overlying rocks may be removed. Depending on the size of the cave and the thickness of the overlying rock layers, the cave roof may collapse, resulting in the formation of surface depressions called sinkholes. Such surface subsidence can not only result in the damage or destruction of surface structures but may also significantly alter the utilization of the surface.

Sediments and Sedimentary Rocks as Conduits of Water Water moving underground passes through sediments and sedimentary rocks called aquifers. In many cases, domestic and industrial pollutants that have entered the groundwater from the surface are removed as the water makes its way through the layers of sediment or rock. The sedimentary materials act, therefore, like a purifier. The sediments surrounding a septic field, for example, remove noxious materials by a combination of filtration, adsorption to the clay minerals, and retarding the movement of the water so that chemical processes such as oxidation are given sufficient time to destroy harmful components.

There is, however, a limit to the ability of sedimentary materials to cleanse groundwater. For example, if the movement of the water through the sediments is impeded, effluents from a septic system may rise to the surface before they have undergone adequate treatment and create both an environmental and a health problem. A similar situation may arise where septic systems are located on steep slopes. Unless measures are taken to route the drainage parallel to the contours of the slope in order to increase the time of contact between the water and the sediments, the effluents may come to the surface downslope before they have been properly treated.

CONCEPTS AND TERMS TO REMEMBER

Sedimentary rocks
Sediment
Classification of sedimentary rocks
  detrital
  nondetrital
Sorting
  roundness versus sphericity
Kinds of detrital rocks
  breccia
  conglomerate
  sandstone
    quartz sandstone
    arkose
    graywacke
  siltstone
  shale versus mudstone
Kinds of nondetrital rocks
  chemical rocks
    chemical limestone
    dolostone
  biochemical rocks
    chert
    biochemical limestone
      coquina
  evaporite rocks
    gypsum
    rock salt or halite
  organic rocks
    coal
Depositional environments
  continental environments/deposits
    fluvial
    glacial
    eolian
    lacustrine
    paludal
  transitional environments
    delta
    beach
    barrier island
    coastal wetland and mudflat
  marine environments
    continental shelf
    continental slope
    abyssal floor
Processes of Lithification
  compaction
  cementation
Sedimentary features
  beds
  ripple marks
    asymmetrical/current
    symmetrical/oscillation
  cross-bedding
  graded bedding
  mud cracks
  animal trails
Walther's law

REVIEW QUESTIONS

1. Which of the major minerals found in sedimentary rocks is usually nondetrital in origin?
   a. quartz c. feldspar
   b. calcite d. clay
2. A well-sorted rock is composed of
   a. one dominant mineral.
   b. a narrow range of particle sizes.
   c. a mixture of particle sizes.
   d. particles that all show the same degree of roundness.
3. A cube would be described as showing
   a. high sphericity but low roundness.
   b. both high sphericity and high roundness.
   c. low sphericity but high roundness.
   d. both low sphericity and low roundness.
4. The difference between a conglomerate and a breccia is
   a. composition. c. particle shape.
   b. bedding thickness. d. particle size.
5. Fluvial deposits are laid down
   a. by the wind.
   b. on the ocean bottom.
   c. by streams.
   d. in wetlands such as marshes and swamps.
6. An arkose in a sandstone that contains an appreciable concentration of
   a. feldspar. c. clay minerals.
   b. calcite. d. rock fragments.

REVIEW QUESTIONS *continued*

7. Compaction is the major lithification process in the formation of
   a. limestones. c. conglomerates.
   b. sandstones. d. shales.
8. Compare the various processes of erosion in terms of their relative ability to sort clastic materials.
9. Why are arkoses not very abundant?
10. What specific sedimentary features would enable you to determine each of the following:
    a. the paleoclimate
    b. the direction of water or wind flow
    c. whether the materials accumulated on land or on the ocean bottom
    d. the specific kind of continental deposit

THOUGHT PROBLEMS

1. In what ways would the assemblage of sedimentary rocks that forms in a hot, dry climate differ from the assemblage that forms in a warm, moist climate?
2. Which detrital sedimentary rock could best be used to determine the kinds of rock that were present in the highland area supplying the sediment?
3. Assume that you are observing a sequence of marine sedimentary rocks in an extensive cliff exposure such as the Grand Canyon. What kind of evidence would you seek to determine the direction to the ancient shoreline, the depth of water at any point in time, and the possible presence of hiatuses in the record?

FOR YOUR NOTEBOOK

Most of us live in areas underlain by sedimentary rocks. The distribution of the various rock types can be obtained from a geologic map of your area. Relate the topography in your area to the distribution of the various kinds of sedimentary rock. Identify the individual rock layers that are responsible for the hills and valleys in your area.

Sedimentary rocks are commonly used for various construction purposes from building stone to crushed stone for road construction. Make a survey of your area and indicate whether sedimentary rock is being quarried and, if so, the purpose for which it is being used. At the same time, a field trip through your town may discover a number of examples of sedimentary rocks being used for construction. Pay particular attention to the kinds of sedimentary rock used for decorative and trim stone.

You might also investigate the various kinds of depositional environments in your area where new sediments are accumulating. Attempt to determine what kinds of rocks will most likely result from these materials.

KEY TERMS		
Arkose	Clastic Sediment (Clast)	Mudstone
Bioclastic Rock	Conglomerate	Oolitic Limestone
Breccia, Sedimentary	Coquina	Shale
Calcarenite	Dolostone	Siliciclastic Rock
Calcirudite	Fissile	Siltstone
Calcisiltite	Graywacke	Wentworth Size Scale
Chert	Micrite	

SEDIMENTARY ROCK IDENTIFICATION

Sedimentary rocks, one of the three major rock categories (Figure 8.32), are formed in one of several ways:

1. **Compaction** of loose sedimentary particles with or without cementation
2. **Direct precipitation** of crystalline materials during the evaporation of water
3. The **accumulation of organic debris** (shells, wood, etc.)
4. Cementation by one or more chemical cements

In the first case, loose particles of sand or mud can undergo simple compaction until the grains are locked together forming a rock.

The same rock material could be cemented together by the growth of calcite or quartz crystals bridging between the grains and filling the pore spaces. The second case generates minerals like

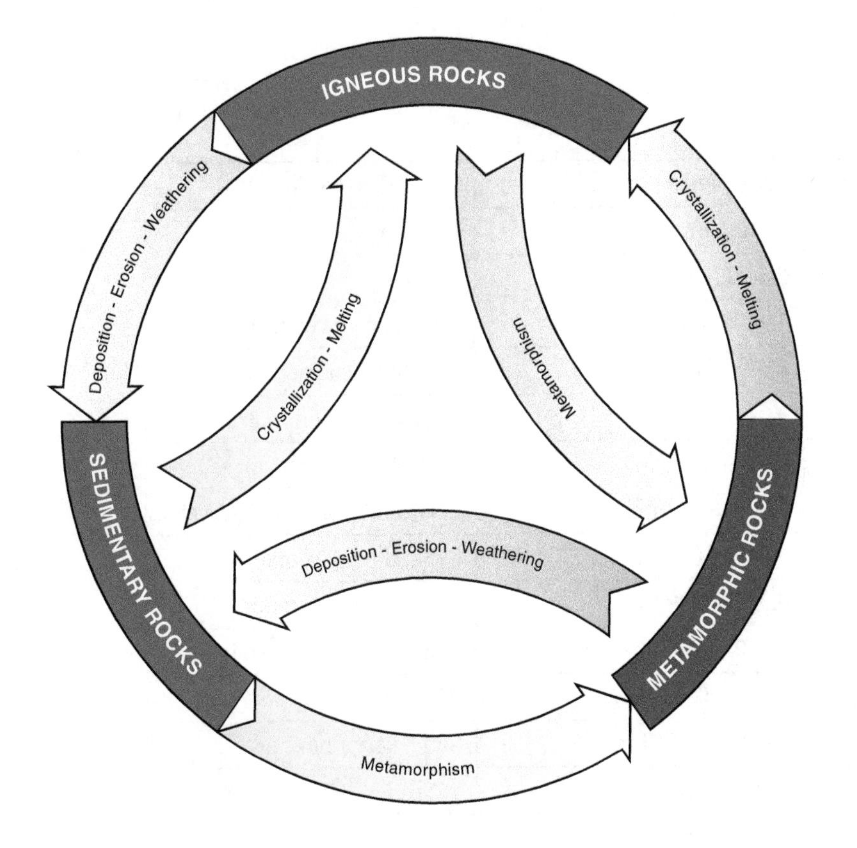

FIGURE 8.32 *Geologic rock cycle.*

gypsum, halite, and dolostone and chemical limestones where the evaporation process produces a supersaturated condition, forcing the less soluble chemical compounds to form crystals and settle out of the water.

Sediment itself can consist of many different things. Simply stated, sediment often consists of loose and fragmented materials that may either have formed in place or may have been transported from some remote point of origin, its **provenance.** The sedimentary particles were originally produced either by **chemical weathering, mechanical weathering, biologic activities** producing or altering some parent material, or some combination of these processes.

Much sediment is transported, and the physical process involved produces effects known as **rounding** and **sorting.** Rounding is the mechanical result of grain-to-grain impact, mechanical weathering, or chemical weathering. Because sharp edges and corners are more easily removed than smooth surfaces, grains become more rounded with further weathering or distance traveled from the source area. This is the main reason that much sand and river rock is round or oval in shape.

In rounding, angular grains retain sharp edges and represent materials close to the primary source material, while subangular and subrounded grains have been altered by weathering and transport to an ever more rounded state. Rounded grains may approach spherical shapes and represent abraded mature sediment transported far from the source terrane.

Sorting is the result of different amounts of depositional energy being available within subdivisions of a transport system. High depositional energy (fast currents and turbulent water) can move large particles while lower energy levels (quiet or slow-moving water) only move the smallest grains. Thus, a river will concentrate coarse sand and gravels in one area while depositing mud elsewhere. Poorly sorted sediments, those containing a mixture of grain sizes and shapes, are generally found close to the source area while well-sorted sediments may consist of a single grain type and shape due to winnowing and transport.

Another important feature of sediments is the grain size. Geologists use the ***Wentworth Scale*** to describe the sizes present in a sample (Figure 8.33).

Size Range (metric)	Size Range (approx. inches)	Wentworth Class
> 256 mm	> 10.1 in	Boulder
64–256 mm	2.5–10.1 in	Cobble
32–64 mm	1.26–2.5 in	Pebble, very coarse gravel
16–32 mm	0.63–1.26 in	Pebble, coarse gravel
8–16 mm	0.31–0.63 in	Pebble, medium gravel
4–8 mm	0.157–0.31 in	Pebble, fine gravel
2–4 mm	0.079–0.157 in	Granule, very fine gravel
1–2 mm	0.039–0.079 in	Sand, very coarse
1/2–1 mm	0.020–0.039 in	Sand, coarse sand
1/4–1/2 mm	0.010–0.020 in	Sand, medium sand
125–250 microns	0.0049–0.010 in	Sand, Fine sand
62.5–125 microns	0.0025–0.0049 in	Sand, very find sand
3.90625–62.5 microns	0.00015–0.025 in	Silt Mud
< 3.90625 microns	< 0.00015 in	Clay Mud

FIGURE 8.33 *Non-graphical Wentworth Scale.*

MAJOR CLASSES OF SEDIMENTARY ROCKS

Clastic Sediments and Detrital Rocks

Clastic materials are those primarily broken away from some parent material by the process of mechanical weathering, and a *clast* is one such fragment (sand, rocks, plant debris, shells or bones).

We have already discussed **pyroclastic** rocks (see Chapter 2), but there are two additional types geologists commonly recognize: ***siliciclastic*** and ***bioclastic.*** Siliciclastic sediments consist of the weathering debris from silicate rocks—that is, clay, silt, sand, gravel, or fragments of silicate rocks. Bioclastic sediments are those composed of fragments of once-living organisms. Bioclastic sediments can consist of such diverse materials as fragments of wood and leaves, animal bones, and shells.

Detritus is a collective term for loose rock and mineral material worn off and removed directly by mechanical means such as sand, silt, and clay derived from older rocks and removed from its place of origin. It is normally considered to consist of clay minerals, fresh and abraded mineral grains, and fragments of preexisting rocks. Rocks formed from these sediments are considered detrital rocks (or clastic rocks).

Clastic rocks are further classified mainly on their grain size and shape. Common classes are taken largely from the **Wentworth Scale** and are: clay, silt, sand, gravel, and larger particles.

Rock names are determined by the size of their particles and to some degree by the process of their formation. Hence, a rock composed of very fine clay-sized particles would be a ***mudstone*** if it is poorly compacted and breaks into rough blocky chunks, but the same material would be a ***shale*** if it broke into flat sheets and could then be said to be ***fissile*** (Figure 8.34).

FIGURE 8.34 *Mudstone and shale.*

Other siliciclastic rocks are named according to their proximity to the source area and according to the presence of key minerals. Examples of these are the rock type ***arkose,*** which is dominated by unweathered and weathered orthoclase feldspar grains; **lithic sandstone,** which contains grains that are recognizable rock fragments; and ***graywacke*** (also referred to as **wackes** and **wackestones**), which are quartz sandstones containing abundant, generally dark-colored muds and clays (hence a "dirty sandstone").

Shape of the rock particles is also important. **Breccia** is defined here much as it was in the last chapter (see volcanic breccia in Chapter 2). Rocks composed of irregular, angular, unsorted debris can form a ***sedimentary breccia,*** while ***conglomerates,*** which may have similar sized fragments, are composed of rounded grains (Figure 8.35).

Chemical and Biochemical Sedimentary Rocks

Chemical sedimentary rocks are those that are entirely produced by chemical means. These are crystalline materials produced directly from water (calcite, aragonite, dolomite, halite, gypsum, chert, opal, etc.). The resulting rocks are termed chemical limestone when the minerals are calcite or aragonite, dolostone when the carbonate is dolomite, rock salt

> **INSIGHTS**
>
> Some geologists have named such fine-grained rocks as **claystones,** but this usage is only proper when the materials are known to be clays—that is, a clay is one of a suite of closely related sheet-silicate minerals. This fact is seldom known and the fine-grained rock may well be composed of very fine particles of other siliciclastic debris. Current usage suggests that fine-grained rocks be called **"mudstones"** when the chemical nature of the sediment is unknown, and that the use of the term "claystone" should be restricted to rocks where clay minerals are known to be the principal siliciclastic material present (clay making up 80 to 90 percent of the rock mass).

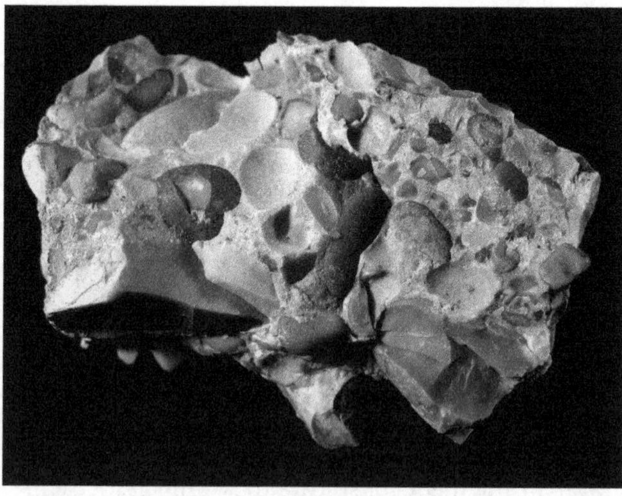

FIGURE 8.35 *Sedimentary breccia (left) and conglomerate (right).*

when the material is halite, rock gypsum when it is gypsum, and chert when the material is microcrystalline quartz.

The rock known as ***dolostone*** is formed from the mineral dolomite, either directly precipitated from seawater or by later modification of chemical limestones by the addition of magnesium, altering the formula from $CaCO_3$ (calcite) to $CaMg(CO_3)_2$ (dolomite).

The sedimentary limestone properly termed **travertine** is an important chemical sedimentary rock often seen in building interiors as floors and wall facings, in stone "eggs" imported from Mexico, and in carved "alabaster" bookends. In the trade, this form of travertine is often called "Mexican onyx" or "alabaster." True alabaster is a rock formed entirely from the mineral gypsum, referred to in this manual as **rock gypsum.**

Biochemical sedimentary rocks consist largely or entirely of material that was once living. Examples of these materials include coal, ***coquina*** (comprised nearly entirely of whole and broken shells), and ***calcirudites*** (rocks containing coarse sand to gravel-sized particles, with large percentages of broken or whole shells) (Figure 8.36).

Miscellaneous Sedimentary Rocks

A few common rocks do not fall under the clastic or carbonate classification schemes, and these are listed in Figure 8.39. If your sample does not effervesce (fizz) and will not scratch glass, it probably belongs in Figure 8.39. The only exception is ***chert,*** which is microcrystalline quartz and will easily scratch glass. Chert may be found in many colors, but it always has the look and feel of porcelain.

By now you should be able to easily recognize halite and gypsum, while peat and coal will be dark and obviously organic in nature. **Peat** is the partially decomposed remains of dead plants, and sedimentary coal, termed **bituminous coal,** is the result of increased pressure and age on peaty materials. Peat is generally brown, but bituminous coals are shiny and black.

> INSIGHTS
>
> Nearly all clastic sedimentary rocks will scratch glass, since they generally contain quartz, but with the exception of chert, almost no chemical sedimentary rocks are as hard as glass.

OTHER ACTIVITIES

Internet Links

Check the Internet for links to sites discussing sedimentary rocks. Use these Internet sources to find information on useful sedimentary rocks in

FIGURE 8.36 *Coquina (left) and fossil-rich calcirudite (right).*

your locality. Several Internet sites have numerous high-quality illustrations of both common and rare types, plus much information on the formation of these interesting rocks.

Urban Field Trips

Your instructor can also be helpful in setting up an urban field trip like that discussed in the last chapter. Most cities have numerous public and private buildings that use stone in their construction. In many cases, these stones are igneous or metamorphic and are composed of minerals you have studied, but many are also constructed of sedimentary rocks. Visit your city center or public buildings and see how many types you can identify from this chapter's study.

INSIGHTS

The commonly used term "flint" should to be reserved for implements and weapons made from the rock chert. Flint is an object name as used by archaeologists, NOT a rock name. Example: "This flint arrowhead was crafted from chert."

Lab 6: Sedimentary Rocks Identification

IDENTIFICATION KEY TO SEDIMENTARY ROCK

Methods:

Identification Procedures and Techniques
The keys presented here are arranged in three parts: (1) siliciclastic sedimentary rocks; (2) carbonate sedimentary rocks; and (3) miscellaneous sedimentary rocks. The steps below should allow you to quickly identify your unknown rock samples:

1. Attempt to scratch a glass plate with the sample. If it scratches the glass, it WILL either be a **siliciclastic** rock or chert. Use Figure 8.37 if the sample is granular; use Figure 8.39 if the sample scratches glass easily but has a smooth, nongranular texture.
   a. These samples will scratch the glass only if the grains are sand sized or larger. Examine the grains, determine their size and shape, and move to the siliciclastic chart.
   b. Some **siliciclastic** rocks are too fine grained to produce easily observed scratches (***siltstones*** and mudstones). Check these in step 4 below.
2. If the rock will not scratch glass, put a drop of dilute hydrochloric acid on the sample. If it **effervesces** (fizzes), it will be EITHER a **carbonate rock** or a **carbonate cemented siliciclastic** rock (if the latter, it should have scratched glass). Use Figure 8.38.
   a. Determine if the rock is well-cemented and dense or poorly cemented and porous.
      - If well cemented and dense, with visible fossil fragments present, determine the grain size and consider **calcirudite, *calcarenite, calcisiltite,*** or ***micrite.***
      - If well cemented and dense, coarsely crystalline without evident fossil fragments being present, determine the grain size and consider **crystalline limestone,** or **micrite.**
      - If poorly cemented and earthy, with visible fossil fragments present, consider **coquina** or **chalk,** depending on the grain size present.
   b. If the rock is composed of small spherical laminated grains, consider an ***oolitic limestone*** (small, spherical, concentrically laminated grains).
   c. **Dolostone** can be tricky. Either scratch the surface before applying the acid or pulverize a small bit of the rock. If the acid is applied to the surface of your sample, it may take as long as 5–10 seconds for the effervescence to begin.
3. Figure 8.39 covers those miscellaneous, common chemical sedimentary rocks. These samples will not effervesce and, with the exception of chert, will not scratch glass. Use the same tests here that you used for those minerals in Chapter 2.
4. Very fine-grained sedimentary rocks can be a challenge. If grain sizes are silt-sized or smaller, test with acid. If they effervesce, they are **PROBABLY** micrites, chalks, or calcisiltites. If they **DO NOT effervesce,** they **MUST** be siliciclastics and will therefore be mudstones, siltstones, or shales.

   Problems arise because some siliciclastics are cemented by calcite and will effervesce when the acid acts on the cement, but not on the rock components.

Once you have decided that the sample is a fine-grained siliciclastic rock, you can easily determine whether its grain size is silt- or clay-sized. With the assistance of your instructor, remove a sample NO LARGER than a pinhead and crush it gently between your incisors [front teeth]. If it crushes to a smooth paste with a few nibbles, the rock was a mudstone or muddy (argillaceous) shale. If it remains slightly gritty, it was a siltstone or silty shale.

Silly as this test may sound, it is the only really useful field technique for a geologist trying to determine the particle size of fine-grained rocks. Don't be afraid of "eating rocks." Give it a try! Just nibble gently and use a SMALL fragment!

Composition	Grain-size Class and Diameter		Comments	Name	
Mainly quartz, feldspar, rock fragments, and clay minerals	gravel (>2mm)		rounded grains	CONGLOMERATE	
			angular grains	BRECCIA	
	sand (0.0625-2.00, or 1/16-2, mm)		mostly quartz grains	QUARTZ SANDSTONE	SANDSTONE
			mostly feldspar grains	ARKOSE	
			mostly rock fragments	LITHIC SANDSTONE	
			mixed with much silt and clay	GRAYWACKE	
	mud and silt	silt (0.0039-0.0625, or 1/256-1/16, mm)	nonfissile (compact)	SILTSTONE	MUDSTONE
			fissile (splits easily)	SILTY SHALE	
		mud (<0.0039, or 1/256, mm)	nonfissile (compact)	MUDSTONE	
			fissile (splits easily)	MUDDY SHALE	

FIGURE 8.37 *Chart I: siliciclastic rocks.*

Composition	**Distinctive properties**	**Grain-size**	**Name**	**Rock type**
Calcium carbonate $CaCO_3$	Shell debris coral fragments and/or calcareous microfossils; reaction with diluted HCl	> 2 mm mostly gravel-sized shells or coral fragments	COQUINA	BIOCLASTIC
		Mostly sand-sized shell fragments	FOSSILIFEROUS LIMESTONE OR CALCARENITE	
		Silty, composed of the microscopic shells of calcareous plankton	CHALK	
		No visible grains; may contain visible fossils	MICRITE	BIOCHEMICAL
	Calcite spheres with concentric laminations; reaction with diluted HCl	< 2mm Mostly spherical grains	OOLITIC LIMESTONE	CHEMICAL
	Calcite crystals, may have cavities or pores, color banding is common; reaction with diluted HCl	Visible crystals or microcrystalline	TRAVERTINE	
	Calcite crystals formed as inorganic chemical precipitates; reaction with diluted HCl	> 2 mm coarse to fine-grained (<0.0039 mm)	CRYSTALLINE LIMESTONE	
		Very fine-grained	MICRITE	
Mainly dolomite $CaMg(CO_3)_2$	Effervesces in diluted HCl only if powdered; usually light colored	Microcrystalline	DOLOSTONE	

FIGURE 8.38 *Chart II: carbonate rocks.*

Composition	Comments	Grain-size	Name
Mainly varieties of quartz, SiO_2 (chalcedony, flint, chert, opal, jasper, etc.)	Commonly occurs as layers, lenses, nodules	microcrystalline or amorphous	CHERT
Mainly halite, NaCl	crystals formed as inorganic chemical precipitates	all sizes	ROCK SALT
Mainly gypsum, $CaSO_4 \cdot 2H_2O$	crystals formed as inorganic chemical precipitates	all sizes	ROCK GYPSUM
Mainly plant fragments	brown and porous	all sizes	PEAT
	black and nonporous	all sizes or dense with conchoidal fracture	BITUMINOUS COAL

FIGURE 8.39 *Chart III: miscellaneous sedimentary rocks.*

CAUTION

Don't be in too great a hurry with the acid testing. If the carbonate is **dolostone,** the effervescence will not start immediately, and it will not be as evident as that seen for calcite rocks (see section C).

INSIGHTS

Put the acid in a crack or hole in the rock to increase the surface area in contact with the acid.

INSIGHTS

Put a drop of acid on the unknown rock sample and wait until all fizzing stops. Take a clean paper towel and firmly blot the spent acid (don't wipe or scrub the rock). Examine the wet area of the paper towel. If it is stained by mud the same color as the rock sample, then the rock was cemented and the rock grains were free to move to the paper towel as a pigment.

If the wet area on the paper towel is more or less colorless, the rock was a chemical limestone, and BOTH the cement and the rock grains were dissolved by the acid.

Data Sheets for Sedimentary Rocks

Sample #	Rock Name	Texture	Composition	Other Properies How did you determine this?

Data Sheets for Sedimentary Rocks

Sample #	Rock Name	Texture	Composition	Other Properies How did you determine this?

Data Sheets for Sedimentary Rocks

Sample #	Rock Name	Texture	Composition	Other Properies How did you determine this?

Data Sheets for Sedimentary Rocks				
Sample #	Rock Name	Texture	Composition	Other Properies How did you determine this?

CHAPTER 9

Metamorphic Rocks

INTRODUCTION Simply stated, metamorphism is *change*. We have talked about change before: *crystallization* is the change of magma or lava into an igneous rock, *weathering* is the change from one mineral assemblage to another resulting from the chemical attack of oxygenated or carbonated waters, and *lithification* is the change of sediment into sedimentary rocks.

Two points set the scene for our discussion of metamorphism and metamorphic rocks: (1) in all of the transformation studied so far, the change was a response to the combined effects of *heat, pressure,* and *chemically active fluids,* primarily water; and (2) transformations take place when materials created under one set of conditions are exposed to a new set of conditions.

In the case of sedimentary rocks, calcium carbonate crystallizes from solution as the water temperature rises. Mud, accumulated at Earth's surface, is converted into shale by the increased pressure of burial. Various cements that convert loose sediment such as beach sands into a sandstone precipitate from groundwater.

In the case of igneous rocks, heat, pressure, and water all contribute to the creation of magma. Heat breaks bonds within crystal lattices and allows the rocks to melt. The decrease in pressure as convection-driven mantle rocks approach the top of the asthenosphere results in their melting. The waters provided by the dehydration of hydrous minerals and the hydrothermal solutions released from crystallizing magmas aid in the formation of magma. Thus, to one degree or another, heat, pressure, and chemically active fluids in various combinations play critical roles in the formation of both sedimentary and igneous rocks.

Most minerals are stable unless exposed to conditions different from those under which they formed. Minerals within plutons are stable as long as they remain deep within Earth. When igneous rocks are exposed at the surface, however, they become unstable and are changed by the process of weathering into a new assemblage of materials that are stable. Similarly, the products of weathering remain relatively unchanged at Earth's surface. If they are buried at depths of a few thousand feet or meters, however, where they are subjected to increased temperatures and pressures and exposed to mineral-laden waters, the sediments change into rocks that are stable under the new conditions.

Now we are about to introduce the third kind of rock, **metamorphic rock.** Once again, we will see that heat, pressure, and chemically active fluids play important roles. We will see that the process of metamorphism involves changes that take place when rocks are subjected to conditions *different* from those under which they formed. In this respect, the process of metamorphism can be viewed as the inverse of weathering in which minerals that formed at conditions of high pressure and temperature are exposed to the atmosphere and converted to a mineral assemblage stable at atmospheric conditions.

METAMORPHISM

Metamorphism is a process of *transformation* by which the mineral composition, texture, or both of an existing rock are changed, creating a new rock through the application of (1) *heat,* (2) *pressure,* and/or (3) *chemically active fluids.* Of the three, heat is the most important, being involved even when pressure and chemically active fluids are the dominant agents. For the most part, metamorphism takes place deep within the confines of active mountain belts at the convergence of lithospheric plates, especially during continent-continent collisions.

A requirement of the process of recrystallizataion involved in metamorphism is that it proceed in the *solid state.* During metamorphism, the mineral assemblage of the original rock recrystallizes into a new mineral assemblage without melting. In some cases, the new minerals are identical in composition to the original minerals, differing only in grain size. The best example is the transformation of limestone to marble. Both rocks are made of calcite ($CaCO_3$). The original limestone may have been a chemical sedimentary rock made up of fine-grained calcite or biochemical limestone consisting of a mixture of fossils, shell fragments, and chemically precipitated carbonate cement. The process of recrystallization destroys the fossil remains and the original calcite grains and replaces them with new crystals of calcite, all of which will be relatively uniform in size.

In other cases, the recrystallization process results in both a change in texture and the creation of new and different minerals. A common example is the metamorphism of shale to produce *slate.* During metamorphism, the clay minerals originally present in the shale are changed into minerals such as mica and chlorite.

These examples illustrate a general characteristic of most metamorphic processes, namely, that the transformations take place with little, if any, addition to or removal of material from the original rock. The original rock and the subsequent metamorphic rock has the same *elemental* composition. Only the texture or the mineralogy has been changed.

Some metamorphic transformations do involve changes in elemental composition, however. In some cases, new elements are *added* to the rock while in other cases, elements are *removed* or *replaced* with different elements. In such cases, mineralogy of the metamorphic rock will always be different from that of the original rock.

As these brief descriptions of the process of metamorphism illustrate, metamorphism is simply another example of change involving the basic agents of heat, pressure, and chemically active fluids, which are also needed for the formation of igneous and sedimentary rocks. These three conditions can be considered as lying along an intensity continuum. As we discuss metamorphism, it will become evident that the conditions required to produce metamorphic rocks are usually *more intense* than those involved in the formation of sedimentary rocks and *less intense* than those involved in the generation of a magma.

Most geologists believe that the transformation from sedimentary to metamorphic conditions occurs at temperatures of about 400°F (200°C) and pressures equivalent to 6,000 to 9,000 feet (2,000–3,000 m) of burial. At some point, the conditions of temperature, pressure, and solution activity that convert sediments into sedimentary rocks may result in mineralogic or textural changes within the rock that go beyond those attributable to lithification. A sedimentary rock that shows evidence of having been subjected to metamorphism is called a *metasediment.*

Because of the transitional relationship that may exist between igneous and metamorphic conditions, some rocks may partially melt as they undergo metamorphic changes. The rock that is produced is called a **migmatite** (meaning "mixed rock") and is considered to be both igneous and metamorphic. Some metamorphic petrologists believe that migmatites can also form by the reaction between the host rock and chemically active fluids, a process called *metasomatism,* to be discussed shortly. The existence of migmatites notwithstanding, however, it is important to remember that the conditions cannot result in the melting of the original rock. Should melting occur, the newly formed rock will be igneous rock, not a metamorphic rock. The conditions may cause the rocks to react plastically and flow, but all metamorphic changes must take place in the solid state.

METAMORPHIC TEXTURES AND CLASSIFICATIONS

Metamorphic textures are of two types: (1) *foliated* and (2) *nonfoliated.* Foliated rocks exhibit more or less pronounced parallelism of platy minerals (those having sheet structures) and commonly split along planes of weakness into layers, sheets, or flakes, a property called **rock cleavage**. The rock cleavage responsible for foliation results from the orientation of platy minerals that exhibit excellent mineral cleavage, such as the micas and chlorite.

Foliated Texture

Foliated rocks develop primarily as a result of dynamo-thermal metamorphism (described later in the chapter). In dynamo-thermal metamorphism, stress can be applied in two different ways, by *direct compression* and by *shear*. In direct compression, the forces act toward and directly opposite each other in much the same fashion you would grip a pencil. In shear, the forces act toward each other, but not directly opposite. As the rocks are exposed to the directional stress of direct compression and the platy minerals begin to crystallize, they grow preferentially in the direction of *least* stress, that is, *perpendicular* to the direction of *maximum* stress (Figure 9.1b). The rock develops cleavage with the plane of cleavage parallel to the plane of mineral cleavage and perpendicular to the direction of maximum stress (Figure 9.2). Note in Figure 9.2 that the cleavage cuts across the bedding in the folded strata. When recrystallization takes place under shear, the platy minerals are parallel to each other and at a small angle to the direction of stress, as shown in Figure 9.1c.

Slaty Cleavage

Depending upon the relative abundance and preferred orientation of the platy minerals and the perfection of the cleavage, the rocks can exhibit varying degrees of foliation. The fullest development of foliation is called **slaty cleavage** because it is exhibited by slate where the orientation of the tiny platy minerals is so perfect that the resultant cleavage allows the rock

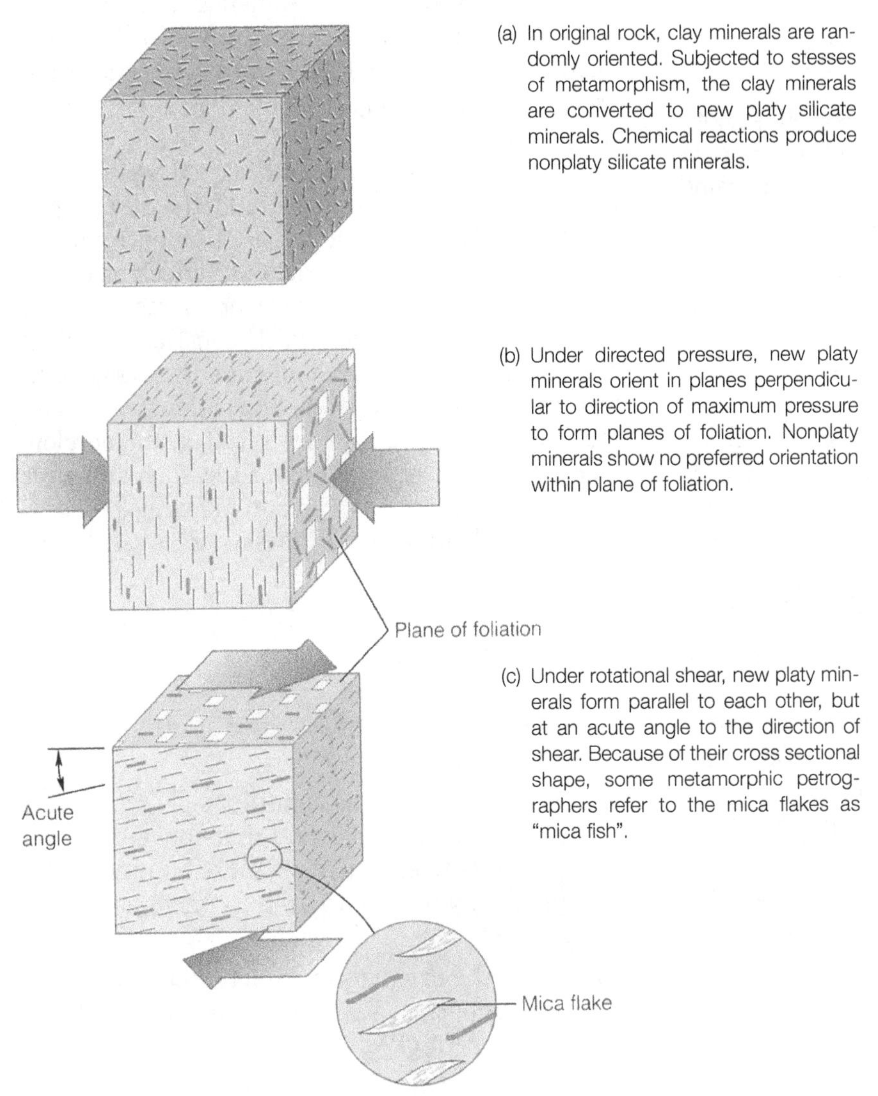

FIGURE 9.1 *When rocks enriched in clay minerals undergo metamorphism, the clay minerals recrystallize to form new types of platy minerals, such as chlorite. The orientation of the new minerals within the rock depends on the kind and direction of the applied force.*

FIGURE 9.2 *As new-formed metamorphic rocks fold, the orientation of the cleavage planes is maintained relative to the forces of deformation. As a result, the cleavage planes cut across the folded beds and are oriented approximately parallel to the axial planes of the folds.*

to be split into slabs of uniform thickness with smooth surfaces. Before the days of painted chalkboards, slate was a common blackboard material as well as being used for roofing tiles and paving stones.

Phyllitic Cleavage Continued metamorphism of slate produces the metamorphic rock, **phyllite.** Phyllite resembles slate, but close examination reveals coarser mica crystals, as well as nonplaty minerals such as garnets, quartz, and feldspars, which serve to decrease the degree of perfection of the cleavage. Rather than splitting in the parallel-faced slabs characteristic of slate, phyllite splits into sheets with more irregular surfaces that exhibit a characteristic sheen. Because it is characteristic of phyllite, this type of cleavage is referred to as **phyllitic cleavage.**

Schistose Cleavage More intense conditions of metamorphism result in the coarsening of micas and nonplaty minerals and the formation of **schist.** Schists can form from a variety of predecessors and, as a result, vary widely in composition. Due to the larger grain size, schist normally splits into fragments with irregular, shiny faces rather than the sheets characteristic of slate and phyllite. This grade of cleavage is referred to as **schistose cleavage** or **shistosity.**

Gneissic Banding The poorest development of cleavage is found in the metamorphic rock called **gneiss** where the minerals have segregated into bands of dark- and light-colored minerals. In fact, some gneisses show no rock cleavage at all.

Gneiss is the most abundant of all metamorphic rocks, forming from a wide variety of premetamorphic rocks and under a wide range of metamorphic conditions. Generally, however, gneisses are the result of high temperatures and pressures. Because gneiss consists largely of nonplaty minerals such as quartz, feldspars, amphiboles, and pyroxenes, it exhibits minimal development of foliation, referred to as **gneissic banding.**

Nonfoliated Texture

Nonfoliated metamorphic rocks consist primarily of interlocking, equidimensional grains. These rocks commonly form under metamorphic conditions where little directional stress is present to orient the newly formed minerals and also where the original rock lacks layered silicate minerals. A well-known example of a nonfoliated metamorphic rock is marble. Sculptors seek specimens of marble that show exceptional uniformity in nonfoliated texture for their work.

In general, monomineralic rocks have less tendency to layer preferentially, and the layering in those that do is difficult to see. Examples include *metaquartzites,* which consist of interlocking quartz grains; *marble,* which is made up of recrystallized calcite; and *serpentinite,* which consists of various serpentine minerals including antigorite, chrysotile, and lizardite. In some cases, impurities such as clay minerals in the original rock result in the development of a degree of foliation in rocks such as marble and metaquartzites. Any foliation in marble produces zones of weakness and precludes its use for sculpting.

SPOT REVIEW

1. What is a "migmatite" and what is its significance?
2. At what point do the conditions of temperature and pressure that convert sediments to sedimentary rocks become sufficiently intense to produce metamorphic changes?
3. Why is the orientation of platy minerals more important than their abundance in the development of foliation?

KINDS OF METAMORPHISM

There are three primary kinds of metamorphic processes: (1) *contact metamorphism* where *heat* is the dominant agent, (2) *dynamo-thermal metamorphism* where both *pressure* and *temperature* are important, and (3) *hydrothermal metamorphism* where the dominant agents are *chemically active fluids* (largely superheated water) derived from the crystallization of magmas.

Contact Metamorphism

Contact metamorphism is primarily associated with magmatic intrusions. The host rocks in contact with the molten mass are subjected to heat from the magma and are literally "baked," as you would bake a ceramic piece in a kiln. Just as the ceramic undergoes chemical and physical changes as it bakes, changes take place within the host rocks as they adjust to the elevated temperatures to which they have been subjected. The zone of metamorphic rocks that forms around an intrusive body is called a **halo** or an **aureole** (Figure 9.3).

A metamorphic rock formed by contact metamorphism of an argillaceous (containing clay minerals) rock is referred to as a **hornfels.** Hornfelses are usually harder, denser, and darker in color than the original rocks. The extent of the changes that take place and the volume of host rock affected depend upon a number of factors including (1) the volume and temperature of the intruding magma body, (2) the amount of water in the host rock, and (3) the grain size and composition of the intruded rock. For example, if a shale is intruded by a relatively small volume of magma such as the amount that would produce a dike or sill, metamorphism of the host rock will be limited to a relatively narrow zone where the clay minerals are altered to mica and the quartz grains are recrystallized to form larger grains. If the shale in intruded by a volume of magma of batholithic dimensions, however, the amount of heat will be so great that the metamorphic process will involve a relatively thick halo. Within the halo, the clay minerals and the quartz will react with each other to form a totally different assemblage of minerals including micas, garnet, and andalusite/sillimanite. In both cases, the rock is called a hornfels, but the garnet, andalusite, and sillimanite in the latter rock record its formation at significantly higher temperatures. Note that the minerals that form are those that are stable at the new, higher temperatures.

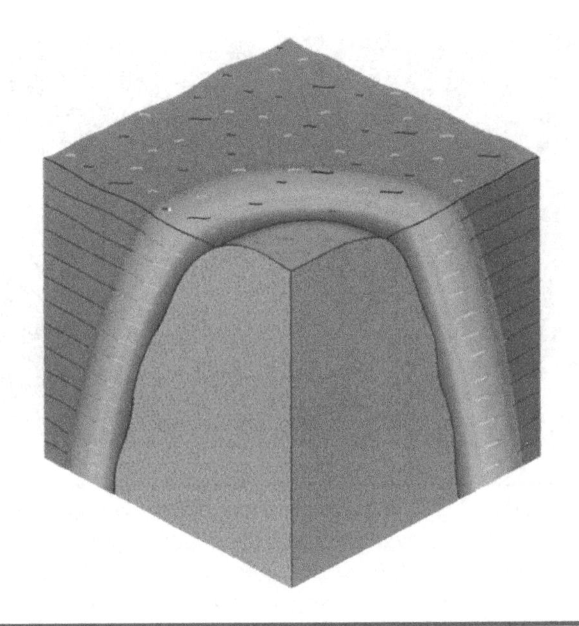

FIGURE 9.3 *The host rock immediately adjacent to an intruding mass of magma is altered to form a zone of metamorphism called a* halo. *The thickness of the halo depends on the amount of heat available from the mass of the intruding magma.*

Water tends to accelerate most metamorphic processes by serving as a transport medium for ions within the rock mass. As a result, the volume of rock within the aureole will be larger if the intruded rock was wet. Because the rate at which rocks undergo metamorphism increases with decreasing grain size, the extent of metamorphism beyond the magma-host rock contact for a given set of metamorphic conditions will be greater for fine-grained host rocks.

Dynamo-Thermal Metamorphism

The primary agents in **dynamo-thermal metamorphism** are pressure and heat. One type of dynamo-thermal metamorphism is **regional metamorphism.** Regional metamorphism is a general term for the metamorphism of rocks that are exposed over an extensive area as opposed to *local* metamorphism where the rocks of a limited area are affected. Metamorphic rocks formed by dynamo-thermal metamorphism are most commonly found in the cores of foldbelt mountains where the rocks have been subjected to the intense pressures generated by plate convergence, especially continent-continent collisions. A variety of foliated rocks is formed by dynamo-thermal metamorphism. A typical and very common example is *schist* (Figure 9.4). Schists are

FIGURE 9.4 *Individual kinds of metamorphic rocks can form from a wide variety of precursors depending on the composition of the original rock and the conditions of metamorphism. A common metamorphic rock that can form from a wide variety of precursor rocks is* schist. *(VU/© Albert J. Copley)*

FIGURE 9.5 *The intricate bedding patterns often seen in metamorphic rocks indicate that the conditions of temperature and pressure were so high that the rock actually flowed plastically. (VU/© Albert J. Copley)*

strongly foliated metamorphic rocks and can be of variable composition.

In some cases, the conditions of pressure and temperature involved in dynamo-thermal metamorphism are so intense that the newly formed rock flows plastically and creates the familiar flowage patterns often seen in marble (Figure 9.5). Metamorphic rocks can also be formed by dynamo-thermal metamorphism at lower temperatures. In major fault zones, for example, a dense, fine-grained metamorphic rock called *mylonite* forms when rocks, caught between the two moving fault blocks, are pulverized and sheared.

Hydrothermal Metamorphism

In both contact and dynamo-thermal metamorphism, the recrystallization usually takes place with little addition or loss of material. With the exception of some loss of the more volatile components such as water, the process simply rearranges the original elements into a new assemblage of minerals. As a result, although the *mineralogy* may change, the *elemental* composition of the metamorphic rock is the same as that of its predecessor.

In **hydrothermal metamorphism**, on the other hand, materials may be both *added to* and *removed from* the original rock composition by the intervention of "chemically active fluids." As the name *hydrothermal* implies, the fluid responsible for the movement of ions is hot water. The water may come from several sources including ocean water drawn down into the zones of subduction and the thermal decomposition of hydrated minerals originally contained within subducted oceanic sediments.

A major source, however, is the water that concentrates during the final stages of magma crystallization. Water is one of the ingredients dissolved in molten rock. As the magma crystallizes, little water is needed for the formation of the major silicate minerals; this is apparent from their formulas, few of which contain either waters of hydration indicated by ($\cdot nH_2O$) or the hydroxyl radical (OH^{1-}). Because water is not needed for the formation of the early formed silicate minerals, it becomes increasingly concentrated in the remaining magma as the minerals precipitate. Eventually, crystallization progresses to the point where what remains is no longer molten rock but rather a hot water (hydrothermal) solution containing the remaining elements. This hydrothermal solution then permeates into and reacts with the host rock to produce new assemblages of minerals.

An example of hydrothermal metamorphism that simply involves the reaction of water with the intruded rock is the conversion of an olivine-rich dunite into serpentine and brucite by the reaction:

$$2Mg_2SiO_4 + 3H_2O \rightarrow Mg_3Si_2O_5(OH)_4 + Mg(OH)_2$$
$$\text{olivine} + \text{water} \rightarrow \text{serpentine} + \text{brucite}$$

Serpentine is a deep green color. When shot through with fracture-filling white calcite, the rock is the beautiful green stone known in the industry as *verde antique,* a rock that is widely used as a decorative stone. Cut and polished, serpentine has been used for years as a decorative trim stone and can be found in many older public buildings. The columns around the rotunda of the National Gallery of Art in Washington, D.C., are spectacular examples.

Metasomatism In a special type of hydrothermal metamorphism, called **metasomatism,** the original mineral assemblage of a host rock is replaced atom-for-atom with a new mineral assemblage.

Another, somewhat more controversial example of metasomatism is *granitization.* In our discussions if igneous rock, we indicated that granite is the most common intrusive igneous rock, emplaced either forcefully or by the process of stoping. However, certain granite bodies can be found where features within the host rock such as bedding can be traced into the granite body. The presence of the residual bedding appears to exclude both forceful entry and stoping as the means of emplacement.

Several possible scenarios could result in relic host rock structures being found within the granite body. The host rock in contact with the magma could have melted in place with the melt moving little before solidification; structures of the host rock would then be preserved in the outer portion of the igneous rock body. On the other hand, the mineral assemblage of the original rock could have been subjected to metasomatic replacement with the original rock being replaced literally atom-for-atom by solutions containing the granitic assemblage of minerals. In this case, the granite would be metamorphic rock, not igneous rock.

Although some geologists once believed that the majority of granite bodies were the result of metamorphic processes, the concept of granitization has now been largely discredited. Most geologists, although agreeing that some granite bodies can form by metasomatism, believe that the vast majority of granites result from the crystallization of magmas.

SPOT REVIEW

1. How do the grain size and the presence or absence of water within the host rock affect the volume of rock modified by contact metamorphism?
2. What are the major sources of water in hydrothermal metamorphism?
3. What is unique about the process of metasomatism?

METAMORPHIC GRADE

When rocks produced by dynamo-thermal or regional metamorphism are exposed by erosion, the regional variation in the mineral assemblages records information about the intensity of the heat, pressure, and deformation experienced by the rocks. Most often, these rocks represent the core of an ancient mountain range, now exposed by erosion. An area showing the effects of regional metamorphism is the *Canadian Shield,* which is the exposed craton of the North American continent. A bewildering array of igneous, metamorphic, and sedimentary rocks that have been subjected to metamorphism are found within the Canadian Shield, some of which date back to some of the earliest mountain-building episodes to occur on Earth. Some of the metamorphic rocks represent rocks that formed deep within collision zones at the extreme metamorphic conditions of heat and pressure. These rocks grade laterally into other assemblages or metamorphic rocks, which, based upon their mineral contents, are interpreted to have formed under less severe conditions (Figure 9.6).

The **metamorphic grade** of a rock is determined by identifying its **index minerals,** which are silicate minerals that form only under specific metamorphic conditions. The *micas* and *chlorite* are indicators of **low-grade** metamorphism. **Intermediate-grade**

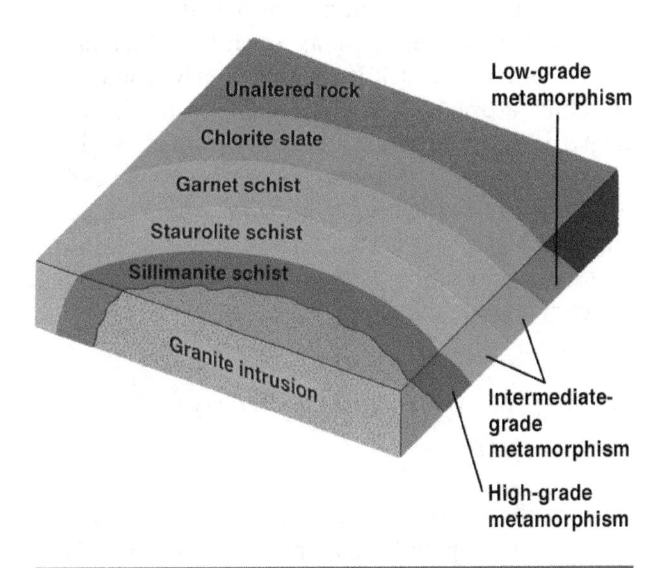

FIGURE 9.6 *In regional metamorphism, the kinds and compositions of the metamorphic rocks record a metamorphic* gradient *with the grade of metamorphism progressively decreasing from a central locale, where conditions were most intense, outward to rocks that were unaffected.*

metamorphism is indicated by the presence of *garnets* and *staurolite.* Fibrous crystals of *kyanite* and *sillimanite* indicate **high-grade** metamorphism, the most intense metamorphic conditions.

METAMORPHIC FACIES

Metamorphic rocks that form under the same conditions of temperature and pressure (depth of burial) are said to belong to a **metamorphic facies.** It is important to remember that most metamorphism does not involve a change in the elemental composition of the original rock. As a result, the kind of metamorphic rock contained within a particular metamorphic facies depends on the composition of the original rock. Although a suite of metamorphic rocks may differ in kind, they will exhibit comparable index minerals if they formed under identical conditions of metamorphism. Thus, a geologist can determine that the rocks formed under the same conditions by examining the index minerals.

It is also important to remember that metamorphism is a dynamic process. Consequently, the mineral assemblage of a rock being subjected to changing conditions of temperature and pressure will change as it approaches equilibrium with the new conditions. When the mineral composition changes in response to *increasingly intense* conditions of temperature and pressure, such as deeper burial, the process is called **prograde metamorphism.** On the other hand, if a metamorphic rock is exposed to new conditions of temperature and pressure that are *less intense* than those under which the rock formed, such as when burial depth decreases as erosion removes the overlying rocks, the original minerals may change to an assemblage that is stable at the lower conditions of temperature and pressure. Such a change is called **retrograde metamorphism.**

METAMORPHISM AND PLATE TECTONICS

Perhaps the major contribution that plate tectonics has made to our understanding of metamorphism is that it has allowed geologists to explain the spatial distribution of metamorphic facies within regional metamorphism. Figure 9.7 illustrates the various metamorphic facies and their temperature/pressure boundaries. Throughout this discussion, the circled

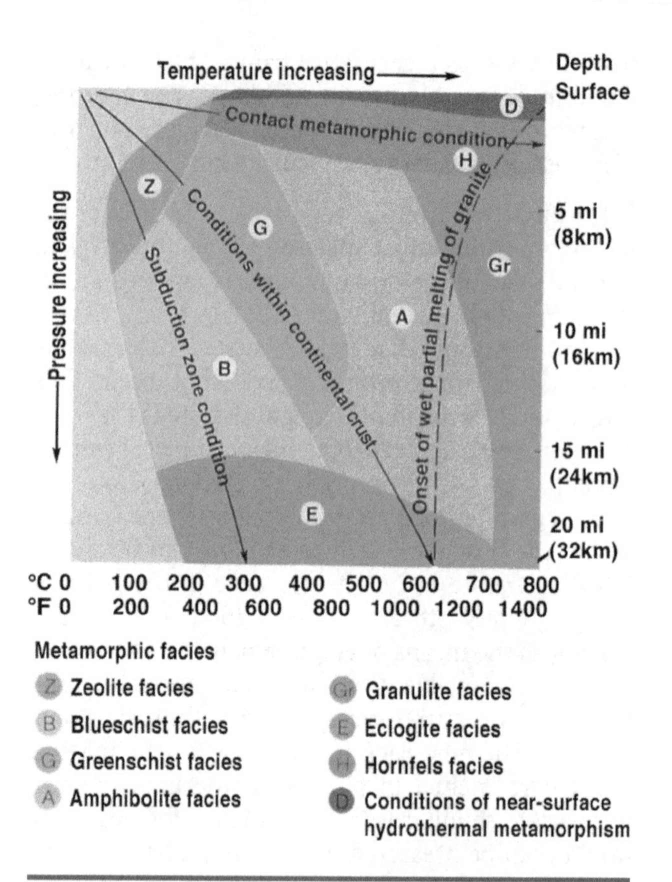

FIGURE 9.7 *Every metamorphic rock belongs to a particular* metamorphic facies *depending on the conditions of temperature and pressure under which it formed.*

letters in the text refer to the corresponding letters in the figure.

Although some metamorphic rocks form at divergent plate boundaries and hot spots where host rocks are intruded by magmas and within some fault zones, most metamorphic rocks are formed at convergent plate boundaries within the overriding plate (Figure 9.8). Within the overriding plate, the temperature, pressure, and availability of chemically active fluids depend on location. For example, near-surface rocks within the continental crust may be modified by hydrothermal metamorphism as they are permeated by solutions being released from crystallizing magma bodies Ⓓ. Many of our rich metal deposits are formed by these processes.

Within a deep-sea trench, land-derived sediment combines with deep-sea sediments and fragments of oceanic basalts stripped from the subducting oceanic plate to form a *mélange* (French for mixture), which may be subsequently transformed into zeo-

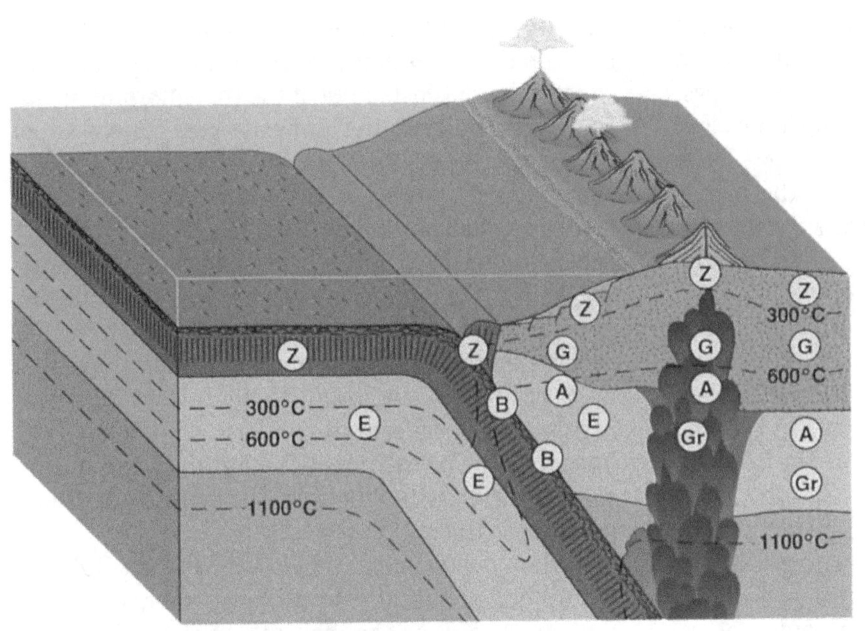

Metamorphic Facies

Ⓩ Zeolite
Ⓑ Blueschist
Ⓐ Amphibolite
Ⓖ Greenschist
Ⓔ Eclogite
(Gr) Granulite

FIGURE 9.8 *Most metamorphic rocks form at the convergence of lithospheric plates where the conditions of temperature and pressure necessary for extensive metamorphism are achieved.*

lite facies metamorphic rocks under relatively low conditions of temperature and pressure Ⓩ. On the other hand, as rocks are subducted to depths of 10 to 20 miles (15–30 km), they are subjected to pressures in excess of 5 kilobars (37 tons/in^2). Because the oceanic rocks are initially cold, however, and warm only slowly as they descend into the mantle, the temperatures at these depths reach only moderate levels Ⓑ. Under these conditions of high pressure and relatively low temperature, the rocks are converted to the blueschist/eclogite facies Ⓑ–Ⓔ. Because of our understanding of plate tectonics, geologists now use the presence of blueschists to identify a sequence of rocks that formed in an ancient zone of subduction. An example of such a rock sequence is the Franciscan Formation that makes up much of the Coastal Range mountains of California.

As the plates converge, the rocks of the continental crust are subjected to compression that generates mountains along the margin of the continental crust. As the continental crust thickens, the rocks within the core of the developing mountains are buried deeper and therefore are subjected to increasing pressures. At the same time, temperatures rise as the core rocks are intruded by magmas. As a result of these high temperatures and pressures, the rocks are converted to metamorphic rocks of the greenschist/amphibolite ©–Ⓐ facies. At the present time, vast volumes of greenschist/amphibolite facies rocks are being created within the core of the Himalayas as the collision of the Asian and Indian plates continues.

SPOT REVIEW

1. What is meant by prograde and retrograde metamorphism?
2. How can a geologist determine the intensity of conditions under which a particular metamorphic rock formed?
3. Where on Earth are the greatest volume of metamorphic rocks formed?

ENVIRONMENTAL CONCERNS

Like igneous rocks, most metamorphic rocks undergo relatively high rates of chemical weathering due to the large difference between the conditions of temperature and pressure under which they form and atmospheric conditions. As a result, thick layers of weathering products quickly develop on exposed surfaces. Should the rocks crop out on steep slopes, the products of weathering are prone to slumps and landslides. Should the rocks exhibit foliation, rock slides are common as water penetrates the planes of foliation, promotes weathering that subsequently decreases the cohesion between the foliation surfaces, and generates planes of slippage.

A good example of the type of engineering problems that are commonly associated with metamorphic rocks, and an excellent example of how preconstruction geologic investigations can prevent deadly and costly structural failures, is the collapse of the St. Francis Dam near Saugus, California.

The St. Francis Dam was built in a narrow section of the San Francisquito Canyon. The geology of the site is illustrated in Figure 9.9. A worse geologic setting for the construction of a dam could not be imagined. The floor and east wall of the canyon were underlain by a schist with planes of foliation oriented parallel to the length of the canyon and dipping westward toward the canyon floor. Numerous landslide scars on the slopes of the eastern canyon wall attested to the inherent instability of the metamorphic rocks. The west wall of the canyon was composed of what appeared to be coherent sedimentary rocks. Unfortunately, strength tests conducted *after* the failure of the dam showed that, like many rocks that appear competent under arid conditions, the sedimentary rocks became unstable when continuously exposed to water. To make the situation worse, the sedimentary and metamorphic rocks were separated by a fault zone containing about 5 feet (1.5 m) of fault breccia. Ironically, the presence of the fault was known before construction began, but the builders apparently chose to ignore it. The setting was ready-made for disaster.

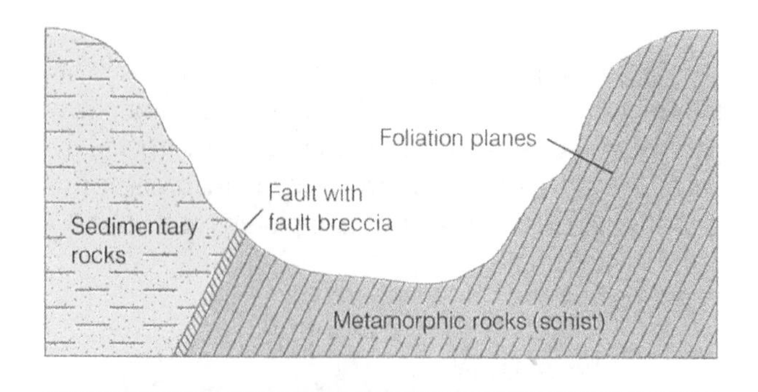

FIGURE 9.9 *The geology of the construction site of the St. Francis dam in San Francisquito Canyon could not have been more unfavorable for the location of a dam.*

When completed in 1927, the dam was 190 feet (63 m) high, 650 feet (214 m) long, and held 62 million cubic yards (47 million m^3) of water. The pressure at the base of the dam was about 5.7 tons per square foot.

Within one year of the completion of the dam, water had weakened the sedimentary rocks anchoring the west end of the dam. Water penetrated the fault, moved under the dam, and resulted in the removal of crushed rock from the dam foundation. Some believed, however, that the major cause of the dam failure was the development of slip planes in the schists anchoring the east end of the dam as water permeated the rock foliation planes.

No one agreed on the exact sequence of events that preceded the failure of the dam on March 12, 1928. The torrent of water that surged down San Francisquito Canyon killed more than 500 people and caused an estimated $10 million in property damage. Had modern preconstruction geologic investigations been applied to the dam site, the St. Francis Dam would never have been built.

CONCEPTS AND TERMS TO REMEMBER

- Metamorphic rock
- Metamorphism
  - migmatite
- Metamorphic textures
  - foliated texture
    - rock cleavage
    - slaty cleavage
    - phyllitic cleavage/phyllite
    - schistose cleavage or schistosity/schist
    - gneissic banding/gneiss
  - nonfoliated texture
- Kinds of metamorphism
  - contact metamorphism
    - halo or aureole
    - hornfels
  - dynamo-thermal metamorphism
    - regional metamorphism

CONCEPTS AND TERMS TO REMEMBER *continued*

- hydrothermal metamorphism
  - metasomatism
- Metamorphic grade
  - index minerals
  - low/intermediate/high grade
- Metamorphic facies
  - prograde metamorphiism
  - retrograde metamorphism

Important Minerals

calcite	$CaCO_3$
chlorite	$(Mg, Fe^{2+}, Fe^{3+})_6AlSi_3O_{10}(OH)_8$
micas	$(K, Na, Ca)\ (Mg, Fe, Li, Al)_{2\text{–}3}(Al, Si)_4O_{10}(OH, F)_2$
garnet	$(Ca, Mg, Fe^{2+})_3(Al, Fe^{3+}, Mn, V)_2(SiO_4)_3$
andalusite	Al_2SiO_5
sillimanite	Al_2SiO_5
magnetite	Fe_3O_4
staurolite	$(Fe, Mg)_2Al_9Si_4O_{23}(OH)$

REVIEW QUESTIONS

1. Heat is the major agent of metamorphism in
   a. contact metamorphism.
   b. dynamo-thermal metamorphism.
   c. hydrothermal metamorphism.
   d. metasomatism.
2. The most abundant metamorphic rock is
   a. schist.
   b. gneiss.
   c. migmatite.
   d. hornfels.
3. Foliation forms primarily as a result of
   a. dynamo-thermal metamorphism.
   b. contact metamorphism.
   c. hydrothermal metamorphism.
   d. metasomatism.
4. Most metamorphic rocks form in association with
   a. divergent plate margins.
   b. convergent plate margins.
   c. volcanic hot spots.
   d. rift zones.
5. In what ways are the processes of metamorphism and weathering similar?
6. How can a geologist determine the relative intensity of the metamorphism that created a particular assemblage of metamorphic rocks?
7. Explain how gneiss can form from rocks as diverse as granite and argillaceous sandstones.
8. Although the theory is now discredited, what led geologists to consider that most granites had formed by hydrothermal metamorphism?
9. Why do metamorphic rocks such as marble show no foliation?

THOUGHT PROBLEMS

1. How would you distinguish between the conditions of burial that result in the conversion of sediments into sedimentary rocks and those that produce metamorphic rocks?
2. Although the conditions of temperature and pressure required for metamorphism are usually found only at depths of tens of hundreds of miles (kilometers) below the surface, what surface or near-surface situations could result in the metamorphism of rocks?
3. Consider all of the examples of landslides used in the text. Can you ascertain certain pre-failure conditions that seem to be associated with most rock mass movements?

FOR YOUR NOTEBOOK

The first step is to investigate a geologic map to see whether metamorphic rocks are exposed in your area. If they are, note the kinds of rocks and the metamorphic grade. From these data and the general geology of your area, attempt to determine the type and extent of deformation that produced the regional geology. Because such a study could be quite complex, you may want to consult your instructor for some help and advice.

For the greater percentage of us who live in areas where metamorphic rocks are not exposed, a field trip through town will often produce examples of metamorphic rocks. As the chapter text noted, metamorphic rocks are favorite materials for decorative stone. Historically, for example, banks have often utilized various kinds of metamorphic rock in their interior decor. Once again, a short conversation with a local architect may prove useful, not only about the use of various kinds of metamorphic rock but also in locating specific examples of their use. A visit to the local cemetery will also provide examples of metamorphic rocks.

Lab 7: Metamorphic Rock, Processes and Metamorphic Environments

Student Name: ____________________

Definition: Metamorphic rocks are rocks changed from one form to another (metamorphosed) by intense heat, intense pressure, or the action of waterly hot fluids.

Part 1: Mineral Identification

Like any other rocks, metamorphic rocks are made of minerals. As a first step in this lab, let's get familiar with common metamorphic minerals.

Use Mineral Identification Charts and tools (glass plate, streak plate, HCl etc.) to identify minerals common in metamorphic rocks:

Sample #	Mineral Luster	Mineral Hardness	Mineral Color	Color of Streak	Other Physical Properties	Mineral Name
1						
2						
3						
4						
5						
6						
7						
8						
9						
10						

Part 2: Rock analysis:

Define metamorphic textures:

Foliated:

Nonfoliated:

Use Metamorphic Rock Identification Chart to analyze the rocks projected on the screen:		
	Photo 1	**Photo 2**
What is the rock texture?		
What is the rock's mineral composition?		
What is the metamorphic rock name?		
What was the parent rock?		

Part 3: Rocks Identification.

1. Use Metamorphic Rock Identification Chart to identify rock samples provided:

Sample #	Texture	Mineral Composition	Rock Name	Metamorphic Environment, Grade, Facies	Uses
1					
2					
3					
4					
5					
6					
7					
8					
9					
10					
11					
12					

2. Refer to Metamorphic Rocks Chapter and Lecture notes to find out the metamorphic environment (contact, regional, hydrothermal) and temperature and pressure conditions (low, intermediate, high grade) at which each identified rock formed and feel in appropriate column in the table above.

3. Using the Internet resources find what metamorphic rocks are being used in industry and economic activities. Feel in the last column of the above table.

TABLE 9.1

METAMORPHIC ROCKS IDENTIFICATION

Metamorphic Rock Identification Chart					
Rock Texture	**Rock's Mineralogical Composition and/or other distinctive features**	**Metamorphic Rock Name**	**Parent Rock**	**Metamorphic Grade**	**Uses**
Foliated	Dull luster; breaks into hard, flat sheets along slaty cleavage (chlorite, muscovite, quartz, plagioclase)	SLATE	Mudstone or shale	Very low	Roofing slate, table tops, floor tile, blackboards.
Foliated	Breaks along wrinkled or wavy foliation surfaces with shiny, metallic luster (Biotite, muscovite, quartz, plagioclase).	PHYLLITE	Mudstone, shale or slate	Moderately Low	Construction stone, decorative stone.
Foliated	Visible sparkling crystals of platy minerals (chlorite, biotite, muscovite; breaks along scaly, foliated surfaces).	SCHIST Garnet schist Chlorite schist Muscovite schist Biotite schist Kyanite schist Amphibole schist Tourmaline schist Sillimanite schist	Mudstone, shale, slate or phyllite	Moderately High	Construction stone, decorative stone.
Foliated	Visible crystals of two or more minerals in alternating light and dark foliated layers (Quartz, plagioclase, biotite, muscovite, garnet)	GNEISS	Mudstone, shale, slate, phyllite, schist, granite or diorite	Very High	Construction stone, decorative stone, source of gemstones.

Metamorphic Rock Identification Chart

Rock Texture	Rock's Mineralogical Composition and/or other distinctive features	Metamorphic Rock Name	Parent Rock	Metamorphic Grade	Uses
Nonfoliated	Mostly visible crystals of amphibole (usually glossy black hornblende, plagioclase, biotite, plagioclase, quartz)	AMPHIBOLITE	Basalt, Gabbro, Ultramafic igneous rocks	Variable	Construction stone
Nonfoliated	Black glossy rock composed of organic carbon that breaks along uneven or conchoidal fractures.	ANTHRACITE COAL	Peat, Lignite, Bituminous coal	Low	Highest grade coal for clean-burning fossil fuel.
Nonfoliated	Usually a dull dark color; very hard.	HORNFELS	Any rock type	Very Low	Decorative stone.
Nonfoliated	Serpentine; Soft, color usually shades of green.	SERPENTINITE	Basalt, Gabbro, or Ultramafic igneous rocks	low	Decorative stone.
Nonfoliated	Talc; can be scratched with your fingernail; shades of green, gray, brown, white.	SOAPSTONE	Basalt, Gabbro, or Ultramafic igneous rocks	Low to Medium	Art carvings, electrical insulators, talcum powder.
Nonfoliated	Sandy texture; Quartz and sand grains fused together; grains will not rub off like sandstone, usually light colored	QUARTZITE	Sandstone	Variable	Construction stone, decorative stone.
Nonfoliated	Calcite (or dolomite) crystals of nearly equal size and tightly fused together; calcite effervesces in dilute HCl; dolomite effervesces only if powdered	MARBLE	Limestone	Variable	Sculptures, monuments, construction stone, decorative stone, source of lime for agriculture.

ESS 101

Lab 8: Geologic Walk Around Campus and Visit to Geology Museum

Group #: ______________ Student Names: ______________________, ______________________

______________________, ______________________

Stop 1: Corner of South High Street and Rosedale:

1a. Carefully examine the wall on both side of steps.

1b. What type of rock was used to build the wall?

Igneous Sedimentary Metamorphic

1c. What is rock name? (Example: sandstone, limestone, granite, etc.) ______________________

1d. What are some recognizable minerals? ______________________

Stop 2: Southwestern corner of Philips Memorial Building:

2a. Look at the right hand corner of the building. What type of rock was used to construct Philips Memorial hall? ________________

2b. Describe the rocks' texture: ________________

2c. Recognizable minerals: ________________

2d. What is the rock name? ________________

Stop 3: Boulder on the side of the path halfway between Philips Memorial Building and Main Hall:

3a. What type of rock is the boulder?

Igneous sedimentary metamorphic

3b. This rock (before it became a boulder) was intruded by igneous dike.

Identify igneous texture of the dike:

Aphanitic Porpheritic Pegmatitic Phaneritic

3c. Identify visible minerals: ________________

3d. What is the dike's rock name? ________________

Stop 4: Entrance to Recitation Hall:

Recitation Hall is the oldest building on campus. It was built from a local metamorphic rock. The color of the rock comes from its mineral composition.

4a. What is the rock's color? ____________________

4b. What is a dominant mineral in this rock? ____________________

4c. What is the rock name? ____________________

Stop 5: Entrance to Old Library:

At this stop use the tools provided to you by instructor.

Using tools provided identify rocks that were used to build Old Library:

Bottom step ______________________________

Pillars, steps, and door frame ______________________________

Walls ______________________________

Door step ______________________________

Stop 6: Sculpture (new location is behind Bull Center)

At this stop use the tools provided to you by the instructor.

5a. What types of rocks were used to create this sculpture? (Circle all correct answers):

Igneous Sedimentary Metamorphic

5b. What minerals can you identify in these rocks? (use some of the tools provided) ____________________

5c. Name at least two different rocks used for this composition:

____

____

Visit to Museum:

Examine the geologic map of Chester County.

1. What is the origin (igneous, sedimentary, metamorphic) of rocks that are exposed in Chester County?

2. Why are minerals such as kyanite, talc, garnet, and graphite common in Chester County?

3. Asbestos is a typical metamorphic mineral. (bottom left case)

3a. How is it being used?

3b. What are the physical properties of this mineral that make it useful?

3c. Why is asbestos bad for human health?

4. Serpentine is one of the most distinctive local rocks.

4a. Did you find any buildings on campus that are made of serpentinite?

4b. A lot of buildings in Chester County were built from this rock but not many of them are still around. Why?

4c. What is the parent rock (the one that was metamorphosed) of serpentinite.

4d. What does the presence of serpentinite and its original parent rock tell you about the geologic history of the area? (former environments, tectonic setting, tectonic activity)

C H A P T E R 10

The Age of Earth

INTRODUCTION The construction of monuments such as *Stonehenge* indicates that the builders were aware of time. Some historians believe that the builders of Stonehenge not only kept records of passing events but used them to predict future occurrences of certain celestial events such as eclipses. These civilizations measured the passage of time with the same kind of celestial observations we use today. The rising and setting of the Sun defined the day, the phases of the Moon determined the month, and the passage of the seasons established the year. To these natural cycles, we have added the artificial subdivisions of weeks, hours, minutes, and seconds in an attempt to make our record keeping more precise.

The idea of an ancient Earth is not a recent concept. The *Brahmins* of India considered the world to be eternal. The early Greek philosophers were also aware of the antiquity of Earth. The Greek philosopher *Xenophanes* (c. 570–470 B.C.) correctly concluded that areas where fossiliferous rocks were exposed had once been covered by the sea and that significant amounts of time must have passed since the land was a part of the ocean floor. In c. 450 B.C., the Greek historian *Herodotus* watched the Nile delta slowly build up with each yearly flood and realized the enormity of time needed to amass the entire structure. Unfortunately, with the fall of the Roman Empire, most of the peoples of Europe were more concerned with protecting themselves from roving marauders than with intellectual pursuits. Questions about topics such as the age of Earth and the mysteries of the heavens were seldom raised until the end of the Middle Ages.

EARLY ESTIMATES OF EARTH'S AGE

The first real attempt to establish the age of Earth was made in 1644 by *John Lightfoot,* the vice chancellor of Cambridge University, who claimed that Earth was created at exactly 9:00 A.M. on October 26 in the year 3926 B.C. In 1658, *James Ussher,* the archbishop of Armagh, Ireland, determined that Earth was created on October 23 in the year 4004 B.C. The two churchmen proposed such similar dates because both used the Bible as their source, in particular, the Old Testament Book of Numbers, which relates the genealogy of the tribes of Israel. Both men started with Adam and Eve and then attempted to calculate how long it would take to evolve all the tribes of Israel. Although today some may scoff at such attempts to establish the age of Earth, one must keep in mind that both men were serious scholars of theology who used what they considered to be the most reliable source of information, namely, the Bible. Unfortunately, in 1701, *Bishop Lloyd* inserted the 4004 B.C. date for the creation of Earth into a footnote to the Great Edition of the English Bible, and for nearly a hundred years thereafter, to deny a 6,000-year age for Earth was tantamount to heresy.

Limited to a short 6,000 years, many early geologists were forced to use extraordinary means to explain the origin of Earth's rocks and landforms. Of these theories, the most widely accepted was **catastrophism.** According to catastrophists, most features of Earth's crust were created by short-lived, violent processes. For example, catastrophists claimed that all mountains were born when violent, convulsive eruptions lifted huge masses of rock out of the ocean depths. Although some volcanoes do arise in such spectacular fashion, we now know that most of Earth's mountains rise slowly over geologic periods of time, driven by forces generated by the convergence of the lithospheric plates. Similarly, catastrophists would have explained features like the Grand Canyon as having been born as the result of a cataclysmic wrenching and tearing of Earth's crust rather than by millions of years of stream erosion during and after slow uplift. But with only 6,000 years at their disposal, early students of Earth had little choice but to propose such origins, especially since they also wanted to avoid contradicting the teachings of the church.

The first real challenge to catastrophism and to the short 6,000-year history of Earth came in 1785 with the efforts of *James Hutton.* Regarded as the founder of modern geology, Hutton was a wealthy physician and landowner who spent much of his time roaming the countryside studying the rocks of his native Scotland. Hutton saw rocks being slowly reduced to rock fragments and minerals, shifted downslope by almost imperceptible movements, and eventually carried off by streams. He watched the streams deposit the materials into the ocean and correctly surmised that the materials were picked up by ocean currents, spread out on the ocean floor, and eventually transformed into sedimentary rocks. He visualized that the newly formed sedimentary rocks were then uplifted from the ocean floor to create bedded rocks similar to those he observed on land. Most importantly, it was obvious to Hutton that given the extremely slow rates of the processes he saw going on about him, the rocks and landforms could not possibly have been created in only 6,000 years.

Hutton's opinion of the extreme age of Earth was reinforced in 1788 when he first saw the rock exposure at *Siccar Point* along the southeastern coast of Scotland (Figure 10.1). As Hutton viewed the near-horizontal sedimentary layers of the Old Red Sandstone resting on the vertical beds of the underlying sedimentary rocks, he immediately grasped the significance of the formation. Hutton recognized that before the sediments that eventually formed into the upper rocks were deposited, the lower rocks had been uplifted, tilted, subjected to a long period of erosion, and resubmerged beneath the sea. As he looked at this single outcrop, he understood that it had required many millions of years to accomplish. *John Playfair,* a friend who accompanied Hutton on the historic visit to Siccar Point, was to write a few years later, "The mind seemed to grow giddy by looking so far back into the abyss of time...."

Hutton published his ideas in *Theory of the Earth.* His concepts eventually gave rise to one of the most important tenets of geology, the concept of **uniformitarianism.** In direct contradiction to the then-popular tenets of catastrophism, uniformitarianism states that the processes that shape Earth's surface today are the same as those that acted in the past. The concept is commonly summarized in the statement "the present is the key to the past." In support of uniformitarianism, Hutton argued that although catastrophic events were not excluded as natural

FIGURE 10.1 *It was the exposure at* Siccar Point, *along the North Sea coast of Scotland, that convinced James Hutton of the great age of Earth. The Devonian Old Red Sandstone resting on the upturned edges of the folded Silurian sedimentary rocks below is an example of an angular unconformity.*

processes, they were not required to produce Earth's landforms. It was Hutton's view that given sufficient time, very slow-acting, seemingly insignificant processes could create all the landforms attributed to catastrophism. One of the most important results of Hutton's work was the general acceptance of a great age for Earth. Hutton summed up his view of Earth's antiquity in a paper published by the Royal Society of Edinburgh in 1788: "The results, therefore, of our present inquiry is that we find no vestige of a beginning no prospect of an end." With Hutton's ideas firmly entrenched, others soon attempted to determine the true age of Earth.

Because Hutton's writings were very difficult to read, James Playfair published *Illustrations of the Huttonian Theory* in 1802 in which he championed Hutton's ideas in a much more readable form. In 1830, *Sir Charles Lyell* published his *Principles of Geology,* which expanded and built on Hutton's basic concepts. Lyell's book not only laid the groundwork for the early development of the science of geology but also established the basis for modern geology.

DATING METHODS

The age of an object or the timing of an event can be determined by either (1) *relative dating* or (2) *absolute dating.* **Relative** (or **sequential**) **dating** only requires that one object or event be determined to be younger or older than another. For example, based upon physical appearance alone, one could conclude with a degree of certainty that the average high school student is younger than the average retiree. **Absolute** (or **chronologic**) **dating** of these two individuals would entail determining how many years have passed since each was born. The availability of other information, such as the number of months, weeks, days, hours, and minutes that have passed, would allow the absolute age to be determined with increasing precision. Although absolute dating has obvious advantages, most geologists employ relative dating techniques to establish the ages of rocks and events because absolute dating techniques are quite expensive and cannot be applied to all rocks.

Relative Dating

The procedures used to determine the relative age of rocks and events date back to 1669 when *Nicolas Steno,* a Danish physician, published the results of studies that he had made of rocks in the vicinity of Tuscany, Italy. In his publication, Steno introduced two of the most important concepts upon which relative dating of rocks is based. The first is the **principle of superposition,** which states that in a sequence of layered rocks, unless it has been overturned by folding or faulting, younger layers are positioned on top of older layers (Figure 10.2). Although this may seem obvious to us now, keep in mind that Steno made this deduction nearly a hundred years before the monumental works of Hutton.

Geologists employ the principle of superposition every time they observe an outcrop of rocks for the first time. The first determination that must be made is whether the sequence of rocks is "right side up" or "overturned." According to Steno's principle of superposition, if the rock layers are right side up, the layers will become progressively *younger upward* while if they are overturned, they will become progressively *older upward.*

In Chapter 8, you learned a number of features that can be used to establish the original position of sedimentary layers. For example, *graded bedding* is a common sedimentary feature that forms when coarse particles accumulate, followed by particles of progressively smaller sizes. To find a bed with coarse grains at the top and progressively finer particles toward the bottom could be evidence of overturning.

The *cross-bedding* that exists within some sedimentary layers is another sedimentary feature used to determine the orientation of beds. During the formation of cross-bedded layers, newly deposited layers commonly erode the uppermost portions of the underlying bed as they are laid down, producing

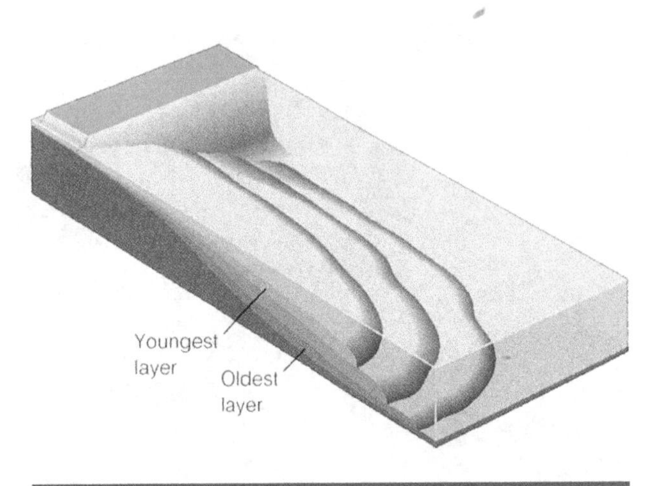

FIGURE 10.2 *This schematic drawing of a delta illustrates the relative ages of the individual layers of the* principle of superposition.

a truncation of the cross-beds at the top of the layer (Figure 10.3). Truncated cross-beds located at the tops of sedimentary beds are therefore evidence of a "right side up" orientation.

Ripple marks are formed on the upper surface of layers of sand-sized deposits as they are deposited by water or wind. When originally formed, the peaks of the waveforms point upward (Figure 10.4). To find a bed in a sequence of rocks with the peaks of the ripples pointing downward would be evidence that the sequence was overturned.

Steno's second major contribution was the **principle of original horizontality.** Steno correctly surmised that sediment layers are generally horizontal or near-horizontal when first deposited. To observe layered rocks tilted from the horizontal means that some force moved the rocks from their original horizontal state (Figure 10.5).

As is often the case with radical new ideas, Steno's colleagues rejected his work. The significance of Steno's contributions to the development of the science of geology would not be recognized and appreciated until the work of Hutton nearly a century later.

In the mid-eighteenth century, the work of *William Smith,* an English engineer, resulted in the formulation of a concept called **correlation** that allowed the age equivalency of distant rock layers to be established. Smith was involved in the construction of waterways throughout England where most of the sedimentary beds are relatively horizontal. He soon was able to recognize certain rock layers based upon their physical appearance and composition and noted further that they always appeared in a definite vertical (stratigraphic) sequence. Once familiar with the rock sequence in a given area, he was able to predict which rock layers would be found from one site to the next. Having identified the rock unit exposed at the surface, Smith was also able to predict the sequence of rock layers that would be found below.

Because most of the rocks Smith encountered in his work were very fossiliferous, he was also able to identify individual rock layers by their diagnostic fossil contents. He demonstrated that over relatively short distances, the fossil content of individual rock layers remained reasonably constant. This allowed Smith to correlate the rock layers from one locale to another. Over long distances, he observed that similar fossil assemblages were found in different lithologies (kinds of sedimentary rocks). At the time, Smith could not explain the cause of these observations. We now know that the rocks Smith was observing formed from sediments that accumulated in a near-shore marine environment, which consists of different subenvironments usually depending on the distance from the shore. Individual rock types represent different depositional environments. Each environment is inhabited by a certain assemblage of plants and animals that is favored by the conditions that exist within that environment. The result is a particular lithology with a diagnostic fossil content. However, because the individual species and

(a)

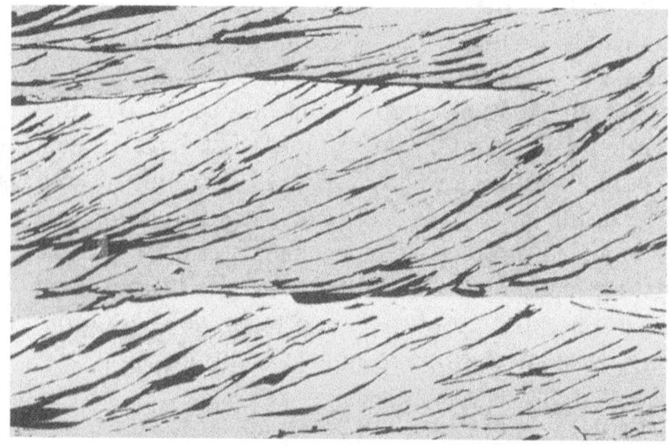
(b)

FIGURE 10.3 *The lateral movement of each new bed of cross-bedded sand removes the upper portion of the underlying bed. As a result, the cross-beds at the top of the underlying bed are truncated. (Courtesy of E. D. McKee/USGS)*

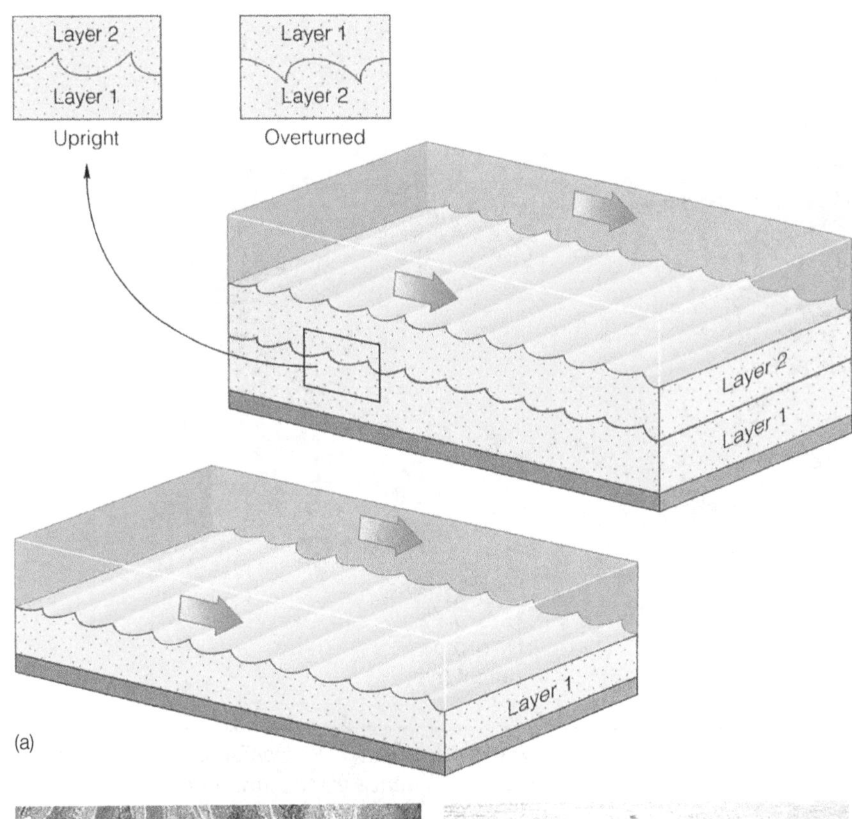

FIGURE 10.4 *(a) The presence of truncated cross-beds at the top of a bed indicates that the bed is right side up. (b)* Symmetrical ripple marks *commonly form on the surface of layers of sand subjected to* oscillatory *water movement such as in the shallow water between low and high tide. (Courtesy of Joe Donovan) (c)* Asymmetrical ripple marks *form where the surface of a layer of sand is being subjected to* directional *water currents such as on the beach at Vancouver Bay, British Columbia, Canada. (Courtesy of Joe Donovan) (d) If a ripple-marked surface is subsequently covered by sediments and the entire sequence is buried and undergoes lithification, the ripple marked surface may be preserved on the bedding surface of a sandstone. (Courtesy of E. D. McKee/USGS)*

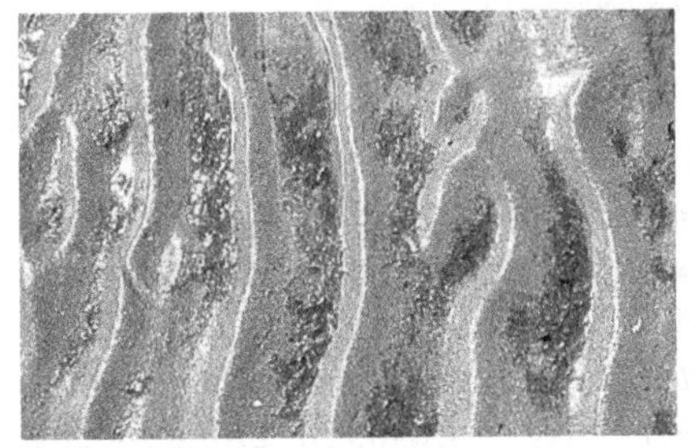

(b)

(c)

(d)

FIGURE 10.5 *When folded rocks are visible in an outcrop, one must keep in mind that folding can only occur deep within Earth. Only after the overlying rocks are removed by millions of years of erosion are the folded rocks exposed at the surface.*

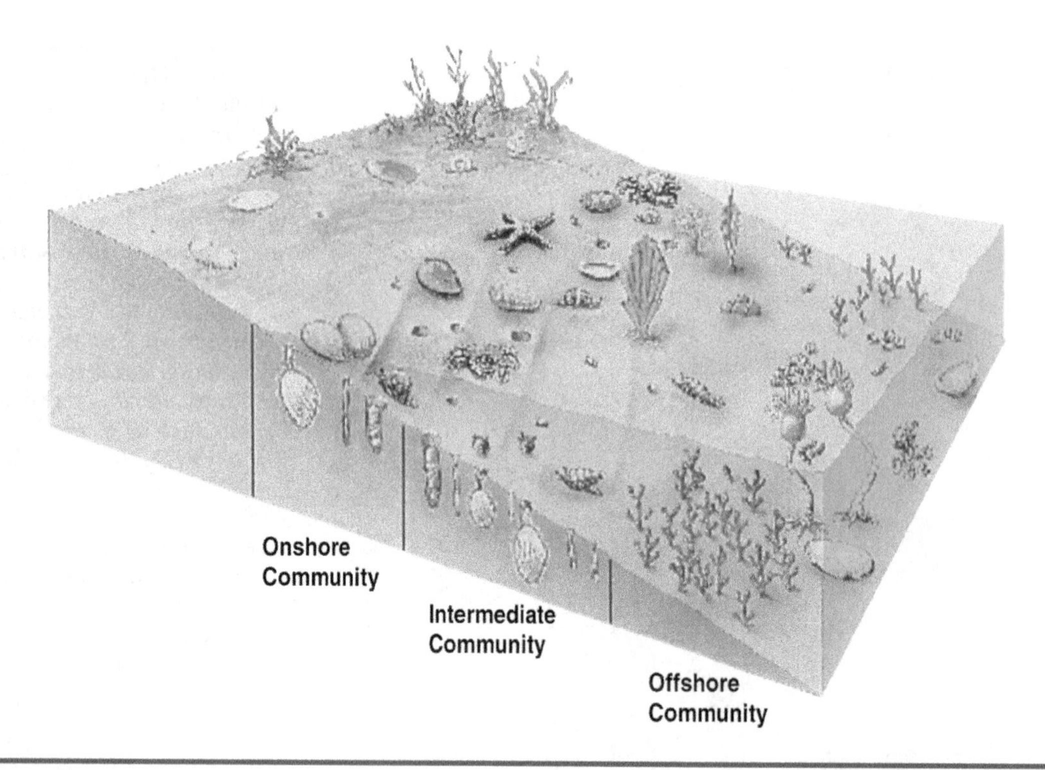

FIGURE 10.6 *Modern plant and animal assemblages change with the depth of water. Recognizing such changes in the fossil record allows geologists to determine the relative depth of water, or distance from shore, in which the sediments forming a particular sedimentary rock were deposited.*

communities of plants and animals evolve over time, rocks deposited during any given period of time will contain a fossil assemblage characteristic of that time interval and different from the assemblages in rocks deposited earlier or later (Figure 10.6). As a result, certain fossil assemblages can be found in rocks of different lithologies.

Smith's inability to understand the reasons for the fossil distributions he observed was primarily due to the fact that the basis for our understanding of the spatial and temporal variation in fossils was not to come until more than a half century later with the publication of Charles Darwin's *On The Origin of Species.*

As an understanding of plant and animal evolution developed, Smith's work was eventually formulated in the **principle of faunal succession.** This concept states that the fossil assemblage found in rocks accumulated during any particular period of geologic time is unique to that time interval and is fundamentally different from the assemblages found in rocks that accumulated earlier or later. More than any other single tool at the disposal of the geologist, fossils are used to correlate rocks or similar or dissimilar lithologies and to establish the age equivalence of isolated exposures anywhere on Earth.

Another important tool that allows geologists to establish the relative ages of rocks was proposed by Charles Lyell in 1830 with the publication of his *Principles of Geology.* Lyell introduced a concept that has become known as the **principle of cross-cutting relations,** which states that a geologic feature or rock is younger than any rock or geologic feature that it cuts across (Figure 10.7). An excellent example is the relationship that Hutton observed at Siccar Point, Scotland, where near-horizontal beds of the Old Red Sandstone overlie the truncated vertical beds of the sedimentary rocks below. According to the principle, the Old Red Sandstone is younger than the truncated rocks below because the sandstone layers cut across the layers of the underlying beds.

The Siccar Point exposure is an example of a relationship known as an **unconformity.** (Figure 10.8) The term was first used in 1805 by *Robert Jameson* to describe surfaces of nondeposition and/or erosion that represent breaks in the geologic record. The time required to develop a surface of unconformity is not recorded either because the sediments needed to record time were not accumulating or because rocks that recorded the passage of geologic time have been stripped away by erosion. The first

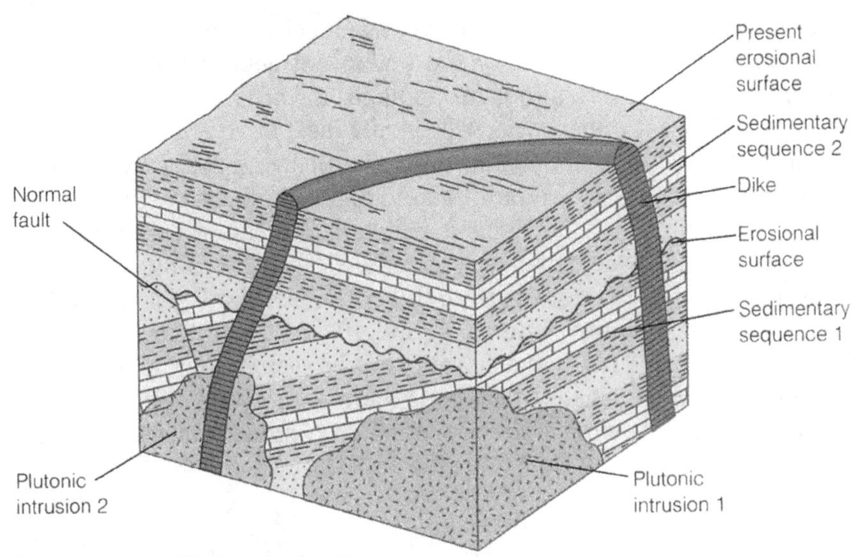

FIGURE 10.7 *The relationship between the boundary of one rock body or geologic feature and another allows a geologist to determine which of the two is the younger or older by applying the* principle of cross-cutting relations.

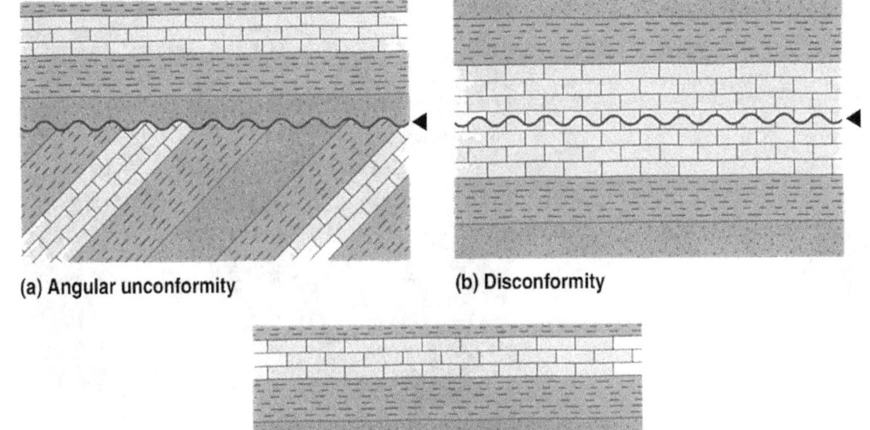

FIGURE 10.8 *The various types of* unconformities *are determined by the combination of rocks and the attitude of the sedimentary beds involved. The most difficult type of unconformity to recognize is the* disconformity *that exists between parallel layers of sedimentary rocks.*

case is comparable to not recording several days' events in a personal diary while the latter case is comparable to tearing out several pages from the completed portion of the diary. In both cases, a gap exists in the record. The interval of unrecorded time between the accumulation of the older sequence of rocks and that of the younger sequence is called a **hiatus.**

The most difficult type of unconformity to detect is a **disconformity,** which is an unconformable surface between parallel bedded sedimentary rocks (refer to Figure 10.8b). The parallelism of the beds indicates that the older rocks were uplifted without deformation or that there was a period of nondeposition (a hiatus) after which sedimentation was resumed. Disconformities are usually identified by recognizing significant differences in the fossil assemblages contained in the rocks on either side of the surface of disconformity. There are several major disconformities in the upper sedimentary sequence within the Grand Canyon (Figure 10.9).

Usually, an unconformity implies deformation, uplift, and erosion of the older rocks before the accumulation of the younger rocks. If the older rocks are sedimentary and underwent deformation by folding or faulting during uplift, the beds of the younger and older rocks will usually meet at an angle, giving rise to an **angular unconformity** such as the exposure at Siccar Point (refer to Figure 10.8a and Figure 10.10).

An unconformity where sedimentary rocks are deposited on the surface of an eroded plutonic igneous or metamorphic rock mass is called a **nonconformity** (refer to Figure 10.8c). A well-known example of a nonconformity can be seen in the Grand Canyon where the Tapeats Sandstone overlies the erosional surface developed on the Vishnu Schist and associ-

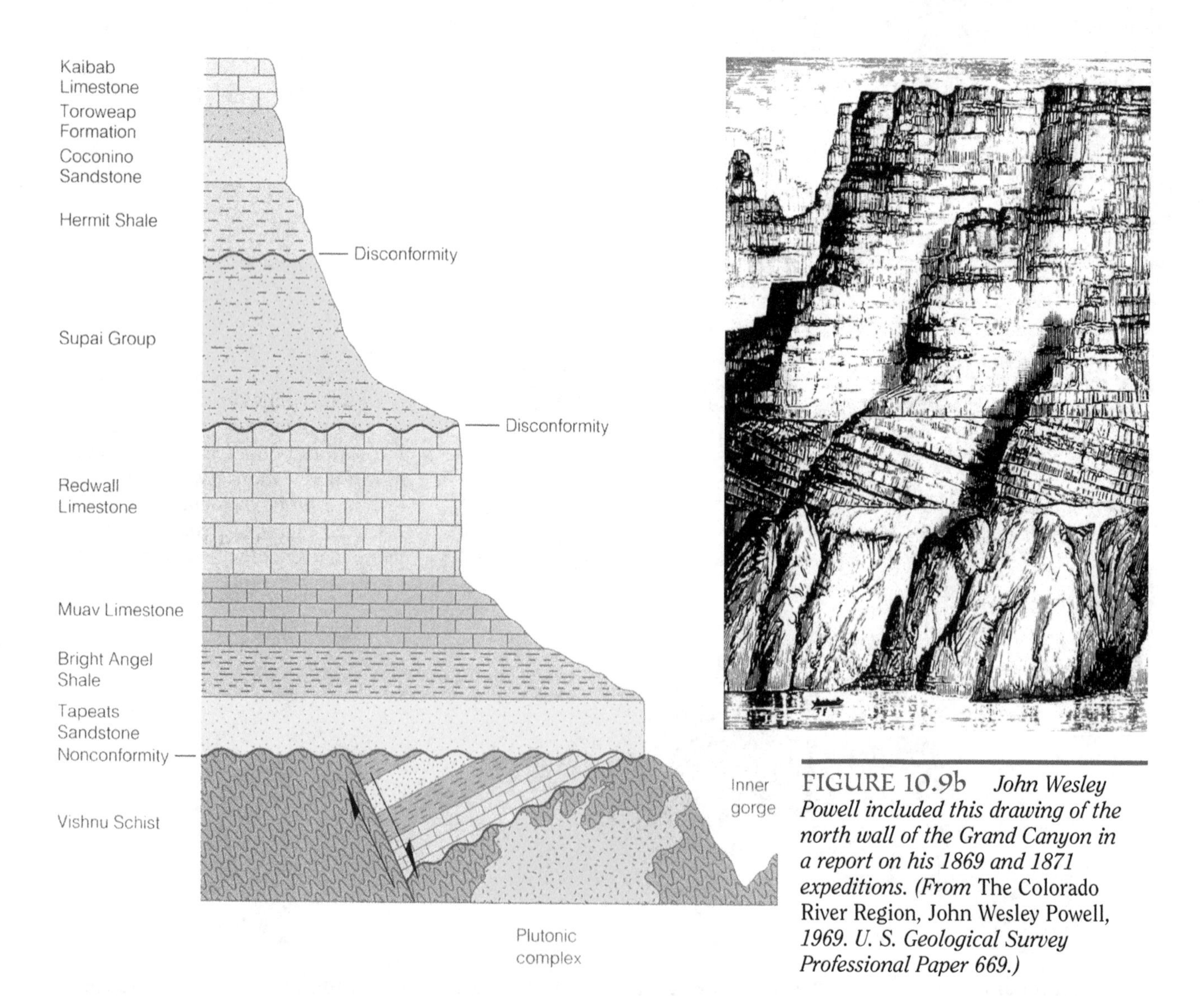

FIGURE 10.9a *The rocks exposed in the Grand Canyon represent a tremendous span of geologic time, but this record is not continuous. Unconformities within the rocks represent vast amounts of time that have been lost to the geologic record.*

FIGURE 10.9b *John Wesley Powell included this drawing of the north wall of the Grand Canyon in a report on his 1869 and 1871 expeditions. (From* The Colorado River Region, John Wesley Powell, *1969. U. S. Geological Survey Professional Paper 669.)*

ated intruded granites and pegmatites (Figure 10.9). It is important to note that the type of unconformity may change from place to place as the kinds of geologic structures and rock types under the erosional surface change.

A number of other cross-cutting relationships between rock bodies can also be observed in the Grand Canyon. The continuity of the contacts of the granitic and pegmatitic bodies within the Vishnu Schist indicates that the igneous bodies are younger than the Vishnu Schist. Similar relationships exist where dikes and sills intrude and cut across host rocks.

The principle of cross-cutting relationships can also be applied to determine the relative ages of geologic features other than rock bodies. Note in Figure 10.9 that the unconformity below the Tapeats Sandstone cuts across major fractures, called normal faults, in the underlying rocks. The principle of cross-cutting relationships indicates that the faults are older than both the erosional surface and the Tapeats Sandstone but younger than both the Vishnu Schist and the sediments of the Grand Canyon Series that they displace.

Absolute Dating

The objective of absolute (chronologic) dating is to determine the age of an event or an object in years. For example, the absolute age of a relatively young object such as a tree can be determined by counting growth rings, each of which represents a year's addition of new wood to the outer surface of the stems. Because the growth rate of plants is influenced by the prevailing climatic conditions including temperature, moisture, and available light, growth rings can be used to correlate the ages of individual trees growing within a specific region by identifying similar patterns of growth ring thickness. In the southwestern United States, a procedure called **dendrochronology** (Greek *dendros* = tree and Greek *khronos* = time), which uses wood growth rings to date events of the recent past, has enabled archaeologists to date human habitation sites back about 6,000 years.

Another example of a feature that can be used to date relatively young materials is the **varve.** Varves are very thin (1–2 mm) sedimentary beds or laminations that represent yearly cycles of deposition in lakes located in temperate-humid regions (Figure 10.11). Some of the best-developed varves have been found

(a)

(b)

FIGURE 10.10 *Angular unconformities, such as those seen at (a) Siccar Point and (b) within the deep gorge of the Grand Canyon, represent a great interval of unrecorded time. Before the sediments from which the younger rocks formed had even accumulated, the older rocks had been deformed, uplifted, eroded, and resubmerged beneath the sea. Several cross-cutting relations exist within the rocks of the Grand Canyon, each of which can be used to determine the relative ages of rocks and geologic events. For example, the unconformity at the base of the Tapeats Sandstone cuts across the metamorphic structures within the Visnu Schist, normal faults, the sequence of sedimentary rocks titled by the faulting, and the unconformity that exists below the titled sedimentary rocks. Thus, according to the principle of cross-cutting relations, the unconformity at the base of the Tapeats Sandstone, and all the rocks above it, are younger than any of the features it cuts across. Can you find other examples where the principle of cross cutting relations can be used to establish relative ages of rocks or events within the rocks of the Grand Canyon? (Courtesy of E. D. McKee/USGS)*

FIGURE 10.11 *In temperate climates, lakes commonly deposit a yearly graded bed. Called* varves, *these graded beds can be used to determine information such as lake age, relative age of the sediments, and climatic changes that have taken place over the lifetime of the lake. If the same event can be recognized in the varves of two lakes, the lake sediments can be correlated. (Courtesy of P. Carrara/USGS)*

in sediments of the lakes that formed as the ice caps of the Great ice Age waned. If a sequence of varves includes the currently forming layer, the varves can be used for absolute dating. However, without the present layer to establish the absolute age of the youngest varve, the varves can only be used for relative dating.

As with tree growth rings, regional climate affects the thickness of varves that accumulate within a particular year, allowing the correlation of varves from different lakes that have been subjected to common climatic influences. Periods of high rainfall, for example, will result in increased runoff and erosion and, subsequently, a thicker yearly varve. A period of drought will have the opposite effect.

Dating by Sediment Accumulation, Ocean Salinity, and Earth's Heat

Dating events and rocks with ages that span geologic time requires the use of more complex methods. In the late 1800s and early 1900s, some scientists attempted to determine the age of Earth based upon the rate of accumulation of sedimentary rocks. The researchers assumed that if they could determine the average rate at which sediments accumulate and accurately estimate the total thickness of the sedimentary rocks accumulated since the creation of Earth, the age of Earth could be calculated by simple division. Obviously, such an approach entails numerous pitfalls. For one thing, the rates at which sediments accumulate vary tremendously from the thousands of years required to accumulate to a few millimeters of deep-sea sediments to the hours or minutes needed to amass several feet of postflood debris. Secondly, there is no single place where sediments have been accumulating continuously since the formation of Earth. Therefore, estimating the total thickness of sedimentary rocks that have accumulated since the formation of Earth would require summing thicknesses of individual sedimentary rock sequences from all over the world that were assumed not only to have accumulated sequentially at different sites but to have done so with no interruption of time. Finding such rock sequences is highly improbable. To complicate the situation further, even if geologists found sequences of sedimentary rocks that recorded an unbroken sedimentary history, they would have no way to estimate either the amount of time that had lapsed between the accumulation of the individual sedimentary beds or the length of time represented by any unconformities that might exist within the individual rock sequences. Nevertheless, despite these potential problems, researchers estimated that 100,000 to 300,000 feet (30,000–90,000 m) of sedimentary rock had accumulated since the formation of Earth at an average rate of accumulation of about 1 foot (30 cm) per 1,000 years. Based on these data, the age of Earth was estimated at about 500 million years. As we will see later, although the estimate of 500 million years falls far short of the actual age of Earth, it was useful in convincing geologists that 6,000 years was not enough time to accumulate the thick layers of sedimentary rocks that exist at Earth's surface.

In 1899, an Irish chemist and geologist, *John Joly* (perhaps the first *geochemist*), suggested that the age of Earth could be determined based upon the salinity (total of all dissolved solids) of the ocean. Joly assumed that the oceans were originally filled with fresh water and became salty as a result of water-soluble materials being washed from the land. Joly made his calculations based upon estimates of (1) the volume of the oceans, (2) the yearly volume of water added to the oceans from the land, and (3) an average value for the salt content of oceanic and stream waters. Estimating that the oceans contained 16 quadrillion tons of salt (16 followed by 12 zeros) and that salt was being added at a rate of 160 million tons per year, Joly calculated the age of Earth at about 100 million years.

Like the researchers trying to determine the accumulation of sedimentary rocks, Joly made some erroneous assumptions and consequently significantly underestimated the age of Earth. For one thing, there is no reason to believe that the oceans were ever fresh. Salts began to be delivered to the ocean basin as soon as the rocks were subjected to weathering. In addition, some salt is contributed as a result of sea-floor spreading. Secondly, accurate estimates of ocean volumes did not exist in Joly's time, nor could he obtain much data on the world-wide salt content of either oceanic or stream waters. A large portion of Joly's error lay in his assumption that salt, once deposited in the ocean, stays in solution; in fact, large volumes of salt, including the sources of most of our domestic and industrial salt, are periodically removed from the ocean and deposited within sedimentary rock sequences.

At about the same time Joly was attempting to determine the age of Earth, *Lord Kelvin,* one of the leading physicists of the day, suggested a procedure that he believed would solve the dilemma once and for all. An expert in heat and heat flow, Lord Kelvin assumed that, in the beginning, Earth was a completely molten sphere of rock. He then calculated how long it would take for a mass of molten rock with the diameter of Earth to cool to the point where the release of heat from its surface equaled the measured heat flow at Earth's surface. Kelvin determined that Earth was no older than 70 million years, an estimate significantly shorter than the ages calculated from sedimentation rates or the accumulation of oceanic salt. When geologists suggested that his estimate might be too low, Kelvin intimidated them with mathematics that "proved" that he was correct and that they must be mistaken in thinking that Earth was any older.

But as with the other attempts at age estimation, Kelvin's error lay in his basic assumptions. For one thing, there is no reason to believe that Earth was ever a completely molten blob of rock. Most damaging, however, was his assumption that all of the heat radiating from Earth's surface was heat remaining from its initial fromation. We now know that most of the heat radiating from Earth's surface is generated within the crust by the disintegration of radioactive elements. In all fairness to Kelvin, he could not have been aware of the radioactive source of heat in-as-much as radioactivity was not discovered until 1896 when *Henri Becquerel* observed that photographic plates darkened when exposed to uranium-bearing minerals.

In 1903, *Maria Sklodowska Curi* and her husband, *Pierre,* became the first to isolate a measurable quantity of a radioactive element, radium—an achievement for which they received the Nobel Prize. Within a few years, *Lord Rutherford,* another well-known scientist determined that the rate at which radioactive **isotopes** disintegrate is a physical constant measured by the **half-life** of the isotope. Isotopes are atoms whose nuclei contain the same number of protons but different numbers of neutrons. As a result, the atoms have the same **atomic number** (determined by the number of protons) but different **atomic mass** (the sum of the number of protons and neutrons in the nucleus). We will discuss the significance of the isotopic half-life in the next section.

In 1905, Lord Rutherford was the first to determine the age of a rock by using the rate of radioactive breakdown of uranium. With that experiment, modern age-dating procedures were born. For the first time, it was possible to determine the age of a rock with reasonable accuracy and, eventually, to determine the age of Earth.

Radiometric Dating

The use of radioactive isotopes to determine the age of rocks is based upon the disintegration of radioactive **(parent)** isotopes into stable **(daughter)** isotopes. (Should you not be familiar with the structure of the atom, it might be well for you to read the appropriate section in Chapter 2 before continuing with the following discussion.) The disintegration of radioactive isotopes involves the release of one or both of two atomic particles from the nucleus: (1) *beta particles* or (2) *alpha particles* (Figure 10.12). **A beta particle** is an electron lost from a neutron. The emission of a beta particle therefore converts a neutron into a proton and increases the atomic number of the daughter isotope by one unit over that of the parent isotope. Since the sum of protons and neutrons has

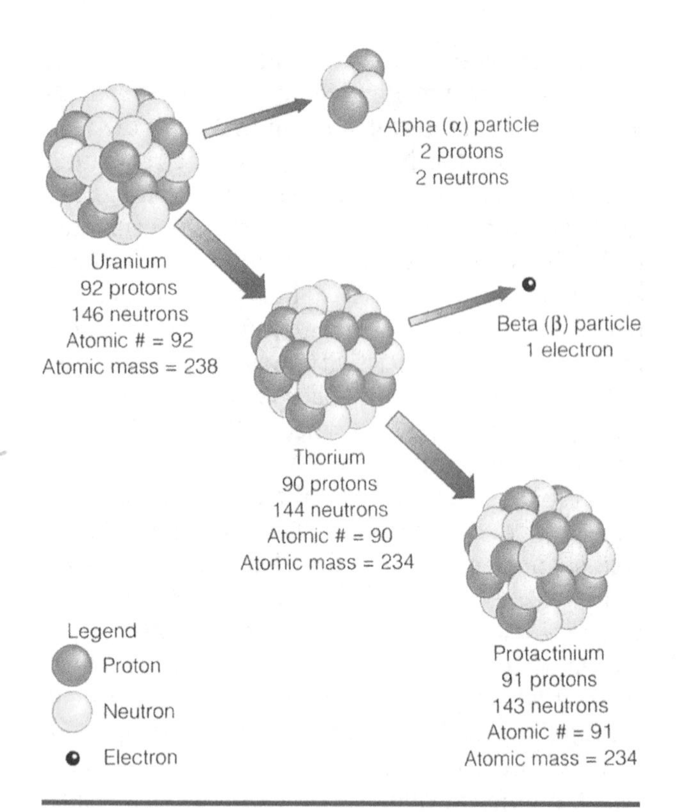

FIGURE 10.12 *Radioactive isotopes disintegrate by releasing subatomic particles from the nucleus.*

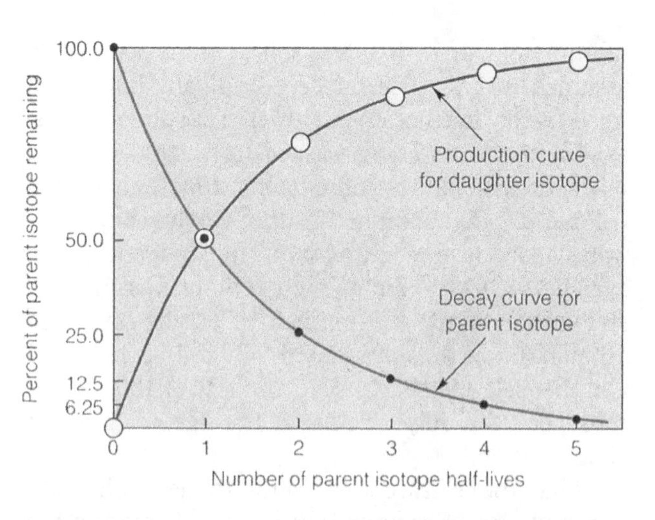

FIGURE 10.13 *The rate of disintegration of radioactive isotopes is measured by the isotope's* half-life. *During one half-life, one half of the existing mass of the* parent *isotope disintegrates into the* daughter *element. Therefore, after one half-life, 50% of the original mass of the parent will remain; after two half-lives, 25% will remain, and so forth.*

not changed, however, the atomic mass is unaffected. An example would be the disintegration of thorium (atomic number 90, atomic mass 234) into protactinium (atomic number 91, atomic mass 234).

An **alpha particle** is the combination of two protons and two neutrons. With each alpha particle released from the nucleus, the atomic number of the daughter isotope will decrease by two, and the atomic mass will decrease by four relative to the parent isotope. An example would be the disintegration of uranium (atomic number 92, atomic mass 238) into thorium (atomic number 90, atomic mass 234).

Another pair of isotopes used in radiometric dating, potassium 40 (^{40}K) and argon 40 (^{40}Ar), involves a different process. The change of ^{40}K to ^{40}Ar involves the capture of an electron from the innermost electron shell by a proton in the nucleus of the potassium atom. As a result, the atomic number of the daughter isotope (^{40}Ar) is one unit less than that of the parent isotope (^{40}K). Potassium 40 may also disintegrate to calcium 40 (^{40}Ca) by the process of beta decay. The radiation released from the nucleus of the parent isotope during disintegration affects neither the atomic number nor the atomic mass.

The rate of disintegration of a parent isotope is measured by the physical constant called the half-life. Defined by Lord Rutherford, the half-life of the parent isotope is the amount of time required for one-half of the parent atoms of the isotope to disintegrate into atoms of the daughter isotope. For example, the half-life of the isotope uranium 235 (^{235}U) is 713 million years. Every 713 million years, one-half of any number of ^{235}U atoms will disintegrate into atoms of the daughter isotope, lead 207 (^{207}Pb). The disintegration of a radioactive isotope is graphically portrayed in Figure 10.13, which relates the atomic fraction of the radioactive isotope to the number of expired half-lives.

To date a rock by a radiometric technique, the number of atoms of parent and daughter isotopes in the rock are counted using an instrument called a **mass spectrometer**. The sum of the number of parent and daughter isotopes in the rock equals the number of parent isotopes originally present in the rock. The atomic fraction of the original parent isotope concentration represented by the remaining parent isotopes is calculated by dividing the number of parent isotopes remaining by the sum of the existing parent and daughter isotopes. The number of half-lives required to reduce the parent isotope concentration to the determined level is then calculated by using a plot such as the one shown in Figure 10.13. If the calculations show, for example, that only 25% of the original concentration of parent isotopes remains, two half-lives have passed since the rock formed. If the parent isotope was ^{235}U with a half life of 713 million years, the age of the rock is 2 × 713 million years or 1.4 billion years.

As with any age-dating technique, the accuracy depends on the validity of certain basic assumptions. One assumption is that no atoms of the parent or daughter isotopes have been removed from or added to the rock by any process other than radioactive disintegration. If the isotopic concentrations have been modified by any outside process such as weathering, the calculated ratio of parent to daughter isotopes and any subsequent age determination will be in error. To minimize such errors, geologists analyze specific minerals within the rock that are known to be especially resistant to external modification. For example, when the $^{235}U/^{207}Pb$ technique is used, the dating is usually based upon the analysis of *zircon* crystals ($ZrSiO_4$) painstakingly removed from the rock. Because zirconium and uranium have similar atomic radii (0.74Å for Zr and 0.97Å for U), uranium atoms commonly substitute for zirconium atoms within the zircon crystal structure during its formation. Atoms of lead, on the other hand, with atomic radii of 1.20Å cannot enter the growing crystal structure of the zircon grain because of the size restrictions. Any lead atoms found in the zircon crystal can therefore be assumed to have formed from the radioactive disintegration of uranium.

The half-life of the parent isotope limits the use of any particular isotopic parent/daughter pair for dating. If, for example, a parent isotope with a very long half-life is used to date a relatively young rock, the number of atoms of daughter isotopes created since the formation of the rock may be too few to be counted even with the highly sensitive mass spectrometer. Because the concentration of the daughter isotope will be evaluated as zero, the atomic fraction of the parent isotope will be calculated as 1.0, and according to the chart in Figure 10.13, no half-lives will have passed since the formation of the rock.

If, on the other hand, a parent isotope with a very short half-life is used to date a very old rock, the number of parent isotopes remaining within the rock will be too few to count. In this case, the atomic fraction of the remaining parent isotope will be vanishingly small, and the curve showing the half-lives in Figure 10.13 will be so close to the horizontal axis that the number of half-lives cannot be calculated, again precluding an age determination.

To allow the dating of rocks with a wide range of ages, a number of isotopic techniques have been devised that utilize parent isotopes with differing half-lives (Table 10.1). The older the rock, the longer the parent isotope half-life must be to determine the age accurately. For example, because the isotope rubidium 87 (^{87}Rb) has a half-life of 47 billion years, the isotopic combination used to determine the ages of the very oldest rocks is rubidium/strontium ($^{87}Rb/^{87}Sr$). The most widely used isotopic parent/daughter pair is uranium/lead ($^{235}U/^{207}Pb$).

Because of the abundance of potassium in Earth's crust (2.6% by weight), another isotopic pair commonly used to date rocks is potassium/argon ($^{40}K/^{40}Ar$). A potential problem that must be recognized in using the $^{40}K/^{40}Ar$ pair, however, is that argon is a gas. Consequently, some of the argon may have escaped from the rock, thereby introducing the possibility of error into the age determination. This is especially true if the rock has been heated by volcanic or metamorphic activity. If fact, the disintegration of radioactive potassium in the crust and the subsequent "leaking" of the argon from the rocks exposed at the surface are responsible for argon being the third most abundant gas in Earth's atmosphere.

The age of human remains and artifacts is determined by using carbon 14 (^{14}C) or **radiocarbon dating.** Carbon 14 is generated in the atmosphere when cosmic rays impact nitrogen atoms (^{14}N). The ratio of ^{14}C to ^{12}C isotopes in the atmosphere is constant with ^{12}C being the more abundant. Because all isotopes of an element have identical chemical responses, both the ^{12}C and the ^{14}C atoms react with oxygen to form carbon dioxide, which mixes throughout the atmosphere. The carbon then enters the tissues of plants during photosynthesis and the tissues of animals when they eat the plants. As long as the organism is

TABLE 10.1

ISOTOPES USED FOR RADIOMETRIC DATING

PARENT ISOTOPE	DAUGHTER ISOTOPE	HALF-LIFE (IN YEARS)
Carbon 14	Nitrogen 14	5,730
Uranium 235	Lead 207	713 million
Potassium 40	Argon 40	1.3 billion
Uranium 238	Lead 206	4.5 billion
Thorium 232	Lead 208	14 billion
Rubidium 87	Strontium 87	47 billion

alive, the ratio of ^{14}C to ^{12}C within the plant and animal tissues is maintained at a constant value. When the organism dies, however, the ^{14}C with a half-life of 5,730 years disintegrates by giving off beta particles and reverts back to nitrogen, thereby changing $^{14}C/^{12}C$ ratio in a predictable way. Because ^{14}C has a short half-life, radiocarbon dating is restricted to materials with maximum ages of about 50,000 to 60,000 years.

With the exception of the radiocarbon techniques used to date plant or animal remains, most radioactive dating techniques are used to date igneous rocks. It is assumed that all the minerals contained within an igneous rock formed in a relatively short period of time. If the assumption that the abundances of parent and daughter isotopes have not been altered by any outside process is valid, the radioactive date will give the time the rock crystalized. If an igneous rock is metamorphosed at a grade sufficiently high to form new minerals, however, the atomic "clock" may be reset, and rather than dating the age of the original rock, the radioactive procedure will date the time of metamorphism. Because sedimentary rocks are made up of the remains of other rocks, dating a sedimentary rock will not give the time the rock formed but rather the time of formation of the rocks from which the fragments were derived.

Since Lord Rutherford's first successful dating of rocks in 1905, a large number of crustal rocks have been dated. Combining these results with ages obtained by dating meteorites and rocks brought back from the Moon, we have now arrived at an age for Earth of 4.6 billion years.

With radioactive techniques available to determine the absolute ages of rocks, why do most geologists use relative dating techniques? As previously discussed, not all rocks can be successfully dated using radiometric techniques. One of the major reasons, however, why most geologists do not use radiometric dating is because the mass spectrometer used to determine isotopic abundances is an expensive instrument, both to purchase and to maintain on a day-to-day basis, and highly trained individuals are required to prepare the samples, operate the instrument, and interpret the data. For these reasons, the absolute dating of rocks is very expensive and is usually far beyond the financial capabilities of most geologists. As a result, although they would prefer absolute dating, most geologists, taking advantage of the ages of rocks that *have* been dated by radiometric methods via the methods of correlation, usually establish the *relative* ages of rocks utilizing the principles set forth by Steno, Hutton, and Lyell more than a hundred years ago.

SPOT REVIEW

1. Why are estimates of Earth's age based on criteria such as the total accumulation of sedimentary rocks and the salinity of the oceans since Earth's creation always in error?
2. Why does the particular isotopic parent/daughter pair used for dating a rock depend on the age of the rock?
3. Why are most radioactive dating techniques used only to determine the age of igneous rocks?

ROCK TIME SCALES

Utilizing relative dating techniques, geologists attempt to place rocks in their proper chronological order. One of the earliest attempts to establish the chronology of Earth's rocks was a subdivision proposed by *Giovanni Arduino* in 1759 based on his studies in the Alps. Arduino subdivided all rocks into three groups. He assigned the oldest rocks to the *primitive group,* which included all the igneous and metamorphic rocks that he observed in the core of the Alps. Younger than the primitive group was the *secondary group,* which consisted of the sedimentary rocks that he observed overlying the crystalline core of the Alps and making up most of the landscape. Arduino then assigned surface accumulations of unconsolidated materials such as regolith and soil to the *tertiary group.* Later, in 1830, a fourth group, the *quaternary,* was added to include the most recent materials such as glacial and lake deposits.

The Modern Time Scale

The modern time scale was first developed during the nineteenth century as the knowledge and understanding of both organic evolution and the record of evolution represented by fossils expanded. The geologic time scale that we use today was created by assembling data from rock sequences from localities around the world that contained fossil assemblages that were both distinctly different from those in the adjacent rock sequences and showed evolutionary changes (Figure 10.14).

Within limited continental areas, geologists can often determine the relative age of a rock sequence simply by recognizing certain rock layers or sequences of lithologies based on their physical appearance. Within the Appalachian Mountains, for example, the Tuscarora Sandstone, which is responsible for most of the topographic relief of the area, can be easily recognized by its white quartzose composition and its location above the red sandstones and shales of the Juniata Formation. Other rock units with similar lithologies may be identified and distinguished by diagnostic fossils or sedimentary features. The identification of a particular rock unit such as the Tuscarora Sandstone allows a geologist to immediately establish the position of the rocks in the relative age scale.

On the other hand, between distant regions within a continent such as between the eastern and western United States and certainly between continents, the lithologies of the rocks that accumulated during any interval of time will be different, and lithologic similarities cannot be used to establish the relative ages of rocks. Nevertheless, even though the lithologies of the rocks exposed in different regions of a continent or between continents may vary, the similarity of the fossil assemblages will usually allow the age equivalency to be established.

Eons of Time The longest period of geologic time is the *eon.* All geologic time is subdivided into two eons, the **Cryptozoic Eon** and the **Phanerozoic Eon** (Figure 10.14). The *Cryptozoic Eon* is subdivided into the **Archean** and **Proterozoic.** The subdivision is primarily based on the formation of large, granitic continents; during the older *Archean,* there were not well-established granitic cratons. Another important development that accompanied the formation of continents in the *Proterozoic* was the development of the shallow water marine environments that, in turn, allowed the development of shallow-water life forms. The importance of fossils in establishing relative age is clearly illustrated by the names given to the various subdivisions of geologic time. The stem-*zoic* is from the Greek word meaning "life," in recognition of the importance of life as a basis for subdivision. The prefix *crypto-* comes from the Greek word meaning "hidden" and refers to the fact that although the most ancient rocks may be devoid of fossils, we realize that life did exist. Because hard body components such as shells, bones, and teeth that are the basis of nearly all fossil remains had not yet widely evolved, indications of their presence are *hidden* from observation. In some cases, a wispy film of carbon on the bedding surface of a sedimentary rock or on the cleavage surface of a metamorphic rock may be all that remains to indicate the presence of former life.

The Cryptozoic Eon includes the vast interval of time from the formation of Earth 4.5 to 5.0 billion years ago until recognizable fossils began to appear consistently in the sediments about 600 million years ago. Attempts to subdivide the Cryptozoic Eon have met with little success due to the low abundance and diversity (the number of different kinds of organisms in an assemblage) of fossils and the deformation to which most of Earth's most ancient rocks have been subjected over the billions of years since their formation.

The appearance of fossils in sedimentary rocks about 600 million years ago is used to subdivide Earth history to Cryptozoic and Phanerozoic Eons in much the same fashion as the appearance of writing about 5,000 years ago is used to establish the "prehistoric" and "historic" portions of human existence. The Phanerozoic Eon beings the "historic" portion of Earth's existence. The prefix *phanero-,* from the Greek work meaning "apparent," refers to the presence of diverse and abundant identifiable fossil remains. A span of about 200 million years prior to the transition and the 20 to 40 million years between the two eons of time is the period during which animals (and to a lesser degree plants) evolved body parts that were sufficiently resistant to decomposition so as to be capable of entering the fossil record.

Eras of Time Eons of time are subdivided into *eras.* Based upon the overall characteristics of the lifeforms, the Phanerozoic Eon has been subdivided into three eras: (1) the *Paleoszoic,* (2) the *Mesozoic,* and (3) the *Cenozoic.*

The **Paleozoic Era** extends from approximately 570 million to about 245 million years ago. The prefix *paleo-,* from the Greek word palaios meaning "ancient" or "old," refers to the dominance of marine invertebrates. Although the life of the Paleozoic was dominated by the invertebrates, the era was a time of spectacular evolutionary change. The Paleozoic had not been long underway when all of the major animal phyla had evolved. Familiar forms of fish were present in the lakes and streams, and simple plants carpeted the land. During the late Paleozoic, dragonflies with three-foot wingspreads buzzed through the extensive swamps that characterized the time. The peat that accumulated in these swamps gave rise to a large percentage of Earth's coal deposits.

EON		ERA	Duration in millions of years	Millions of years ago
PHANEROZOIC		CENOZOIC	66	66
PHANEROZOIC		MESOZOIC	179	245
PHANEROZOIC		PALEOZIOC	325	570
CRYPTOZOIC	PROTEROZOIC	LATE	330	900
CRYPTOZOIC	PROTEROZOIC	MIDDLE	700	1,600
CRYPTOZOIC	PROTEROZOIC	EARLY	900	2,500
CRYPTOZOIC	ARCHEAN	LATE	500	3,000
CRYPTOZOIC	ARCHEAN	MIDDLE	400	3,400
CRYPTOZOIC	ARCHEAN	EARLY		4,600

Era	Period	Epoch	Duration in millions of years	Millions of years ago		Distinctive occurrences
CENOZOIC	Quaternary	Pleistocene	1.6	1.6	Age of Mammals	Humans
CENOZOIC	Tertiary	Pliocene	3.7	5.3	Age of Mammals	Mammals become dominant
CENOZOIC	Tertiary	Miocene	18.4	23.7	Age of Mammals	
CENOZOIC	Tertiary	Oligocene	12.9	36.6	Age of Mammals	
CENOZOIC	Tertiary	Eocene	21.2	57.8	Age of Mammals	Extinction of many species
CENOZOIC	Tertiary	Paleocene	8.6	66.4	Age of Mammals	
MESOZOIC	Cretaceous		78	144	Age of Reptiles	First flowering plants, maximum development of dinosaurs
MESOZOIC	Jurassic		64	208	Age of Reptiles	First birds and mammals, abundant dinosaurs
MESOZOIC	Triassic		37	245	Age of Reptiles	First dinosaurs
PALEOZOIC	Permian		41	286	Age of Amphibians	Extinction of many marine animals
PALEOZOIC	Carboniferous: Pennsylvanian		34	320	Age of Amphibians	Great coal swamps; abundant insects, first reptiles
PALEOZOIC	Carboniferous: Mississippian		40	360	Age of Amphibians	Large primitive trees
PALEOZOIC	Devonian		48	408	Age of Fishes	First amphibians
PALEOZOIC	Silurian		30	438	Age of Fishes	First land plants
PALEOZOIC	Ordovician		67	505	Age of Marine Invertebrates	First fish
PALEOZOIC	Cambrian		65	570	Age of Marine Invertebrates	First shelled organisms
PRECAMBRIAN	Contains Proterozoic, and Archean					First multicelled organisms; First one-celled organisms; Approximate age of oldest rocks; Origin of Earth

FIGURE 10.14 *The* geologic time scale *is largely based on the chronological evolution of life as recorded in the fossils contained in sedimentary rocks.*

The swamps of the early Devonian were the sites where amphibians evolved from air-breathing fish that struggled for life in the oxygen-poor waters. Although the amphibians were the first animals to adapt to the land, they never made a complete break with the marine environment. Because of their fragile, unprotected eggs, they had to return to the water to spawn. As a result, the amphibians were never able to completely colonize the land. Always having to remain near the water, they were unable to venture into the drier or more upland parts of the land.

By the end of the Pennsylvanian, one group of amphibians had evolved into the reptiles by evolving the shelled egg. The new, protected egg not only provided a better chance of survival, but it also allowed the reptiles to more successfully and completely invade the land.

The shrinking of the vast wetlands and the onset of the drier climate toward the end of the Paleozoic spurred rapid reptilian evolution. At the close of the Pennsylvanian and into the Permian, a "mammal-like" reptile evolved that was to foretell the coming of the mammals. The animal had changed its stance from the characteristic reptilian sprawl by developing strong, muscular legs that lifted and supported it's body off the ground. The shape of the teeth had changed from the spikelike teeth of the reptile to more diversified shapes better designed to eat a

FIGURE 10.15 *The* dinosaurs, *which dominated every ecological niche during the Mesozoic Era, went to extinction at the close of the Cretaceous Period. Exactly what caused the worldwide extinction is still the topic of debate.*

variety of food. An enlarged skull indicated a larger and better developed brain.

The **Mesozoic Era** extends from 245 million years ago to 65 million years ago. The prefix *meso-*, from the Greek word *mesos* meaning "middle," refers to the middle position the dominant Mesozoic reptilian life-forms represent between the marine invertebrates of the Paleozoic and the mammalian species of the Cenozoic Era. During the Mesozoic, the reptiles occupied all of Earth's major environments as they spread across the land, returned to the water, and took to the air (Figure 10.15). Huge marine reptiles such as *Elasmosaurus* (whose descendants are still thought by devout Scots to live in the black depths of Loch Ness) dominated the sea (Figure 10.16). *Pteranodon,* the great "flying" reptile, gained access to the air by taking advantage of folds of skin that stretched between its fingers and rear limbs not unlike those possessed by the modern bat (Figure 10.17).

Best known, however, was a new class of animal called the *dinosaur* that evolved early in the Mesozoic and was destined to dominate the land during that era. The name means "terrible lizard," and although it was coined with the large carnivorous kinds in mind, most dinosaurs did not fit the description.

As the era closed, birds evolved from the biped dinosaurs, which, according to some, had already developed internal body thermostats that allowed their blood to warm. The end of the Mesozoic Era saw a mass extinction of marine and flying reptiles, flightless dinosaurs, and many marine invertebrates. The scene was set for the mammals.

Although mammals had evolved early in the Mesozoic, their small size indicates that they were unable to compete with the reptiles and dinosaurs. Most early mammals were the size of mice with the largest being no bigger than a cat. They probably survived by keeping out of the way of the larger and more numerous reptiles and dinosaurs and subsisted on seeds an insects. As the Mesozoic came to a close and the numbers of reptiles declined and the dinosaurs vanished, the mammals underwent rapid evolution and soon filled the vacated ecological niches.

The **Cenozoic Era** extends from 65 million years ago to the present. The name *Cenozoic* is derived

FIGURE 10.16 Elasmosaurus, *a well-known Cretaceous marine reptile, was over 36 feet (12 m) long. Devout Scots believe that the creature supposedly living in the depths of Loch Ness may be an elasmosaur.*

FIGURE 10.17 *Not all Mesozoic reptiles were tied to the land or dwelt in the sea.* Pteranodon *is an example of a reptile that was able to overcome the force of gravity and achieve soaring flight. Although its body was only about the size of a goose,* Pteranodon *had a wingspan of over 20 feet (7 m).*

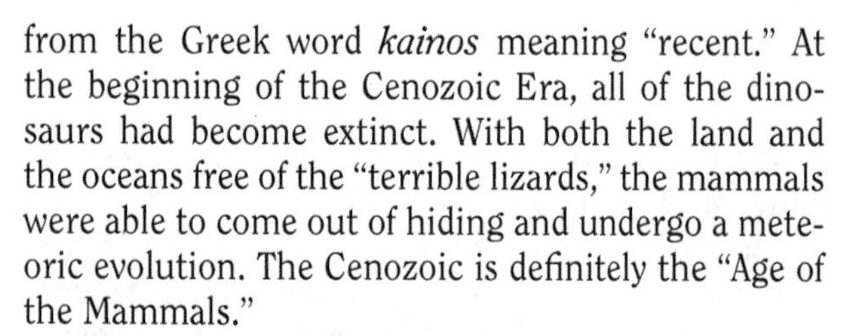

from the Greek word *kainos* meaning "recent." At the beginning of the Cenozoic Era, all of the dinosaurs had become extinct. With both the land and the oceans free of the "terrible lizards," the mammals were able to come out of hiding and undergo a meteoric evolution. The Cenozoic is definitely the "Age of the Mammals."

All of the major features of the modern landscape also evolved during the Cenozoic. Continents moved to their present positions and took on their present sizes and shapes. The great mountains (the Rockies, the Alps, the Himalayas, the Andes, and the Appalachians) were sculpted to their present appearance by the processes of erosion during the Cenozoic. In fact, most of the landforms we see today are the result of the activity of geological processes during the last 2 million years.

Periods of Time An era is subdivided into **periods** of time with the sequence of rocks deposited during a period being referred to as a **system** of rocks. The subdivision of the rock column into periods and systems is based upon both fossil and rock contents and includes the missing time represented by the unconformities.

The present geologic time scale was largely established by European geologists of the eighteenth and nineteenth centuries who began to recognize the repetition of particular rock types from place to place and the sequences of life's evolution represented by their fossil contents. They began to assign the time of formation of the rocks to periods and the rocks themselves to systems. The geographic locality where the rocks of a particular system were first described is called the **type locality.** Whoever first described the rocks was afforded the privilege of assigning the name, usually of local origin. Rocks of similar age are recognized throughout the world by referring back to the rock sequence exposed in the type locality and in particular to the diagnostic fossil assemblage that the sequence contains.

Periods of the Paleozoic Era

The oldest rocks of the Paleozoic Era belong to the **Cambrian System,** first described in 1831 by *Adam Sedgwick,* one of the foremost geologists of the day. At the type locality of the Cambrian System in northern Wales, the rocks are a sequence of sparsely fossiliferous sedimentary and volcanic rocks. Sedgwick called the system the Cambrian after *Cambria,* the Roman name for Wales. Because the Cambrian is the first period of the Phanerozoic Eon, the rocks belonging to the Cryptozoic Eon are usually referred to as **Precambrian.**

While Sedgwick was studying and describing the rocks of northern Wales, a colleague and friend, *Roderick Murchison,* was studying a sequence of fossiliferous rocks in southern Wales that he named the **Silurian System** after an ancient Welsh tribe, the *Silures.* In 1835, Sedgwich and Murchison presented a joint paper describing the Cambrian and Silurian Systems in England and Wales.

In later years, a dispute broke out between the two friends that alienated them for the rest of their lives. As they continued to study the rocks of England and Wales, it became evident that the upper portion of Sedgwick's Cambrian System overlapped with the lower portion of Murchison's Silurian. Because Sedgwick did not provide the detailed description of the fossil evidence needed to prove that the fossils were of Cambrian age, Murchison suggested that only Sedgwick's lower unfossiliferous rocks were Cambrian in age and that the rocks of Sedgwick's upper Cambrian belonged in his own Silurian System. Needless to say, Sedgwick did not agree.

The dispute that subsequently developed between Sedgwick and Murchison not only divided the two men but also split the geologists of Europe into two camps until a solution to the problem was suggested in 1879 by another geologist, *Charles Lapworth.* Lapworth proposed that the disputed sequences of rock be combined into a new system that he proposed to name the **Ordovician,** after another ancient Welsh tribe, the *Ordovices.* Although generally accepted by most geologists as a period of geologic time, the Ordovician Period is not universally recognized because it is not based on rocks exposed in a type locality.

In 1839, before the beginning of their feud, Sedgwick and Murchison had studied a section of rock in Devonshire in southwestern England and tentatively identified it as either Cambrian or what was later to be described as Ordovician. However, based on fossil evidence, *William Londsdale* showed the rocks were intermediate in life-forms between the Silurian and the Carboniferous Systems and named the system **Devonian.**

In 1822, *William Coneybeare* and *William Phillips* described a section of rocks in northern England that they named the **Carboniferous System** because of its coal content. In North America, because the coals are largely restricted to the upper portion of the system, the Carboniferous System is subdivided into two systems, the older **Mississippian** and a younger **Pennsylvanian,** which have a widespread unconformity between them. Although the Mississippian System in North America does contain some coal, the dominant lithology of the Mississippian is limestone. The Mississippian System was named after rock exposures along the Mississippi River that were first described and subdivided as the Lower Carboniferous by *Alexander Winchell* in 1870 about the time that he became the first chancellor of Syracuse University. In 1891, *Henry Shaler* proposed that the rocks represented a separate system and assigned the name Mississippian. The upper portion of the Carboniferous in North America was named the Pennsylvanian System in 1891 by *H. S. Williams* after excellent exposures in Pennsylvania where the rock sequence contains rich deposits of coal.

In 1841, the tsar of Russia invited Murchison to study a sequence of rocks west of the Ural Mountains. Murchison found the rocks contained fossils of a younger stage of biological succession than those of the Carboniferous. On that basis, he designated the sequence as a separate system that he called the **Permian** after the Perm Province in which the rocks were described.

Periods of the Mesozoic Era

The Mesozoic Era is subdivided into three periods of time with their respective systems of rock, the *Triassic, Jurassic,* and *Cretaceous.*

The **Triassic System** was named by a German geologist, *Frederich von Alberti,* in 1834 after a sequence of rocks in Germany. The name refers to the threefold lithologic subdivision of the sequence in the type locality (not characteristic of the system elsewhere) where a marine rock sequence is sandwiched between two continental sequences. In eastern North America, some of the best known Triassic rocks come from the down-faulted troughs called the Triassic Basins. Extending from North Carolina to Nova Scotia, the Triassic rocks of the eastern United States are known for their content of fossil dinosaur tracks. In the western United States, Triassic rocks are responsible for much of the beauty of the southwestern desert including the Painted Desert, Zion National Park, and the Petrified Forest (Figure 10.18).

The name **Jurassic** was originally applied in 1795 by a German geologist, *Alexander von Humboldt,* to

(a)

FIGURE 10.18 *The rocks of the Triassic provide the Southwest with some of its most spectacular scenery and information about the environment. (a) The sandstones of the Wingate and Navajo Formations exposed in the cliffs of* Zion National Park *are fossil dunes that record a vast desert. (b) The colored rocks of the* Painted Desert *record the presence of streams and lakes. (Courtesy of National Park Service) (c) The silicified logs that characterize the* Petrified Forest National Park *were trees similar to modern redwoods that became trapped in logjams during floods, were buried by stream deposits, and eventually were preserved as silica replaced the wood. (Courtesy of E. D. McKee/USGS)*

(c)

(b)

a sequence of rocks exposed in the Jura Mountains, located between France and Switzerland. At the time Humboldt described the rocks, the concept of periods of time and systems of rocks had not yet been developed. Later in 1839, the rocks were redefined as a system by another German geologist, *Leopold von Buch.*

The richly fossiliferous rocks of England that inspired William Smith were Jurassic in age. The Jurassic rocks are known worldwide for their abundance of dinosaur remains. In the United States, for example, the fossil content of the *Morrison Formation* in Colorado and Wyoming has provided a rich picture of the life of Jurassic time. One of the single most famous fossil species, the remains of the first bird, *Archaeopteryx,* was found in the Jurassic-age rocks within the lithographic stone quarries near Solenhofen, Bavaria.

The **Cretaceous System** was established based upon studies of the rocks in the chalk cliffs along the Strait of Dover, England, and within the structural basins surrounding London and Paris. The name, which was first applied by a Belgian geologist, *Omalius d'Halloy,* in 1822, is derived from the Latin word *creta* for "chalk," a common lithology within the Cretaceous System around the world. The rocks of the Cretaceous record a burst of evolution among the mammals as the Age of the Reptiles began to decline, but not before the evolution of the most fearsome of the "terrible lizards," *Tyrannosaurus rex.*

In addition to its spectacular fossil content, the Cretaceous is important as a source of economic deposits of coal throughout the world, including most of the major coal deposits of the western United States.

Periods and Epochs of the Cenozoic Era

The Cenozoic Era is subdivided into two periods, the older **Tertiary** and the younger **Quanternary.** Although the names are carryovers from Arduino's original classification of rocks, they no longer have their original connotations.

A period of time is subdivided into *epochs.* In some cases, periods are subdivided into *lower, middle,* and *upper* time units while in other cases, such as the Tertiary Period, specific names are applied to the individual epochs.

The subdivision of the Tertiary Period into *epochs* of time represents the first application of statistics to geology (Table 10.2). The names applied to each of the epochs of the Tertiary were based upon detailed studies of the fossil assemblages in the rocks of the Paris Basin. Originally studied by the French geologist *Gérard-Paul Deshayes,* the assemblages attracted the attention of Lyell who noticed that as the individual beds became younger, they showed a greater percentage of living species. Lyell then proposed a classification scheme, based upon the percentage of still-living shelled invertebrates, which subdivided the Tertiary Period into the **Eocene, Miocene,** and **Pliocene Epochs** (refer to Figure 10.14). The stem *-cene* is from the Greek *kainos* meaning "recent" while the prefixes *eo-*, *mio-*, and *plio-*, referring to the fossil content, are from the Greek words for "dawn," "less," and "most," respectively. The other epochs, the **Paleocene** (Greek *palaios* = ancient) and the **Oligocene** (Greek *oligos* = little), were added later.

The *Quaternary Period* is the shortest of the periods of time and consists of a single epoch, the **Pleistocene.** Although it represents only the last 2 million years of Earth history, it was during the Pleistocene that much of the modern world developed. During this short period of time, the continental glaciers gripped the northern continents, the continents took on their modern shapes and sizes, and the finishing touches were applied to the modern landscape.

Not only was the land taking on its modern appearance, but so were the life-forms. Forests were essentially modern by the beginning of the Pleistocene. Grasslands carpeted the plains, giving rise to herds of grazing animals including elephants, camels, and horses. Familiar (toothless) birds dominated the skies while a few large flightless birds served as predecessors to modern flightless birds such as the ostrich.

Perhaps the most significant evolutionary development of the Pleistocene was the arrival and meteoric evolution of *Homo sapiens.* Beginning with *Australophithecus* in the Pliocene and culminating

TABLE 10.2

THE SUBDIVISION OF THE TERTIARY BASED ON THE PERCENTAGE OF MODERN SPECIES REPRESENTED IN THE FOSSIL ASSEMBLAGE OF THE PARIS BASIN

Period	Epoch		Percentage of Modern Species
Quarternary	Pleistocene	(Greek *pleistos,* most + *kainos,* recent)	90–100
	Pliocene	(Greek *pleios,* more + *kainos*)	50–90
	Miocene	(Greek *meios,* less + *kainos*)	20–40
Tertiary	Oligocene	(Greek *oligos,* little + *kainos*)	10–15
	Eocene	(Greek *eos,* dawn + *kainos*)	1–5
	Paleocene	(Greek *palaios,* ancient + *kainos*)	0

with *Homo sapiens,* the human was to become the life-form that has had the greatest impact upon Earth since the blue-green algae changed the composition of the atmosphere from predominantly carbon dioxide to 21% oxygen and 78% nitrogen.

SPOT REVIEW

What is the basis for subdividing geologic time into eons, eras, periods, and epochs?

CONCEPTS AND TERMS TO REMEMBER

- Early estimates of Earth's age
  - catastrophism
  - uniformitarianism
- Dating methods
  - relative dating
    - principle of superposition
    - principle of original horizontality
    - correlation
    - principle of faunal succession
    - principle of cross-cutting relations
    - unconformity
    - hiatus
    - disconformity
    - angular unconformity
    - nonconformity
  - absolute dating
    - dendrochronology
    - varves
  - radiometric dating
    - isotope
    - half-life
    - atomic number
    - atomic mass
    - parent isotope
    - daughter isotope
    - beta particle
    - alpha particle
    - mass spectrometer
    - radiocarbon dating
- Geologic time scales
  - eons of time
    - Cryptozoic
    - Archean
    - Proterozoic
    - Phanerozoic
  - eras of time
    - Paleozoic
    - Mesozoic
    - Cenozoic
  - periods of time
    - system of rocks
      - type locality
    - Cambrian
      - Precambrian
    - Silurian
    - Ordovician
    - Devonian
    - Carboniferous
      - Mississippian
      - Pennsylvanian
    - Permian
    - Triassic
    - Jurassic
    - Cretaceous
    - Tertiary
    - Quaternary
  - epochs of time
    - Eocene
    - Miocene
    - Pliocene
    - Paleocene
    - Oligocene
    - Pleistocene

REVIEW QUESTIONS

1. The age of Earth is about ______ years.
   a. 500 million
   b. 1 billion
   c. 5 billion
   d. 10 billion
2. A surface of unconformity between parallel sedimentary beds is called a (an)
   a. angular unconformity.
   b. nonconformity.
   c. hiatus.
   d. disconformity.

REVIEW QUESTIONS *continued*

3. The person credited with originating the principles of superposition and original horizontality is
   a. Lord Kelvin.
   b. Nicolas Steno.
   c. James Hutton.
   d. Charles Lyell.
4. The percentage of radioactive isotopes that disintegrate during each half-life period
   a. increases at a constant rate.
   b. decreases at a constant rate.
   c. depends on the atomic number of the isotope.
   d. is constant.
5. When 75% of a parent isotope has disintegrated to the daughter isotope, ______ half-lives have passed.
   a. 1 c. 3
   b. 2 d. 4
6. The radioactive isotope used to date human remains and artifacts is
   a. carbon 14.
   b. uranium 235.
   c. potassium 40.
   d. rubidium 87.
7. The greatest portion of Earth history is included in the
   a. Paleozoic Era.
   b. Phanerozoic Eon.
   c. Cryptozoic Eon.
   d. Mesozoic Era.
8. The appearance of the present surface of Earth evolved during the
   a. Miocene Epoch.
   b. Pleistocene Epoch.
   c. Pliocene Epoch.
   d. Eocene Epoch.
9. What assumptions are made in using (a) sediment thickness, (bocean water salinity, and (c) radioactivity to determine the age of rocks?
10. How are the principles of superposition, original horizontality, and cross-cutting relations used to determine the relative age of rocks?
11. Why can't a radioactive isotope with a short half-life be used to date very old rocks?
12. What is the basis behind the prefixes *paleo-*, *meso-*, and *ceno-* used with the various geologic eras of time?

THOUGHT PROBLEMS

1. The concept of uniformatarianism seems to preclude the existence of castrophism, yet there are ample examples of natural catastrophes such as floods, earthquakes, and volcanic eruptions. How can this dilemma be resolved? Can there be "catastrophic uniformitarianism"?
2. The oldest rocks yet found are only about 4.1 billion years old, although the age of Earth is estimated at 4.5 to 5.0 billion years. How can this discrepancy be explained? Do you think older rocks exist? If so, where do you think they are most likely to be found, and how old do you think they will be?
3. One of the justifications given for the space program and visits to distant celestial bodies is that the information obtained will shed light on the age and origin of Earth. How valid do you think these arguments are?

FOR YOUR NOTEBOOK

Geologic maps give information about the kind, distribution, and age of the rocks in any area. A geologic map of the United States has been included in the appendix of this text. Acquire a geologic map of your area from your instructor or from the state geological survey, and investigate the ages of rocks that occur in your area in more detail. Most people are intrigued by fossils. Investigate the fossil assemblage that are found in the rocks in your area, and attempt to determine the kind of environment that existed when the rocks were formed or accumulated.

ESS 101

Lab 9: Radiometric Dating

Student Name: ______________________________

Background info useful for this assignment:

Radiometric dating is used by geologists to estimate the age of the rock based on radioactive decay of unstable isotopes present in minerals. During the decay atoms of unstable isotopes (parent) convert into stable atoms of other elements (daughter). For example atoms of 40Potassium (common element in Potassium feldspar and thus in many igneous rocks) is unstable and after it was crystallized from cooling magma, atoms of 40Potassium (parent) begin to change in to a stable element 40Argon (daughter) (see Chapter 10 for details). Each radioactive isotope has its own rate of decay which is measured by a physical constant known as half-life (T ½). For example T ½ of 40Potassium (40K) disintegration is 1.3 billion years. That means that it will take that long for half (50 percent) of 40K atoms present in rock to convert into 40Argon (40Ar). Examples of measured decay rates of other isotopes commonly present in rocks are presented in Table 10.1 in Chapter 10.

To date a rock applying radiometric technique, the number of atoms of parent and daughter isotopes in the rock is counted using the instrument known as a mass spectrometer. Either atom counts or percentage of atoms are compared by ratio (dividing amount of Daughter atoms by amount of parent atoms). This ratio is related to the length of time period (measured in half-life) during which the radioactive decay is going on. Let's use an example of 40K/40Ar decay common for igneous rocks such as granites; results of spectrometric analysis of a rock sample report that rock sample contains 1000 atoms of 40K and 1000 atoms of 40Ar. Remember when the rock formed only atoms of 40K were present since 40Ar is a product of radioactive decay. During the time of rock existence 1000 atoms of 40K converted to 40Ar, while another 1000 atoms of 40K is still present in the rock sample. At the time of the rock formation the sample contained total of 2000 atoms of 40K. Now only half (50 percent) of the original amount is still present. How old is the rock sample? Lets calculate the ratio: 1000 atoms of 40Ar/1000 atoms of 40K = 1 or 50%/50% = 1, 1 half-life elapsed since the analyzed rock was formed. Half-life (T ½) of 40K equals 1.3 billion years. The rock is that old! What if mass spectrometer counted only five atoms of 40Ar? Applying the same principle lets figure out how many half-lives elapsed since rock was formed 5 atoms of 40Ar/1995 atoms of 40K = 0.0025 half-lives elapsed. What is that in real years? 0.0025 × 1.3 billion years = 3,250,000 years old.

To determine the age of geologic material, it must contain atoms of a radioactive isotopes (parent atoms) that originated when the material formed. When atoms of the parent (P) isotope decay to a stable form, they have become a daughter (D) isotope. A parent isotope and its corresponding daughter are called a decay pair (see Table 10.3). Note that T ½ is different for different isotopes (see Table 10.1).

TABLE 10.3

DECAY PARAMETERS FOR RADIOACTIVE PAIRS

Decay Parameters for All Radioactive Decay Pairs			
Percent of Parent Atoms (P)	**Percent of Daughter Atoms (D)**	**Half-lives Elapsed**	**Age**
100	0	0	0.000 × T ½
98.9	1.1	1/64	0.015 × T ½
97.9	2.1	1/32	0.031 × T ½
95.8	4.2	1/16	0.062 × T ½
91.7	8.3	1/8	0.125 × T ½
84.1	15.9	1/4	0.250 × T ½
70.7	29.3	1/2	0.500 × T ½
50	50	1	1.000 × T ½
35.4	64.6	1 1/2	1.500 × T ½
25	75	2	2.000 × T ½
12.5	87.5	3	3.000 × T ½
6.2	93.8	4	4.000 × T ½
3.1	96.9	5	5.000 × T ½

1. Astronomers think that the Earth probably formed at the same time as all of the other rocky materials in our solar system, including the oldest meteorites. The oldest meteorites ever found on Earth contain nearly equal amounts of both Uranium-238 and Lead-206. Based on Figure 10.13 and Table 10.1, what is Earth's age? Explain your reasoning.

Let us come up with our own approach. Since we know the relationship between D/P and half-lives we can use a graphical approach to estimate rock ages. Let's create a graph which can be used in application to any pair of isotopes.

First complete the following table referring to Fig. 10.13

2. Complete the following Table:

Half-Lives elapsed	P Parent Fraction	D Daughter Fraction	D/P ratio
0			
1			
2			
3			
4			
5			

3. Using data from the table, construct a graph in which the ordinate is "D/P ratio" and the abscissa is "Half-Lives Elapsed." The point representing one half-life is already plotted. Plot the rest, and draw a smooth curve connecting the data points; do not connect the points with straight-line segments, but estimate the curvature between points as best you can so that the entire curve bends smoothly.

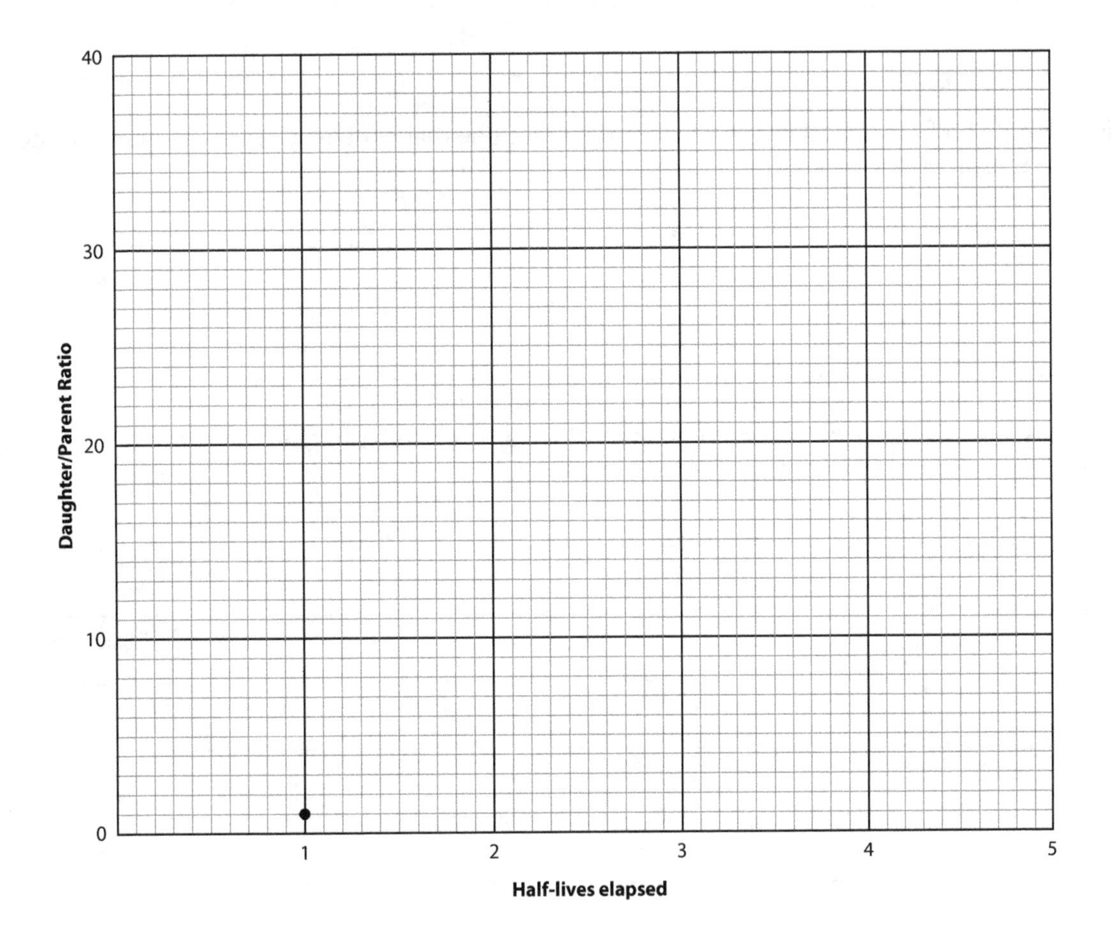

4. Uranium-235 counts for three rock samples reported in the table below. Using your graph, determine the number of half-lives that elapsed for each sample and write your answer in the "Half-Lives Elapsed" column. Refer to Table 10.1 and calculate the age of each sample and record it in the last column.

Sample Number	Atoms of Parent (P)	Atoms of Daughter (D)	D/P Ratio	Half-lives Elapsed	Age in Years
1	2135	3203			
2	4326	10,815			
3	731	14,620			

5. If you assume that the global amount of radiocarbon (formed by cosmic-ray bombardment of atoms in the upper atmosphere and then dissolved in rain and seawater) is constant, then decaying Carbon-14 is continuously replaced in organisms while they are alive. However, when an organism dies, the amount of its Carbon-14 decreases as it decays to Nitrogen-14. The carbon in a buried peat bed has about 6 percent of the Carbon-14 of modern shells. What is the age of the peat bed? Explain.

____

____

____

6. About what percent of modern ^{14}C remains in a piece of coral that is 11,640 years old?

____

7. Using your graph from Question 3, how old is the limestone that contained sea shells in which the ratio of $^{14}N/^{14}C$ was estimated to be equal to 5?

____

____

____

8. Layers of sand on a New Jersey beach contain common zircon crystals. Could the zircon crystals be used to date exactly when the layers of sand were deposited? Explain.

____

____

____

____

____

____

GRAND CANYON NATIONAL PARK NORTHERN ARIZONA	
Area: 1,217,158 acres; 1902 square miles	**Address:** P.O. Box 129, Grand Canyon, AZ 86023
Proclaimed a National Monument: January 11, 1908	**Phone:** 928–638–7888
Established as a National Park: February 26, 1919	**E-mail:** deanna_prather@nps.gov
Designated a World Heritage Site: 1979	**Website:** www.nps.gov/grca/index.htm

The incomparable Grand Canyon of the Colorado River is ten miles wide and over a mile deep. The national park encompasses 277 miles of river, plus adjacent uplands. Weathering and erosion have produced colorful cliffs, slopes, spires, buttes, and mesas. Exposed in these features are rock formations that illustrate a vast span of geologic history. Two sequences of Precambrian rocks record early mountain-building episodes. A classic Paleozoic section includes carbonates, shales, and sandstones deposited in five geologic periods.

The Indians of the Southwest had legends to explain the origin of the Grand Canyon. The stories reveal an awareness of natural causes and effects. The Navajos, for example, believed that rain for many days and nights caused a great flood that covered the land with water that rose higher and higher. Finally an outlet formed, and as the rushing waters drained away, the Grand Canyon was cut deeply into the earth.

More than 500 Indian sites in the park, both on the rim and within the canyon, indicate that as recently as 600 years ago large numbers of Indians lived in the Grand Canyon vicinity. Havasupai Indians still live in a branch canyon watered by Havasu Creek in the western section of the park.

In 1540, a band of Coronado's conquistadores were the first white men to see the Grand Canyon, after being guided by Hopi Indians to the rim. More than 300 years later, Lt. Joseph C. Ives, a U.S. Army surveyor, came upriver from the southwest and reached the lower Grand Canyon. After an unsuccessful attempt to reach the rim, he concluded that surveying the Grand Canyon of the Colorado River was impractical.

John S. Newberry, a member of the Ives party and a geologist, thought otherwise. He convinced Major

FIGURE 10.19 *The Grand Canyon is so large that it can be seen from outer space. The canyon was formed around two million years ago as the Colorado River eroded down as rapidly as the Colorado Plateau was uplifted. Erosion exposed one of the most complete geologic section of rocks in the world, from the Precambrian to the Cenozoic. The river developed along a fault zone where rock was weaker. The canyon rims are from 9 to 18 miles apart. Photo by John Van Hosen.*

John Wesley Powell, a fellow geologist (who had lost a hand in the Civil War), that an expedition by boat down the Colorado River through the Grand Canyon would be worth the risk in order to complete the survey. With scientific help from the Smithsonian Institution, the Powell expedition of four boats and nine men left Green River, Wyoming, in May 1869, and began their hazardous journey downstream. One boat was smashed; scientific instruments and food were lost. Most of the time the party was wet, tired and hungry, and some were convinced that they would never get out of the canyon alive. Three men climbed out of the canyon but were killed on the plateau by hostile Indians or Mormons. The men who stayed with Powell survived and reached the settlements near the western end of the Grand Canyon. Two years later, Powell went down the Colorado River again in order to collect more data for his geographical and geological report on the region.

Theodore Roosevelt, after a trip to the Grand Canyon in 1903, was determined that it should be preserved. He was able to have part of the area placed under government protection in 1908, but opponents blocked the establishment of a national park until 1919.

The Colorado River carries tons of sediment past any one point each day. The rate of erosion is not constant throughout the course of the river, but it is estimated that every thousand years the drainage basin of the Colorado River system is lowered by an average of 6 1/2 inches. It was his observation of the sediment load that convinced Powell that he would not be washed over high falls (as some people had predicted) on his initial expedition. He reasoned correctly that a river carrying so much sediment would have scoured its bed down to a more or less even grade, or slope.

The Colorado River no longer rampages every spring or falls to lowest low-water stage in summer. The Colorado has been a regulated stream since 1966, when Glen Canyon Dam, just upstream from Lees Ferry, impounded river water to fill Lake Powell in Glen Canyon National Recreation Area. The tremendous sediment loads that used to scour the canyon during spring floods now settle to the bottom of Lake Powell, and the pre-1966 average load of half a million tons per day has been greatly reduced. The curtailment of the sediments has resulted in the removal of sand bars and beaches, to the detriment of the environment. A "test flood" in 1996 lasted a week, and another in 1997 lasted two days. Both deposited sand bars and beaches, but they disappeared within a year once the source of the sediments was removed. The river below the Grand Canyon has been controlled since 1935 by Hoover Dam, which backed up water in Lake Mead to Grand Wash, where the river leaves the national park.

Grand Canyon National Park extends from the mouth of the Paria River, which joins the Colorado River just below Glen Canyon Dam, to the eastern boundary of Lake Mead National Recreation Area, a river distance of 277 miles. The Grand Canyon is over a mile deep, with the distance from rim to rim ranging from 9 to 18 miles. The elevation of the North Rim (8900 feet) averages some 1200 feet higher than the South Rim (6900 feet).

Dimensions of such magnitude have biological as well as geological significance. Life forms on the two canyon rims evolved differently because of physical separation over time. At the higher altitude of the North Rim, under conditions of cooler average temperatures and greater annual precipitation, plants and animals characteristic of Canadian life zones became established. Trees and shrubs such as blue spruce, ponderosa pine, fir, and aspen, that are darker in color, took over. A typical North Rim animal is the Kaibab squirrel, which has a black body, black stomach, and white tail.

On the South Rim, vegetation that is lighter in color and typical of the Upper Sonoran life zone (junipers, pinon, etc.) predominates. The Abert squirrel is found here, with its grizzled gray body, white stomach, and gray tail. The common ancestor of these two modern species was a black and white squirrel with tufted ears that lived on the plateau before the Grand Canyon was incised. When the squirrels were no longer able to migrate back and forth across the physical barrier of the canyon, and as differences in climate became pronounced between the North Rim and South Rim, new squirrel species developed due to evolutionary adaptation—the darker Kaibab squirrel on the North Rim and the lighter Abert on the South Rim.

GEOLOGIC FEATURES

In terms of geologic time, the Grand Canyon of the Colorado River is surprisingly young. The river accomplished the major part of the downcutting during the last two million years. It is the rocks that are ancient, not the canyon itself. Exposed by this downcutting is one of the most nearly complete

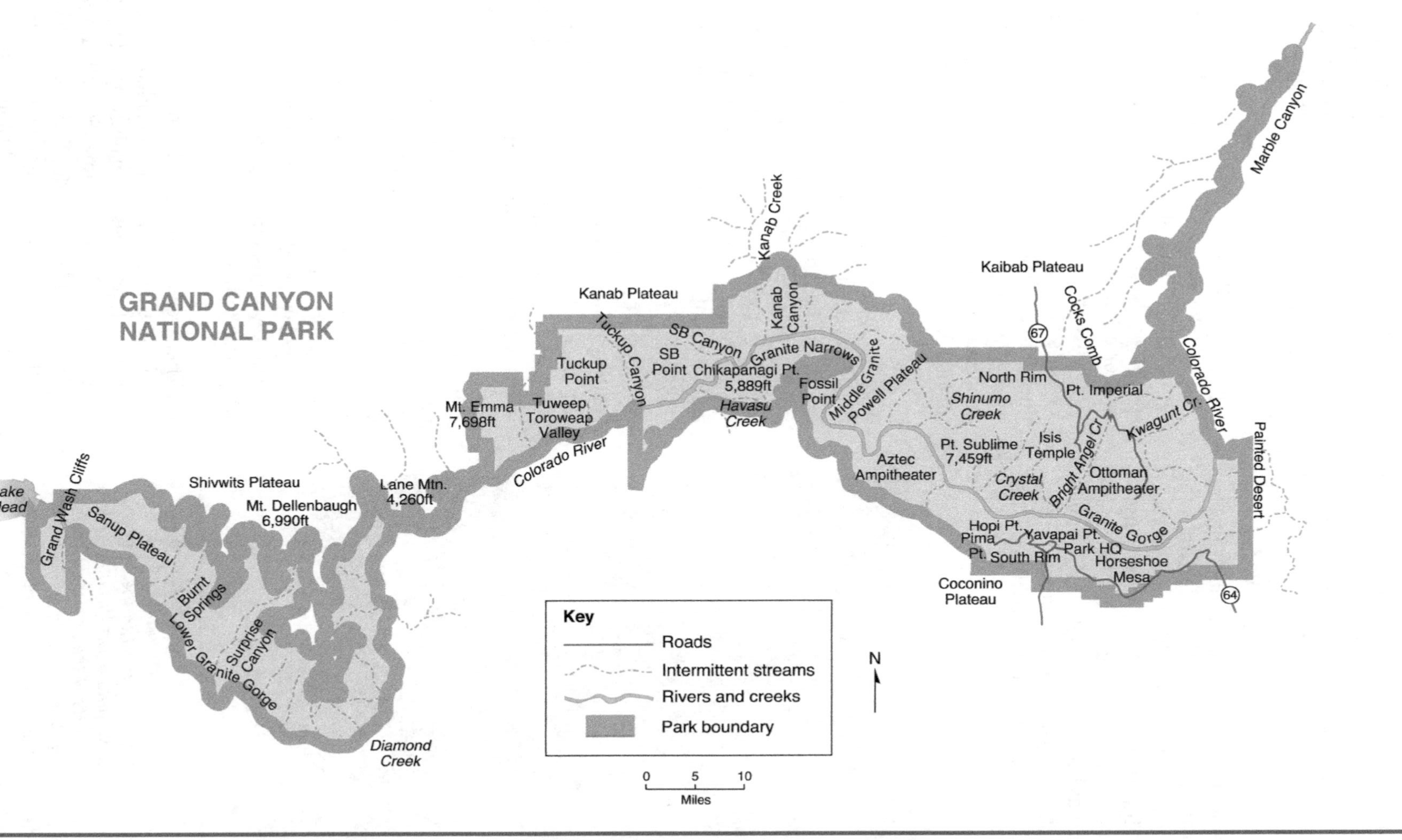

FIGURE 10.20 *Grand Canyon National Park, Arizona.*

geologic columns on earth, encompassing some two billion years of geologic history. The layers of sedimentary rock seen in the Grand Canyon are essentially flat-lying; many of these beds extend throughout the Colorado Plateaus region.

How Does a River System Evolve?

The Colorado River is a major through-flowing stream with a total length of 1450 miles from the Rocky Mountains to the Gulf of California. Like other major streams, the Colorado developed in response to environmental variables over time; in the Colorado's case, its history has been more complicated. For over a hundred years, geologists have argued about possible courses and directions of flow of tributaries and "ancestral" streams that may have had some part in the Colorado's evolution. So far, no hypothesis has met with overall acceptance. The details of the problem are fairly technical, but some general principles that apply to stream development are outlined below.

FIGURE 10.21 *Natural cave and a flowing spring in the Redwall Limestone. Major Powell named the area Vasey's paradise after the botanist in the 1869 expedition. Ground water, seeping down through rock layers, over time dissolves parts of the limestone and enlarges openings in the rock. Prehistoric animals and humans found shelter in the caves in the canyon walls. Photo by Michael Little.*

Base Level

The limiting surface to which a stream can cut down is called *base level.* Sea level is the ultimate base level for through-going streams like the Colorado, but sometimes local base level, such as a lake, a reservoir, or an interior basin without an outlet, controls a stream's downcutting ability. The Colorado's base level has been affected by uplift and also by sea level fluctuations, especially during the Pleistocene Ice Ages. Moreover, segments of the river, probably before it became through-flowing, may have emptied into inland lakes that functioned as temporary base levels.

Gradient

A stream whose *gradient* (rate of descent, or slope) is low does not have much erosive power. But if an area is uplifted, the gradient steepens, increasing the velocity of stream flow. When this happens, a stream downcuts rapidly, especially if its volume of water also increases. Sometimes, during uplift, a stream is able to maintain its original course and cut a narrow trench down through bedrock.

Headward Erosion and Stream Capture

While a trunk stream is downcutting or entrenching itself, its tributaries lengthen themselves by *headward erosion,* which means that they cut back into a plateau or upland at the head of each valley. A drainage system that is favorably situated, perhaps flowing on weaker rock, for example, is able to erode headward faster and eventually intersect and *capture* the headwaters of a neighboring stream or drainage system. Geologists are sure that stream capture occurred numerous times while the Colorado was evolving into a major stream; but the details of how, when, and where these events happened is not clear.

The key factors in the Colorado River's evolution are probably these:

1. The river's source area in the Rocky Mountains, in addition to the Colorado Plateau, has undergone a long series of relatively rapid uplifts since the end of Mesozoic time (65 million years ago). Overall, the Colorado Plateaus rose 5000 to 10,000 feet. Stream gradients became steeper, increasing stream velocity and erosion capability.

2. More rain and snow fell in the Rocky Mountains as the elevation increased. In the colder Pleistocene climates, glaciers formed and grew. As more meltwater and runoff became available, streams gained volume. The streams that drained the Rocky Mountains were powerful forces in erosion as they poured out on the plateaus.
3. The climate of the plateaus was (and is) drier than the mountain climate. Therefore, the high-volume, swift rivers cut narrow, deep canyons. Because rainfall is low on the uplifted plateaus, runoff is not ample enough to widen the valley side slopes.
4. When the Gulf of California opened more than 6 million years ago, this lowered base level for the Grand Canyon region and caused rapid headward erosion. The modern course of the through-flowing Colorado River was probably established after this time. Previously to the opening of this arm of the sea, the Colorado—or an ancestral stream that drained the Rockies—flowed into the ocean at a different location.
5. The east-west Grand Canyon segment of the Colorado's course across the Kaibab Plateau is asymmetrical; that is, the river flows closer to the south side due to a slight tilting that occurred during uplift. The fact that the North Rim receives more moisture and sheds more sediment also tends to push the river toward the south side of the canyon. The rate of erosion on the Kaibab Plateau is greater because all of the precipitation falling on the north side of the river drains to the south and into Grand Canyon. On the South Rim, rain and meltwater also drain to the south; since the runoff goes away from the canyon, it has less erosional effect.
6. How the Colorado River is regulated will determine how it evolves in this century and the next. Lake Powell became filled to capacity in 1980. How long will it be before silt accumulating in the lake bottom will cause the river to run over the dam? If such overflow is allowed to happen, the dam will inevitably wash out.

The controlled average release of water by Glen Canyon Dam is 12,200 cubic feet per second. Contrast this with the range of flow before dams were built above the Grand Canyon: a low of 1000 cubic feet per second during the driest time of year and a recorded flood high of 325,000 cubic foot per second. The Paria River, which joins the Colorado just below Glen Canyon Dam, brings some sediment, seasonal high-water flow, and occasional flash-flooding to the Grand Canyon. To a lesser extent, the Little Colorado River, with its confluence just above the big bend where the main river turns westward, does the same. But these contributions, added to sediment-free water released by the dam, are not powerful enough to keep the canyon bottom free of debris. Meanwhile, beaches on bank areas that used to be replenished by sediment during seasonal floods are being eroded due to fluctuating river levels when large amounts of water are released from the power station at Glen Canyon Dam.

Restoring the river's normal flow is not a practical answer because the Colorado River is the only significant source of surface water in the entire Southwest. More of its water is legally allocated to users than the river normally carries. How national parks and recreational areas are affected by the regulation of the Colorado River is only one aspect of a difficult problem of crucial importance to millions of people.

Igneous and Metamorphic Rocks in the Grand Canyon

As the Colorado River sliced its way deeper into the upwarped rocks of the plateau, it lets itself down onto ancient rocks and structures beneath the flat-lying Paleozoic beds. These igneous and metamorphic rocks, exposed in the Inner Gorge of the Grand Canyon, are Precambrian, and over 600 million years old. The oldest is the Vishnu Schist at the very bottom of the canyon. The rock originally accumulated as silts, muds, and volcanic material and lithified about 2 billion years ago. During mountain-building episodes, the layered rocks were highly deformed and metamorphosed. On dark-colored, contorted Vishnu outcrops alongside the river, little, rounded garnets stick out from the rock surface like a scattering of seeds.

A *pluton* (a body of cooled magma), the Zoroaster Granite, intruded the metamorphosed Vishnu rocks deep underground and gradually cooled. The granite shows up as light-colored, irregular bands and blotches in the Vishnu Schist.

Later in Precambrian time, many layers of shales, sandstones, and other sedimentary rocks—interbedded with countless lava flows—covered the region. These rocks, which comprise the Grand Canyon Supergroup, were also metamorphosed by heat and pressure during tectonic activity. Some of the lavas show up as dark ribbons or bands between or cutting through the old sedimentary beds. Geologists call such features *dikes,* when they cut through older rocks, and *sills,* if they are squeezed between rock layers (Figure 10.22). A long period of erosion removed a great portion of the Precambrian rocks.

In recent geologic time, around a million years ago, dark (basaltic) lava erupted from fissures and flowed down into the Grand Canyon in the western section of the park. Volcanic ash, thrown out from small cinder cones on the canyon rim, fell on the land. These flows and ash falls made the youngest rocks in the park.

Geologic Structures

Unconformities

Before erosion planed off the igneous and metamorphic rocks of the Inner Gorge, the rock units had been broken into large blocks and tilted about 15° by mountain-building. But the surface was relatively flat by the time an ocean covered the region and the sedimentary beds of Paleozoic time began to accumulate. The surface or contact between Paleozoic and Precambrian rocks, which is called an *unconformity,* represents a gap in the geologic record. In this case, it is an *angular unconformity* because the bedding planes of the rock units above and below the erosion surface are *not* parallel; specifically, the younger overlying sediments rest upon the eroded surface of tilted or folded older rocks (Figure 10.24).

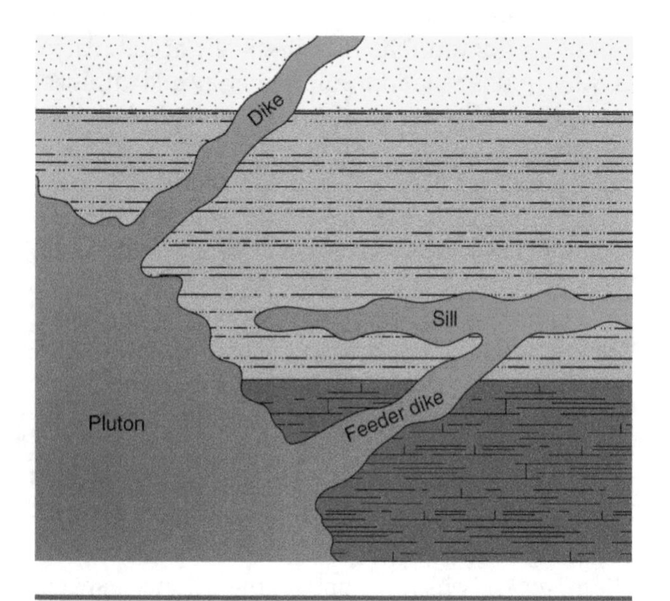

FIGURE 10.22 *Magma from a small pluton intrudes sedimentary rock beds. Dikes cut across rock layers, while a sill squeezes between bedding planes.*

What causes gaps in the rock record? The marine sedimentary rocks in the walls of the Grand Canyon cover a large portion of geologic history, but some geologic periods are not represented. For example, a major erosional unconformity exists between horizontal Cambrian and Devonian strata in the canyon sequence. Were the marine beds of the missing geologic periods completely eroded away? Or did marine sedimentation not occur during those periods?

Tectonic forces generated within the earth have resulted in uplift, movement, and deformation of portions of the earth's crust. Continents have stood at times higher and at times lower during geologic history. Ocean levels have fluctuated over millions of years.

When sea level rises, the oceans *transgress* onto the land; and when sea level falls, they *regress* from the land. During the long and slow process of transgression or regression, streams carrying sediment down to the ocean continue to erode. If the ocean is regressing, the streams may strip and redeposit (at lower elevations) most or all of the sediment layers that were put down during an earlier period of transgression. When the seas finally come back and again cover an area that has been stripped of sediment, a gap, shown as an unconformity in a geologic column, occurs in the geologic record. The rock above an unconformity is considerably younger than the underlying rock.

Geologists who try to reconstruct ancient changes in elevation and sea level find unconformities to be useful indicators. Typically, an unconformity can be recognized by the presence of a *basal conglomerate* that rests on the old erosion surface and is composed of fragments of old eroded sediment incorporated with younger sediment deposited by a transgressing sea.

As a shoreline shifts landward or seaward (depending on whether the sea is transgressing or regressing), beach deposits accumulate in this narrow depositional environment. A *time-transgressive rock unit,* with the rocks becoming younger in the direction in which the sea was moving, may show the shifting location of the old shoreline. The Tapeats Sandstone, which is on the upper side of the angular unconformity between the Precambrian rocks and the Paleozoic beds, is an example of a time transgressive rock unit (Figure 10.25).

Folds and Faults

Layered sedimentary rock units, such as those exposed in the canyon walls above the igneous and metamorphic rocks of the Inner Gorge, have other interesting clues about geologic conditions in the past. The presence of marine fossils leads to the obvious conclusion that these rock layers have been uplifted more than a mile from their original position beneath an ancient sea. Uplift must have been essentially uniform over a large area in order to keep the layers so nearly horizontal.

What happens to layered bedrock when the uplifting forces are not uniform? Keep in mind that most tectonic activity takes place at considerable depth when rocks are subjected to great heat and pressure. Under such conditions, layered rock that would fracture or break in a surface environment may become *plastic,* or malleable, and bend into *folds* rather than break. The sides, or arms, of a fold are called *limbs.* Most folds are either *anticlines* (upfolds) or *synclines* (downfolds) (Figure 10.26). If the limbs are parallel (as with a hairpin), the fold is called an *isocline* or *isoclinal fold,* and we know it must have been formed by intensely compressive forces (i.e., squeezed together). Folds of this type are found in some of the Precambrian metamorphic rocks in the Grand Canyon.

A type of fold more characteristic of the Colorado Plateaus is the monocline, which usually shows up in flat or slightly tilted beds. *Monoclines,* or *monoclinal folds,* have one limb and a gentle bend, or flexure, connecting rock layers at one level (elevation) with the same layers at another level (Figure 10.26). The East Kaibab monocline, which outlines the eastern edge of the Kaibab Plateau, is an example.

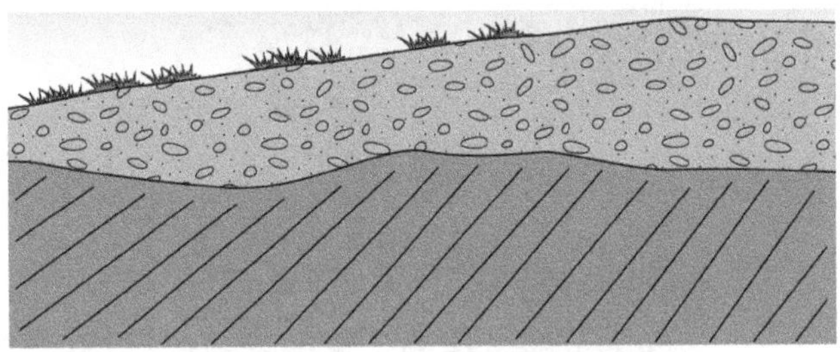

FIGURE 10.24 *Unconformities represent a gap in the geologic record. They are old erosional surfaces separating younger rock units from older rocks exposed to erosion before being covered over by younger rock layers.* A. *This* unconformity *may be due to an interruption in sedimentation or to uplift and erosion. Deformation did not occur while processes of weathering and erosion operated because beds above the unconformity are approximately parallel to layers below the contact.* B. *This* angular unconformity *shows that the older rocks below the contact were folded ("deformed") and subsequently eroded. Younger beds were deposited on the old erosional surface.*

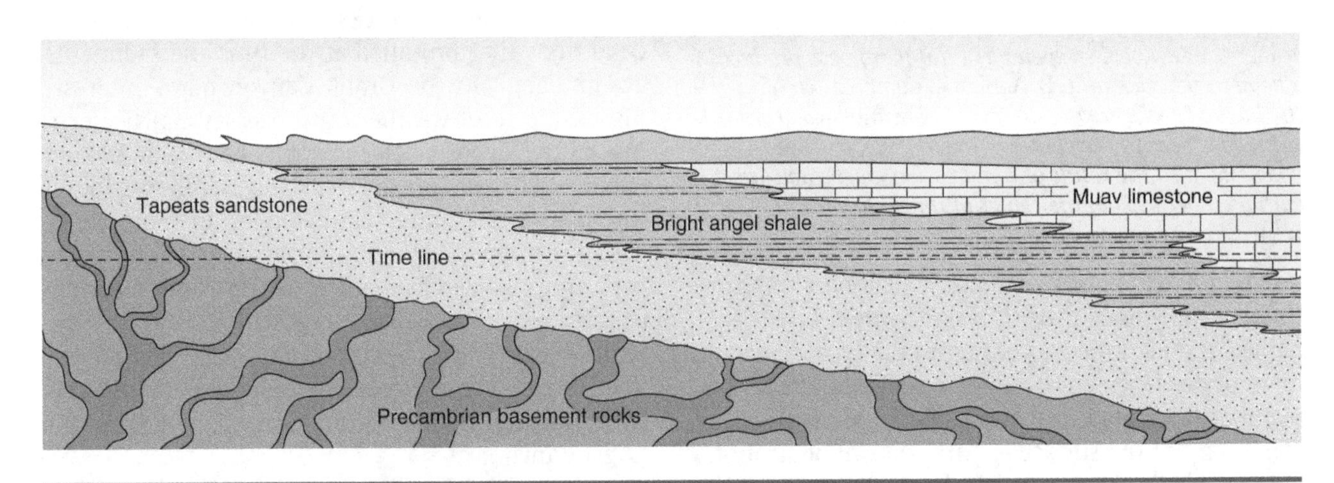

FIGURE 10.25 *Time-transgressive rock layers may be deposited when a shoreline shifts laterally, causing the age of a formation to become progressively younger across an area. Beach deposits of the Tapeats Sandstone accumulated as an ocean advanced over Precambrian rocks; mud and silt settled in shallow water offshore and became the Bright Angel Shale; the Muav Limestone formed in deeper water. The time line in the cross section represents a theoretical equal age across the formations shown. From D. Nations and E. Stump,* Geology of Arizona. *© 1981 Kendall/Hunt Publishing Company.*

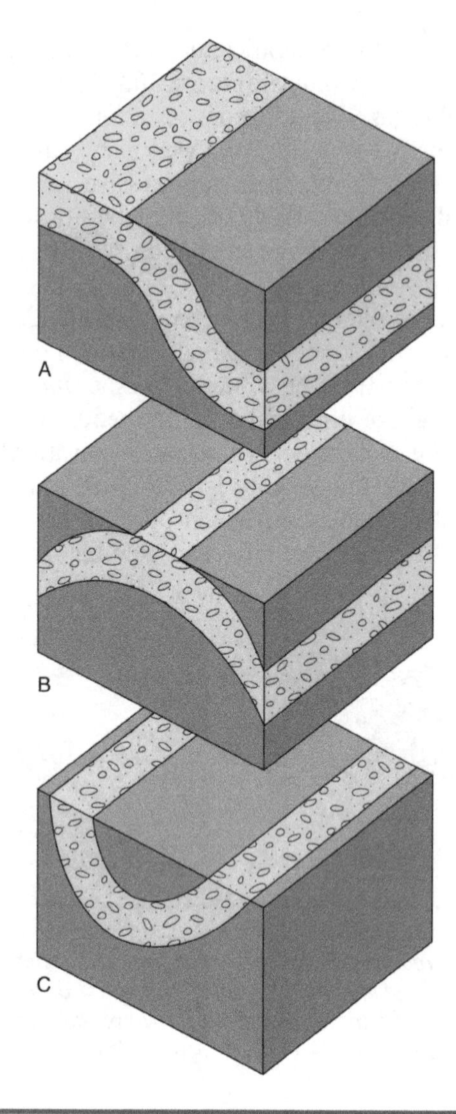

FIGURE 10.26 *Block diagrams of simple folds.* A. Monocline. *Local steepening of flat-lying beds produces a bend in the rock unit.* B. Anticline. *Compression of the layered rocks creates an upfold, opening downward, with older beds in the center of the fold.* C. Syncline. *A downfold, also the result of compression, opens upward, with younger beds in the center of the fold.*

A *fault* is the result of movement along a break or fracture in bedrock causing *displacement.* Faults are named and identified according to the type of movement or displacement that occurred. Faulting may or may not be related to the topography of an area. In some regions the surface features give no indication of the geologic structure of the bedrock below, but in the Basin and Range Province, just west of the Grand Canyon, great mountains and valleys display evidence of faulting on a grand scale.

Normal faults separate blocks that have been either raised or dropped, mainly as a result of tensional and vertical stresses affecting the earth's crust. Tectonic forces that stretch the crust cause normal faults. The normal faults in the Precambrian rocks in the Grand Canyon indicate block-faulting during ancient geologic times and are unrelated to present surface features.

Faults involve relative movement between two blocks, one of which is called the footwall block and the other the hanging wall block. If a person were able to stand on a fault surface within the earth (as if in a mine), his feet would be supported by the *footwall,* while the *hanging wall* would be the rock surface over his/her head. In a normal fault, the hanging wall has moved *down* in relation to the footwall.

If the hanging wall has moved *up* in relation to the footwall, it is called a *reverse fault.* A special type of reverse fault, in which the fault plane has a low angle (about 10° or less), is referred to as a *thrust fault* (Figure 10.27). Reverse faults are produced when tectonic forces compress crustal rocks, causing them to break vertically or at an angle. In the Grand Canyon, some of the Precambrian normal faults (produced by tensional forces) were changed millions of years later to reverse faults by compressive forces.

Sometimes compressive forces cause blocks to slide past each other, producing a *strike-slip fault* (Figure 10.27) that shows lateral displacement.

Sedimentary Environments

Studying the sizes, arrangement, and composition of sediments can reveal a great deal about the origin of sedimentary rocks and the environmental conditions that prevailed at the time the sediments accumulated. In the Grand Canyon many interesting examples of weathering processes can be seen. *Brecciated* deposits of rock that had been broken up and crushed into angular fragments are found where a cave ceiling has fallen or in fault zones where movement along fault surfaces has crushed rock. Rock broken by faulting weathers more rapidly because more surfaces are exposed to water and the atmosphere. Tributaries of the Colorado River tended to form along fault zones where the rock is susceptible to weathering and erosion. An example is Bright Angel Canyon.

Some of the Grand Canyon limestones have *karst* features, which are the result of slightly acidified underground water dissolving soluble rock. Because limestone consists principally of the mineral calcite (calcium carbonate), which is soluble under humid conditions, water trickling down through cracks and

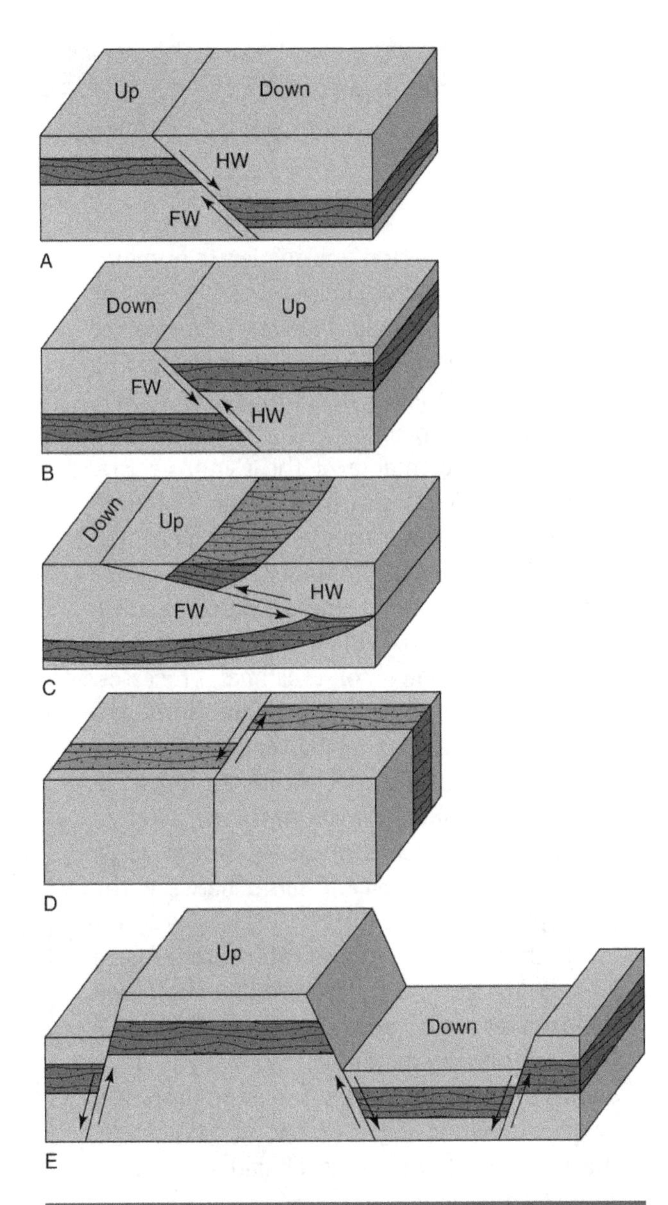

FIGURE 10.27 *Block diagrams of types of faults. Stresses that create faults may cause one block to move up, down, obliquely, horizontally, or even to rotate, in relation to another block.* A. Normal fault. *The hanging wall side of the fault moves down in relation to the footwall side.* B. Reverse fault. *The hanging wall side moves up in relation to the footwall side.* C. Thrust fault. *A reverse fault in which the fault plane dips at a low angle.* D. Strike-slip fault. *One side of the fault moves horizontally by the other side.* E. Block-fault structure. *The upthrown block and the downdropped block are bounded by normal faults. Extensional forces stretching the crust produce normal faults. Reverse faults are the result of crustal compression.*

crevices in rocks tends to enlarge any voids into hollows or caverns. The caves in the Red Wall Limestone developed during an earlier time when the climate was more humid than now.

In the semiarid climate of the present, the limestones of the Colorado Plateaus tend to be resistant to weathering. The Kaibab Limestone, which forms the caprock on the rim of the Grand Canyon, is an example. However, visitors approaching the North Rim can see circular pits and hollows, some containing water, beside the highway. These are *sinkholes,* which are depressions dissolved beneath the surface in the Kaibab Limestone. The Kaibab Plateau is flat enough so that rain and meltwater tend to collect in pools that over time eat away the soluble rock.

Precipitation that falls on the Kaibab Plateau also percolates down through the sedimentary beds of the North Rim (Figure 10.21). When the water reaches an impervious rock unit, it emerges from the canyon wall as seeps or natural springs. Roaring Springs, one of the larger springs supplied in this way, gushes from a chasm high on the north wall. Roaring Springs provides water (with the aid of a pumping station) for the Park Service facilities on the South Rim.

When water containing dissolved calcium carbonate from the limestone evaporates, deposits of finely crystalline, pale *travertine* are precipitated. Along Havasu Creek, travertine deposits have encrusted the canyon walls and formed low, sinuous dams in the creek bed.

Sandstones may be formed from old dunes, such as the *eolian* (windblown) deposits that make up the Coconino Sandstone. Other sandstones were formed from water-laid deposits, either in freshwater lakes or in marine environments. The degree of sorting and rounding of sand grains, the presence or absence of fossils, the type of bedding, and the kind of cementing (or matrix) holding the grains together can all provide clues as to the geologic history of an area.

GEOLOGIC HISTORY

1. Accumulation of sediments early in Precambrian time.

Around 2 billion years ago, sands, silts, muds, and volcanic ash were deposited in a shallow marine basin associated with an orogenic belt. Fossil evidence of life has not been found in these marine beds.

AN OUTLINE OF THE GEOLOGIC HISTORY OF THE COLORADO PLATEAU

1. **Precambrian Time**
   The Precambrian rock units exposed in the Grand Canyon do not crop out in the other national parks of the Colorado Plateaus, presumably because erosion and stream downcutting have not reached low enough depths.
2. **Lower Paleozoic: Cambrian to Devonian Periods**
   The Colorado Plateaus were a tectonically stable area for millions of years, probably as the trailing, subsiding edge of a *continental plate* (a segment of continental crust). It is believed that continental rifting (the splitting up of a continent) had occurred, possibly before the end of Precambrian time. Shallow seas spread over a vast area of the continental plate that would eventually become North America. In these seas, thousands of feet of sediment were deposited.
3. **Upper Paleozoic and Mesozoic: Mississippian through Cretaceous Periods**
   A series of *orogenies* (mountain-building episodes) deformed parts of western North America surrounding the Colorado Plateaus. Instead of breaking up or pulling apart, some land masses were coming together and closing up former oceans. Sand dunes overspread vast plains. The abundance of quartz-rich sands that built the great Jurassic dune belts in coastal deserts were derived from the weathering and recycling of quartz grains from older rocks.

   In Cretaceous time, the last great inundation by ocean waters spread over western North America. Thick units of richly fossiliferous, marine rocks were laid down. The Mesozoic Era drew to a dramatic close as accelerated orogenic activity lifted the land and drained the shallow ocean. This great event, which began 65 million years ago and lasted for millions of years, drastically changed the face of western North America. It is called the Laramide orogeny.
4. **Cenozoic Era**
   Minor uplifts and adjustments, accompanied by gentle deformation and moderate outpourings of dark lava, characterized early Cenozoic time in the Colorado Plateaus. Nonmarine sediments, washed down from surrounding highlands and ranges, accumulated in basins, and spread over plains.

   Toward the middle of the Cenozoic Era, tectonic activity resumed. The Colorado Plateaus region was uplifted still higher and tilted. Streams began to dissect the landscape into plateaus, benches, mesas, buttes, arches, and deep canyons. Each successive uplift increased the power of erosional processes. Uplift was uneven, causing some faults and open folds to develop in the rocks. Upwarping and stretching of the continental crust in the Basin-and-Range region to the west of the Colorado Plateaus produced several large faults, such as Hurricane fault, between the two regions.

   In the meantime, the elevation of the Sierra Nevada to the west deprived the Colorado Plateaus of rain, and the climate became increasingly arid. (When moisture-laden clouds from the ocean are forced to drop rain or snow when rising over a mountain range, the far side of the range is said to be in a *rain-shadow zone.)*

   The Pleistocene Epoch was a time of colder climates, increased precipitation, and intensified erosion. Mountain glaciers formed at higher elevations, and ice and frost broke up rocks everywhere on the plateaus. The results of colder temperatures, and more rain show up in the present Colorado Plateaus landscapes as dry valleys that once held streams, abandoned lakebeds, landforms eroded by fluvial sculpturing, and stream deposits unrelated to today's creeks and rivers.

2. Ancient mountain-building and igneous intrusions.

About 1.7 million years ago, tectonic activity, resulting from plate collision, converted the accumulation of sedimentary beds and volcanic material into the metamorphic Vishnu Schist. Then molten rock pushed into the schist, slowly cooling to form Zoroaster Granite. Some of the rock became metamorphosed to gneiss. In Trinity Chasm near Isis Temple, light-colored gneiss (containing appreciable amounts of quartz, feldspar, and micas) shows up as veins in the ancient rocks.

3. Uplift and erosion of the early Precambrian rocks.

We know little about the Mazatzal Mountains that were constructed; only their roots are left. For millions of years, sediments were stripped from highlands and deposited in lowlands or oceans.

4. Deposition of the Grand Canyon Supergroup (1.2 billion to 800 million years ago).

The rock units of the *supergroup* (an assemblage of vertically related formations and groups) consist of sedimentary beds that were later subjected to varying degrees of metamorphism. (Since they have retained the layering and certain other characteristics of sedimentary rocks, most have been given sedimentary rock names.) The Unkar Group, the older section of the supergroup, was deposited in an offshore environment. Waves worked over and smoothed out the early Precambrian land surface. Larger fragments accumulated as gravel, creating a basal conglomerate known as the Hotauta Member of the Bass Limestone, a formation. (A *group* consists of two or more geographically associated formations with notable common features. A *formation* is a rock unit containing one or more beds with distinctive physical and chemical characteristics. A *member* is a minor unit within a formation.) Ranging from 120 to 340 feet in thickness, the Bass Limestone accumulated in a shallow sea as a mixture of limestone, sandstone, and shale.

The Hakatai Shale, the next younger formation of the Unkar Group, is composed of units of thin-bedded shales, mudstones, and sandstones. Since most of the shale is nonmarine, the formation represents a temporary retreat of the sea.

The Shinumo Quartzite, a marine sandstone before it was metamorphosed, was so resistant to erosion that it juts up as "fossil islands" into overlying Cambrian beds. Apparently cliffs of Shinumo Quartzite withstood the wave action of the Cambrian ocean until the islands were completely engulfed. The Dox Sandstone, mostly sandstone but containing some shale layers, is another accumulation of shallow ocean deposits.

The flows of dark (basaltic) Cardenas Lava, the youngest rock unit of the Unkar Group, intruded zones of weakness in the sedimentary rock. Lava flows up to 1000 feet thick erupted from fissures and spread over older rocks.

The beds of the Nankoweap Formation, which are shallow-ocean deposits, lie unconformably over the eroded surface of Cardenas Lava and separate the Unkar Group from the younger rocks of the Chuar Group. Three formations—Galeros, Kwagunt, and Sixty Mile—make up the Chuar Group. Evidence of algae has been found in Kwagunt rocks, but fossils are rare in rocks this ancient.

5. Block-faulting of the Grand Canyon Supergroup.

Remnants of the Grand Canyon Supergroup crop out in scattered sections of the Inner Gorge and the deeper tributary canyons. The pockets of rock are isolated from each other because when tilting and blockfaulting occurred, some rocks were elevated and some were dropped down (Figure 10.27E). Tensional tectonic forces were stretching and rifting crustal segments at that time. Movement was along fault lines, such as the Bright Angel fault, creating north-south trending fault-block mountain ranges.

6. "The Great Unconformity."

During the long erosional period that followed the uplifting and tilting of the fault-block mountains, most of the Chuar beds and some of the Unkar Group were stripped from the mountain slopes. This exposed some of the more resistant beds, such as the Shinumo Quartzite, that stood as islands in the encroaching ocean of Cambrian time. Most of what is left of the Grand Canyon Supergroup is preserved in the down-faulted blocks. The horizontal contact between the Precambrian and Paleozoic beds is the Great Unconformity. In 1869, when John Wesley Powell recognized the angular relationship between

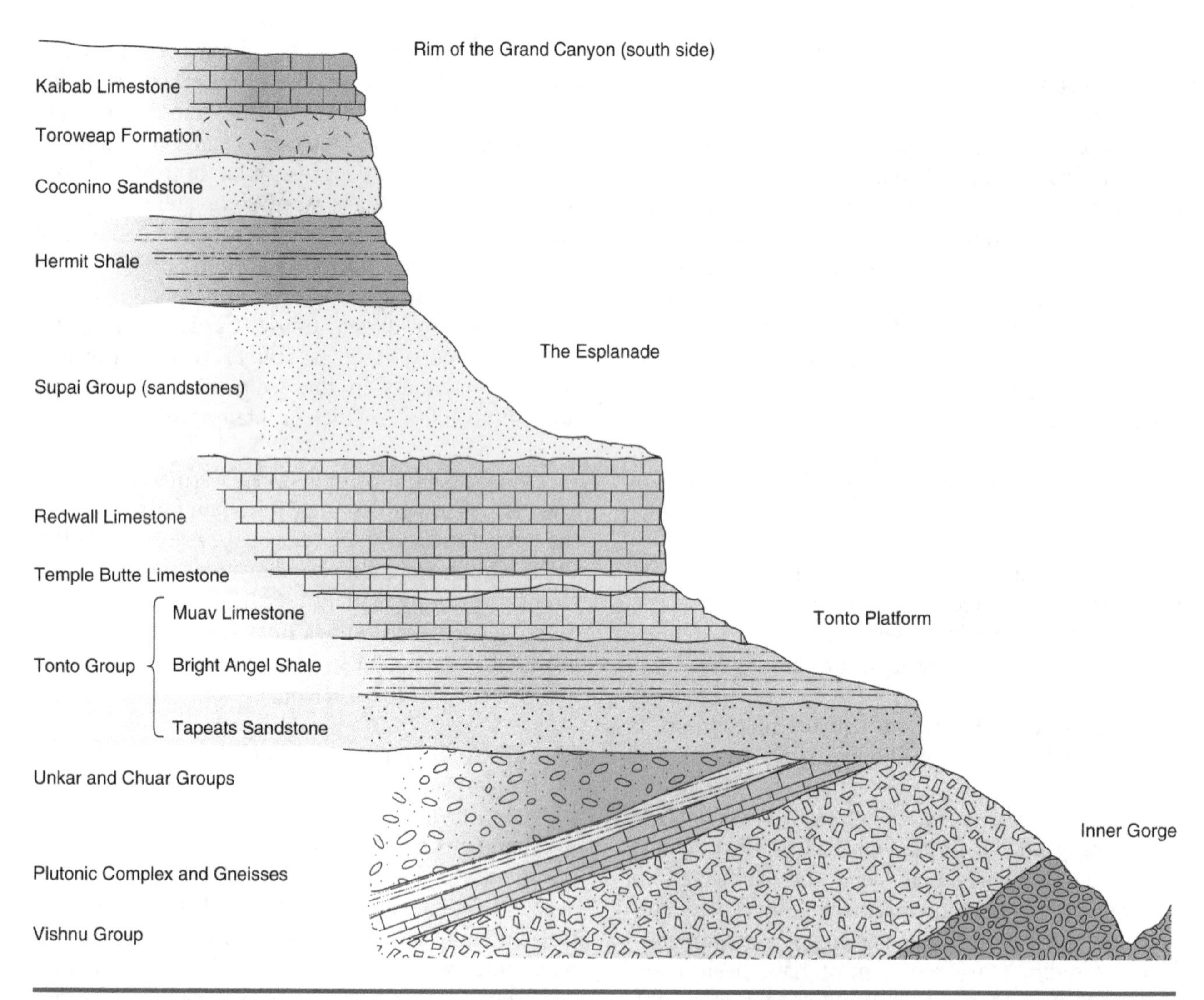

FIGURE 10.28 *Schematic profile of rock units exposed on the south wall of the Grand Canyon. Relative thickness of formations are approximate, but slope angles are greatly exaggerated. In the semiarid climate, limestones tend to be cliff-formers, while shales form gentle slopes. The resistance of sandstones depends on their purity and cementation. Adapted from Billingsley and Breed 1980.*

the older tilted rocks and the flat-lying formations above, he named the contact the Great Unconformity.

7. Deposition of Cambrian rock units.

About 550 million years ago, a western ocean began to spread over the eroded coastal lowlands. Three formations were deposited, each representing a different environment. The way in which the sediments were laid down illustrates the principle of time transgression (Figure 10.25).

Imagine yourself standing on a sandy beach that would eventually become the Tapeats Sandstone. Pebbles, cobbles, and a few boulders are on the sand after falling from the cliffs along the shore. This is a turbulent coast. Pounding storm waves attack the cliffs; cobbles and sand are rolled back and forth in the surf and tossed up on the shore. These are the constituents of the basal conglomerate at the bottom of the Tapeats. Also preserved in the Tapeats are the shapes of ancient coastal sand dunes.

Farther out in the water, the sea bottom becomes muddy. These muds, which contain the remains of primitive brachiopods, trilobites, and worms, formed the Bright Angel Shale. Offshore, calcium carbonate precipitates from seawater and accumulates as a limy ooze on the bottom. From this ooze was formed the Muav Limestone.

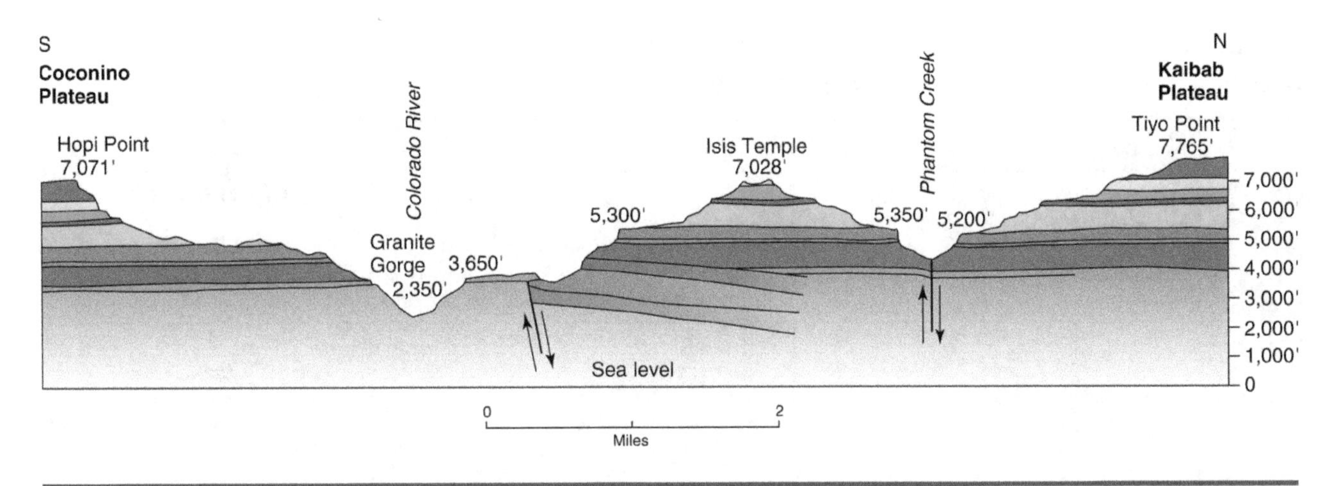

FIGURE 10.29 *South Rim to North Rim cross section through the Grand Canyon from Hopi Point to Tiyo Point. Canyon features are shown in their relative sizes and correct proportions. Uplifting of the Colorado Plateau enabled the Colorado River, with the help of mass wasting, to remove immense amounts of rock and carve the Grand Canyon in a few million years. Adapted from F. Matthes, U.S. Geological Survey Bright Angel quadrangle, 1906.*

Because sea level is gradually rising, the shoreline and the beach are shifting eastward, and the pattern of sediment deposition is also moving eastward without a break in continuity. Thus when the Tapeats Sandstone and the Bright Angel Shale were first laid down, it was early in Cambrian time; but by the time the last grains were being deposited some distance to the east, the time was Middle Cambrian, about 30 million years later. Evidence that the sea was transgressing is shown by finer-grained shale deposited over coarser-grained sandstone. Had the sea been regressing, coarser sediment would have overlaid the finer.

The Tapeats Sandstone, Bright Angel Shale, and Muav Limestone belong to the Tonto Group and are the beds that make up the Tonto Platform. These beds lie in their original horizontal position, virtually undisturbed since they were deposited on top of the Precambrian erosional surface. The Tapeats Sandstone and the Muav Limestone are consistent cliff-formers in the walls of the Grand Canyon, while Bright Angel Shale is a slope-former (Box 10.3).

8. Unconformity between the Cambrian and Devonian beds.

Beds from the Ordovician and Silurian periods are missing from the Grand Canyon geologic column. There is no evidence to indicate whether beds were deposited in the region during these time periods and subsequently removed by erosion, or whether deposition did not occur. However, deep channels were carved by streams (or possibly by marine scour) on top of the Muav. Devonian Temple Butte Limestone was deposited in these depressions. The channel deposits are especially well displayed in Marble Canyon in the eastern section of the park. Toward the western end of the park, near Grand Wash Cliffs, Temple Butte is a cliff-former, in some places a thousand feet thick. Because the limestone in Temple Butte has largely been replaced by dolomite, only a few fossil fish teeth remain in the formation. Erosion, producing an unconformity, brought the Devonian Period to a close.

9. Deposition of the Mississippian Redwall Limestone.

The visually striking cliffs of the Redwall in the Grand Canyon today are stained red by rainwater dripping down from the overlying Supai and Hermit Shale redbeds. The Redwall is actually a bluish gray limestone containing chert nodules. It was deposited in a shallow tropical sea in early to middle Mississippian time, some 330 million years ago. Thickness averages 500 feet. Fossil remains of crinoids, brachiopods, bryozoans, and other marine creatures are abundant in the formation. Added to the list of fossils are the chambered nautiloid fossils that were discovered in the streambed of a branch of Marble Canyon. Following the deposition of the Redwall Limestone, the region was gradually uplifted

TABLE 10.2

GEOLOGIC COLUMN, GRAND CANYON NATIONAL PARK

Era	Period	Group & Supergroup		Formation	Geologic Events
Cenozoic	Quaternary Tertiary			Lavas Slump and slide deposits, terrace gravels, river deposits, travertine.	Minor lava flows Development of Grand Canyon by uplifting and tilting, accompanied by down-cutting Uplift, erosion, minor deformation Erosion
Mesozoic	Cretaceous Jurassic Triassic			Moenkopi	Flood-plain deposition Withdrawal of seas Erosion
Paleozoic	Permian			Kaibab Limestone	Transgression of shallow seas; regression Erosion
				Toroweap Coconino Sandstone	Transgression of shallow seas, flattening sand dunes Flood-plain deposits covered by migrating sand dunes Erosion
				Hermit Shale	Deposition on flood plain, lagoons Erosion
	Pennsylvanian	Supai		Esplanade Sandstone Wescogame Manakacha Watahomigi	Flood-plain deposition Retreat of seas Marine sedimentation Erosion
	Mississippian			Surprise Canyon	Channel deposits; uplift Erosion
				Redwall Limestone	Karst development Marine sedimentation Erosion
	Devonian			Temple Butte Limestone	Marine deposits in channels Erosion
	Cambrian	Tonto		Muav Limestone Bright Angel Shale Tapeats Sandstone	Deposition in a transgressing sea, forming a facies change in these formations Erosion
Late & Middle Precambrian time (Proterozoic)		Grand Canyon	Chuar	Sixty Mile	Stream deposition Erosion
				Kwagunt Galeros	Volcanic activity; Grand Canyon orogeny, block-faulting, thrust-faulting, folding Erosion
				Nankoweap	Deposition in a shallow sea Erosion
			Unkar	Cardenas Lava Dox Sandstone Shinumo Quartzite Hakatai Shale Bass Limestone (including Hotauta Conglomerate)	Faulting, igneous intrusions, lava flows Shallow sea deposits Deposits as a marine sandstone Temporary retreat of the sea Deposition in warm, shallow sea Transgression of seas; major unconformity Erosion
Early Precambrian time (Archean)				Trinity and Elves Chasm Gneisses	Block-faulting; uplift. Igneous intrusions; metamorphism Erosion
				Zoroaster Granite	Uplift of Mazatzal Mountains; orogeny, folding, faulting, intrusions, metamorphism
				Vishnu Schist	Deposition of marine sediment; volcanic activity

Source: Modified after Huntoon et al., 1981, 1982, 1986.

and the beds underwent erosion late in Mississippian time. Because limestone is highly susceptible to solution by ground water, some natural caves developed.

10. A geological "Surprise" in the Grand Canyon.

In very latest Mississippian time (and possibly very earliest Pennsylvanian time), a singular depositional event occurred in what is now the Grand Canyon region (Beus 1986). A sedimentary rock, heretofore unrecognized, was first discovered by McKee and Gutschick, but was not considered as a separate formation until 1985 by Billingsley and Beus. This "new" unit has been appropriately named the Surprise Canyon Formation. The exposures, which are in discontinuous lenses a few hundred feet thick at the top of the Redwall Limestone, record the evolution of tidal estuaries on a low coastal plain that gradually subsided, drowning the valleys. At the end, a short period of erosion was followed by deposition of the lowermost beds of the Supai Group. Fossils in the Surprise Canyon Formation—from abundant logs and plant materials mixed with coarse sands in the bottom beds to marine shells in beds of sand and lime deposits—reflect the changing environments that produced this unusual sequence.

11. Deposition of the Supai Group during Pennsylvanian and early Permian time.

The Supai Group, a prominent ledge- and cliff-former in the canyon walls, is 600 to 700 feet thick. Mostly nonmarine, red siltstones and sandstones accumulated in a swampy environment. Distinctive crossbedding may be an indication of deltas and coastal dunes. Fossils include footprints of amphibians and an abundance of plant materials. The Esplanade Sandstone of early Permian age has been oxidized to a bright red color.

12. Early and middle Permian deposition.

The Hermit Shale, a red slope-former, accumulated in an environment similar to the underlying Supai Group. The beds contain ripple marks, mud cracks, footprints, plant fossils, and a few insect remains.

Migrating sand dunes that buried the swampy flood plain formed the Coconino Sandstone. This formation is the whitish cliff-former near the top of Grand Canyon. The uniformly sized, perfectly rounded, frosted quartz grains, arranged in a cross-bedded pattern characteristic of dunes, testify to a wind-blown (eolian) origin. The dunes must have been large because the Coconino is 400 feet thick. Reptile tracks are preserved in the sandstone.

The Toroweap Formation, with its nearly 200 feet of sandstones and limestones interbedded with gypsum, records marine advances and retreats. Brachiopods, mollusks, corals, and other organisms lived in the ancient ocean. The Toroweap is a ledge- and cliff-former.

Forming the rim of the Grand Canyon and the edge of the Kaibab Plateau is the massive, thick-bedded Kaibab Limestone of middle Permian age. More than 300 feet thick, the Kaibab Limestone is a prominent cliff-former. Some zones in the Kaibab have sand grains and chert nodules. Fossils include organisms such as crinoids and brachiopods that lived in warm, shallow seawater.

13. Withdrawal of the Permian Sea; Mesozoic deposition.

As the Mesozoic Era began, the continent was rising. Streams carved wide, low valleys. Sediment was transported from higher land to lower slopes and basins. These deposits became the Moenkopi Formation of early Triassic age, a mixture of brightly colored sandstones and shales with interbedded gypsum layers. The number and variety of fossils indicate an abundance of life forms.

Presumably the early Triassic rock units, such as those now exposed along the Colorado River above Marble Canyon, may have overspread the uplands of the Grand Canyon region. The soft Triassic shales, less resistant to erosion than the Permian Kaibab and Toroweap rock units, were probably stripped off long ago. Younger Mesozoic and early Cenozoic beds, identified in nearby areas, may also have been removed by erosion.

14. Mesozoic uplift and mountain-building.

The compressive forces generated by the Laramide orogeny had profound effects on western North America at the end of the Mesozoic and in the early Cenozoic. Nevertheless, the horizontal beds of the Colorado Plateaus remained stable and relatively undeformed while still being elevated like a great platform. All around the Colorado Plateaus, rock units were folded, faulted, and uplifted, and great mountain ranges were being pushed up. The

THE TERRACED WALLS OF THE GRAND CANYON

Observers who look down into the Grand Canyon from either the South Rim or the North Rim are usually fascinated by the "stairstep" appearance of the canyon walls (Figure 10.28). The upper part of the Grand Canyon is terraced while the lower part, or Inner Gorge, is narrower and V-shaped.

Four geological rock units are responsible for the most prominent cliffs or terrace risers. They are, from the top, the Permian Kaibab Limestone and the Supai Group, the Mississippian Redwall Limestone, and the Cambrian Tapeats Sandstone. Each cliff has at its base blankets of loose rock fragments, or *talus,* overlying gentle slopes. On top of the Tapeats Sandstone, a broad terrace top, the Tonto Platform, has been created by removal of less resistant rock units. A similar platform, called the Esplanade, in the western part of the park, has developed on the resistant Esplanade Sandstone. These slope segments, formed as the Grand Canyon was eroded down through successive geological formations, illustrate how some rock units are more resistant to weathering and erosion than others. This phenomenon is called *differential erosion.* The narrow Inner Gorge is not terraced or stepped because the metamorphic and intrusive igneous rocks have fairly uniform hardness and are more resistant to weathering and mass wasting than the Paleozoic formations.

Because the canyon is more than 10 miles wide from rim to rim, it becomes evident that the processes of valley widening and cliff recession are also significant. Except on the faces of the steepest cliffs, great sheets of broken rock on the valley sides prove the effectiveness of weathering, which produces the fragments, and mass wasting, which moves loose rock downslope.

North of Grand Canyon National Park, stepped topography on a larger scale is produced by the retreat of the margins of Mesozoic and Cenozoic formations in the form of the Chocolate Cliffs, Vermilion Cliffs, White Cliffs, and Pink Cliffs. These cliffs, which make up "the Grand Staircase" of the Colorado Plateaus, trend generally east-west, and step up toward the north. The most prominent features in Zion National Park are formed in the White Cliffs. The Pink Cliffs are strikingly displayed in Bryce Canyon National Park.

East Kaibab monocline (and other monoclinal folds) developed along the edge of the Kaibab Plateau during this time. (Presumably a southflowing segment of the Colorado River began to erode Marble Canyon along the lower flank of the monocline, long before the stream cut across the Kaibab Plateau.)

15. Middle Tertiary uplifting and faulting.

About 20 million years ago, as the Colorado Plateaus were uplifted and warped by tensional forces, local faulting occurred—sometimes along old fault lines—along with moderate volcanic activity. The stress of crustal stretching that was going on to the west in the Basin and Range province created several major north-south trending faults between the Colorado Plateaus and the Basin and Range. The Grand Wash fault, which intersects the Colorado River at the western edge of the park, is on the boundary between the two provinces. The Hurricane fault extends north into Utah. By late in Tertiary time, the Colorado Plateaus region had attained most of its present elevation.

16. The connecting of Colorado River segments and establishment of its present course.

The uplift and tilting of the plateaus accelerated the development of the drainage system. With its gradient increased and base level lowered by the opening of the Gulf of California, the young Colorado River downcut rapidly, captured the headwaters of other streams, and carved out the configuration of the Grand Canyon.

FIGURE 10.30 *View from Bright Angel Point on the North Rim. The Coconino Sandstone stands out as a cliff-former near the top of the cliff wall on the opposite side, overlying the Toroweap Formation. Both are Permian in age. Note the resistant dike (center bottom) that stands out in stark relief as it cuts across the horizontal beds. Photo by Ann G. Harris.*

17. Intensification of weathering and erosion during Pleistocene glacial episodes.

The colder, wetter climates of Pleistocene time increased erosional rates in and around the Grand Canyon. Most of the work was done by mighty floods of meltwater in the spring and by sudden flash floods from violent summer storms. In the last two million years, most of the downcutting of the Grand Canyon was accomplished.

18. Late Cenozoic volcanism.

Only a million years ago, flows of fluid basaltic lava covered parts of the plateau. Some poured down side canyons and dammed the Colorado River in the western section of the Grand Canyon. These lava flows are still being eroded by the Colorado River. The remains of small cinder cones on the rim mark the sites of ash eruptions.

19. Continued deepening of the Grand Canyon by downcutting; widening of the Canyon by mass wasting.

The climatic patterns of spring and summer floods, along with the age-old processes of weathering and erosion, continued into Holocene time, although at a somewhat slower pace. Since the Colorado River has become a regulated stream system from its source area to its outlet, its power to incise its channel has been reduced. However, as weathering loosens particles, mass wasting—powered by gravitational energy—moves materials downslope. In an arid climate these processes operate slowly except when quantities of loose sediment are transported rapidly during flash floods or landslides.

SOURCE

Baars, D. L. 1983. *The Colorado Plateau, a Geologic History.* Albuquerque, New Mexico: University of New Mexico Press. p. 3–48.

Beus, S. S. 1986. A Geologic Surprise in the Grand Canyon. *Fieldnotes* 16 (3): 1–4. Tucson, Arizona: Arizona Bureau of Geology and Mineral Technology, University of Arizona.

——, and Morales, M. (eds.) 1990. *Grand Canyon Geology.* Oxford University Press, 582 p.

Collier, M. 1980. *An Introduction to Grand Canyon Geology.* Grand Canyon Natural History Association.

Hamblin, W. K. 1994. *Late Cenozoic Lava Dams in the Western Grand Canyon.* Geological Society of America, 144 p.

——, and Murphy, J. R. 1980 (revised edition). *Grand Can_yon Perspectives.* Provo, Utah: Brigham Young Univer_sity Geology Studies, Special Publication No. 1, 48 p.

Hunt, C. B. 1969. Geologic History of the Colorado River, in *Colorado River Region: John Wesley Powell.* U.S. Geological Survey Professional Paper 669, pp. 59–130.

McKee, E. D. 1969. Stratified Rocks of the Grand Canyon, in *Colorado River Region: John Wesley Powell.* U.S. Geological Survey Professional Paper 669, pp. 23–58.

Nations, D. and Stump, E. 1981. *Geology of Arizona.* Dubuque, Iowa: Kendall/Hunt Publishing Company.

Powell, J. W. 1981 reprint (1875 *Scribner's Monthly* articles) W. R. Jones, editor. *The Canyons of the Colorado.* Golden, Colorado: Outbooks. 64 p.

Rigby, J. K. 1977. *Southern Colorado Plateau,* K/H Geology Field Guide Series. Dubuque, Iowa: Kendall/ Hunt Publishing Company.

CHAPTER 11

Rock Deformation and Geologic Structures

INTRODUCTION In earlier discussions, we saw that sedimentary rocks are originally deposited as horizontal layers. If you see horizontal sedimentary rocks in a hillside or road cut, it means that the rocks were not tilted as they were uplifted. Unless they are exposed in an outcrop such as the Grand Canyon, horizontal sedimentary rocks are not exceptionally awe inspiring, even to a geologist. The surrounding scenery is likely to be more eye-catching, whether it is the rolling hills of the Appalachian Plateau or the rugged beauty of the Colorado Plateau. When tipped on end, however, soaring vertically for hundreds of feet, or intricately folded and contorted, sometimes broken and thrusted one over another, sedimentary rocks capture the imagination like no other kind of exposure. How did they get that way? What force could bend and twist these "rock-hard" materials like so much taffy?

These bend and broken rocks exhibit the basic geologic structures of *folds, faults,* and *joints.* The forces that are responsible for making geologic structures and deforming the rocks are the topics of this chapter.

STRESS, DEFORMATION, AND THE STRENGTH OF MATERIALS

This chapter is about *stress, deformation,* and the *strength* of materials. **Stress** is *force per unit area,* whether it is the stress you apply to mold a piece of modeling clay or the stress that creates mountains. The response to stress, called **strain** or **deformation,** is the *change* in either *shape* or *volume* of the material undergoing stress. Deformation may or may not be permanent. The breaking of a window by the stress of an impacting rock is permanent deformation while the flattening of a tennis ball by the impact of a racket is not.

Every solid material has an inherent **strength,** which is the ability of the material to withstand stress. Once stress *exceeds* the strength of the material, the object will *deform;* that is, it will *change* in either shape or volume. It is important, however, to realize that the amount of deformation may not be sufficiently great to be detected by the unaided eye. For example, the deformation in rocks before they rupture to cause an earthquake can only be detected by very sensitive instruments.

Kinds of Stress

There are two fundamentally different kinds of stress: (1) *compression* and (2) *tension.*

In **compression,** the forces act *toward* each other. In some cases, the forces are *directly opposed* to each other, and the material is squeezed (Figure 11.1). An example would be the stress a pencil experiences as you hold it between your fingertips. In other cases, the forces are directed toward each other, but along *parallel, nonopposing* paths (Figure 11.2). These **shear** forces, as they are called, can be modeled by pushing on the diagonal edges of a deck of cards or by turning a doorknob (Figure 11.3).

The stresses involved in **tension** also act directly opposite each other and along parallel paths, but they act *away* from each other (Figure 11.4). In other words, tension is the *opposite* of compression. The stretching of a rubber band is an example of a tensional extension.

The stresses that deform the rocks of Earth's crust have many sources. The greatest stresses are associated with the plate margins. Tensional stresses are generated at divergent margins while enormous compressional stresses are created at convergent

FIGURE 11.1 *The forces involved in* compression *act* toward *and* directly opposite *each other. With few exceptions, this is the force you apply whenever you pick up an object.*

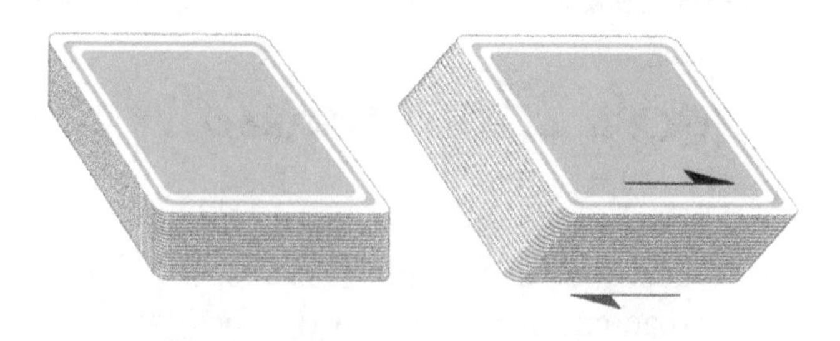

FIGURE 11.3 Shear Forces *can be modeled using a deck of cards. Applying forces directed toward but not directly opposite each other causes the cards to move over one another (shear), all in the direction of the applied force.*

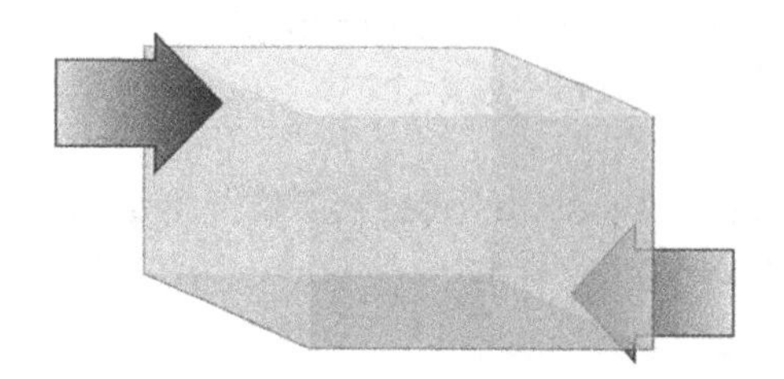

FIGURE 11.2 *The forces in* shear *also are directed* toward *each other, but are* not directly opposed. *This is the kind of force you apply when you take the cap off a bottle or turn a doorknob.*

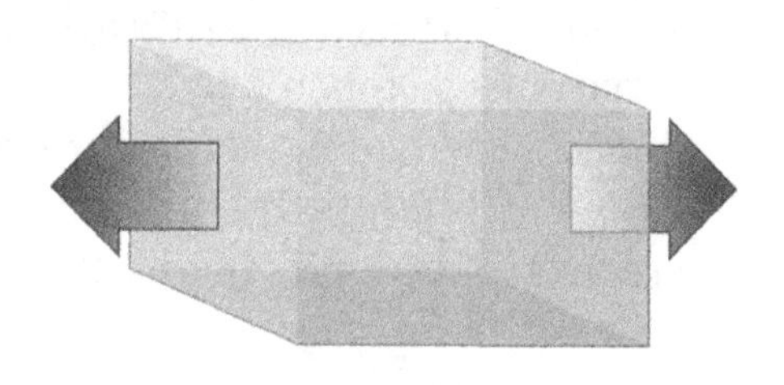

FIGURE 11.4 *The forces involved in* tension *operate* directly opposite *and* away *from each other. Anytime you pull on an object, be it a drawer or a rubber band, you are applying tensional forces.*

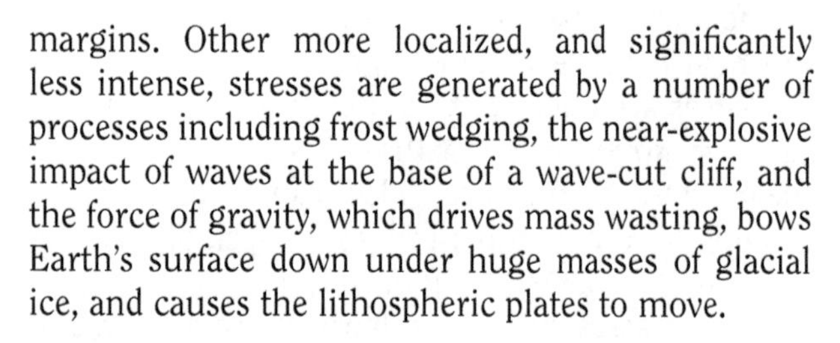

margins. Other more localized, and significantly less intense, stresses are generated by a number of processes including frost wedging, the near-explosive impact of waves at the base of a wave-cut cliff, and the force of gravity, which drives mass wasting, bows Earth's surface down under huge masses of glacial ice, and causes the lithospheric plates to move.

Kinds of Deformation

We will consider three different kinds of deformation: (1) *elastic*, (2) *plastic*, and (3) *brittle*. A fourth type of deformation, *viscous*, is experienced by liquid rock such as intruding magmas and mobile rocks such as those of the asthenosphere. Because our discussions will involve only deformation of the crust, we will not consider viscous deformation further.

Elastic Deformation A generalized stress-strain relationship is illustrated in Figure 11.5. The first response to stress is **elastic**. The characteristic of elastic deformation is that when the stress is *released*, the object will *return* to its original shape; that is, it is not permanently deformed. Think of squeezing a rubber ball or stretching out a rubber

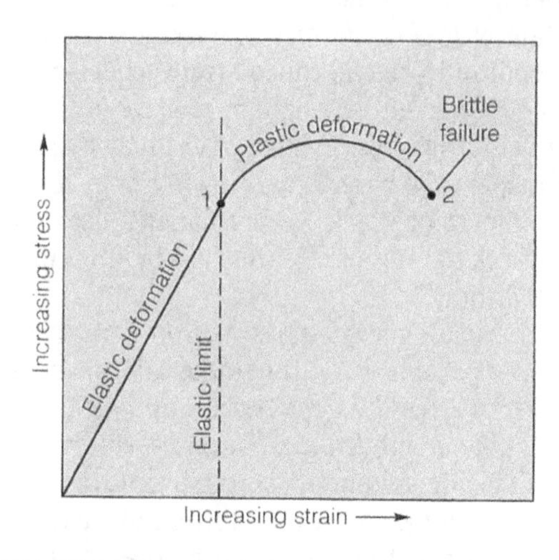

FIGURE 11.5 Strain *or* deformation *is the response to* stress. *These are responses that are part of our everyday life. When subjected to tensional forces, a rubber band deforms by* elastic deformation *and stretches, but returns to its original shape when the stress is released. The butter you spread on your toast deforms by* plastic deformation. *Once it is spread, it stays spread. Finally, we are all familiar with the common example of* brittle deformation, *the breaking of a glass as it hits the floor.*

band. The importance of the elastic response is that while the material is being stressed, some of the applied energy is *stored* within the material and is *released* when the stress is removed. It is also important to realize that the released energy can be used. For example, the impact of a tennis ball and a racket results in the elastic deformation of both the ball and the racket strings. As the tennis ball and the racket strings return to their original shape, the combined energy is released and drives the tennis ball over the net.

At some point during elastic deformation, deformation will begin to increase nonlinearly with stress, signaling that the material has reached its **elastic limit** (point 1 in Figure 11.5). If the material is stressed beyond the elastic limit, it will undergo permanent deformation, either plastic or brittle.

Plastic Deformation

If stresses are applied slowly, especially under conditions of high confining pressure and high temperatures, the material will begin to undergo **plastic deformation.** During plastic deformation, the energy applied to the object is *absorbed* internally by friction and is used to rearrange the makeup of the material as it flows to take on a new *shape.* An example would be the response of a ball of modeling clay as you squeeze it in your fist. The modeling clay *remains* deformed when the stress is released. The clay cannot go back to its original shape because there is no stored energy to generate a stress to make it go back. Structural geologists would say that the material "has no memory." Except for the tiny amount of energy involved in elastic strain, nearly all of the applied energy was absorbed and *used* to deform the clay.

Brittle Deformation

When some materials are subjected to high rates of stress application, especially under low confining pressures and low temperatures, they break with *little* or *no* plastic deformation beyond the elastic limit (point 2 in Figure 11.5). Such materials are said to undergo **brittle deformation.** As the material fails, the energy stored during elastic deformation is released. Following brittle failure, the material may once again undergo elastic deformation.

A few comments concerning the significance of breaking are in order. Solid materials release energy from their surface. The sound of a ringing bell or a crystal goblet struck with the fingernail is the response of your ear to the energy released from the surface as the bell or glass deforms elastically. Consider now striking the glass a little too hard or perhaps dropping it. The amount of energy that can be released during any interval of time is limited by the surface area of the object. Because the amount of energy applied to the glass by the impact is more than can be absorbed and released from the *existing* surface of the glass, the glass breaks in order to generate the *additional* surface area needed to release the excess energy. The number of pieces into which the glass will break depends upon the amount of energy that must be released. In contrast to the melodious ringing of the crystal goblet carefully struck with a fingernail, the familiar, discordant sound of breaking glass is the result of the mixing of the sounds from a number of different-sized and odd-shaped "bells."

The Effect of Time

Whether a solid material deforms as a plastic or as a brittle solid is often determined more by the *rate* at which the energy was applied than the *amount* of energy. Consider, for example, the response of glass to stress. Most individuals would consider glass to be a brittle solid as evidenced by the ease with which it breaks when impacted by a rock. As a rock strikes a pane of glass in a window, the rate of energy application is very high. Within a fraction of a second, the strength of the glass is exceeded, the applied energy is far beyond the amount that can be absorbed by elastic deformation, the glass responds as a brittle solid, and it breaks.

Consider, on the other hand, the same pane of glass standing on the end in the door frame of an antique china closet, subjected to the *same* amount of energy but this time in the form of gravity pulling downward on the glass over a period of a hundred or more years. The rate of energy application will be very low. During a century or more of confinement within the frame of the china closet door, the glass will have sufficient time to absorb the energy, make internal molecular rearrangements, deform plastically, and flow downward. Variations in thickness in the glass will be created in the form of horizontal wavy lines across the pane while vertical shortening of the pane will be evident at the upper portion of the door frame. Antique hunters look for these results of plastic deformation as evidence for the age of glass in considering potential purchases.

An example of a natural material that may respond by brittle failure under one set of conditions but deform plastically under another is ice. In our discussions of glaciers, we pointed out that ice, although a brittle solid when responding to a short-term stress as a jab of an ice pick, deforms plastically when it

THE STRAIN ELLIPSOID

A helpful tool used by geologists to portray the response of rocks to stress (strain) and to illustrate the various strain orientations represented by the geologic structures is the **strain ellipsoid.** An ellipsoid is a geometric figure where the maximum, intermediate, and minimum diameters are mutually perpendicular and the corresponding sections are ellipses (Figure 11.B1). A sphere is a special case where all axes are equal in length and the corresponding sections are circles.

To illustrate the use of the strain ellipsoid, consider Figure 11.B2. In Figure 11.B2a, the sphere represents the distribution of strain within a mass of buried rock that is being subjected to compressive forces equally from all directions. Such forces would be the confining forces to which rocks are subjected at depth. Now consider subjecting the rock mass to a *directional* compressive stress. The sphere will be deformed into an ellipsoid as the three axes change length (refer to Figure 11.B2b). The axis parallel to the direction of the maximum compressive stress experiences the greatest shortening (strain)

FIGURE 11.B1 *An* ellipsoid *is a geometric figure with three mutually perpendicular axes of different lengths and corresponding elliptical cross sections. A sphere is a special case where the three mutually perpendicular axes are equal in length and all sections are circles.*

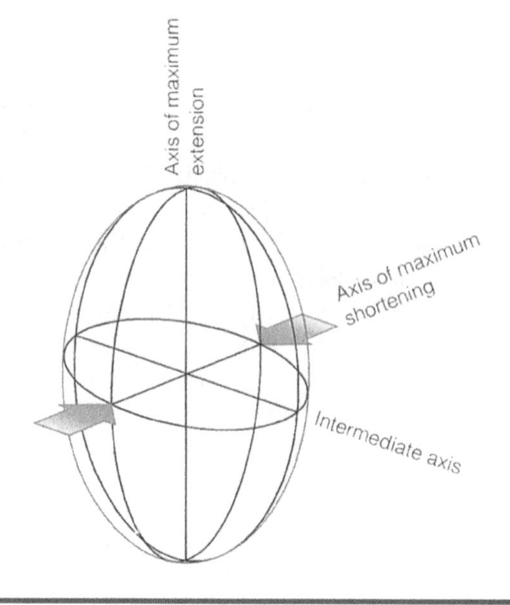

FIGURE 11.B2 *An* ellipsoid *forms as one axis of a sphere becomes the direction of* maximum directional compressive stress *and* shortens. *In order to maintain a constant volume and remain symmetrical, one of the other axes becomes elongated while the length of the third remains essentially* unchanged. *In terms of the strain* ellipsoid, *the three axes are called the* axis of maximum shortening, *the axis* of maximum extension, *and the* intermediate axis, *respectively.*

accumulates to thicknesses in excess of 150 feet (46 m) and begins to flow under the force of gravity.

The Effect of Temperature and Pressure

Increasing temperatures usually favor the plastic response of solids, an example being the ease with which glass rods bend when heated to redness over a Bunsen burner. Increased confining pressure increases the ultimate strength of most solids by extending the range of plastic deformation. As a result of the combined effects of temperature and pressure, rocks under stress at or near Earth's surface usually behave as brittle solids and break, while with increased depth of burial and the resulting increase in both tempera-

and is therefore called the **axis of maximum shortening.** Of the two remaining axes, one will undergo maximum extension (strain) and is called the **axis of maximum extension.** The remaining axis is the **intermediate axis.**

The response of rocks to stress can be represented by three different orientations of the strain ellipsoid (Figure 11.B3). Note that the strain ellipsoids in Figures 11.B3a and 11.B3c both characterize the response of rocks to horizontal compressive forces such as those that exist at convergent plate margins. The axis of maximum extension of the strain ellipsoid shown in Figure 11.B3b, on the other hand, is horizontal, and the axis of maximum shortening is vertical. This is the strain orientation characteristic of areas undergoing extension, such as the margins of divergent plates.

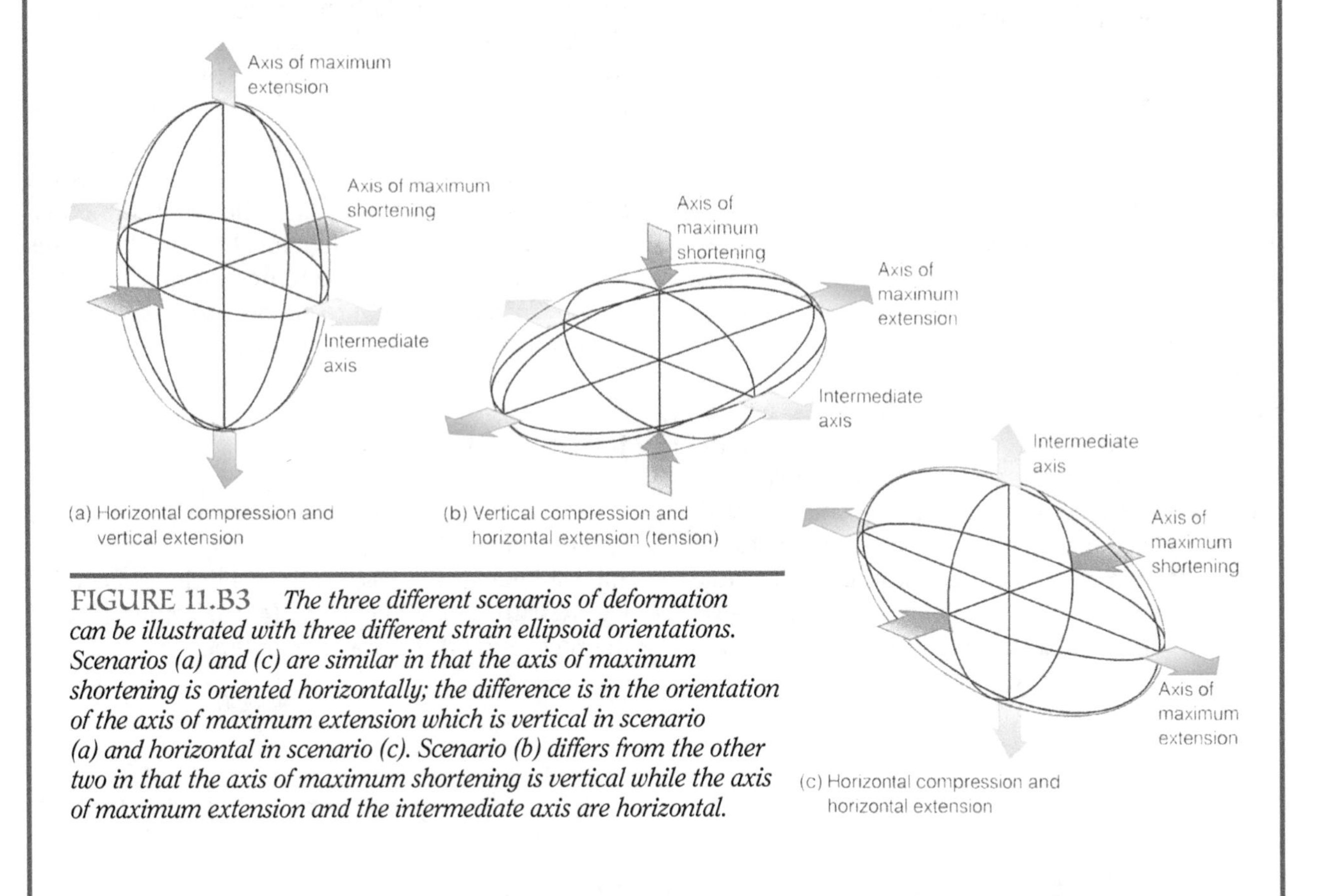

FIGURE 11.B3 *The three different scenarios of deformation can be illustrated with three different strain ellipsoid orientations. Scenarios (a) and (c) are similar in that the axis of maximum shortening is oriented horizontally; the difference is in the orientation of the axis of maximum extension which is vertical in scenario (a) and horizontal in scenario (c). Scenario (b) differs from the other two in that the axis of maximum shortening is vertical while the axis of maximum extension and the intermediate axis are horizontal.*

ture and pressure, rocks become increasingly plastic and tend to fold or flow.

Effect of Rock Composition The composition of rocks may critically control their response to stress. For example, under similar conditions a sandstone composed of brittle quartz grains may fail by brittle fracture while an adjoining shale made up primarily of clay minerals may deform plastically and flow. Water contained within rocks usually promotes a plastic response by enhancing the dissolution of the more water-soluble minerals while at the same time reducing cohesion and friction at the contacts of others.

SPOT REVIEW

1. What is stress?
2. How are the strength of a material, stress, and deformation related?
3. What is the difference between compression and shear?
4. What is the elastic limit of a material?
5. How do plastic and brittle deformation differ? In what way(s) are they similar?
6. How do time, pressure, and temperature affect the deformation response of a material?

THE STRENGTH OF ROCKS

Rocks are very strong under compression but weak under tension. Perhaps a historical example would serve to demonstrate. When one compares the architectural styles of the ancient Greeks and Romans, the major difference is that the Greeks made prolific use of **columns** whereas the Romans used **arches** (Figure 11.6). The Greeks were master column makers, not so much because they wanted to be, but because they never discovered how to overcome the basic weakness of rocks under tensional forces. The Greeks spanned the distance between columns with flat rock slabs. They found that when they attempted to span too great a distance, the rock slab would sag and eventually break when loaded from above. The rock slabs failed due to *tensional* forces that are generated within any slab supported at its ends. A load placed on the slab increases the amount of tensional stress. As a result, the Greeks had to limit the distances to be spanned by using many columns spanned with short rock slabs, all of which provided a lot of work for column makers.

The Romans, on the other hand, invented the Roman arch. The secret of the Roman arch is the **keystone**. When loaded from above, the wedge-shaped keystone directs the forces outward against the first blocks within the arch. The contact between the keystone and the first block of the arch is under compression, as are all the contacts between successive blocks down to the supporting columns and continuing through the columns to the ground surface. No parts of the arch or the supporting columns are under tension. Because rocks are *very* strong under compression, the arch and supporting columns can support enormous weight. The concept of the Roman arch has been responsible for all the beautiful vaulted arches constructed throughout history. It is this inherent strength under compression that allows a volume of rock at depth to support the weight of all the overlying rocks.

ROCK DEFORMATION

Once the strength of a rock is exceeded, its response depends upon (1) the kind of stress, (2) the rate of application of the stress, (3) the conditions of confining pressure and temperature, (4) the composition of the rock, and (5) the presence of fluids.

FIGURE 11.6 *One reason the ancient Greeks used columns so prolifically was that they never learned how to overcome the inherent weakness of rocks under tension. (Courtesy of M. M. Reddy/USGS)*

Elastic Deformation

The initial response of rocks to stress is elastic. Some examples of elastic deformation involve large volumes of Earth's crust. During the Great Ice Age, for example, a large portion of the North American continental crust centered over what is now Hudson Bay and a comparable region underlying the Scandinavian Peninsula were depressed thousands of feet by the weight of the overlying ice mass, displacing the underlying asthenospheric rocks. Since the ice melted about 10,000 years ago, these two areas have been rising at the rate of about 3 feet (1 m) per century. Because the lithosphere is physically joined to the asthenosphere, the rate of elastic rebound is being controlled by the rate at which the asthenospheric rocks flow back under the crust. As the crust returns to its original elevation, the water will drain from Hudson Bay and the Baltic Sea and return to the North Atlantic Ocean. As is the case with all elastic deformation, once the crust returns to its pre-Ice Age position, no indication will remain that the rocks were ever deformed.

Another example of the elastic response of large volumes of rocks, but on a totally different scale in terms of both time and amount of rock movement, involves the passage of *seismic* (earthquake) waves through Earth's interior. As seismic waves propagate, the rocks are subjected to stress for a period of a few seconds or minutes and are moved millimeters or fractions of millimeters. The rocks respond elastically and are neither broken nor permanently deformed but return to their pre-earthquake positions after the seismic wave passes.

The elastic response of relatively small volumes of rock, such as that exposed in a single cliff face or road cut, is so small that it can be detected only by very sensitive instruments called strain gauges. Yet, the combined amount of energy stored within the rocks may one day be released as a destructive earthquake.

Plastic Deformation

In our discussion of plate tectonics, the asthenosphere was described as being *plastic.* You will remember that it is the heat-driven plastic flow of rocks within the asthenosphere that reduces the frictional contact between the asthenosphere and the lithosphere to the point where the lithospheric plates can move over the underlying asthenosphere. Because of the high temperatures and pressures within the asthenosphere and mantle, the rocks tend to respond to stress as plastic solids and do not break.

As foldbelt mountains are created by the collision of lithospheric plates, the conditions of temperature and pressure within the cores of the mountains are oftentimes sufficient to metamorphose and cause the plastic deformation of rocks. The swirling patterns of color commonly seen in marble are an example.

Brittle Deformation

The outermost layer of Earth, the lithosphere, is made up of the *brittle* crust and the outermost *brittle* portion of the mantle. When stress is applied at rapid rates or in great amounts, the lithosphere will tend to break. Indeed, it has. The lithosphere is broken into the *plates* that move relentlessly over the surface of Earth.

Not all rock deformation within the lithosphere is by brittle failure, however. Within the volume of rock involved in the creation of a mountain range, for example, localized stress conditions generated by the movements of the lithospheric plates may result in either plastic deformation or brittle failure of the rocks. The folds that characterize the great mountain ranges of Earth are examples of the plastic deformation of the lithosphere (Figure 11.7) while the generation of faults is an example of brittle deformation (Figure 11.8).

FIGURE 11.7 *The folds seen in this satellite photograph of the northern Appalachian Mountains are a spectacular example of plastic deformation of the lithosphere. For reference, note the Potomac River in the center near the bottom and Harrisburg, Pennsylvania to the middle right. (Courtesy of NASA)*

FIGURE 11.8 *This linear fault trace is an example of the breaking of the brittle crust in response to deformational forces. (Courtesy of R. E. Wallace/USGS)*

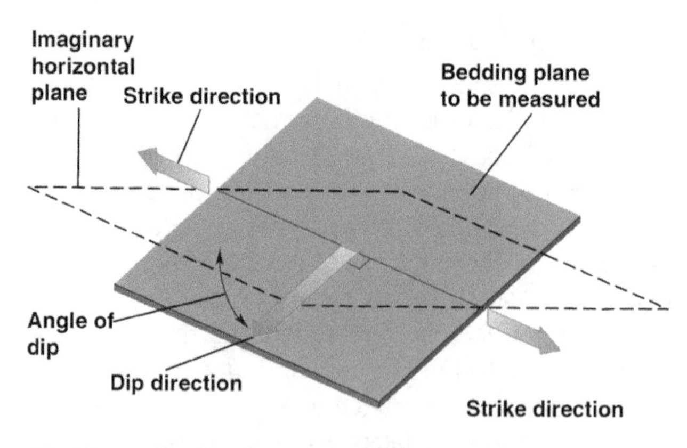

FIGURE 11.9 *The two measurements of* strike *and dip allow a geologist to describe a planar surface in three-dimensional space. Note that the directions of strike and dip are mutually perpendicular.*

Largely because no effect of the elastic response to stress remains after the stress is released, most geologists, except for seismologists (those who study earthquakes), tend to be concerned only with brittle or plastic deformation.

THE CONCEPT OF STRIKE AND DIP

The features we will soon discuss, folds, faults, and joints, all have *planar* (two-dimensional) components. The method used by geologists to describe the three-dimensional orientation of these planar surfaces employs two parameters: (1) *strike* and (2) *dip* (Figure 11.9).

Strike is defined as the *direction* of the *line* of intersection between the plane and a horizontal plane. The definition makes two important points: (1) strike is a *direction* and (2) strike is a *horizontal line.* Although the direction of either end of the line will suffice to describe the direction of the line, we may want to specify the use of one end rather than the other. By convention, for example, the direction of strike is always noted relative to the *north* magnetic pole.

Dip is defined as the *angle* the plane makes with the horizontal and is always measured *downward* from the horizontal. Although not implicit in the definition, the *direction* of the dip must also be given because a plane may dip in two different directions (to either side of the line of strike).

Strike and dip directions are mutually *perpendicular* (refer to Figure 11.9). Once the strike of a plane has been determined, the direction of dip can be only in one of two directions. For example, if the strike of a plane is to the north, the dip must be to the east or west.

GEOLOGIC STRUCTURES

Geologic structures are (1) *folds,* (2) *faults,* and (3) *joints*. The plastic response of rocks to stress produces folds. Faults and joints form by brittle failure.

Folds

Folds form primarily by compressional stress. This can be easily demonstrated with a sheet of paper. You can form folds easily by pushing any two opposite edges of the sheet *directly* toward each other. Folds may also form by shear forces where compressive forces act along parallel, nonopposing paths. This can be demonstrated by moving opposing edges of a sheet of paper toward each other while simultaneously moving them in opposite directions. Although folds can be formed by nonopposing compressive forces, most folds form by directly opposing forces.

FIGURE 11.10 *The simplest kind of fold, the* monocline, *is illustrated by the inclined outcrop of rocks that surrounds the San Rafael Swell, Utah.*

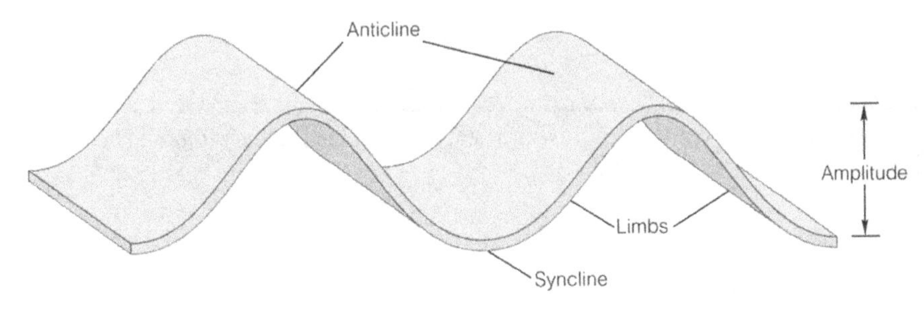

FIGURE 11.11 *The most commonly observed folds are anticlines and synclines. For simplicity, only one bed has been shown in the drawing. Anticlines and synclines are common sights in road cuts in areas of foldbelt mountains.*

Basically, there are three different kinds of folds: (1) *monoclines,* (2) *anticlines,* and (3) *synclines*. The simplest fold is the **monocline** in which the layers are tilted in *one direction.* Monoclines commonly form around the margins of local or regional uplifts and in the strata above faults (Figure 11.10).

Rocks may be arched either *upward* to form an **anticline** or *downward* to form a **syncline.** Normally, anticlines and synclines form together as is illustrated in Figure 11.11.

Anticlines and synclines are described using (1) the strike and dip of the *limbs* relative to the *axial plane,* (2) the direction of the fold *axis,* (3) the *plunge* of the fold axis, and (4) the *amplitude* of the fold.

The **axial plane** is an imaginary plane that parallels the length of the fold and divides the cross section of the fold in half. The **limbs** are the flanks of the fold, which make an angle with the axial plane. The **axis** of the fold is the line of intersection between the axial plane and the limbs. The **plunge** is the angle between the axis and the horizontal. The **amplitude** of the fold is the distance from the top of the anticline to the bottom of the adjacent syncline (refer to Figure 11.11).

Based on the angle and direction of the dip of the limbs relative to the axial plane, folds are classified as (1) *symmetrical,* (2) *asymmetrical,* (3) *overturned,* or (4) *recumbent*. If the limbs dip *away* from the axial plane in *opposite* directions at *equal* angles, the fold is described as **symmetrical** (Figure 11.12a). Folds whose limbs dip *away* from the axial plane in *opposite* directions but at *different* angles are called **asymmetrical** (Figure 11.12b). If both limbs and the axial plane dip in the *same direction*, the fold is referred to as being **overturned** (Figure 11.12c). When rocks are subjected to extreme compression, the folds may assume the attitude shown in Figure 11.12d. where the axial plane *approaches the horizontal.* Such folds are called **recumbent.** Symmetrical, overturned, or recumbent folds where the limbs are parallel are described as being **isoclinal.**

Along most of its length, the axis of a fold is approximately horizontal; that is, the plunge is zero. Eventually, however, every fold comes to an end by plunging as the axis dips away from the horizontal and the amplitude of the fold diminishes (Figure 11.13).

Plastic deformation of rocks usually requires such high confining pressures and temperatures that they cannot be achieved anywhere near Earth's surface. Because folded rocks are a common sight at Earth's surface, we tend to forget that we see them only after they have been uplifted from their place of origin deep within Earth and exposed by millions of years of weathering and erosion.

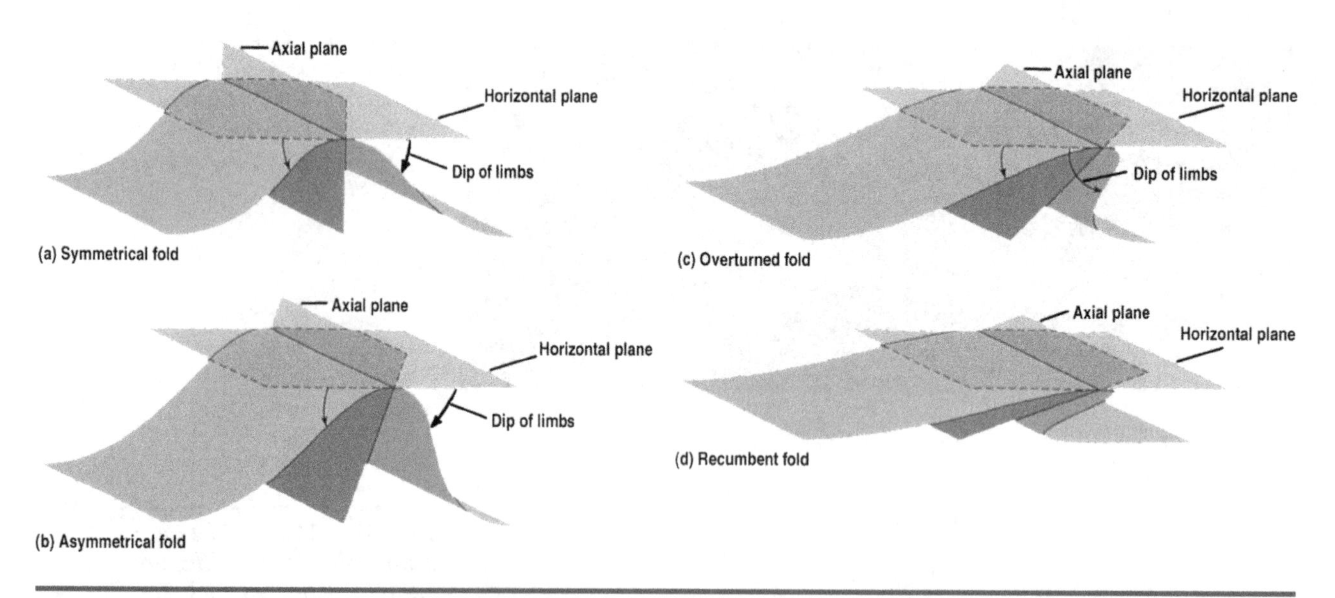

FIGURE 11.12 *Folds are described as* symmetrical, asymmetrical, overturned, *or* recumbent *depending on the relationship of the* axial plane *and the* limbs *relative to the* horizontal.

SPOT REVIEW

1. Give an example of rocks undergoing elastic deformation.
2. Why would you not expect rocks to undergo plastic deformation at Earth's surface?
3. What is strike? How are strike and dip related?
4. What is the axial plane of a fold? How does the orientation of the axial plane differ among the various kinds of folds?
5. How do the forces that produce symmetrical folds differ from those that produce nonsymmetrical folds?
6. What orientation of the strain ellipsoid is common to all folds?

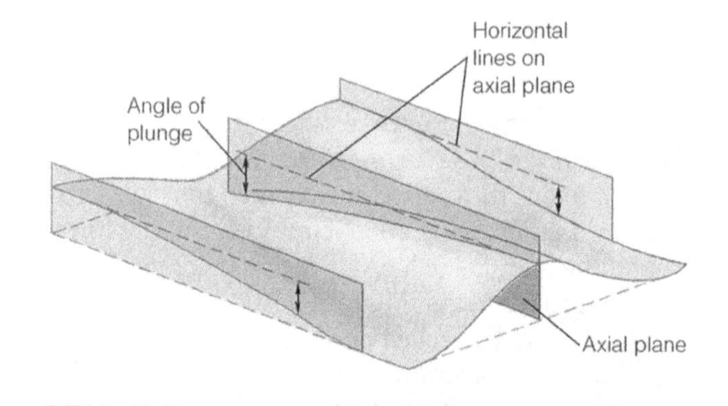

FIGURE 11.13 *Folds eventually come to an end by* plunging *as the amplitude of the fold decreases along its length. As erosion removes the axial region of an anticline, the eroded limbs dip away from the axis of the fold and wrap around the plunging structure as is shown in the photo.*

Faults

Faults are relatively planar breaks in the lithosphere along which the rocks have been *offset*. In other words, all faults form as a result of *shear*. The frequency of faulting is highest near Earth's surface where the rocks are most brittle and decreases with depth as the rocks become increasingly plastic due to the increase in temperature, pressure, and the presence of fluids. Deep-focus earthquakes disappear with depth along subducting plates, indicating that below a depth of about 450 miles (725 km), rocks respond totally as plastic solids and faults do not occur. Within continental crust, the brittle-plastic transition occurs at a depth of about 10 miles (15 km).

Most earthquakes result either from movement along existing faults or from a rupture that initiates new faults. Before rocks can be deformed by faulting, they must first undergo elastic deformation. The energy that destroys buildings and triggers the mass wasting during an earthquake is some of the energy that was stored in the rocks as they underwent elastic deformation over a period of perhaps hundreds or thousands of years. Because the elastic response of rocks under localized conditions is *very* small, there is generally no visual evidence that energy is being stored within the rock. Yet all the while the rock is silently undergoing strain. When the ability of the rock to store additional energy is exceeded, it breaks or slips along a new or preexisting break, and the stored energy is released. Some of the energy is used to move the rocks on opposite sides of the fault. The remaining energy is released as seismic waves. A seismic wave has many of the same characteristics as the waveform that is generated by a breaking stick and is converted by our ear to a sound.

The three kinds of stress conditions mentioned earlier result in the generation of different kinds of faults.

Reverse or Thrust Faults Figure 11.14 shows a block of rock that has failed under compression where the forces were acting directly opposite each other. Upon fracturing, the rock is displaced (sheared) along a planar surface that is inclined to the direction of compression and at an angle to the horizontal. Some of the energy stored in the rock during the elastic phase of deformation is used to move the blocks of rock on opposite sides of the fault plane in opposite directions. The block *above* an inclined fault plane is called the **hanging wall**, and the one *below* the fault plane is called the **foot wall.** Miners coined these terms many years ago as a result of following mineral veins that are commonly emplaced along faults. When the miners were at work, one block *hung* over their heads while the other was below their feet (refer to Figure 11.14).

The fault shown in Figure 11.14, which formed by compression with the hanging wall moving *up* relative to the foot wall, is a **reverse fault.** A reverse fault where the dip of the fault plane is less than 45° is often referred to as a **thrust fault.**

Three measurements are made on the most faults: (1) *displacement,* (2) *throw,* and (3) *heave* (refer to Figure 11.14). The **displacement** is the amount of offset measured along the fault plane. The **throw** and **heave** are the apparent *vertical* and *horizontal* components of movement, respectively. In the case of a reverse or thrust fault, the throw is the amount of crustal *uplift* generated by the fault, and the heave is the amount of crustal *shortening* the fault has created. All of these measurements may vary in magnitude along the fault plane and will diminish to zero at the ends of the fault.

A major thrust fault in the eastern United States underlies the Blue Ridge Mountains of the Appalachians and the rocks to the east. Although the actual displacement is neither known or agreed upon, it is estimated that during the formation of the Appalachians, the rocks that make up the Blue Ridge Mountains and the Piedmont to the east were moved along this thrust fault at least 50 miles (80 km) westward from their original location.

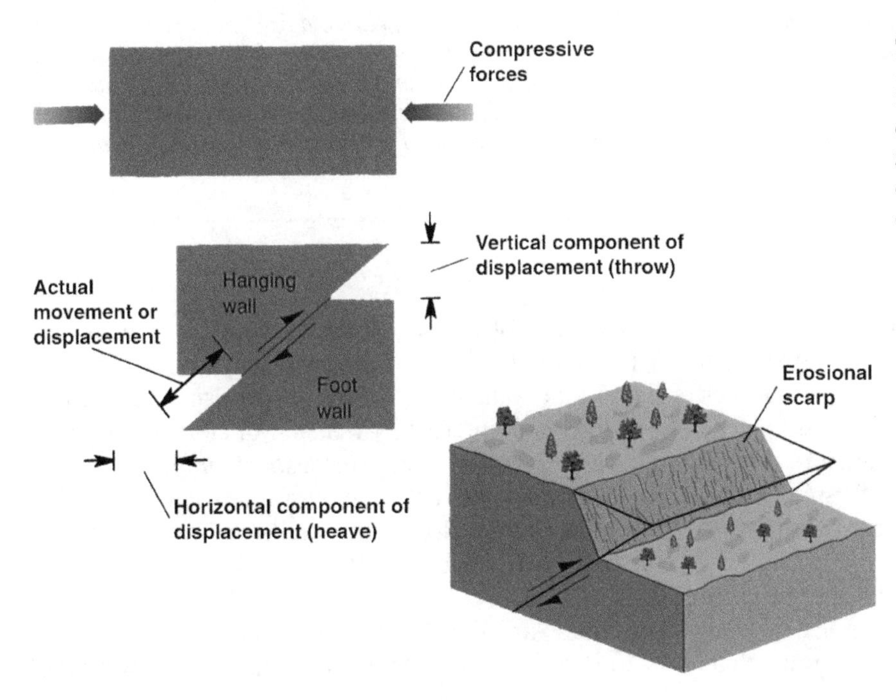

FIGURE 11.14 Reverse *or* thrust faults *form by the brittle failure of rocks under compression. As a result of the movement, the length of Earth's crust is* shortened *and the crust is* locally thickened.

FOLDS AND THE STRAIN ELLIPSOID

Each of the different types of folds has a characteristic orientation of the strain ellipsoid. Consider the symmetrical folds illustrated in Figure 11.B4 and note the orientation of the strain ellipsoids. In both cases, the axis of maximum extension is vertical, the axis of intermediate extension is parallel to the axis of the fold, and the plane of the axes of intermediate and maximum extension is parallel to the axial plane of the fold. The greater amplitude of the fold in Figure 11.B4b is represented by the increased length of the axis of maximum extension while the increased deformation that produced the higher-amplitude fold is represented by the shorter axis of maximum shortening.

Consider now the nonsymmetrical folds in Figure 11.B5. The nonsymmetrical folds form by the application of horizontal, nonopposing, compressive

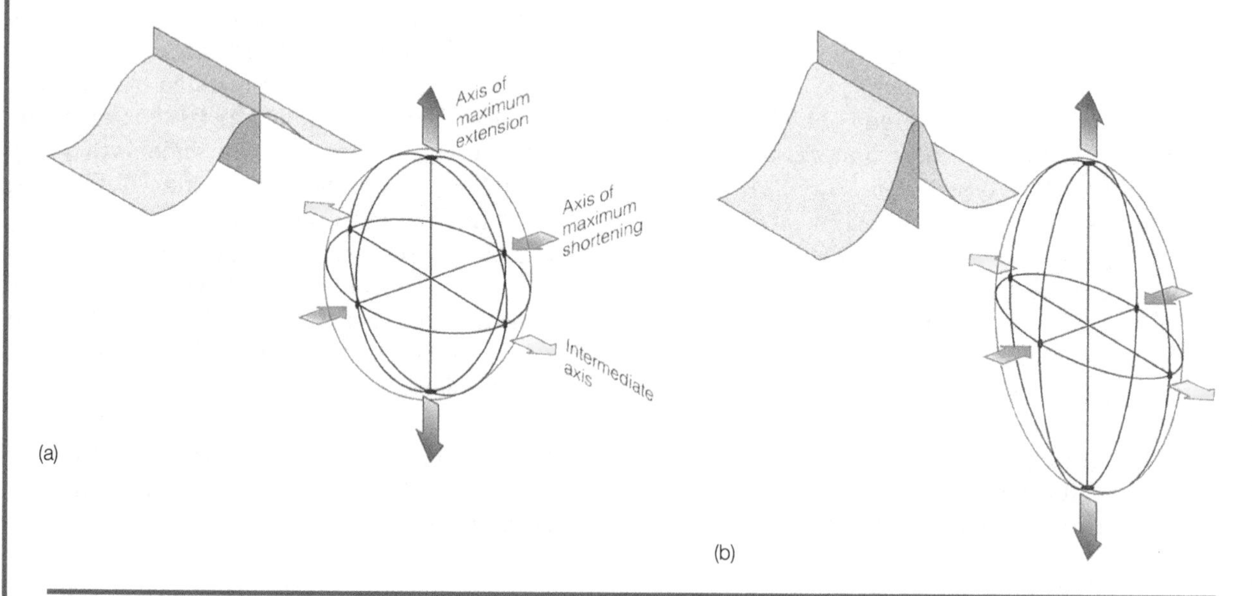

FIGURE 11.B4 Folds *form by the application of force described by the strain ellipsoid orientation shown in scenario (a) of Figure 11.B3. In all cases, the intermediate axis is parallel to the axis of the fold.* Symmetrical folds *form by the application of compressive forces such that the axis of maximum shortening is* horizontal *and the axis of maximum extension is* vertical. *As a result, as stress is applied, the crustal segment is shortened and the rocks arch* upward *into anticlines and/or* downward *into synclines. Continued application of force results in increased fold* amplitude *and additional shortening of the crust.*

Normal Faults As previously indicated, rocks are weakest under tensional stress and will usually fail by brittle fracture. Figure 11.15 shows a block of rock before and after failure under tensional stress. The fault, where the hanging wall moves down relative to the foot wall, is called a **normal fault.** Displacement, throw, and heave for a normal fault are measured in the same way as for a reverse fault.

Note that the orientation of forces that produce normal faults is opposite to the orientations of forces that produce reverse faults. As a result, certain characteristics of the two fault types are also opposites. For example, the relative movements of the hanging and foot walls are reversed. Note also that normal faults result in crustal *extension* or *elongation* while reverse faults produce crustal *shortening.*

In cases where the dip of reverse or normal faults decreases with depth, the faults are referred to as being **listric** (Figure 11.16).

Strike-Slip Faults When a segment of Earth's crust breaks as a result of compression where the forces. The effect of such forces is to rotate the strain ellipsoid around the intermediate axis. Note that, as was the case with the symmetrical folds, the intermediate axis is always parallel to the axis of the fold and the axial plane parallels the plane of the axes of intermediate and maximum extension. The three types of folds differ mainly in the change in the orientation of the axial plane as increased deformation rotates the strain ellipsoid.

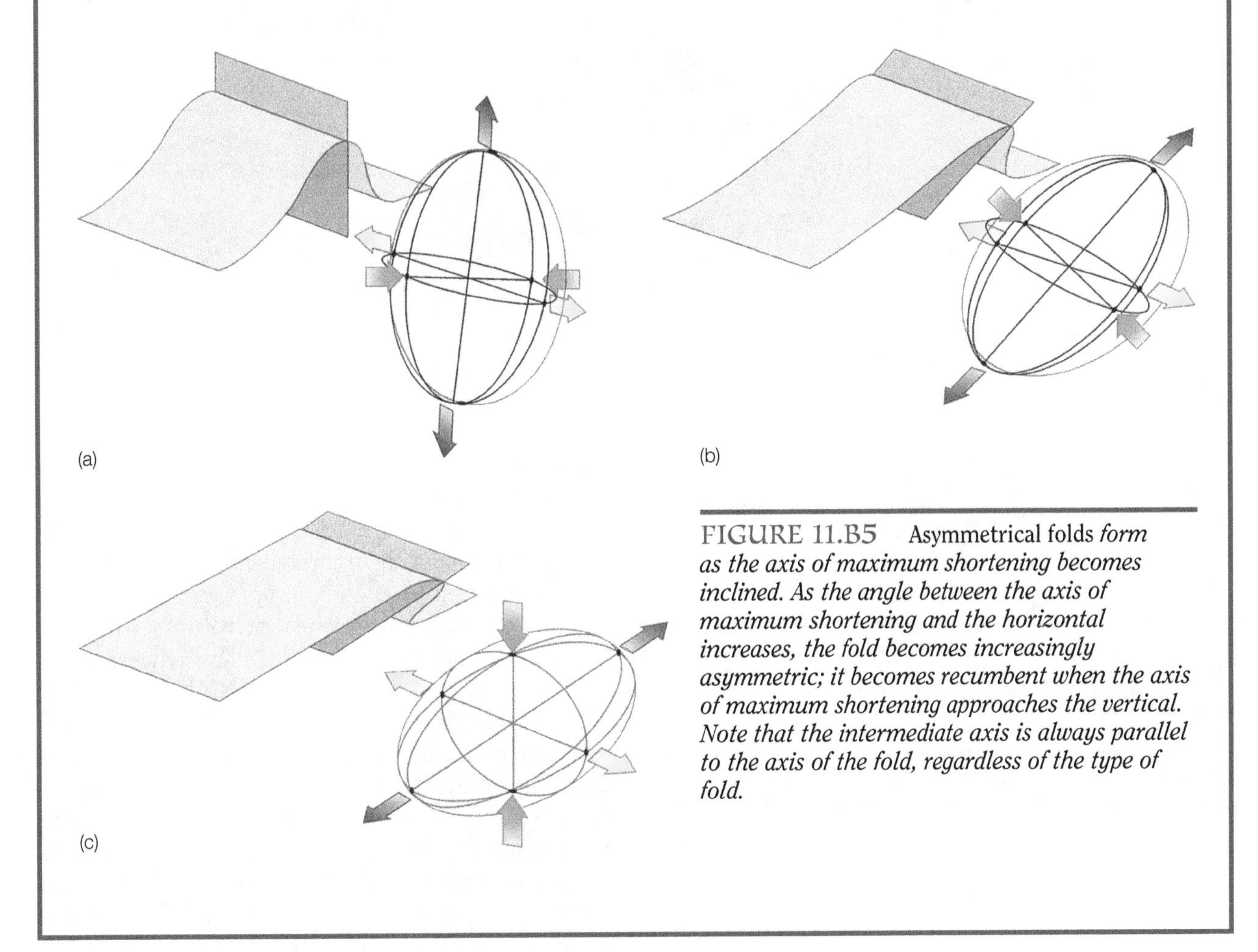

FIGURE 11.B5 Asymmetrical folds *form as the axis of maximum shortening becomes inclined. As the angle between the axis of maximum shortening and the horizontal increases, the fold becomes increasingly asymmetric; it becomes recumbent when the axis of maximum shortening approaches the vertical. Note that the intermediate axis is always parallel to the axis of the fold, regardless of the type of fold.*

shear zone is oriented vertically, the fracture surface will be vertical or near-vertical and parallel to the direction of the compressive stress. The blocks on either side of the fault move horizontally in opposite directions and produce a **strike-slip fault.** The name arises from the fact that the movement (the *slip*) is parallel to the *strike* (the compass direction) of the fault plane (Figure 11.17).

When describing strike-slip faults, the terms foot wall, hanging wall, throw, and heave have no meaning because there is no vertical movement of the rocks. Displacement, however, is measured as before.

Strike-slip faults are described as being either **right-lateral** or **left-lateral** faults depending upon the relative direction of movement of the two blocks (Figure 11.18). The best-known strike-slip fault in the United States, the San Andreas Fault, is responsible for many of the earthquakes that shake the west coast. The San Andreas Fault is also the plate boundary between the North American and Pacific plates.

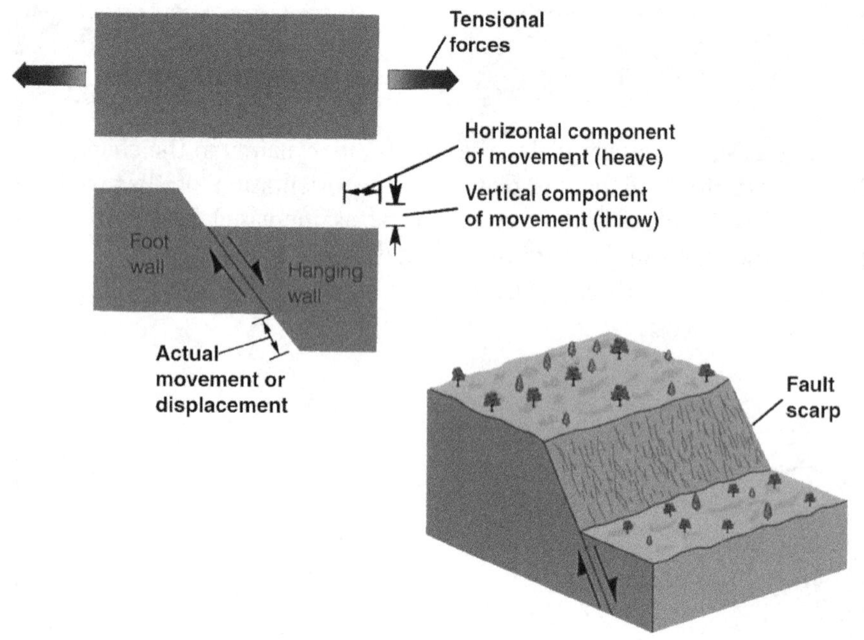

FIGURE 11.15 Normal faults *form by the brittle failure of rocks under tension. As a result of the movement, the length of Earth's crust is lengthened. Even though the forces and rock movements involved in reverse and normal faulting are exactly opposite, the surface scarps may be identical in appearance. Note, however, that one scarp is an actual fault surface while the other is erosional.*

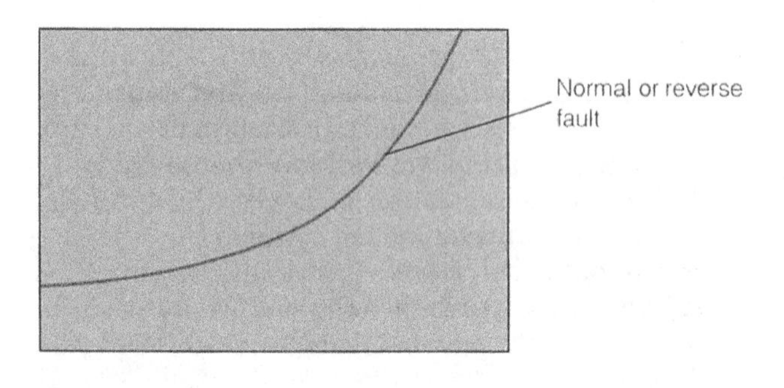

FIGURE 11.16 *The term* listric *refers to normal or reverse faults that decrease in angle with depth.*

Oblique-Slip or Diagonal-Slip Faults In some cases, the stress orientation results in two components of rock movement. One component will be horizontal, within the plane of the fault and *parallel* to the strike of the fault plane, while the other will be within the plane of the fault and *perpendicular* to the strike of the fault plane (Figure 11.19). The resultant movement will angle across the fault surface and produce an **oblique-slip** or **diagonal-slip fault.**

Faulting after Initial Folding

Folding and faulting can develop during the same deformational episode if the rocks break after a period of plastic deformation. The onset of deformation initiates the formation of folds within a horizontal shear zone. As the folds develop to the highly asymmetrical or overturned stage, the volume of rock simultaneously *shortens* and *thickens* in order to maintain a constant volume. Think of squeezing a stick of modeling clay between your hands. As the rock section thickens, the strength of the rock mass increases. Eventually, the rocks begin to resist continued folding and become unable to consume additional energy by plastic deformation. At this point, the rocks will break and develop reverse faults. The faults will invariably break the steeper or overturned limb of the anticlinal folds. Deformation then proceeds with additional shortening and uplift as the folds are thrust one on top of another (Figure 11.20). This type of deformation is common during the formation of foldbelt mountains where thick sequences of sedimentary rocks are involved.

Joints

A **joint** is one of a set of fractures in Earth's crust where only opening displacement has occurred with no appreciable displacement parallel to the fracture surface. Joints are the most common geologic structures, being found in all exposed rocks.

Joints occur in sets within which the individual joints are parallel or subparallel to each other. There

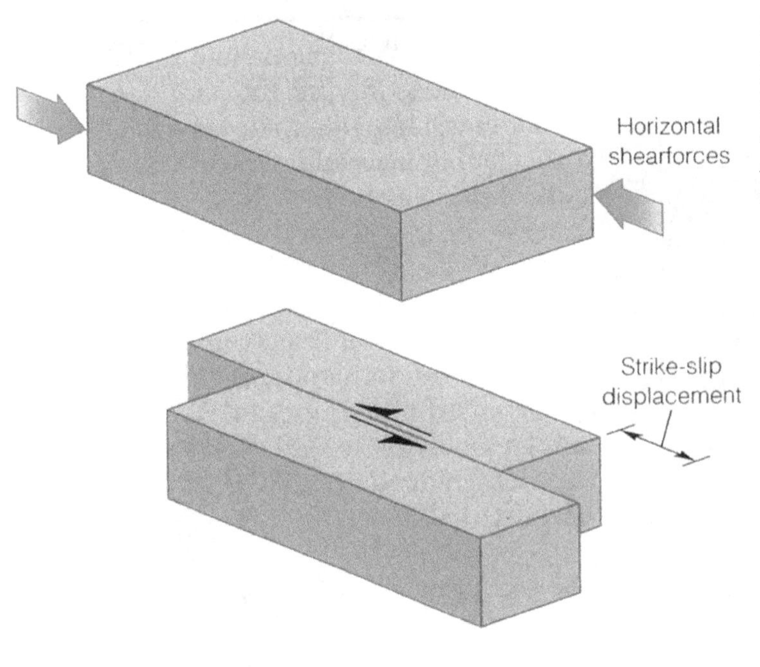

FIGURE 11.17 Strike-slip faults *have only horizontal displacement. The best-known strike-slip fault in North America, the San Andreas Fault, is the boundary between the North American and Pacific lithospheric plates and the source of most earthquakes in the United States.*

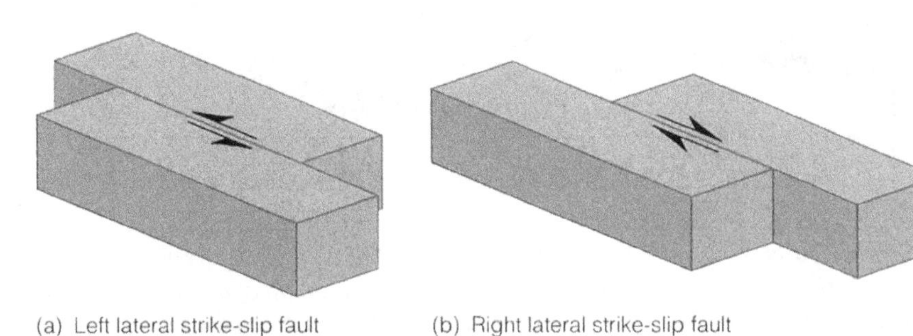

FIGURE 11.18 *Strike-slip faults are described as* right-lateral *or* left-lateral *depending on the relative movements of the rocks on opposite sides of the fault. Picture yourself walking toward the fault and asking, "Which way do I have to turn to reach the same point on the other side of the fault?" The San Andreas Fault is an example of a right-lateral strike-slip fault.*

are two major types of joints depending on the orientation of the fractures relative to the direction of maximum stress (Figure 11.21). Joints that are not parallel but form at an angle to the direction of stress are called **shear joints.** Shear joints usually occur as two equally well developed sets intersecting at an *acute* angle of about 80° that is bisected by the axis of greatest stress (Figure 11.22). In **tensional joints,** the fractures are aligned *parallel* to the direction of maximum compressive stress. They are called tensional joints because the fracture opens by tension. Think of how the cards in a deck of cards open when you compress the deck along the long dimension of the cards.

Certain types of joints are specifically associated with igneous rock bodies. It is common to find joints parallel to the surface of an igneous rock body, apparently formed by the release of pressure as the overlying rocks are removed by weathering and erosion (Figure 11.23). A type of joint found in a variety of igneous rock bodies is called **columnar jointing.** Columnar joints typically break the rock into elongate hexagonal columns, generally perpendicular to the cooling surface; they are formed by the tensional

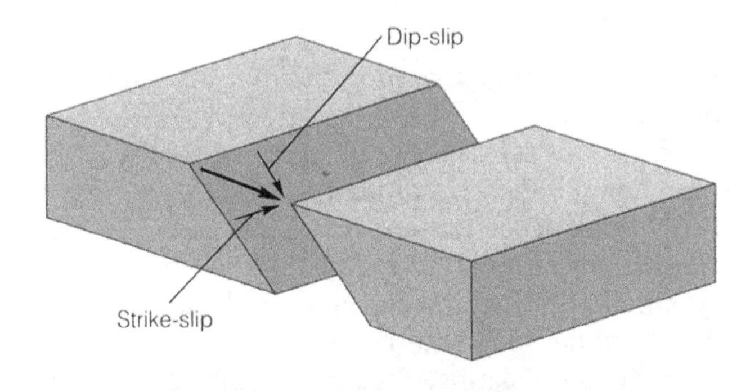

FIGURE 11.19 Oblique-slip *or* diagonal-slip faults *are those that have both a dip-slip and a strike-slip movement.*

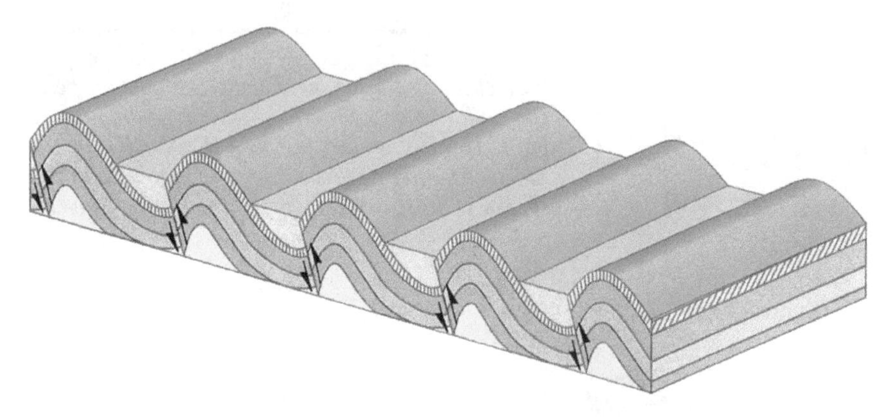

FIGURE 11.20 *A common type of deformation, especially in foldbelt mountains, is asymmetric or overturned folds that have been broken by reverse faults on the oversteepened limb.*

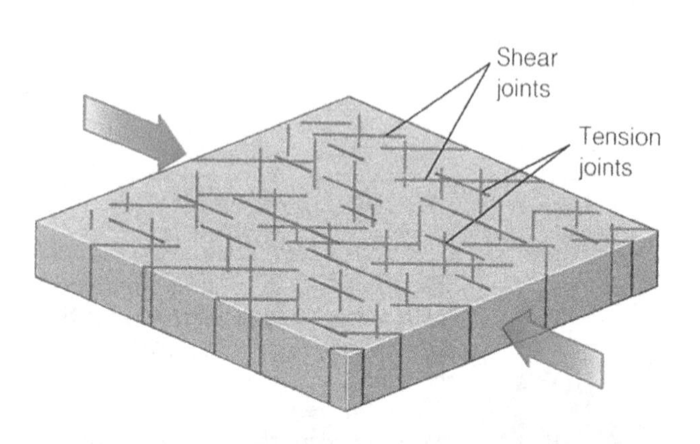

FIGURE 11.21 *Of all the geologic structures, joints are the most common. There are two types of joints;* shear joints *which form at an angle to the direction of applied stress, and* tension joints *which form parallel to the direction of applied stress.*

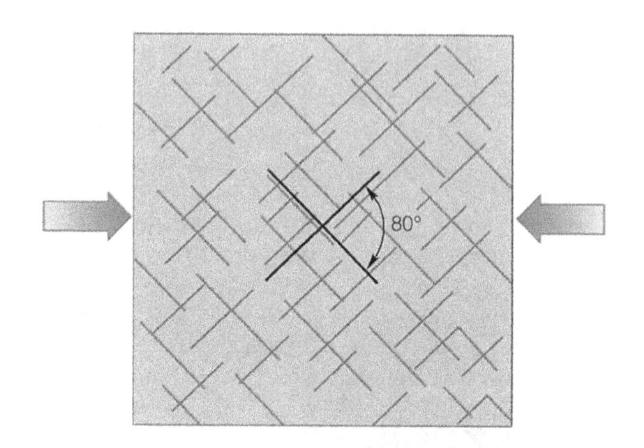

FIGURE 11.22 *Shear joints consist of two sets of fractures that intersect at an angle of about 80° to the direction of applied stress. Geologists use this relationship to determine the direction of applied stress.*

stresses that are generated as the rock mass shrinks upon cooling.

In some cases, geologists agree on neither the time nor the mode of formation of joints. We normally think of the formation of joints as the brittle response of an existing rock. Field evidence, however, indicates that in sedimentary rocks some joints may form not long after burial while the sediments are still consolidating. Studies of jointing in some folded sedimentary rocks show that, regardless of the orientation of the bedding, the fractures are always nearly perpendicular to the bedding, indicating that the joints formed before folding began. Based on our previous discussions that indicated that the initial response to stress is elastic followed in order by plastic and brittle deformation, you might question why rocks would fail by brittle deformation even before they began to deform plastically.

FIGURE 11.23 *The sheeting exhibited by the plutonic rocks exposed in Yosemite Park is the result of joints that formed parallel to the surface of the igneous body when pressure was released as erosion removed the overlying rocks. (Courtesy of N. K. Huber/USGS)*

You will remember, however, that under conditions of low temperature and low confining pressure, as are found near Earth's surface, rocks can go directly from elastic deformation to brittle failure with little or no plastic deformation.

SPOT REVIEW

1. What fault movements are measured by heave, throw, and displacement, respectively?
2. How does the orientation of the strain ellipsoid differ between reverse and normal faults? What do the different strain ellipsoid orientations say about the difference in the forces that produced the faults?
3. How would you determine whether a strike-slip fault was right or left lateral?
4. How do the relative movements of the hanging and foot walls differ between reverse and normal faults?
5. What are the differences between shear and tensional joints?
6. How are the orientations of shear and tensional joints used to determine the direction of maximum compression?

DETERMINING THE DIRECTION OF ROCK MOVEMENT

Figures such as those used earlier to demonstrate folding and faulting might give the impression that the opposing rock masses were both moving; in other words, that the movement of the opposing rock masses is comparable to the head-on collision of two moving automobiles. This, however, is not always the case. As rocks undergo deformation, the stresses may be the result of one rock mass impinging on another in much the same fashion as a car collides with a wall. An example is the collision of India and Asia that gave rise to the Himalayas. In this case, we are fairly certain that India was moving northward while Asia was relatively stationary. In many other cases, however, we are unable to determine the actual movement of two impinging plates based on the observed deformation. Consider that the deformation experienced by the head-on collision of two automobiles each traveling at 30 miles per hour (50 kph) will be the same as that resulting from the collision of an automobile traveling at 60 miles per hour (100 kph) with one that is parked.

Rock Transport

Although we may not be able to determine the *actual* direction of rock movement, we can make judgments

FAULTS AND THE STRAIN ELLIPSOID

As was the case with folds, the stresses involved in the formation of the various types of faults can be portrayed by the strain ellipsoid. When materials are stressed to the point of brittle failure, fractures develop at angles of about 30° to the direction of maximum compressive stress. Consider Figure 11.B6. Figure 11.B6a illustrates the orientation of the strain ellipsoid for a reverse fault. Note that the intermediate axis parallels the strike of the fault, the crustal shortening is in the direction of the axis of maximum shortening, and the crust thickens as a result of the fault movement in the direction of the axis of maximum extension. Note also that the fault makes an angle of about 30° with the direction of maximum shortening.

As we have observed, normal faults are the opposite of reverse or thrust faults. Compare the strain ellipsoid orientations in Figures 11.B6a and in the strain ellipsoid in Figure 11.B6b has been rotated about the intermediate axis by 90°, with the result that the directions of maximum shortening and extension are opposite to those shown for the reverse fault. Note also that the fault plane makes an angle of about 30° with the vertical, which explains why most normal faults are high angle.

Figure 11.B6c illustrates the orientation of the strain ellipsoid for a strike-slip fault. As was the case with the reverse and normal faults, the intermediate axis is within the fault plane. Because both the axes of maximum shortening and extension have a horizontal orientation, however, the intermediate axis does not parallel the strike of the fault.

(continues on next page)

FAULTS AND THE STRAIN ELLIPSOID *continued*

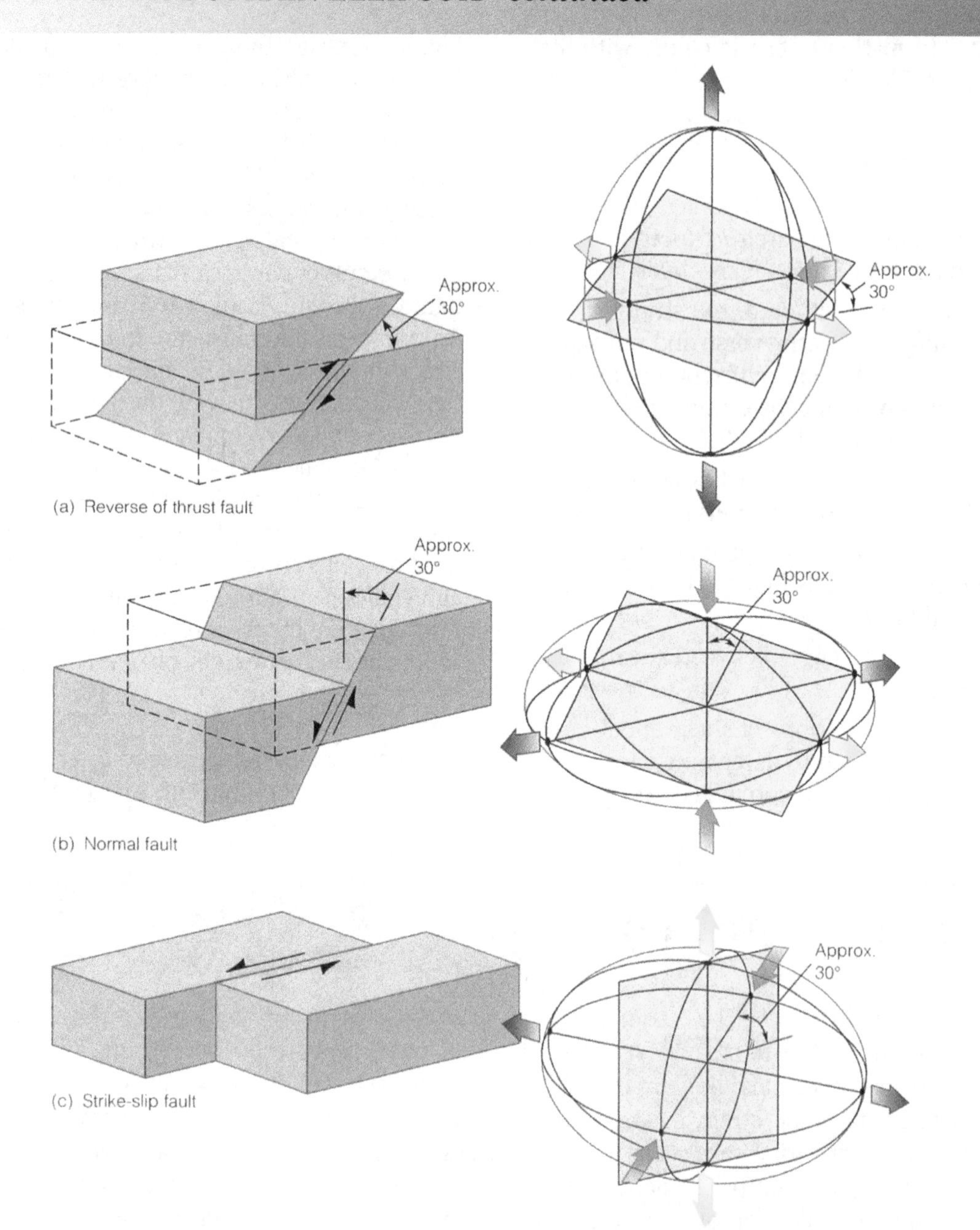

FIGURE 11.B6 *Although the three kinds of faults have different orientations of the strain ellipsoid, studies have shown that in all faults, the fault plane, shown here in blue, makes an angle of about 30° with the axis of maximum shortening. In all cases, the intermediate axis is located within the fault plane.*

Reverse faults *result in the vertical* stacking *of rocks in response to the vertical orientation of the axis of maximum extension.* Normal faults, *on the other hand, result in an orientation of the axis of maximum extension.*

In strike-slip faults, *the intermediate axis is oriented vertically within the fault plane with the axis of maximum shortening oriented horizontally and at an angle of about 30° to the fault plane. The orientation of the axis of maximum shortening relative to the fault plane determines whether the fault will be* right-lateral *or* left-lateral. *The fault shown in Figure 18.B6c is left-lateral. Note, however, that by rotating the axis of maximum shortening 60° clockwise, the fault movement becomes right-lateral.*

as to the *relative* directions of the rock movement initiated by the deformation. This is accomplished by determining the direction of **maximum rock transport** observed in the resultant structures. Figure 11.24 illustrates the generalized structures observed in the Appalachian Mountains. The continental collision that formed the Appalachian Mountains drove the rocks now seen in the Appalachians *westward* by way of thrust faults and by the formation of folds. Note that in Figure 11.24, the folds become increasingly asymmetrical toward the east and that both the fault planes and the axial planes of the nonsymmetrical folds all dip toward the east. As the easternmost rocks were deformed beyond their ability to respond plastically, they broke to form thrust faults with the hanging walls being moved upward and westward. Note that in the development of the faults, the rock mass of the hanging wall actually moves *laterally* westward relative to the foot wall as the hanging walls of the faults move up and over the foot walls. The development of asymmetry in the folds with the axial planes tipping to the west also results in a westward movement of the rock mass.

Structures such as those observed in Figure 11.24 have been reproduced experimentally using models where layers of clay represent the rock layers. These model layers are deformed by the same kind of stress conditions that have been suggested for the formation of the Appalachian Mountains. We can use our belief about how major mountain ranges have formed, coupled with carefully conducted laboratory experiments, to create some basic rules for interpreting the observed geologic structures. In this case,

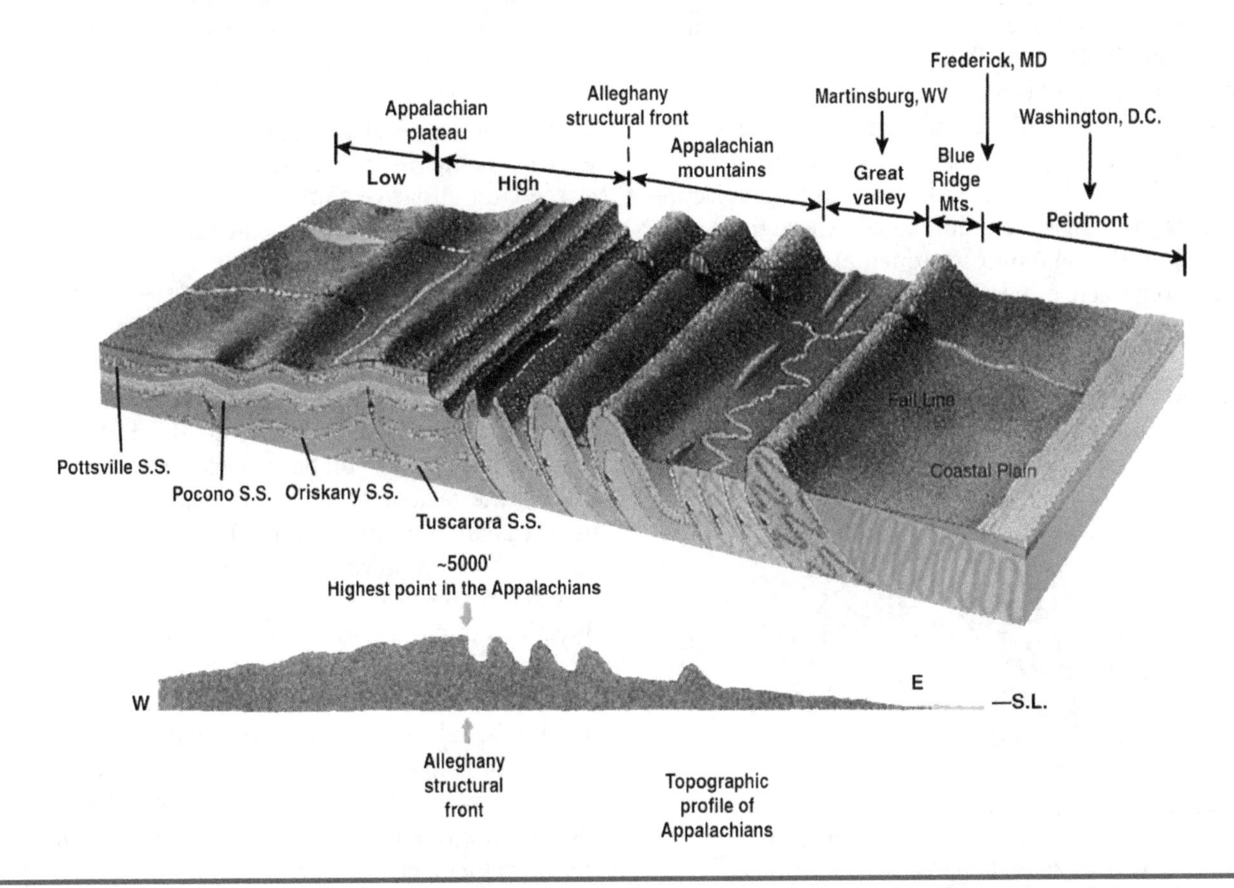

FIGURE 11.24 *The structures of the Appalachians illustrate how geologists determine the direction of* maximum rock transport. *Note that from east to west (a) the folds become increasingly symmetrical and lower in amplitude; (b) the displacements of the thrust faults progressively decrease until, within the Appalachian Plateau, they can no longer be recognized; and (c) the overall degree of rock deformation decreases from the metamorphosed and intruded rocks underlying the Piedmont to the essentially undeformed sedimentary rocks of the Appalachian low plateau. These observations indicate that as the continents collided to form Gondwana, the forces that formed the Appalachians were directed from east to west. In the same fashion, the forces that affected the continent of India were directed from north to south as India collided with Asia to form the Himalaya Mountains. The eastward-dipping thrust faults and steepening of the western limbs of folds indicate that the deformation resulted in the westward movement of rock masses from their original locations.*

both the experimental and the field data indicate that when rocks are deformed by compressive stresses, the axial planes of nonsymmetrical folds and the planes of thrust faults usually dip in a direction *opposite* to both the direction of the maximum applied stress and the direction of maximum rock transport. The structures illustrated in Figure 11.24 indicate that the Appalachian Mountains formed by compressive stresses and that the direction of maximum rock transport was toward the west.

Actual Direction of Fault Movements

The actual movements involved in the formation of faults are not always easily determined. Experiments producing reverse faults by subjecting clay cakes to compressional forces indicate that most of the movement is experienced by the hanging wall as it moves up and over a stationary foot wall. However, this does not preclude the possibility of *underthrusting* in which the foot wall is driven under the overlying rock mass. The same kinds of experiments also show that in normal faulting, most of the movement is experienced by the hanging wall. Consider our discussions of the East African Rift Valley where the valley formed as the adjoining continental masses moved away from each other. A similar scenario was used to describe the downfaulting along the crest of the oceanic ridges as the oceanic plates move away from each other (refer to Figure 11.25). In either case, the formation of the valley is interpreted as the result of the *downdropping* of the rock mass between normal faults located on opposite sides of the rift. It must be pointed out, however, that in both scenarios the valley margins are usually uplifted during rifting due to the buoyancy of the underlying rocks. Did the rift valleys form as the rocks of the valley floor moved *down,* did they form as the rocks along the valley margins moved *up,* or was it a combination of both? Because of the difficulty in determining the *actual* movement along normal faults, the question cannot be answered with certainty.

Evidence of the direction of maximum rock transport involved in strike-slip faults is usually not present. Only in cases such as the right-lateral strike-slip movement of the San Andreas Fault, where we understand the plate movements that are responsible for the creation of the fault, can we make a statement relative to the direction of maximum rock transport. We can say with some certainty that the rock mass west of the San Andreas Fault is moving northwestward while the North American continental plate east of the fault either remains relatively stationary or is moving northwestward at a slow rate.

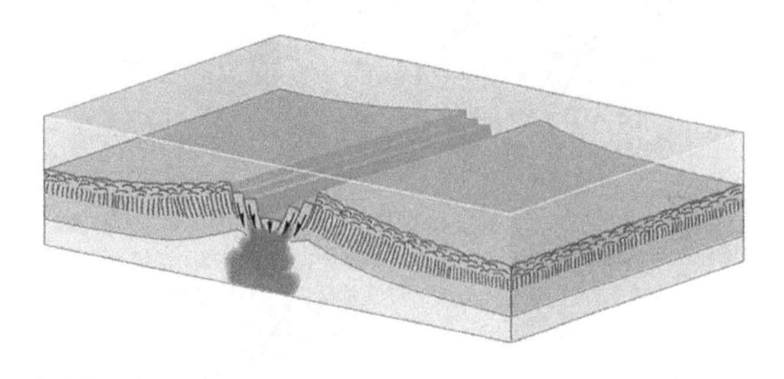

FIGURE 11.25 *Normal faulting is studied experimentally by subjecting clay cakes to tensional forces. The circles impressed into the clay are used to evaluate the degree of elongation of the clay cake under tension, and the stripes are used to simulate bedding. The photo shows the result of an experiment designed to model the formation of a rift valley. Note the surface fractures and the steps along the sides of the main valley formed by normal faulting and the curving of the fault surface with depth. Compare the results of this experiment with the drawing.*

Relative Direction of Fault Movements

A problem commonly confronted by a geologist studying a particular fault is how to determine the *relative* direction of the fault movement. If the fault cuts across the rock layers and there are distinctive layers that can be identified on both sides of the fault, determining the relative direction in which each side has moved is an easy task (Figure 11.26). Without such marker beds, however, the direction of movement may not be apparent. Without some evidence as to the directions of movement, a geologist may not be able to determine whether a particular fault is a normal or reverse fault. Another situation where a geologist may not be able to determine the relative directions of movement is where the fault parallels the rock layers.

Some minor structures, however, may be used to indicate the relative direction of movement. During the movement of one rock mass by the other, less competent rocks such as shales immediately adjacent to the fault surface are deformed into small folds called **drag folds.** The direction in which the broken ends of the beds are bent indicates the relative direction of the movement and allows the type of fault to be determined.

Another useful feature, called a **slickenside,** is commonly generated as a result of the two rock

FIGURE 11.26 *Having a* recognizable bed *such as a coal bed, sill, or distinctively colored sandstone or shale on opposite sides of the fault allows the direction of fault movement to be determined. (Photos VU/© Albert Copley)*

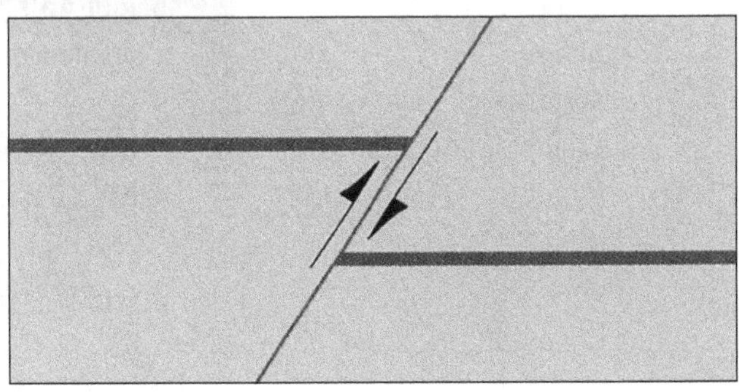

masses grinding against each other. The high polish of the opposing rock surfaces is an indication of rock movement. Close examination of the polished surface will invariably show **striations** that parallel the direction of movement. In some cases, the relative direction of movement of the blocks on opposite sides of the fault can be determined by running one's hand over the slickensided surface parallel to the directions of the striations. Whichever direction feels the smoother is the direction the block *above* the slickensided surface was moving. Field geologists learn to employ a great number of such techniques in the conduct of their everyday work.

SPOT REVIEW

1. In terms of deformation, what is meant by "rock transport"?
2. What kinds of evidence can be used to determine the most probable direction of rock transport within a deformed area?
3. What kinds of features can be used to determine the relative directions of movement of the rocks on opposite sides of a fault?

CONCEPTS AND TERMS TO REMEMBER

Strength
Stress
- compression
  - shear
- tension

Deformation or strain
- elastic
  - elastic limit
- plastic
- brittle

Strain ellipsoid
- axis of maximum shortening
- axis of maximum extension
- intermediate axis

CONCEPTS AND TERMS TO REMEMBER *continued*

- Strength of rocks
  - columns
  - arches
  - keystone
- Orientation of planar features
  - strike
  - dip
- Geologic structures
  - folds
    - monocline
    - anticline
    - syncline
    - descriptive elements
      - axial plane
      - limbs
      - axis
      - plunge
      - amplitude
    - classification of folds
      - symmetrical
      - asymmetrical
      - overturned
      - recumbent
      - isoclinal
  - faults
    - descriptive elements
      - hanging wall
      - foot wall
      - displacement
      - throw
      - heave
    - kinds of faults
      - reverse or thrust
      - normal
      - listric
      - strike-slip
        - right lateral
        - left lateral
      - oblique-slip or diagonal-slip
  - joints
    - shear
    - tensional
    - columnar jointing
- Rock movement
  - maximum rock transport
  - drag folds
  - slickensides
  - striations

REVIEW QUESTIONS

1. The force that you apply when opening a drawer is an example of
   a. rotational compression.
   b. shear.
   c. tension.
   d. nonrotational compression.
2. Which of the following kinds of strain is not permanent?
   a. elastic c. brittle
   b. viscous d. plastic
3. Rocks are weakest under
   a. nonrotational compression.
   b. shear.
   c. rotational compression.
   d. tension.
4. An anticline that has an inclined axial plane with limbs dipping in opposite directions is an example of a (an) _____ anticline.
   a. overturned c. asymmetrical
   b. recumbent d. symmetrical
5. A fault in which the hanging wall is down relative to the foot wall is a ______ fault.
   a. strike-slip c. thrust
   b. normal d. reverse
6. If, in a series of folds, the axial planes all dip toward the east, the direction of rock transport can be interpreted as having been from the
   a. east toward the west.
   b. west toward the east.
   c. north toward the south.
   d. south toward the north.
7. The most common of all the geologic structures are
   a. thrust faults. c. folds.
   b. normal faults. d. joints.
8. How are stress, strain, and strength related?
9. How are the various kinds of strain related to the depth below the surface of the crust?

REVIEW QUESTIONS *continued*

10. How can folds and faults be used to determine the direction of rock transport resulting from deformation?
11. What kinds of evidence can be used to indicate the direction of fault movement?

THOUGHT PROBLEMS

1. How do geologic structures control the processes of weathering and erosion?
2. How can you utilize topography to determine the structures present in subsurface rocks?

FOR YOUR NOTEBOOK

If you live in an area where various kinds of structures are exposed, certainly a photo-taking field trip is in order to document the various kinds. Even though the more spectacular folds and faults may not be exposed, joints are always present. In the case of joints, note the spacing of the joints relative to the bed thickness and to the competence of the individual rock layers.

Should the geologic structures not be readily available, this might be a time to further investigate the various kinds of stress and strain by observing everyday activities and objects. List common activities, from closing drawers to opening twist-top bottles, that are examples of the various kinds of stress.

Simple experiments using sheets of paper or thin layers of clay to reproduce the various kinds of folds and faults could also be performed. Perhaps the geology department has a clay-cake table that you may use, or perhaps structural geology students may be conducting clay-cake experiments that you may be able to observe.

TYPES OF FOLDED STRUCTURES

Folded structures form when rock layers are compressed (pushed together), similar to carpet folds when the sides are pushed together. Folds can take on many different shapes. They can be open or narrow, symmetric or asymmetric, or overturned. Also, folds aren't always horizontal—they can plunge into Earth at an angle. The sides of the folds are referred to as the limbs. The hinge line marks the maximum bend in the fold. Typically, the hinge line is horizontal to the surface of the Earth, however, the hinge line can be inclined downward from horizontal creating a fold that plunges into the Earth. The most common types of linear folds are **anticlines** and **synclines** (Figure 11.27).

An **anticline** is a convex up fold in which the rock layers form an arch with the limbs **dipping** away from the center of the structure. Erosion exposes the oldest rocks in the center.

A **syncline** is a fold in which the rock layers are warped downward and form a "U" shape with the limbs dipping toward the center of the structure. The youngest rocks will occur in the center of a syncline.

There are circular folds called domes and basins (Figure 11.28). A **dome** is geologically similar to an

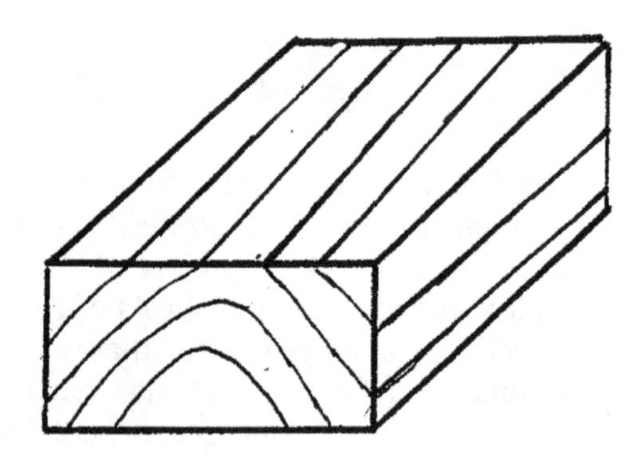

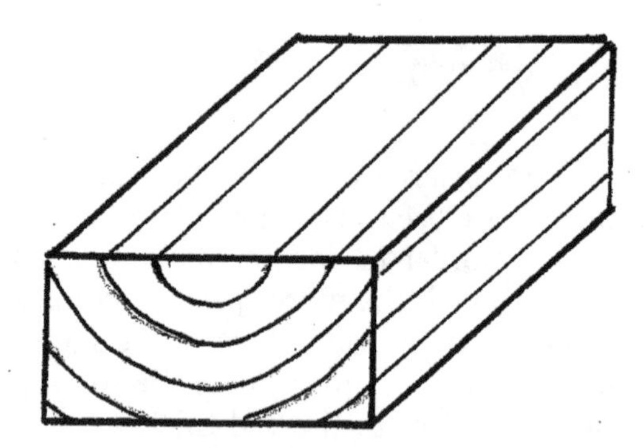

FIGURE 11.27 *Three-dimensional views of an anticlinal fold (left) and a synclinal fold (right).*

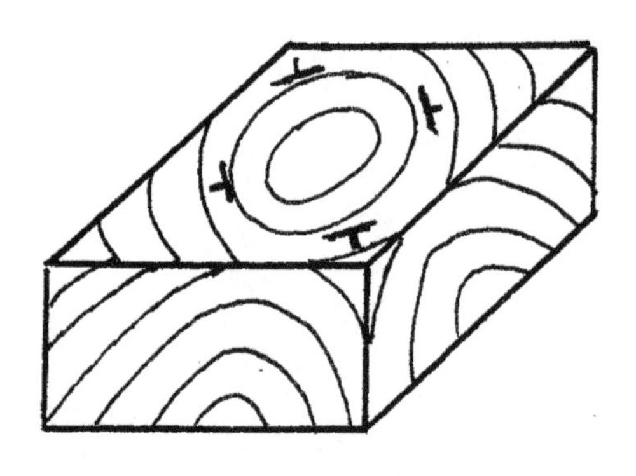

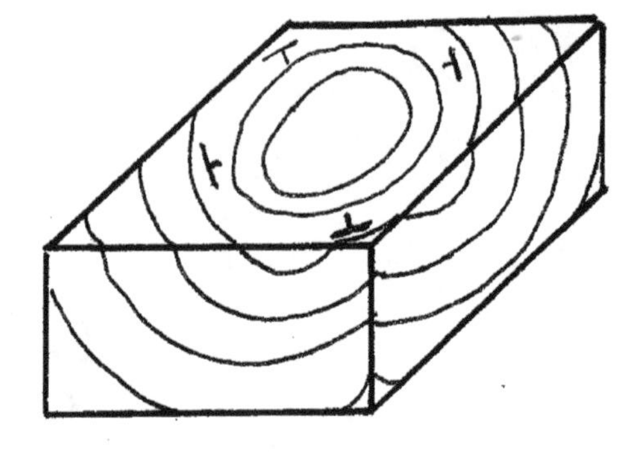

FIGURE 11.28 *A three-dimensional view of a dome (left) and a basin (right).*

anticline in that the oldest rocks are in the center, although this fold is roughly circular in nature. The rock layers dip away from the center in all directions. A good comparison is to picture an overturned bowl.

The opposite is true of a basin. A **basin** is a roughly circular geologic fold in which the rock layers dip in toward the center of the structure.

TYPES OF FAULTED STRUCTURES

Faults are fractures or breaks in rocks along which movement has occurred. Offset is the amount of movement that has occurred. The fault block above the fault is the hanging wall and the fault block below the fault is the foot wall. Geologists recognize three major classes of faults: normal, reverse, and strike-slip.

Normal faults form when tectonic forces stretch the Earth's crust, pulling it apart. This is commonly associated with extensional stress. The hanging wall block (the rocks above the fault) slide down the plane of the fault relative to the foot wall block (Figure 11.29). This movement puts younger rocks against older rocks.

Reverse faults occur when the hanging wall has moved up relative to the footwall (Figure 11.30). This type of fault is common in regions under compression, which is associated with mountain building. In a reverse fault, the hanging wall is pushed up over the footwall so that this places older rocks on top of younger rocks. Shallow-dipping (angles $< 30°$) reverse faults are termed **thrust** faults.

A **strike-slip** fault is common in regions where plates are sliding past one another horizontally (transform margins). The motion of a strike-slip fault can be right-lateral (the opposite side moves to the right) or left-lateral (the opposite side moves to the left).

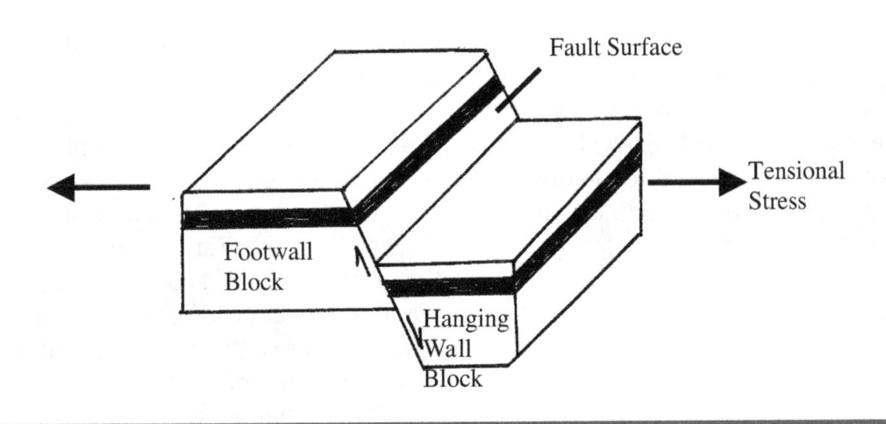

FIGURE 11.29 *A normal fault.*

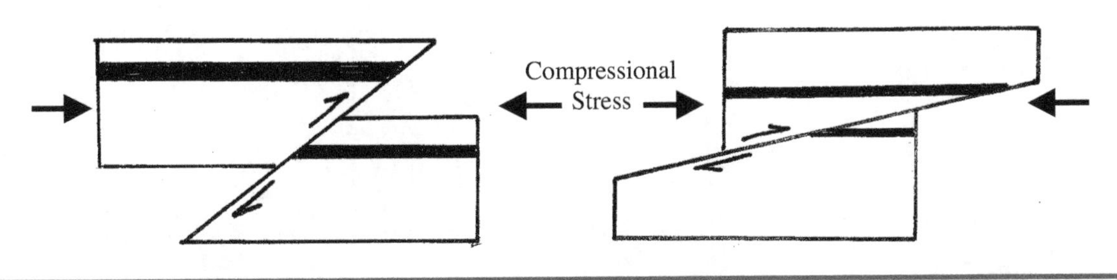

FIGURE 11.30 *Side view of a reverse fault (left) and a side view of a thrust fault (right).*

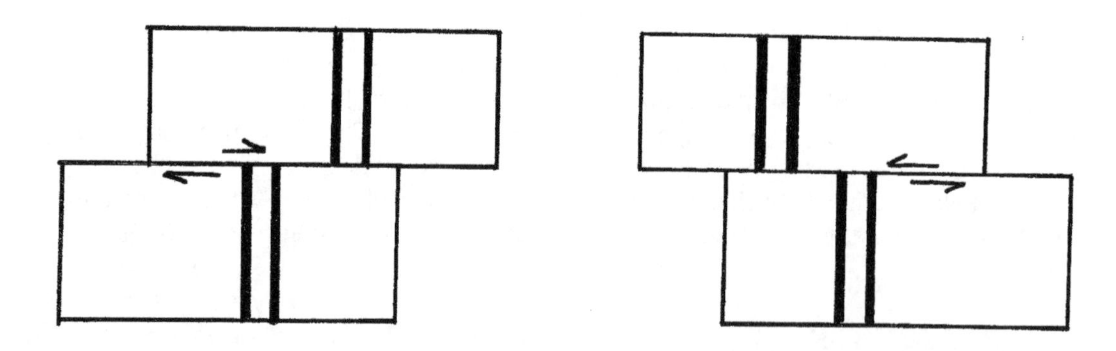

FIGURE 11.31 *A right-lateral strike-slip fault (left) and a left-lateral strike-slip fault (right) in plane view (looking down from above).*

FIGURE 11.32 *A strike/dip symbol oriented to S85°E, 23°SW (left) and N30°E, 37°NW (right).*

ORIENTATION OF STRUCTURES

It is important for geologists to describe how a given rock layer is oriented with respect to compass direction as well as angle of tilt of the rock. Strike and dip are two measurements to describe this orientation.

Strike is the compass direction of a line formed by the intersection of an imaginary horizontal plane and an inclined geologic feature.

Dip is the direction and degree of inclination measured down from horizontal of an inclined planar geologic feature.

Assuming the top of this paper is north, the strike and dip are represented by these symbols (Figure 11.32). The longer line is the strike line measured in degrees from magnetic north and the shorter line points in the direction of dip with the angle quantified.

Geologists use maps and diagrams as tools to explain observations of structural features. **Geological maps** are topographic maps on which different rock types are represented. **Cross sections** are vertical "slices" into the Earth that are used to interpret the geology of an area at depth. **Block diagrams** are a combination of geologic maps and cross sections. They are a three-dimensional representation of observed structural elements.

The orientation of structures and strata depicted on geological maps, cross sections, and block diagrams may be depicted using the symbols in Figure 11.33.

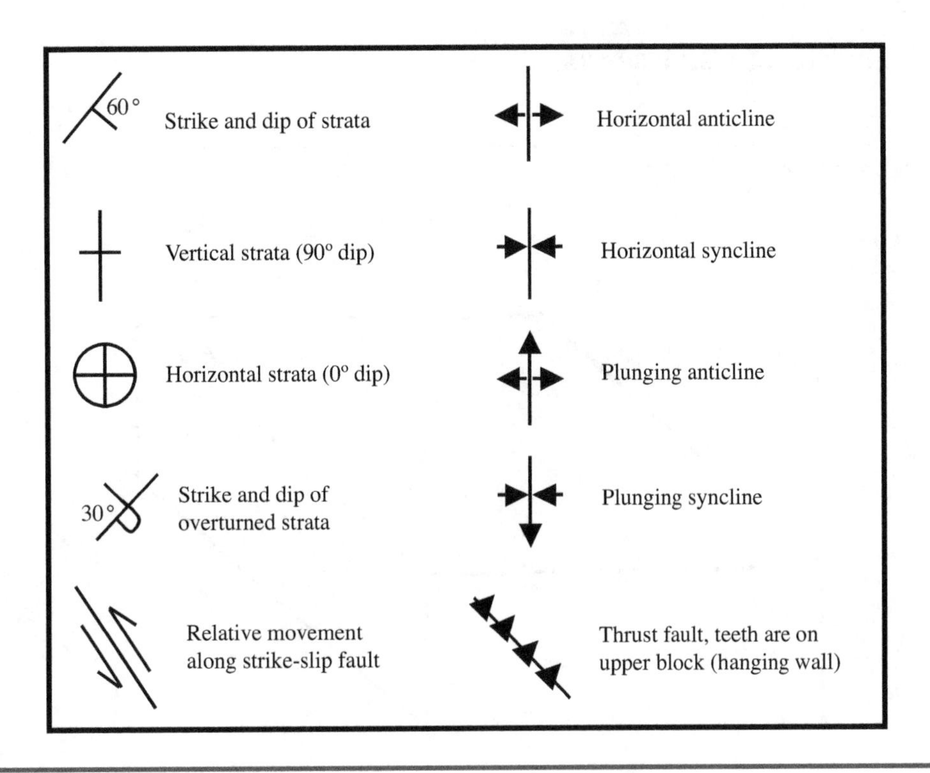

FIGURE 11.33 *Geologic structural symbols.*

Lab 10: Geologic Structures Exercise

Student Name: ______________________________

Follow the instructions below each figure.

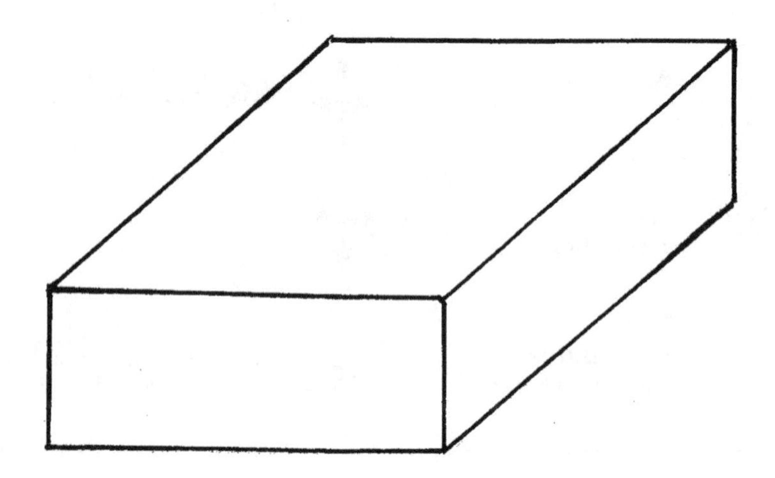

Figure 1. Draw three horizontal layers or beds.

What is the dip of these layers? ____________

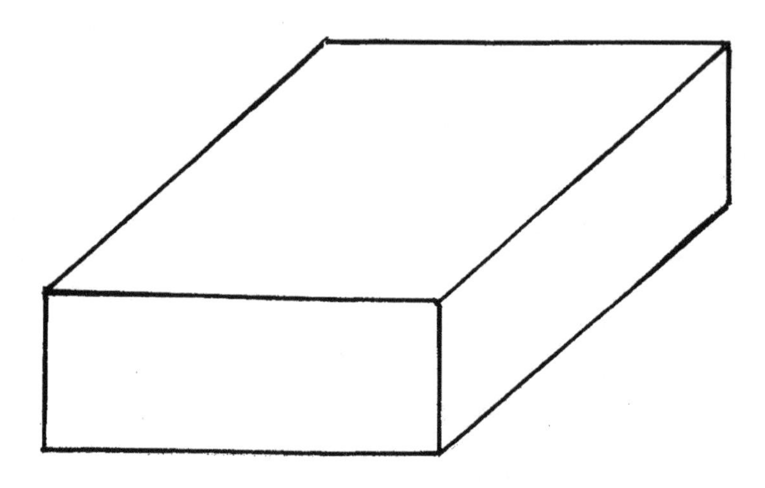

Figure 2. Draw three vertical layers or beds.

What is the dip of these layers? ____________

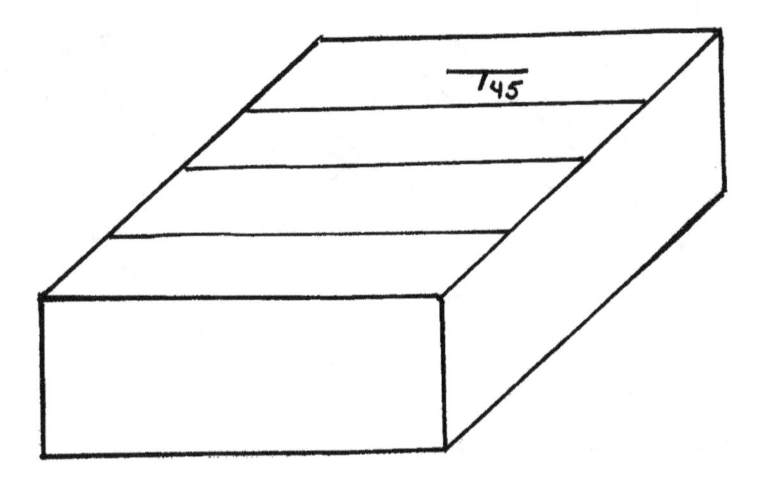

Figure 3. Complete the side and front of the block diagram using 45° for the dip of the beds. Include a strike and dip symbol on the diagram.

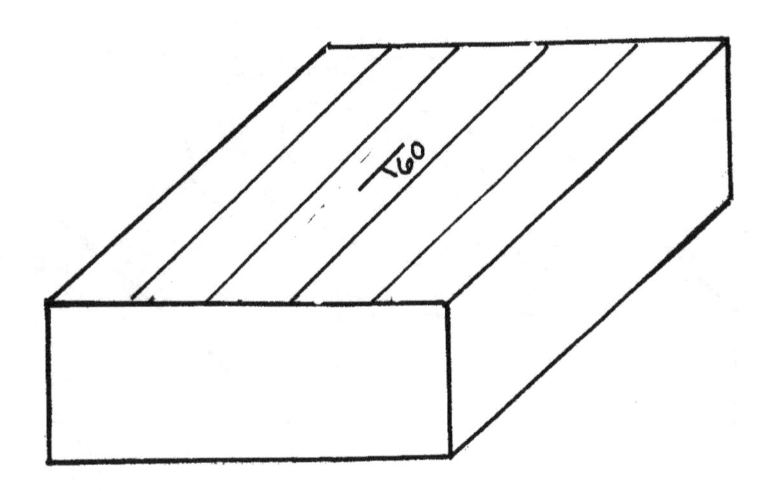

Figure 4. Complete the side and front of the block diagram using 60° for the dip of the beds. Include a strike and dip symbol on the diagram.

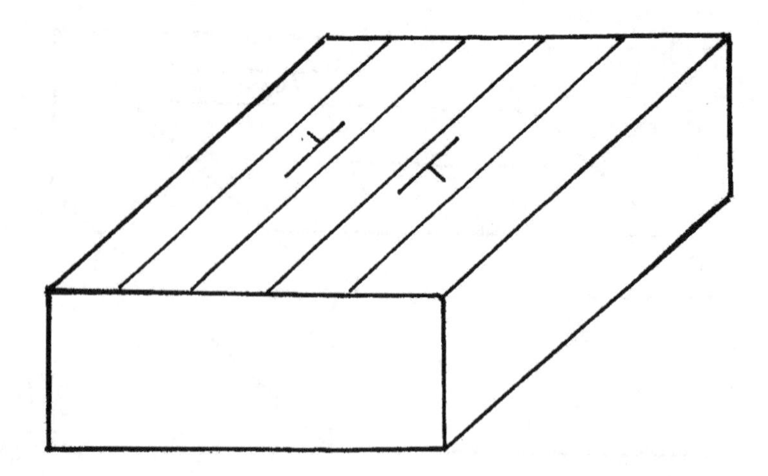

Figure 5. Complete the side and front of the block diagram following the strike and dip information provided.

What geologic structure is shown in this figure? ______________________________

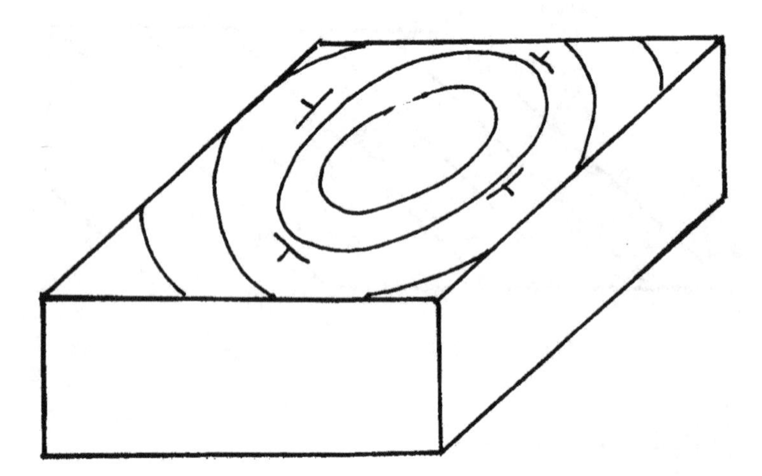

Figure 6. Complete the side and front of the block diagram following the strike and dip information provided.

What geologic structure is shown in this figure? ______________________________

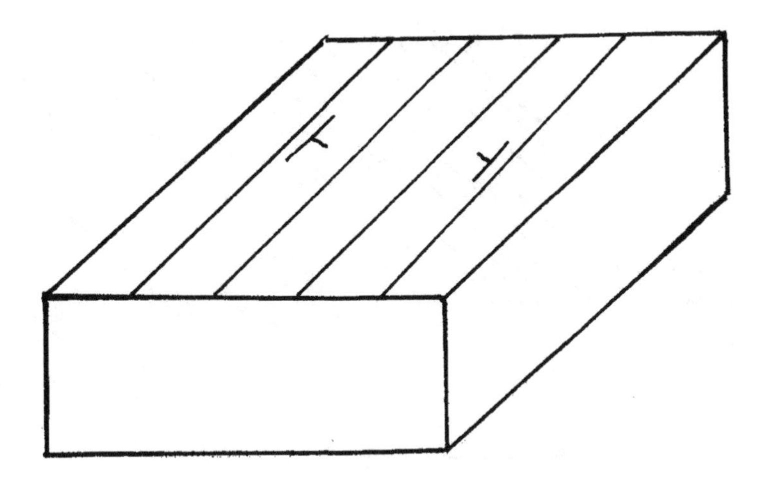

Figure 7. Complete the side and front of the block diagram. Use the strike and dip symbols to guide your interpretation.

What geologic structure is shown in this figure? ______________________________

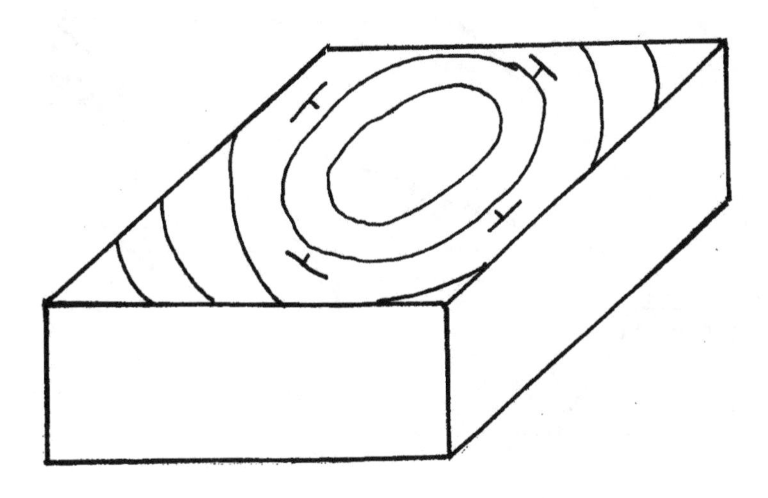

Figure 8. Complete the side and front of the block diagram following the strike and dip information provided.

What geologic structure is shown in this figure? ______________________________

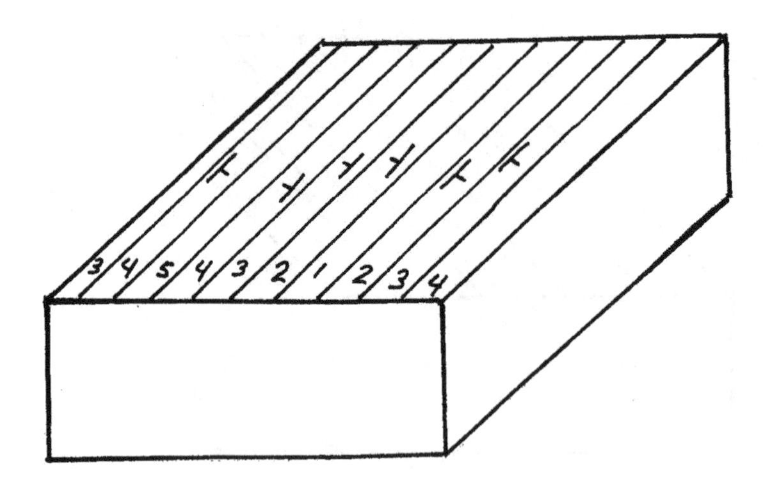

Figure 9. Complete the side and front of the block diagram following the strike and dip information provided.

What geologic structure is shown in this figure? ______________________________

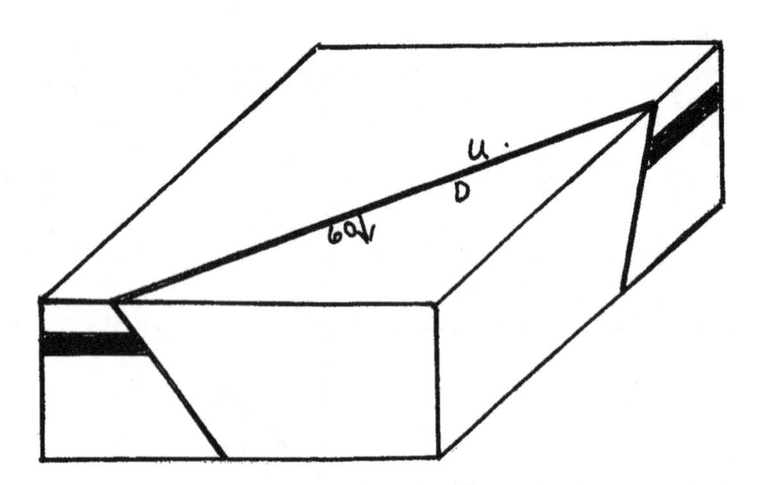

Figure 10. Complete the block diagram drawing in the location of the layers or beds.

What type of fault is shown in this figure? ______________________________

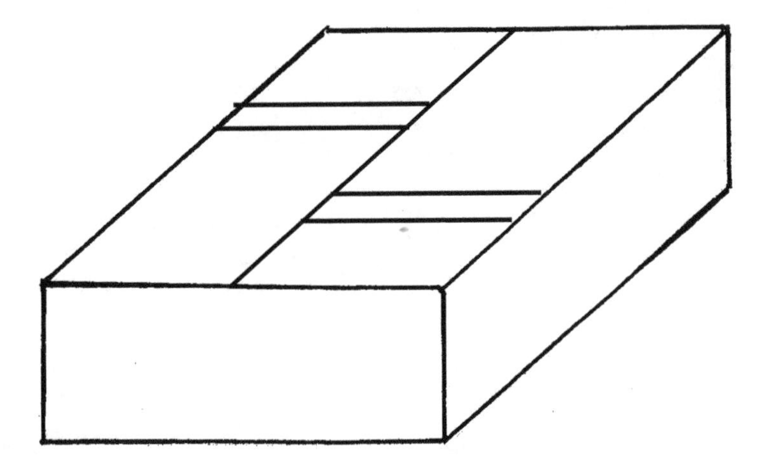

Figure 11. Complete the end of the block diagram. Add arrows indicating the direction of movement.

What type of fault is shown in this figure? ____________________

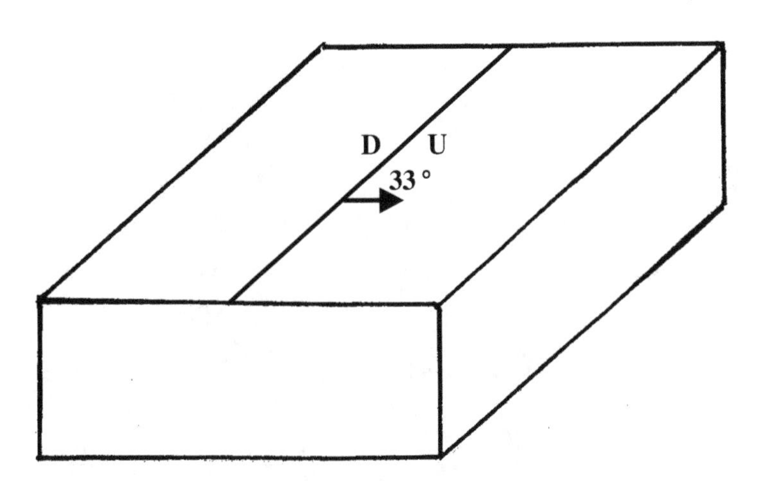

Figure 12. Complete the end of the block diagram. Add arrows indicating the direction of movement.

What type of fault is shown in this figure? ____________________

CHAPTER 12

Earthquakes and Seismology

INTRODUCTION Those of us whose daily lives are unlikely to be touched by earthquakes often find it difficult to imagine how people living in earthquake-prone areas such as parts of California can tolerate knowing that at any moment of any day they may become victims of a major earthquake. Unlike tropical storms and most volcanic eruptions, earthquakes seldom, perhaps never, give any warning of their coming. Within a matter of seconds, a quiet, stable landscape is converted into a quaking surface with fearsome possibilities of the destruction of both property and life. Then, as quickly as it began, the quake is over, except for the damage and suffering that it leaves behind.

Records of earthquakes have been kept for some time. The Chinese records go back nearly 3,000 years. The ancient Greeks were the first to attempt a "scientific" explanation of earthquakes. The philosopher Strabo noted that the frequency of earthquakes decreased as one moved inland from the coast. Aristotle thought earthquakes were caused by strong winds blowing through Earth's interior. (You will remember from our discussions of groundwater that Aristotle thought that the interior of Earth was hollow.) Some scholars have interpreted portions of the Bible as describing the aftermaths of earthquakes. Foremost among these are the collapse of the walls of Jericho about 1100 B.C. and the destruction of the cities of Sodom and Gomorrah.

The first recorded earthquake in the Western Hemisphere occurred in Massachusetts in 1636—reportedly, chimneys were knocked down. Nearly three decades later, in 1663, an earthquake was recorded in the lower Saint Lawrence River Valley, a site of frequent earthquakes even today.

Parts of California have the highest frequency of earthquakes in the continental United States. A relatively complete record of earthquakes for the area is available, dating back to the early 1800s. The record is largely the result of diaries kept by the Franciscan missionaries who established a chain of missions along the west coast. The missions were unknowingly located near faults because of the presence of fault-controlled springs.

THE DISTRIBUTION OF EARTHQUAKES

A map plotting the locations of recent earthquakes shows a pattern similar to that of active volcanoes (Figure 12.1). Before the advent of plate tectonics, the similarity in the distribution of earthquakes and active volcanoes led geologists to believe that one was the cause of the other. Although the rise of molten rock does cause swarms of earthquakes as it shoves rock out of its path, the earthquakes are usually weak and rarely result in damage or loss of life. The spatial relationship that exists between volcanoes and earthquakes occurs not so much because one causes the other as because both of these geological phenomena are related to a common feature, the **plate margins.**

Most earthquakes occur along the margins of the plates with only a few originating within the plates. Ironically, the highest magnitude earthquake ever to occur in North America was an intraplate

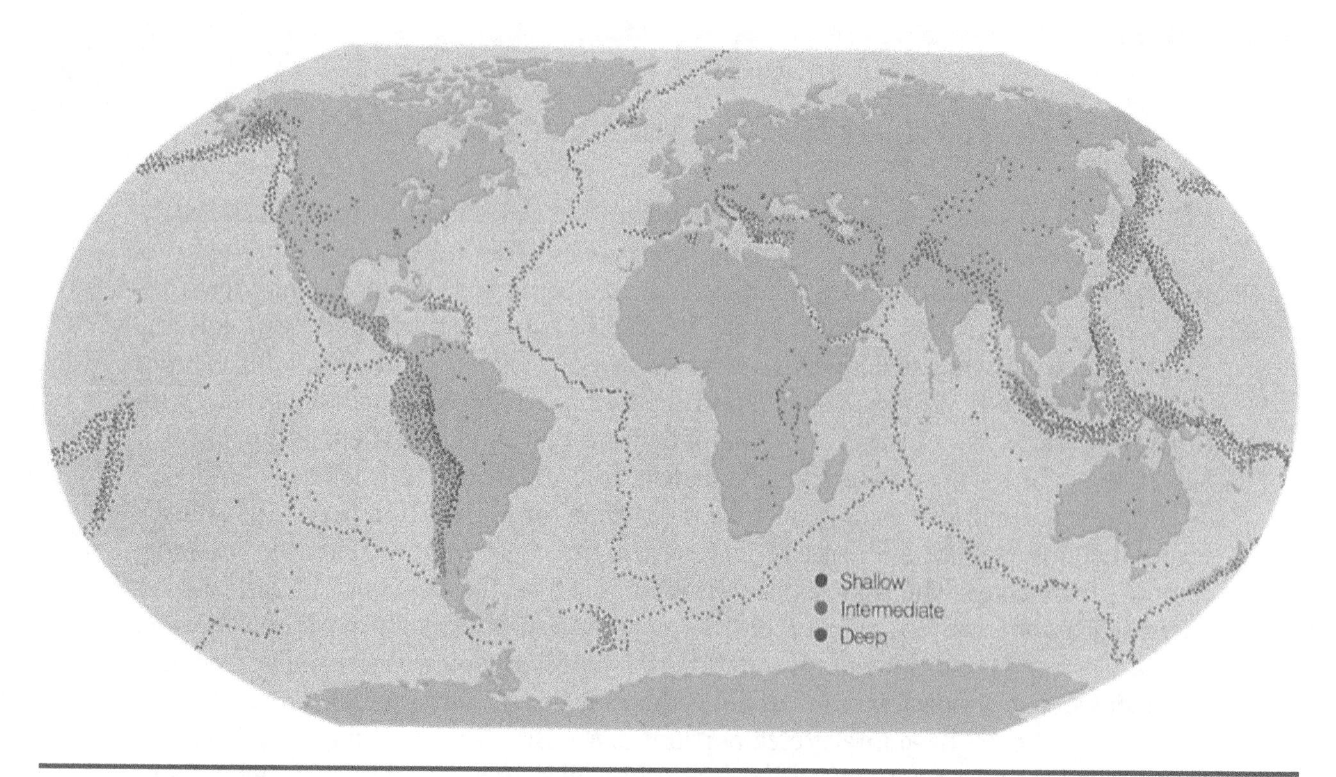

FIGURE 12.1 *With the advent of plate tectonics, the explanation for the* distribution *of earthquakes became apparent. Most of the energy released in the form of earthquakes around the world, and all of the* intermediate- *and* deep-focus *earthquakes that occur, are associated with the convergent plate boundaries.*

earthquake; it occurred in New Madrid, Missouri, on December 6, 1811. Although the Richter Scale did not exist at the time of the earthquake, seismologists have estimated that the New Madrid earthquake would have been between 8.1 and 8.7 on the Richter Scale. We will discuss the Richter Scale later, but an earthquake with a Richter magnitude greater than 8.0 is considered to be major. Geologists believe the New Madrid earthquake was the result of fault movements along a failed rift zone that dates back to the Pre-Cambrian. Two other well-defined earthquake zones within North America are not associated with modern plate boundaries: one is located along the Saint Lawrence River, and the other forms a belt trending from Charleston, South Carolina, to Roanoke, Virginia.

THE FREQUENCY AND LOCATION OF EARTHQUAKES

Earthquakes are caused by the sudden rupture of rocks and the subsequent movement along faults. In our discussions of rock deformation in Chapter 11, we saw that when stressed, rocks will store energy up to the elastic limit; after the limit is reached, they will either react plastically and bend (fold) or react as brittle solids and break. If they bend, the stored energy is consumed as the rock is contorted into various kinds of folds. If the rocks break, however, the stored energy is released. Some of the released energy is used to move the rocks on either side of the fault. The remainder is released as an earthquake or **seismic wave** in much the same way that the energy released from a breaking stick produces a snapping sound. The released energy causes surface rocks to move or "quake." As the seismic wave moves away from the fault into the surrounding rocks, it may spread the impact of the earthquake and its potential for destruction over a large area.

The severity of earth movement depends upon the amount of energy released. Although only a small amount of energy may be stored in any cubic foot or meter of rock during elastic deformation, the total volume of rock involved in an average fault movement may measure miles wide and deep and tens of miles long; thus, the total amount of potential energy available to be released can be awesome.

Because rocks are stronger under compression than they are under tension, a mass of rock undergoing compression can store more energy before it finally fails than a mass of rock being subjected to tension. As a result, 90% of all earthquakes—and

95% of all the seismic energy released—are associated with **subduction zones** where all types of stress occur. This relationship explains why the area around the Pacific Ocean basin experiences the highest frequency of destructive earthquakes.

Most earthquakes occur near the surface, where the rocks are most brittle; the number of earthquakes decreases exponentially with depth as the rocks become increasingly plastic. The deepest earthquakes are recorded at about 450 miles (725 km). At depths greater than 450 miles (725 km), all rocks respond as plastic materials and will not break regardless of the amount or the rate of increase of the applied stress.

The 450-mile (725 km) depth limit of earthquakes represents the deepest penetration of subducting oceanic lithosphere into the mantle. The swarms of earthquakes that are associated with zones of subduction constitute the **Benioff Zone** (Figure 12.2).

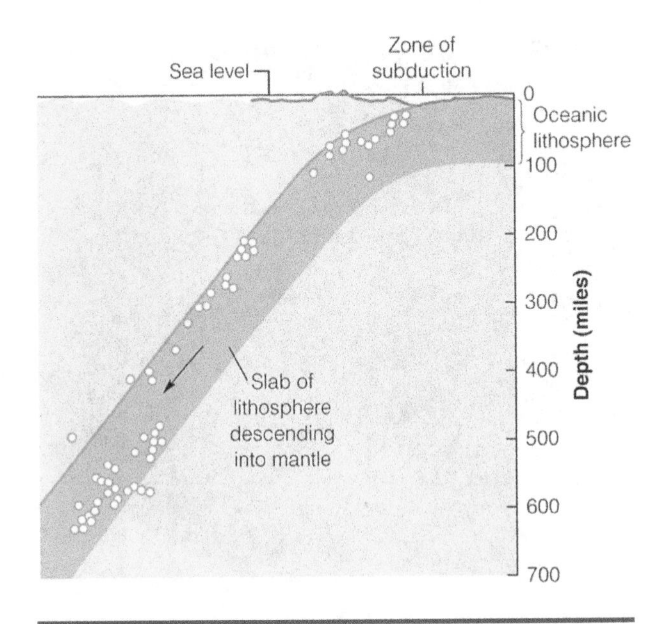

FIGURE 12.2 *Below the zones of subduction and dipping toward the continents at an angle of about 45° is a plane called the* Benioff Zone *along which abundant earthquake foci are located.*

Focus

The initial rupture point causing an earthquake, and also the point where the energy is first converted into a seismic wave, is called the **focus** (or **hypocenter**) of the earthquake. Earthquakes that occur from the surface to depths of about 40 miles (65 km) are referred to as **shallow-focus** earthquakes. As we have noted, the frequency of earthquakes decreases with depth. Nearly 75% of all earthquakes, including all major earthquakes, are shallow focus. **Intermediate-focus** earthquakes occur within the subducting plate from 40 miles down to about 185 miles (300 km) while those that occur from depths of 185 miles down to the maximum depth of about 450 miles (725 km) are called **deep-focus** earthquakes (Figure 12.3). Some geologists believe that rapid changes in the crystal structures of the minerals contained within the rocks rather than faulting may cause some of the deeper earthquakes.

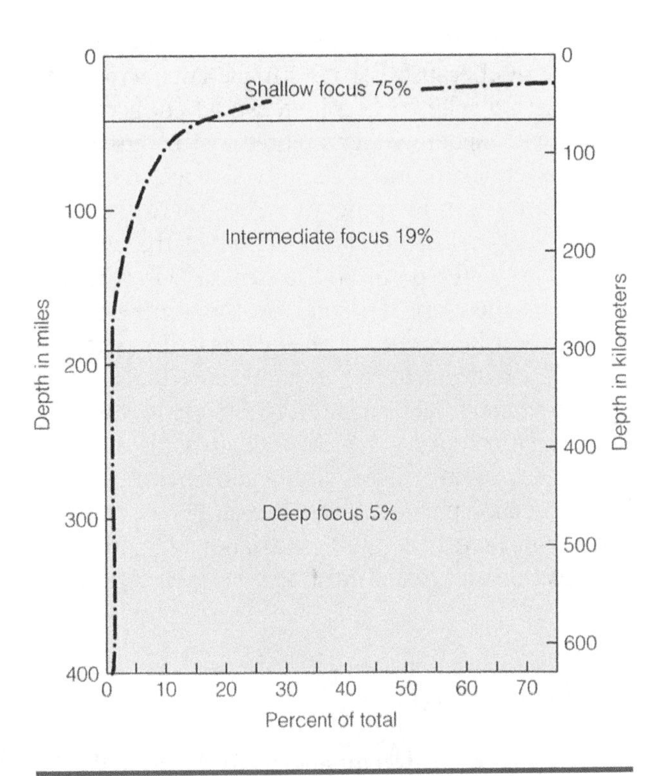

FIGURE 12.3 *The increase in the number of earthquakes above depths of 200 miles (300 km)—and especially the rapid increase above a depth of about 40 miles (65 km)—is due to the increasingly brittle nature of the rocks approaching Earth's surface.*

Epicenter

The point on the surface directly above the focus of the earthquake is called the **epicenter** (Greek *epi* = upon, over). Note that the focus and the epicenter of an earthquake are *not* synonymous. The focus of an earthquake may be located anywhere from the surface down to a depth of about 450 miles, but the epicenter is *always* on the surface.

The significance of the epicenter and its relationship to the focus are illustrated in Figure 12.4. The energy released from the earthquake radiates from the focus in all directions. The decrease in energy away from the focus is represented by a set of nested

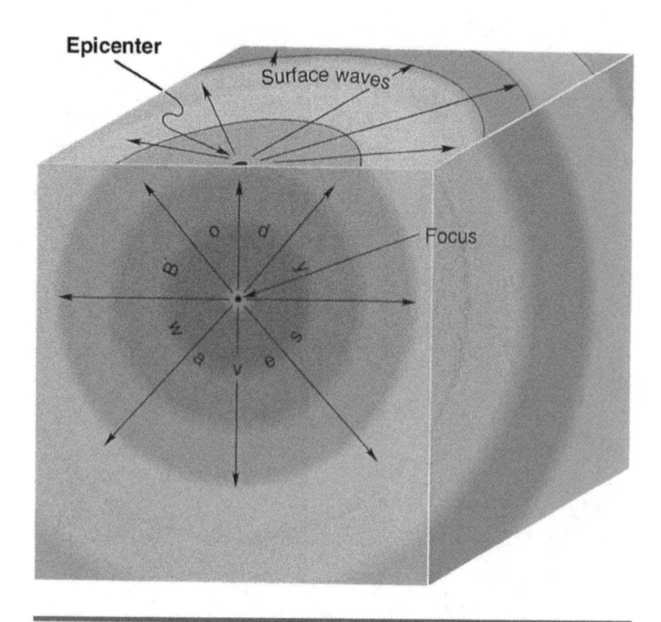

FIGURE 12.4 *The* focus *of the earthquake, the point where the seismic energy was released, can be located anywhere from Earth's surface to a depth of about 450 miles (725 km). The epicenter, on the other hand, is always located at the surface directly above the focus. Note that if the focus is at the surface, it and the epicenter are located at the same point.*

spheres with successive outer spheres representing less energy. Because Earth's surface cuts across the set of nested spheres, as a knife would cut across an onion, each sphere is represented at the surface by one of a set of concentric circles with each successive inner circle representing a higher-energy sphere. The epicenter, located at the center of the concentric circles, is the point on the surface closest to the focus and therefore the point on the surface where the earthquake energy is the greatest. In practical terms, the concentric circles can be considered zones indicating potential damage and destruction resulting from the earthquake, which decrease with distance from the epicenter. Although the concentric zones in Figure 12.4 are perfect circles, in reality their outlines would be irregular because of variations in both rock composition and structure at or near Earth's surface.

SPOT REVIEW

1. What is the source of energy for earthquakes?
2. Why are the distribution patterns for recent earthquakes and active volcanism similar?
3. Why does the frequency of earthquakes decrease with depth?
4. What is the difference between the focus and the epicenter of an earthquake? Can they ever be located in the same place?

MEASURING EARTHQUAKE INTENSITY AND MAGNITUDE

Seismologists use two different measurements to describe the strength of an earthquake. The *intensity* is a qualitative assessment of the effects of the earthquake. The magnitude is a quantitative measurement of the energy released by the earthquake.

Intensity

The **intensity** of an earthquake is determined by the effects observed (such as swinging chandeliers) and the kinds of **damage** caused. The first of several scales used to measure earthquake intensity was devised in 1873 by *M. S. de Rossi* of Italy and *François Forel* of Switzerland. A more detailed intensity scale was introduced in 1902 by an Italian volcanologist and seismologist, *G. Mercalli*. The **Modified Mercalli Scale** presented in Table 12.1 is the result of a revision of the original Mercalli Scale in 1931 by *H. O. Wood* and *F. Neumann* at the California Institute of Technology Seismological Laboratory.

An earthquake intensity scale is in part a stepwise verbal description of what one would expect to experience and observe at the time of the earthquake. This aspect of the scale is based on the comments of individuals who have experienced the particular earthquake combined with the direct observations of the earthquake's effect. Persons who have experienced low-intensity earthquakes typically give such responses as "dishes rattled," "bells rang," and "chandeliers began to swing."

Note, however, that above an intensity of about VII on the Modified Mercalli Scale, the major criterion for establishing the level of earthquake intensity is the amount of *observed* damage. Significant damage, including cracked masonry walls, shattered windows, and the collapse of tall brick smokestacks, begins to occur at intensity readings of VIII. At an intensity of IX, most structures such as buildings, interstate overpasses, and dams could be heavily damaged. At intensities of X, few structures will survive without extensive damage, and mass movements such as rock slides and snow avalanches will be common in areas of high relief. At an intensity of XI, the surface of the ground undulates visibility as the seismic waves pass by. At an intensity of XII, most humanmade structures will be destroyed, trees will be uprooted, and streams will be diverted from their channels. Seismologists estimate that the earthquake that struck New Madrid, Missouri, in 1811, had an intensity of XII at its epicenter.

TABLE 12.1

Modified Mercalli Intensity Scale of 1931 (Abridged)

I. Not felt except by a very few under especially favorable circumstances.

II. Felt only by a few persons at rest, especially on upper floors of buildings.

III. Felt quite noticeably indoors, especially on upper floors, but many people do not recognize it as an earthquake. Vibration like passing of truck.

IV. During the day felt indoors by many, outdoors by few. At night some awakened. Dishes, windows, doors disturbed; walls make cracking sound. Sensation like heavy truck striking building.

V. Felt by nearly everyone; many awakened. Some dishes, windows, etc., broken; a few instances of cracked plaster; unstable objects overturned. Disturbance of trees, poles, and other tall objects sometimes noticed. Pendulum clocks may stop.

VI. Felt by all; many frightened and run outdoors. Some furniture moved; a few instances of damaged chimneys. Damage slight.

VII. Everybody runs outdoors. Damage *negligible* in buildings of good construction; *slight* to moderate in well-built ordinary structures; *considerable* in poorly built or badly designed structures.

VIII. Damage *slight* in specially designed structures; *considerable* in ordinary substantial buildings; *great* in poorly built structures. Fall of chimneys, factory stacks, columns, monuments, walls.

IX. Damage *considerable* in specially designed structures; well-designed frame structures thrown out of plumb; *great* in substantial buildings, with partial collapse. Ground cracked conspicuously. Underground pipes broken.

X. Some well-built wooden structures destroyed; most masonry and frame structures destroyed; ground badly cracked. Considerable landslides from river banks and steep slopes.

XI. Few, if any, (masonry) structures remain standing. Bridges destroyed. Broad fissures in ground. Underground pipe lines completely out of service. Earth slumps and land slips in soft ground.

XII. Damage total. Waves seen on ground surface. Lines of sight and level distorted. Objects thrown upward into the air.

Magnitude

In contrast to the descriptive nature of earthquake intensity scales, the **Richter Scale** is a measure of the **magnitude** of an earthquake, which is an evaluation of the actual amount of **earth movement** and **energy released.** In 1935, *Charles Richter,* a professor at California Institute of Technology, devised a method of calculating the amount of energy released during an earthquake based upon actual measurements of Earth movement made by a sensitive instrument called a **seismograph** (refer to Figures 12.13–16). Richter based his method on experiments that subjected rocks to compressive forces up to the breaking point while he carefully measured the amount of energy applied and the amount of subsequent rock deformation. Each step on the Richter Scale represents a 10-fold increase in the amount of shaking (amplitude) and an approximately 30-fold increase in the amount of energy released. An earthquake of magnitude 7 therefore involves 100 times (10 times 10) the amount of rock movement and 900 times (30 times 30) the amount of energy released as an earthquake of magnitude 5. The Richter Scale is now used throughout the world to evaluate earthquake magnitude.

Table 12.2 shows the frequency of earthquakes at each level of the Richter Scale in any given year. Fortunately, most earthquakes are low magnitude and can only be detected by sensitive instruments. These earthquakes cause no damage of any kind.

TABLE 12.2

FREQUENCY AND ENERGY RELEASE OF EARTHQUAKES OF VARIOUS MAGNITUDES

Richter Magnitude	Number Per Year
Over 8.0	1 to 2
7.0–7.9	18
6.0–6.9	120
5.0–5.9	800
4.0–4.9	6,200
3.0–3.9	49,000
2.0–2.9	300,000

Frequency data from *Earth* by F. Press and R. Seiver. Copyright © 1986 by W.H. Freeman and Company. Reprinted with permission

Note: For every unit increase in richter magnitude, ground displacement increases by a factor of ten, while energy release increases by a factor of thirty. Therefore, most of the energy released by earth quakes each year is released not by the hundreds of thousands of small tremors, but by the handful of earthquakes of magnitude 7 or larger, the so-called major or great earthquakes. For comparison, an earthquake of magnitude 6 releases about as much energy as a 1-megaton atomic bomb.

THE GREAT SAN FRANCISCO EARTHQUAKE AND FIRE

At the turn of the century, San Francisco was the largest and fastest-growing city on the west coast. It was, for its time, a modern city with many innovations to which its citizens pointed with pride. Although "skyscrapers" were being built around steel frameworks, most of the buildings in the city were still built of wood. To make them look more substantial, the buildings were constructed with masonry facades facing the streets.

San Franciscans were also proud of their modern fire-fighting system. The city streets were equipped with a system of fire hydrants fed with water from central storage tanks, which eliminated the need to use horse-drawn wagons to haul water and steam-powered pumps to the sites of fires.

Some of the streets were illuminated by the recently invented electric light, another source of pride for San Franciscans. The main sources of light throughout the city were gas- or kerosene-fueled lamps, however.

Much of the city was built on steep slopes covered with the thick regolith that characterizes the coastal regions of California, setting the scene for disaster. At 5:12 A.M. on April 18, 1906, a segment of the San Andreas Fault to the north of the present location of the Golden Gate Bridge moved about 18 feet (5.5 m) (Figure 12.B1). As is usually the case, buildings with foundations firmly anchored into bedrock survived the quaking with little damage, but many of the buildings and homes built on the unconsolidated regolith experienced substantial movement. Kerosene lamps crashed to the floor and set the wooden buildings on fire. Downslope movement of the regolith broke gas lines, which released gas that fueled the fires. Firefighters attempting to reach

(a)

(b)

FIGURE 12.B1 *The earthquake that devastated San Francisco in 1906 was one of the most severe to occur in North America in historic time. You will recognize the statue with its head stuck in the pavement after being toppled from its perch as being that of Louis Agassiz, the father of the Great Ice Age. Considering the severity of the earthquake, it is surprising that only about 600 people lost their lives. In this classic photo, citizens watch the raging fires that destroyed nearly 90% of the city before they were finally extinguished. The debris generated by the cleanup was dumped into San Francisco Bay where it formed the land on which the Marina was built. Little did the inhabitants realize at the time that they were setting the scene for most of the damage that would occur during the 1989 earth quake.* (a *Courtesy of W. Mendenhall/USGS;* b *courtesy of T. L. Youd/USGS)*

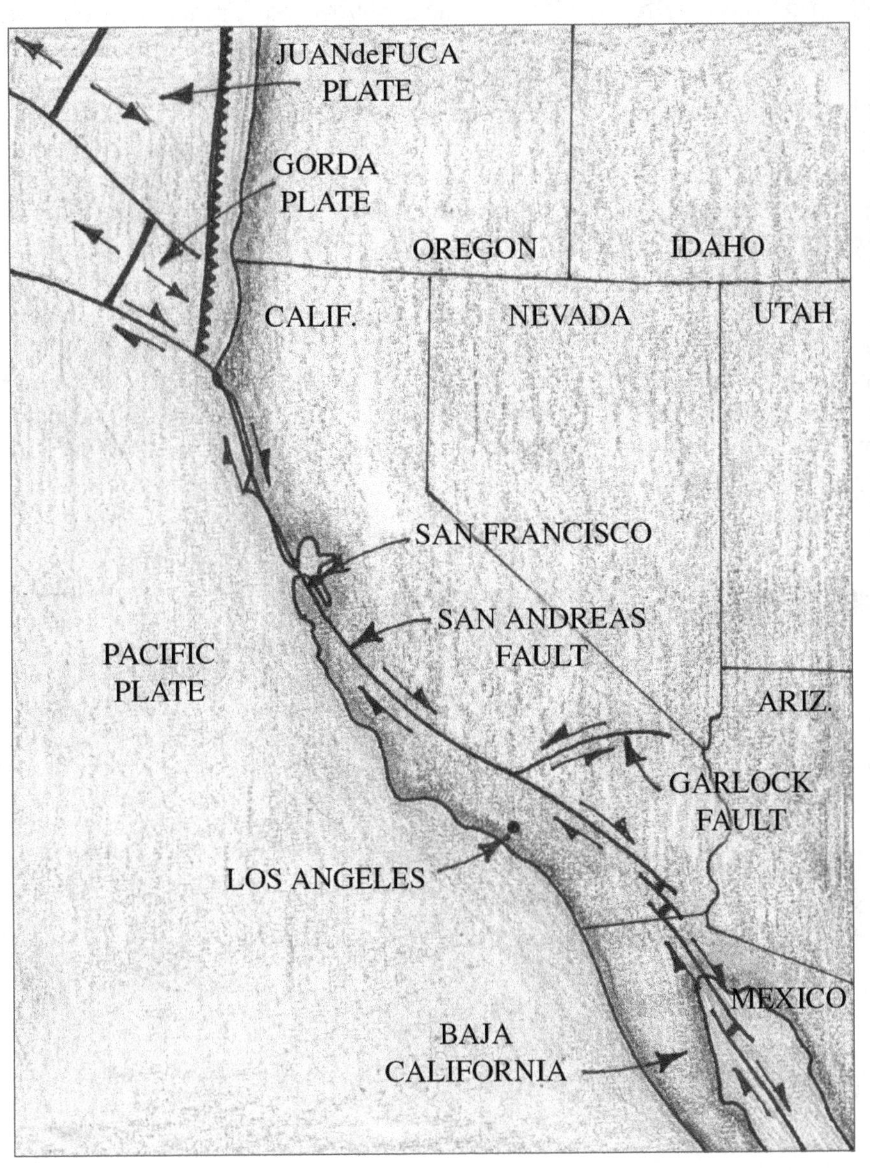

FIGURE 12.B2 *Many of the major earthquakes that occur in California are associated with the* San Andreas Fault. *The San Andreas Fault is a right-lateral strike slip fault and represents the boundary between the Pacific Ocean and North American lithospheric plates. The earthquakes that occurred in San Francisco in 1906 and in Loma Prieta in 1989 were the result of movements along segments of the San Andreas Fault.*

the fires found that many of the masonry building facades had fallen into the streets and made them impassable. When firefighters were able to arrive at the site of a fire, they found that the same earth movements that had severed the broke gas lines had also broken the water mains that fed the hydrants and the water hydrants were dry. The city was soon ablaze (Figure 12.B1). Before the fires were finally brought under control several days later by dynamiting entire city blocks, nearly 90% of the city lay in ruins.

DAMAGE FROM EARTHQUAKES

Although most earthquakes are of low magnitude on the Richter Scale, a single high-magnitude earthquake can kill thousands of people and cause millions of dollars worth of damage. The greatest loss of life due to a single earthquake occurred in 1556 in Shensi Province, China, where an estimated 800,000 people died. As recently as 1976, at least 650,000 people died in Tangshan, China, when an earthquake with a Richter Scale magnitude of 7.6 caused dwellings carved into loess cliffs to collapse, trapping the inhabitants.

Some damage such as the offsetting of roadways, fence lines, and the buckling of pavement is a direct result of the fault movement (Figure 12.5). Most earthquake damage is not directly related to fault movements, however, but results from events caused by the seismic energy, with fire being a major source of damage. An example is the earthquake that destroyed nearly 90% of San Francisco in 1906 where a combination of factors conspired to initiate a fire that caused 10 times more damage than was incurred by the quaking of the earth.

In 1989, San Francisco experienced another severe earthquake. Its epicenter was located in the Santa Cruz Mountains approximately 75 miles (120 km) to the south at Loma Prieta, California. Significantly, the Marina area of San Francisco, which suffered the greatest damage during the 1989 earthquake, was built atop the debris removed from the city after the 1906 earthquake (Figure 12.6). The material had been dumped into San Francisco Bay and partially compacted to provide additional building space. Thus,

FIGURE 12.5 *Although some earthquake damage is due to fault-induced vertical or horizontal offsets of the surface, most earthquake damage is only indirectly the result of fault movements. Fire is actually the most common cause of damage. (Courtesy of W. B. Hamilton/USGS)*

in a sense, the aftermath of the 1906 earthquake laid the foundation for some of the destruction suffered by the city in the 1989 earthquake. Unfortunately, most of the damage experienced by the Marina area occurred for the same reasons as the destruction in 1906. For one thing, the extensive structural damage experienced by buildings was directly the result of having been built on loose, unconsolidated materials. As in 1906, fire fueled by gas leaking from broken gas mains was a major cause of damage in the Marina area until shutoff valves could be located beneath the debris. The collapse of the Nimitz Freeway overpass, crushing motorists in their vehicles, accounted for most of the lives lost during the 1989 earthquake (Figure 12.7). Portions of the roadway failed, at least in part, because they had been constructed on poorly consolidated bay mud.

As the 1989 Loma Prieta earthquake clearly showed, the greatest amount of damage resulting from an earthquake is not necessarily in the area nearest the epicenter. An even better example of this relationship occurred on September 19, 1985, when

FIGURE 12.6 *In general, structures anchored in bedrock have the best chance at surviving an earthquake, while those built on unconsolidated materials generally suffer the most damage. A case in point is the effect of the 1989 earthquake in Loma Prieta, California, on San Francisco. In the downtown area where the foundations of the buildings are in bedrock, there was little damage. Most of the damage occurred in the Marina area where the structures were built on land formed by dumping the unconsolidated debris of the 1906 San Francisco earthquake along the edge of the bay.* (a *Courtesy of J. K. Nakata/USGS;* b *courtesy of C. E. Meyer/USGS)*

FIGURE 12.7 *Most of the lives that were lost in the 1989 Loma Prieta earthquake were the result of the collapse of the Nimitz Freeway. Although the experts still do not agree on the exact cause of the structural failure experienced by the overpass, the fact that portions of the freeway that collapsed were located on unconsolidated bay sediments may have contributed to its failure. (Courtesy of H. Wilshire/USGS)*

FIGURE 12.8 *Another example of indirect earthquake damage is the 1985 Mexico City earthquake where the fault movement occurred more than 200 miles (300 km) away. As was the case in the Marina area of San Francisco, most of the damage was located in areas of the city underlain by unconsolidated materials, in this case, lake sediments. The observed variation in the magnitude of the damage experienced by individual buildings during the Mexico City earthquake has led engineers to hypothesize that the extent of damage could be due to the specific vibrational frequency of the buildings relative to the seismic vibrations of the underlying rocks and sediments. (Courtesy of M. Celebi/USGS)*

movement along the subducting Cocos plate caused an earthquake of Richter magnitude 8.1 on the Pacific coast of Mexico. The greatest amount of destruction, however, was experienced in Mexico City, more than 200 miles (300 km) to the east. The earthquake resulted in more than 8,000 deaths, 30,000 injuries, and the destruction of about 500 buildings. Most of the deaths occurred in ghetto areas where the buildings were poorly constructed and unreinforced. The total damage was estimated to exceed $4 billion (Figure 12.8).

How could an earthquake 200 miles away possibly cause such extensive damage? The answer is not yet fully understood, but certainly the local geology played a major part. Much of Mexico City is built on the bed of ancient Lake Texcoco, which was drained by the Spaniards to allow for the expansion of the city after they defeated the Aztecs. Most of the buildings destroyed during the earthquake were in the portion of the city built on the soft, water-saturated, unconsolidated sediments of the old lake bed. Buildings in adjacent portions of the city constructed on the bedrock surrounding the old lake shoreline experienced little or no damage from the seismic waves even though the amount of earth movement was greater than in the devastated portion of the city. Seismologists and engineers examining the damage concluded that upon reaching the old lake sediments, the seismic waves caused the sediments to begin vibrating like a bowl full of Jello. The vibration of the lake sediments was then transferred to and amplified by the buildings. Buildings with the same natural vibration frequency as the underlying sediments began to vibrate in harmony with the sediments and literally ripped themselves apart. Other buildings whose dimensions and construction rendered them "out of tune" with the vibrating sediments experienced less damage. All the while, buildings with foundations firmly anchored in bedrock survived with little or no damage.

Tsunami

The ultimate in extensive damage caused by a distant earthquake involves the impact of a **tsunami**. Commonly, but incorrectly, referred to as a "tidal wave," a tsunami is a gigantic seawave that has nothing to do with the tides but is usually generated by energy released during a submarine earthquake. Because most major earthquakes are associated with the zones of subduction surrounding the Pacific Ocean, the Pacific Ocean basin suffers more tsunamis than any of the other ocean basins, with Japan and Hawaii experiencing the most tsunamis (Figure 12.9). The worst tsunami to hit Japan occurred on June 15,

FIGURE 12.9 *With major earthquakes occurring around nearly the entire perimeter of the Pacific Ocean basin, the Hawaiian Islands located in mid-Pacific are especially prone to destructive* tsunamis. *(Courtesy of USGS/HVO)*

1896, when a wall of water estimated at 75 to 100 feet (23–30 m) high crashed onto the eastern shore of Honshu, sweeping away more than 10,000 homes and 26,000 people.

The occurrence of earthquakes within the ocean basins is not the only cause of tsunamis. You will remember from our discussions of volcanoes in Chapter 4 that energy released during the eruption of Krakatau in 1883 produced a tsunami that overwhelmed all the low-lying parts of the surrounding islands and carried 36,000 people to their deaths.

With an average amplitude of only a few feet, a tsunami crossing the open sea would pass undetected, even by a relatively small ship. However, with an average wavelength of 60 to 70 miles (100–110 km) and speeds of 500 miles per hour (800 kph) not uncommon, the energy of a tsunami is almost unimaginable! When a tsunami with an amplitude of a few feet at sea moves onto the shoreline, the amplitude may increase to 100 feet (30 m) or more and cause enormous damage.

The **Seismic Sea Wave Warning System (SSWWS)** was established in 1946 to warn the inhabitants of Pacific coastal areas of potentially destructive tsunamis. Earthquakes occurring around the Pacific Ocean basin or on the ocean floor are constantly monitored at the headquarters in Honolulu, Hawaii. Warnings of potential tsunamis are broadcast to stations throughout the Pacific Ocean basin where strategically located sirens warn inhabitants and provide adequate time for protective action to be taken.

SPOT REVIEW

1. What is the difference between the Modified Mercalli Scale and the Richter Scale? In what ways are the two scales similar?
2. What factors have we learned from experience are most important in determining the amount of damage that results from an earthquake?
3. Why are tsunamis so destructive?

EARTHQUAKE (SEISMIC) WAVES

Whether generated by an earthquake, an explosion, or a breaking stick, waves are of two kinds: (1) **shear waves,** in which the material conducting the wave is moved back and forth *perpendicular* to the direction in which the wave is moving, and (2) **compressional-extensional (compression) waves,** in which the material moves back and forth *parallel* to the direction of propagation (Figure 12.10).

A shear wave can be visualized as the ripple that moves from the handle to the tip of a whip as it is snapped and can be illustrated as a sine wave (Figure 12.10a). A compressional-extensional wave is somewhat more difficult to represent graphically but is commonly illustrated as a "Slinky" toy, which, you may remember, is a spring along which a series of coils open and close, forming a wave that moves from one end of the spring to the other and allows the Slinky to "walk" downstairs and do other tricks.

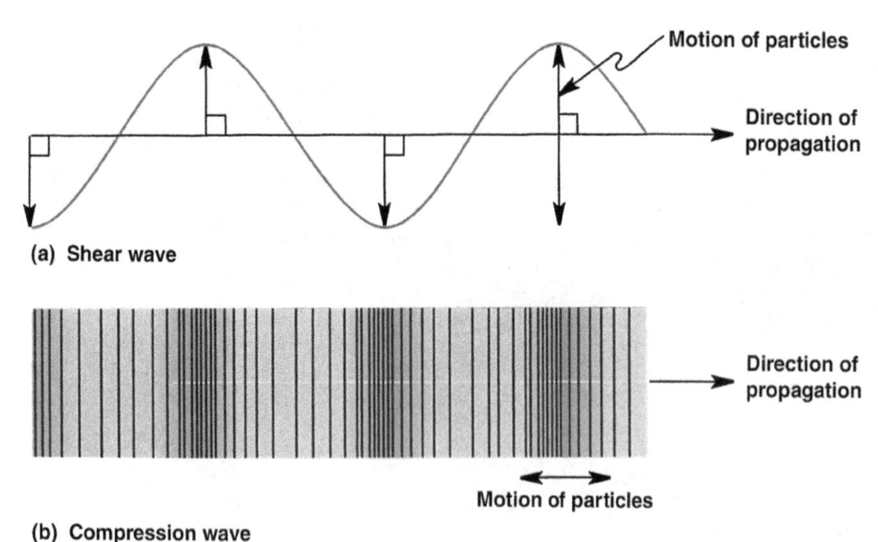

FIGURE 12.10 *Waves are of two kinds:* shear waves *where the movement of the material through which the wave is propagating (moving) is* perpendicular *to the direction of propagation (a), and* compression waves *where the movement of the material is* parallel *to the direction of propagation (b). In the case of shear waves, it is important to keep in mind that the material moves perpendicular to the direction of propagation in* all directions, *not simply in the vertical orientation illustrated in the drawing.*

An important distinction between the two kinds of waves is that shear waves can only be transmitted through *solids* while compression waves can be transmitted through matter in *any* state—solid, liquid, or gas. Shear waves cannot be transmitted through a liquid or a gas because neither has the rigidity necessary to allow it to be sheared (cut). All materials, on the other hand, can be compressed and therefore can transmit compressional-extensional waves.

Body Waves

As movement occurs along a fault and the energy stored in the rock during the elastic deformation is released, shear- and compression-type seismic waves are generated simultaneously and move out in all directions from the focus (refer to Figure 12.4). The seismic waves that move through Earth, called **body waves,** are both of shear and compression type (Figure 12.11). The shear body waves are called **S waves** (*s*econdary waves) while the compression-type body waves are called **P waves** (*p*rimary waves). An important characteristic of both types of body waves is that they are of low amplitude; that is, they do not generate much shaking or vibration in the rock mass as they pass through Earth's interior. Of the two body waves, the P wave has the higher velocity and is therefore referred to as primary.

Surface Waves

Upon reaching Earth's surface, the seismic energy generates another set of waves called **surface waves,** which are of two types: (1) **Love (LQ)** waves and

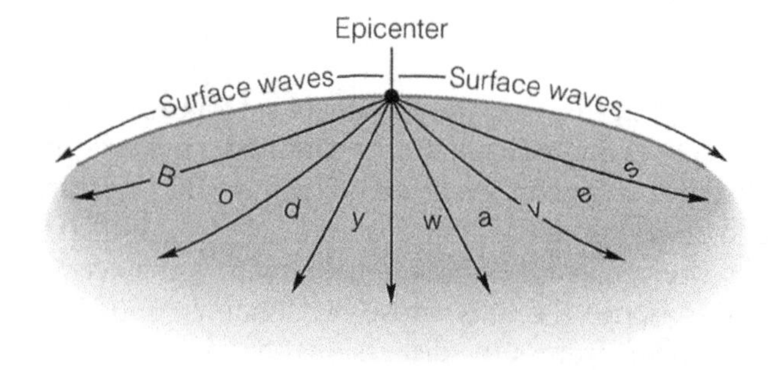

FIGURE 12.11 *The* body waves *that move through the* body *of Earth consist of both shear waves* (S waves) *and compression waves* (P waves). *Body waves are lower in amplitude and travel with higher velocities than surface waves. Of the two types of body waves, the P waves have the higher velocity.*

(2) **Rayleigh (LR) waves.** Love waves are shear waves that move Earth's surface back and forth *horizontally* perpendicular to the direction of propagation (Figure 12.12a) while Rayleigh waves are a combination of compression and shear waves. They produce an elliptical, rolling motion that moves the rocks in a direction opposite to the direction of propagation (Figure 12.12b). The combined effect of these waves results in the *vertical* and *horizontal* movement of Earth's surface. The surface waves radiate from the epicenter of the earthquake.

Surface and body waves differ in several important respects. Surface waves are slower and have much higher amplitudes than body waves. The amplitude of the body waves is too low to cause any damage upon reaching the surface. The surface waves with their

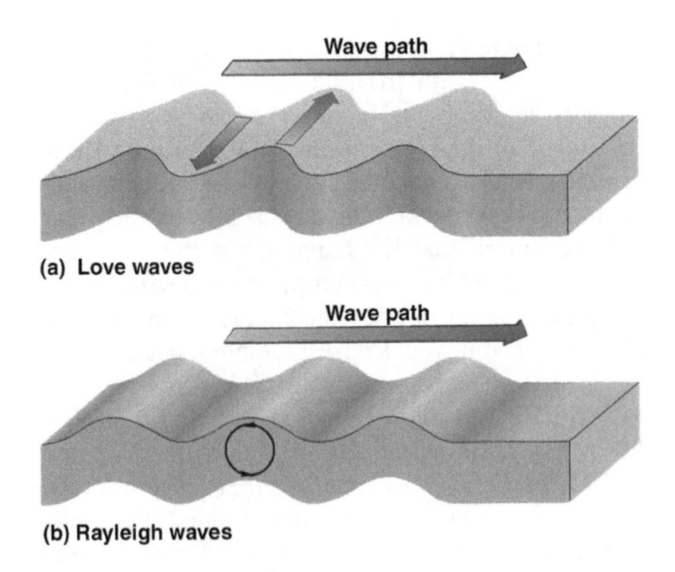

FIGURE 12.12 *There are two types of* surface waves: *(a) the shear-type* Love wave, *which moves the surface in a* horizontal *plane, and (b) the* Rayleigh wave, *which is a combined shear and compression wave that moves the surface in a* rolling *motion opposite to the direction of propagation. Studies have indicated that most of the damage caused by an earthquake is the result of the horizontal motion of the Love waves.*

higher amplitudes are therefore responsible for the destruction resulting from an earthquake.

Engineering studies have shown that the horizontal motion induced by the Love waves is more destructive to structures such as buildings, bridge supports, and dams than the vertical motion caused by the Rayleigh waves. Consider holding a stack of dominoes in your hand. Moving your hand vertically may not disturb the stack, but moving your hand sideways is more than likely to cause the stack to fall.

As we will see later in our discussions of seismology, the magnitude of an earthquake on the Richter Scale is calculated from the largest measured amplitude of the surface waves.

THE SEISMOGRAPH

The **seismograph** is an instrument designed to *record* the movements of Earth's surface generated by a distant earthquake. The Chinese were the first to devise an instrument to *detect* the arrival of a seismic wave. The first true seismic instrument was invented by the Chinese scholar Chang Heng about A.D. 132. The instrument was a large jar with dragon heads evenly spaced around it. An earthquake would cause balanced balls to drop out of a dragon's mouth into the mouth of a waiting copper frog. Assuming that the instrument worked according to plan, the early Chinese seismologists could both detect the arrival of a seismic wave and tell the direction of approach. However, they could tell very little else. The instrument could not, for example, measure or record any information concerning the *amount* of earth movement. Nevertheless, with all of its limitations, many centuries would pass before anyone designed a better instrument.

Consider the difficulty of inventing a *"black box"* that could be set anywhere on the surface of Earth and measure the amount of earth movement. The problem is that as the surface of Earth moves, so does the box. How can the box detect that Earth is moving, let alone the direction and amount of movement?

Any instrument used to measure the movement of another object must possess a component that either (1) moves in a known direction at a known velocity or (2) remains immobile as everything around it moves. It was the latter requirement that was finally satisfied by the first seismograph, which was simply a pendulum suspended from an arm attached to a ringstand. A pen tip extended from the bottom of the pendulum, just barely touching a pad of paper mounted on the base of the ringstand (Figure 12.13). As a seismic wave passed, it moved the base of the ringstand with the attached pad of paper, but the pendulum, because of its large mass and consequent *inertia,* remained *stationary.* The pen attached to the immobile pendulum drew a line on the moving paper that was proportional in length to the amplitude of the seismic wave and was oriented perpendicular to the direction of propagation (Figure 12.14). However, the instrument could not tell from which of the two directions the seismic wave arrived.

The vertical movement of the Earth's surface was measured by a similar instrument in which the pendulum was suspended by a spring rather than by a string (Figure 12.15). With its ability to measure the seismic wave amplitude and determine the direction of the line along which the seismic wave approached, the pendulum seismograph was a definite improvement over the Chinese design. The major shortcoming of both the Chinese and the early pendulum instruments was that neither could record *time*-related parameters. As a result, parameters such as the *frequency* and *time* of arrival of the seismic wave, the *duration* of the quake, or any changes in either wave amplitude or frequency with *time* could not be evaluated.

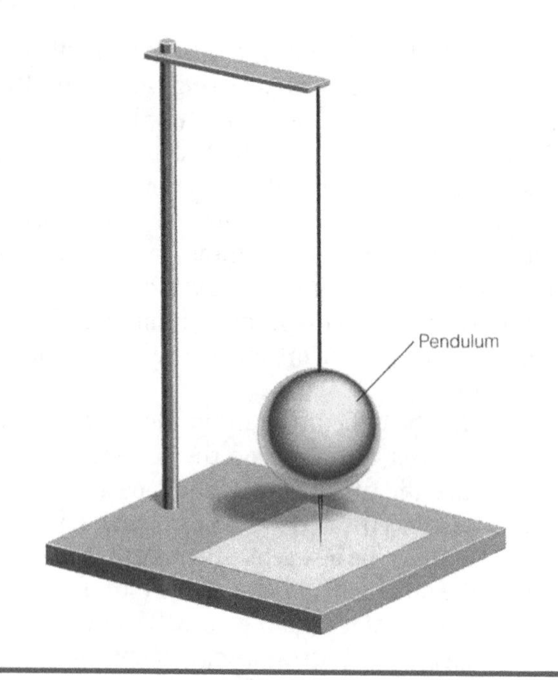

FIGURE 12.13 *The first real improvement on Heng's instrument was a vertically suspended* pendulum. *Because of its* inertia, *the mass of the pendulum would remain stationary as the remaining portion of the instrument was moved by the passing seismic waves. As we will see below, the instrument can generate a* record *of the seismic event called a* seismogram. *Any instrument that generates a record is identified by the suffix* -graph. *As a result, the pendulum instrument is a seismo*graph.

Seismograph design did not undergo another major change until the mid-1800s. John Milne, an English engineer, had gone to Japan shortly after its leaders opened the country to westerners. Milne wanted to make a detailed study of the large number of earthquakes to which Japan is subjected each year but recognized that the hanging pendulum instruments could not record any time-related parameters. He therefore set out to find a way to modify the instrument so it would record temporal data.

Milne's modifications were simple and few. His major design change was to reorient the pendulum horizontally. He placed the pendulum mass at one end of the rigid, horizontal bar that pivoted at the other end, allowing the bar to move in a plane parallel to the base of the instrument. A cylinder was mounted on the instrument base just forward of the pendulum mass with its axis perpendicular to the pendulum bar (Figure 12.16). The cylinder was driven by a clock mechanism that simultaneously rotated the cylinder and moved it parallel to its axis (translation). With the rotation of the drum set at some convenient speed, for example, once per hour, the time at the beginning of each line could be noted and the length of each line subdivided into minutes and seconds.

As a seismic wave moves the base of the instrument and the attached rotating and translating

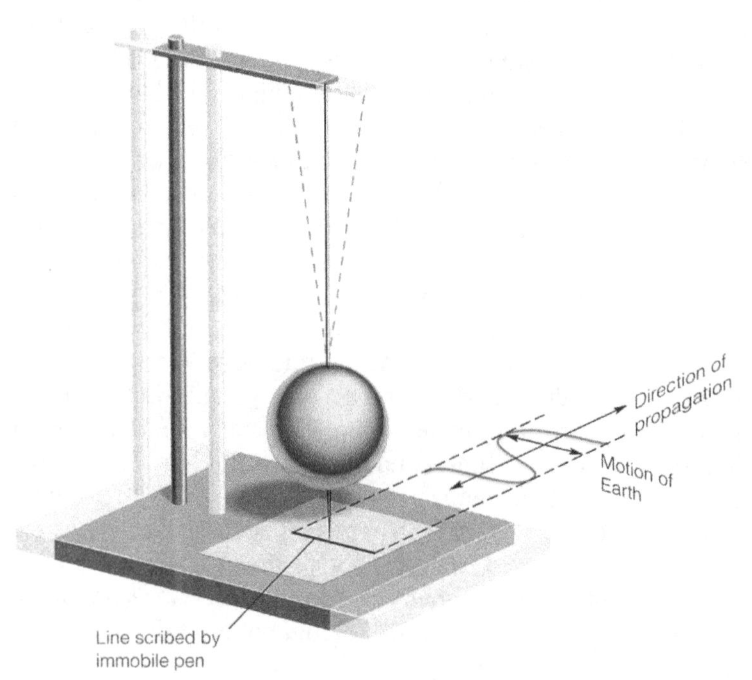

FIGURE 12.14 *As a* Love wave *arrived at the pendulum instrument, the base would be moved horizontally a distance equal to the amplitude of the wave in a direction perpendicular to the direction of propagation. As the base moved, a pen attached to the immobile pendulum drew a line that not only recorded the amplitude of the wave but was oriented perpendicular to the direction of the line of propagation. Note that the wave could have approached in one of two directions. Unfortunately, the instrument is incapable of distinguishing between the two possibilities.*

cylinder moves back and forth, the trace of the pen generates a **seismogram** that records (1) the arrival of the individual types of seismic waves, (2) the change in both wavelength and amplitude of the seismic waves with time, and (3) the end of the quake (Figure 12.17). Although modern instruments are significantly more sensitive and have incorporated all the benefits of modern electronic and computer technologies, the basic instrument design devised by John Milne more than a hundred years ago still underlies each modern seismograph.

Note that a seismic wave approaching the instrument perpendicular to the pendulum bar causes the pendulum to move with the instrument base. Because the pendulum does not remain stationary as the instrument base moves, the instrument cannot record the seismic wave (Figure 12.18). As a result, in order for a seismic station to be able to record seismic waves approaching from all directions, it must

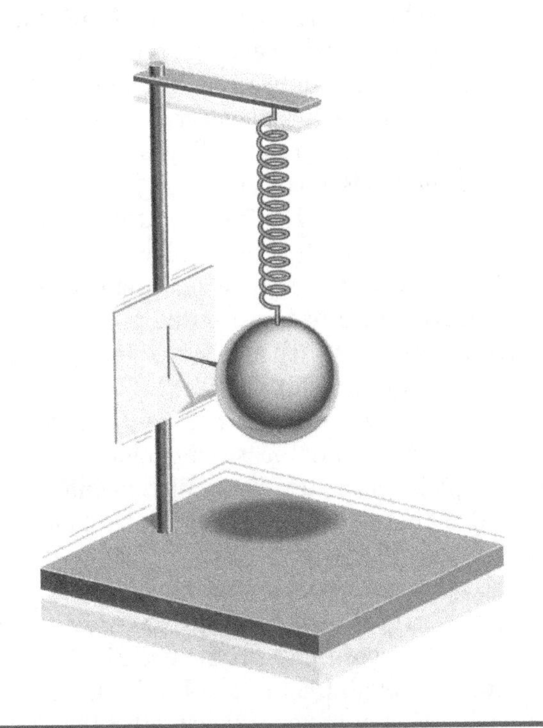

FIGURE 12.15 *The vertical motion imposed by the* Rayleigh *waves is recorded by a similar instrument where the pendulum is suspended by a* spring. *As the instrument is displaced vertically by the wave, a pen attached to the immobile pendulum draws a line whose length records the wave amplitude.*

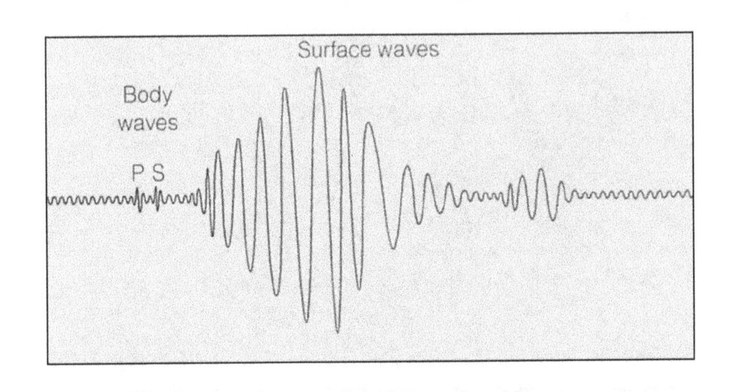

FIGURE 12.17 *Perhaps the most important new capability provided by Milne's improved design was the recording of* time-related *parameters. Because the seismogram is recorded on the surface of each wave, the change of both wavelength and amplitude of each wave with time, and the* end *of the seismic event.*

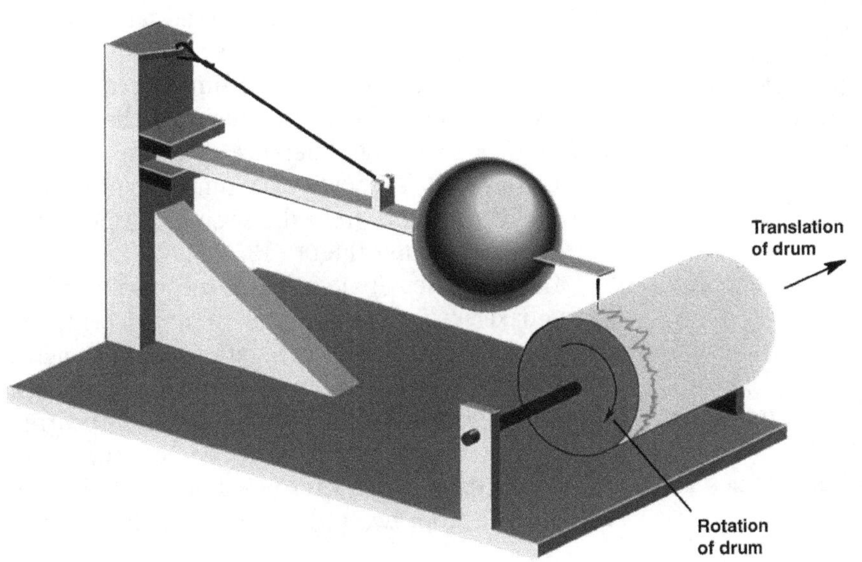

FIGURE 12.16 *John Milne made two major design changes to the seismograph: (a) he mounted the pendulum at the end of a horizontal bar, and (b) he mounted a rotating, translating drum to the base of the instrument on which the pendulum pen rested. Milne's instrument is the basis for all modern seismographs. Subsequent improvements increased the sensitivity of the instrument to allow the recording of all seismic waves including both the low-amplitude body waves and the higher-amplitude surface waves.*

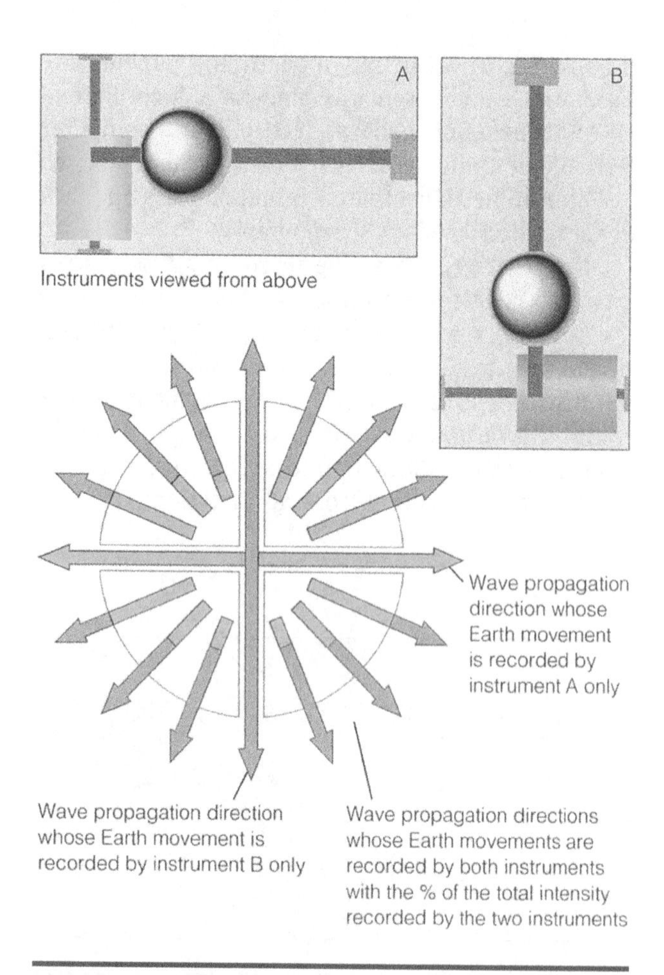

FIGURE 12.18 *A seismic wave approaching a horizontal-pendulum instrument perpendicular to the length of the pendulum bar cannot be detected because the entire instrument, including the pendulum, will be set in motion. Remember that in order for a pendulum instrument to detect movement, the pendulum must remain immobile. As a result, in order to ensure the recording of seismic waves approaching from all directions, two horizontal instruments oriented perpendicular to each other must be employed. Except for waves approaching perpendicular to one of the instruments, waves will be recorded by* both *instruments, with the amplitude recorded for a particular seismic wave depending on the angle of approach.*

have two horizontal pendulum instruments oriented at right angles to each other.

The vertical movements of Earth are still detected and recorded by a spring-mounted vertical pendulum. The only basic difference in design from the earlier seismograph is the inclusion of a rotating, translating drum with its associated electronics to record the earth movement.

LOCATING EARTHQUAKES FROM SEISMIC DATA

Figure 12.19 is an actual seismogram. The seismogram shows the arrival of the body waves with the higher-velocity compressional P wave being recorded first, followed by the shear-type S wave. Note that the amplitudes of both body waves are small. Following the arrival of both body waves, the high-amplitude surface waves arrive with the faster Love waves arriving first, followed by the slower Rayleigh waves.

Assuming this is a typical seismogram, we might ask several questions: (1) Where did the earthquake occur? (2) What was the magnitude of the earthquake? (3) How deep was the focus? Remember that the information recorded on a seismogram describes the seismic waves as they arrive at the seismic station, *not* at the point of origin.

First, let us consider the problem of determining the distance from the seismograph station to the earthquake epicenter. You will remember from Figure 12.4 that both of the body waves follow the same pathways. Consider the specific pathway taken by the seismic waves from the earthquake focus to the seismic station that generated our seismogram. The different velocities with which the two kinds of body waves travel through the rocks of Earth's interior are known with great precision. Knowing that they follow the same path, we can plot time versus distance traveled by the two body waves to produce a **time-distance** (or **travel-time**) curve. A typical time-distance curve, shown in Figure 12.20, graphically illustrates the relationship between the time and distance of travel. In other words, it is a graphic portrayal of the "race" between the P wave and the S wave. As time passes (increases), the distance between the advancing fronts of the two body waves will increase correspondingly. Because every seismogram is a timed recording, it accurately records the arrival times of the P wave and S wave. Once the difference in arrival times of the body waves is known, the time-distance curve provides the distance from the seismic station to the earthquake epicenter (Figure 12.21a).

The next step in locating the epicenter of an earthquake is to draw a circle on a map with the center at the location of the station and the radius equal to the distance determined to the earthquake focus. We now know that the earthquake occurred somewhere on the circle. But where? The exact location is most easily determined by using distance information from two other seismic stations. When

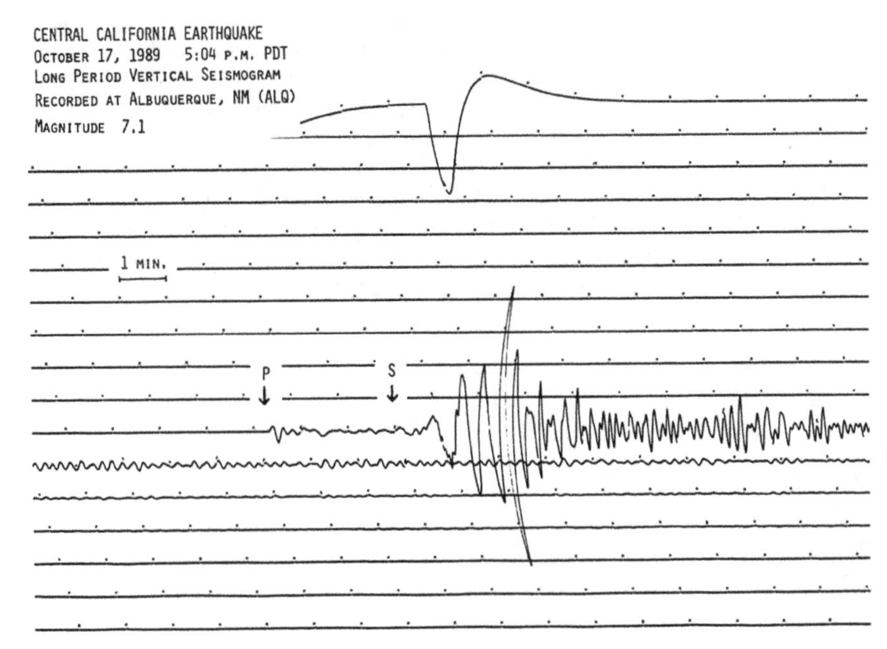

FIGURE 12.19 *An actual seismogram recorded at the Albuquerque, New Mexico seismic station illustrating the arrival of the body and surface waves during the Loma Prieta Earthquake on October 17. 1989. (Courtesy of USGS)*

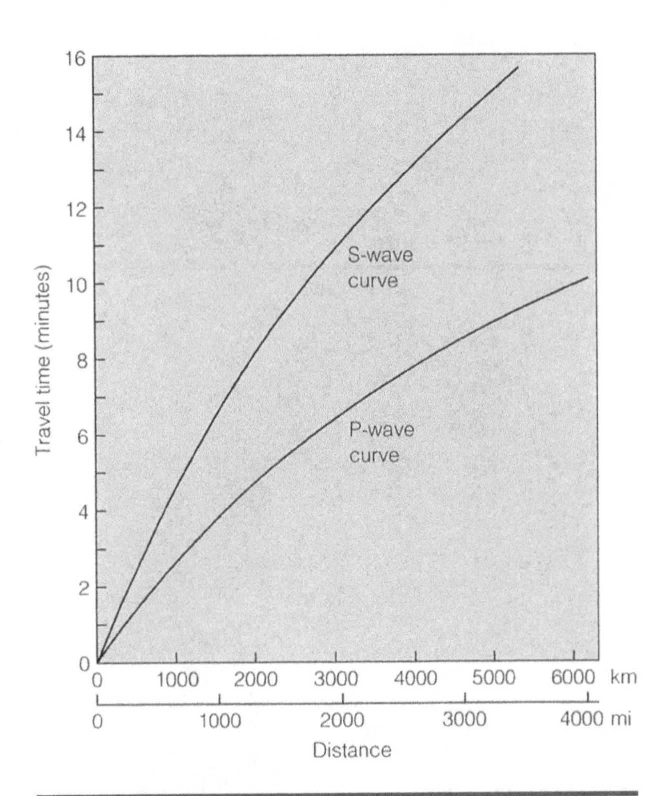

FIGURE 12.20 *Because the P and S body waves travel at different, but known velocities, a graph can be made that portrays the time-distance relationship for each. Called a* time-distance *or* travel-time curve, *the plot is used to determine the distance between the focus of an earthquake and the station recording the event.*

the distance circles determined by three different stations for a specific earthquake are plotted on the same map, the three circles will overlap (refer to Figure 12.21b). The three circles intersect at a single, discrete point on the map, the earthquake focus is located on the surface at the point of intersection (Figure 12.22).

Should the intersection of the three circles produce a spherical triangle as illustrated in Figure 12.23, the epicenter is located at the center of the triangle, and the focus is located some distance below the surface. The depth to the focus is then determined by drawing a large number of station-focus distance data as hemispheres with centers located at the corresponding stations. The point of intersection of all the hemispheres then gives the location of the earthquake focus (Figure 12.24).

Because the amplitudes of the surface waves decrease away from the epicenter, the Richter Scale magnitude of an earthquake at its epicenter can be calculated using the distance data determined from the body wave arrivals and the maximum amplitude of the surface waves recorded on the seismogram. Richter provided equations for the calculation of magnitude, which take into consideration (1) the distance from the earthquake to the seismic station, (2) the kind of rock through which the seismic waves pass, and (3) the physical parameters of the specific seismograph employed at the seismic station. A nomograph

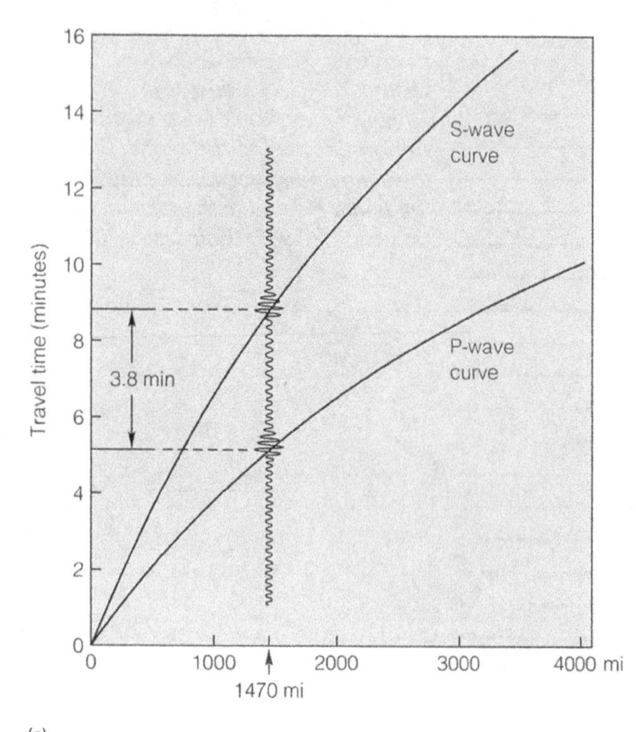

(a)

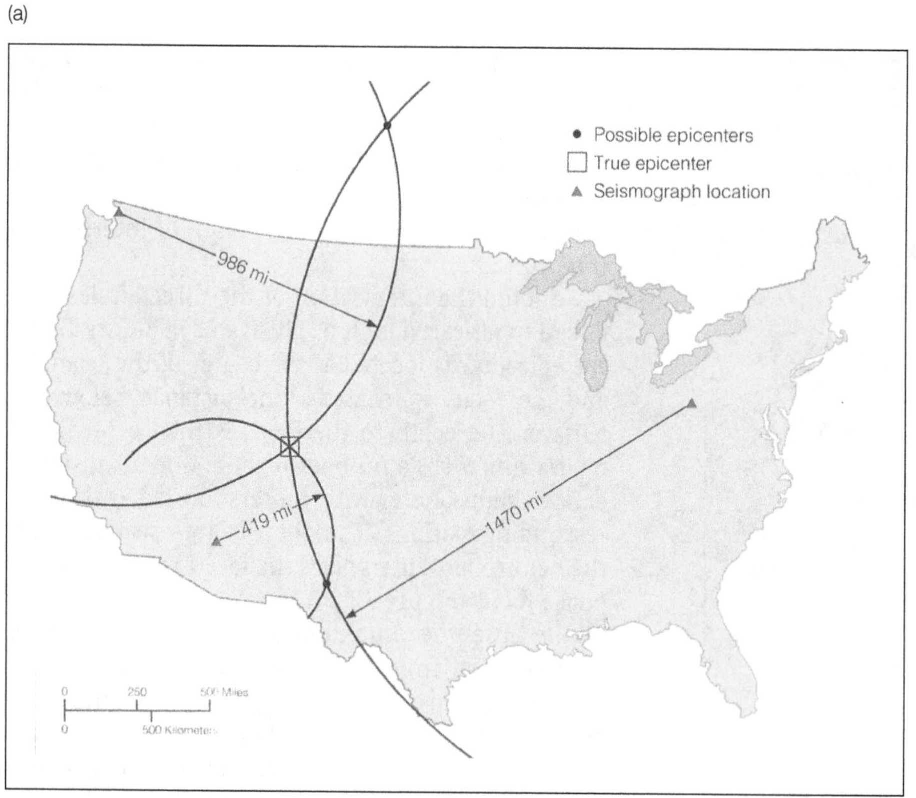

(b) Location of earthquake epicenter by triangulation

FIGURE 12.21 *The location of an earthquake epicenter is accomplished in two steps. (a) First, using the difference in arrival times of the P and S body waves at three different seismic stations, the distance to the epicenter from each station is determined using the time-distance curve shown. With the three distances known, a circle of appropriate radius is then drawn around the map location of each of the three seismic stations (b). The epicenter of the earthquake is located where all three circles intersect.*

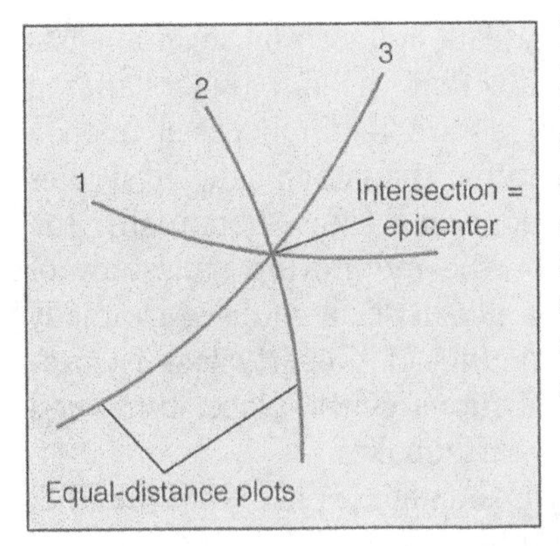

FIGURE 12.22 *The intersection of the three distance circles at a common point indicates that the earthquake focus was located at the surface.*

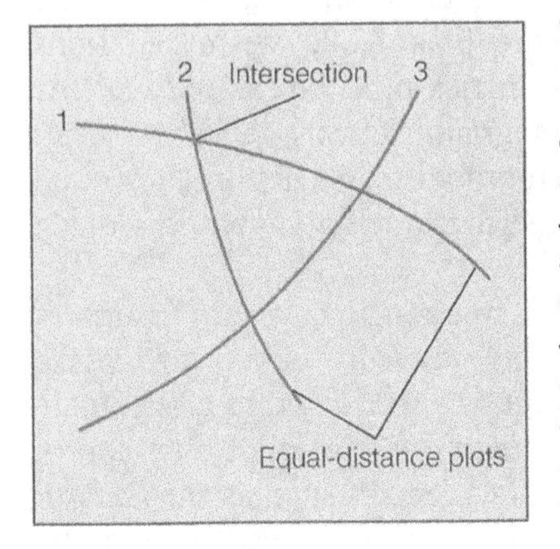

FIGURE 12.23 *The location of an earthquake focus below Earth's surface is indicated when the distance circles from three seismic stations intersect in the form of a spherical triangle*

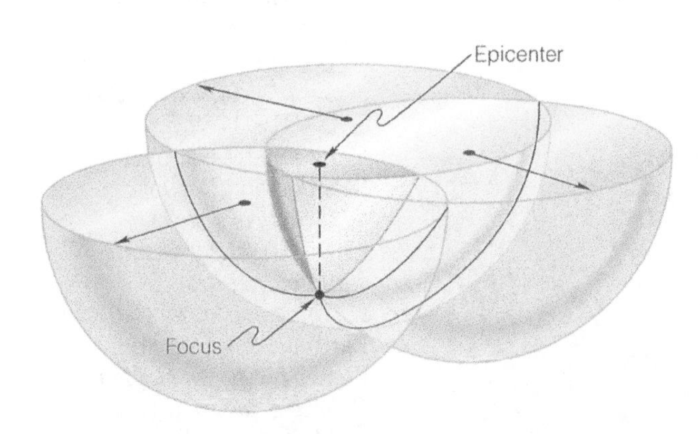

FIGURE 12.24 *In Figure 12.22, the distance was taken as the radius of a circle centered at the seismic station when, in reality, the distance determination provided by the time-distance graph locates the earthquake focus somewhere on the surface of a* hemisphere *centered at the seismic station. The location of the earthquake focus is actually the point of intersection of the three hemispheres, a determination that allows an earthquake to be categorized as shallow, intermediate, or deep focus.*

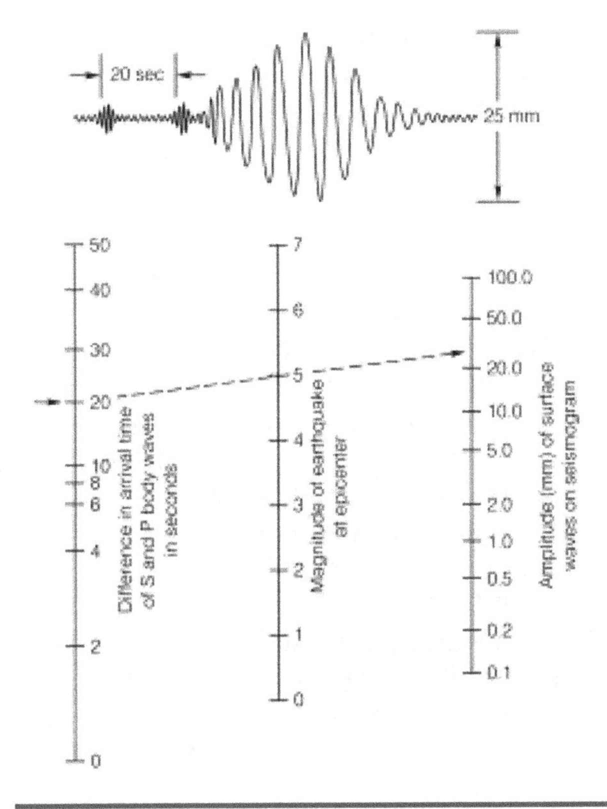

FIGURE 12.25 *Because the* amplitude *of the surface waves at an epicenter is determined by the* magnitude *of the earthquake and* decreases *with* distance *from the epicenter in a regular fashion, a* nomograph *can be used to determine the* magnitude *of the earthquake at the epicenter based on the data recorded on an individual seismogram. A nomograph consists of* three *parallel scales, each graduated for a different variable such that a straight line cutting all three scales intersects at the related values for each variable. Note that the three scales in the nomograph are (1) the difference in arrival times for the body waves (a measure of the* distance *between the epicenter and the seismic station), (2) the* magnitude *of the earthquake at the epicenter, and (3) the* amplitude *of the surface waves on the siesmogram. In our example, by drawing a line that connects the difference in arrival times of the P and S waves (20 seconds) with the amplitude of the surface waves as recorded by our seismograph (25 mm), the nomograph indicates that the magnitude of the earthquake at the epicenter was about 5 on the Richter Scale. (Note that both the arrival times scale and the amplitude scale are logarithmic).*

used to determine the magnitude of an earthquake based upon distance and maximum surface wave amplitude is shown in Figure 20.25.

SPOT REVIEW

1. What are the differences between shear- and compression-type waves?
2. Which seismic waves are responsible for most of the damage resulting from an earthquake?
3. How does the seismograph satisfy the requirements for any instrument designed to measure the movement of another object?
4. How was the seismic instrument designed by John Milne able to record time-related seismic parameters?
5. How are the data recorded on a seismogram used to locate the epicenter, focus, and magnitude of the earthquake?

WORLDWIDE SEISMIC NETWORK

The stresses that build along each plate margin are, at least in part, the result of the combined movement of all the plates. The data amassed by seismologists so far clearly indicate that the full understanding of earthquakes will require research based upon a worldwide database rather than upon detailed studies of specific earthquake-prone areas. Potentially, the knowledge gained by the monitoring and study of seismic activity on a global scale could spare many lives and untold amounts of property loss by providing accurate predictions of the place, time, and magnitude of future earthquakes.

To provide such a database, in 1966 the U.S. Department of the Interior established the **National Earthquake Information Center (NEIC)** at Golden, Colorado, which collects seismic data from all over the world. More than 3,300 seismic stations from 80 foreign countries plus 120 stations of the **U.S. National Seismograph Network (USNSN)**, originally established to monitor Soviet underground nuclear testing, send data to the center in Golden (Figure 12.26).

The center's first mission is to obtain estimates of the magnitude and intensity of an earthquake occurring anywhere in the world and to disseminate the information as quickly as possible to the appropriate federal and state agencies such as the *Federal Emergency Management Agency (FEMA)* and the tsunami early warning system. The center also noti-

FIGURE 12.26 *The* United States National Seismograph Network *(USNSN) consists of 120 cooperating seismic stations that provide seismic data to the* National Earthquake Information Center *(NEIC) at Golden, Colorado, where they are combined with data from more than 3,300 stations worldwide. The center not only conducts research to improve our ability to detect and locate earthquakes worldwide, but also is closely associated with organizations whose mission is to provide disaster relief to areas struck by major earthquakes anywhere in the world.*

fies agencies responsible for providing disaster relief assistance throughout the world.

The second mission of the NEIC is to establish and maintain the national and global network of seismic stations and to serve as a data collection and dissemination center. The data collected by the NEIC are available to anyone.

The third mission is to conduct research that will not only improve the ability to detect and locate earthquakes but will also allow the development of a potential to forecast the occurrence of earthquakes worldwide. Ultimately, the goal of the NEIC is to reduce the incidence of earthquake hazards to all humans.

EARTHQUAKE PREDICTION

Since a major earthquake is capable of destroying an entire city and killing hundreds of thousands of people within a matter of minutes, the ability to accurately predict the time, location, and magnitude of an impending earthquake would be of immense value. Unfortunately, the prediction of earthquakes suffers the same shortcomings as those previously outlined for the prediction of volcanic eruptions. Historic and prehistoric data can be assembled to allow **long-term prediction**; that is, that a major earthquake will take place sometime within the next hundred to a few *hundred thousand* years. Most people responsible for disaster planning and most private individuals find long-range predictions to be of limited use. At the other extreme is **short-term prediction** where the earthquake can be accurately predicted to within days or hours of its actual time of occurrence. In 1975, Chinese scientists used a series of low-magnitude foreshocks (up to a magnitude of 3) to make an accurate short-range prediction of a 7.5 magnitude earthquake that enabled thousands of people to be saved in a city of 90,000 that was nearly destroyed. When the same set of criteria were applied a year later, they failed to predict the Tangshan earthquake that killed several hundred thousand people. It appears that accurate short-range earthquake prediction is still a long way off.

Seismologists have had some success in **medium-term prediction,** which establishes the time of occurrence of an earthquake within a window of a few years to a few months. Medium-range prediction is based on identifying a series of precursory events that have been shown to precede a major earthquake. For example, several major earthquakes have been preceded by a series of low-magnitude foreshocks. Such relationships are then used to devise an empirical model that evaluates the predictive potential of each parameter. The sum total of all events is then applied to a fault zone to predict future fault movements.

One criterion is the historical earthquake data for the fault zone, in particular, the identification of **seismic gaps** or segments of the fault where no recent earthquake activity has occurred. Because these zones have not experienced recent fault movement, stress is still accumulating within the rocks, and, as a result, they are considered to be the most likely sites of future earthquakes. Another observation shown to precede major earthquakes is an as yet unexplained increase in radon gas within deep water wells in or near the fault zone.

One of the most widespread events observed to precede some major fault movements (and other natural disasters) is the erratic behavior of animals. Animals have been observed to run in random patterns, dogs to howl, and burrowing animals to emerge from underground. According to some accounts, the reported use of foreshocks by the Chinese to predict the 1975 earthquake was coincidental, and the prediction was actually based on the behavior of hibernating animals who left their burrows in December (the month the earthquake occurred). Although scientists have had some successes in medium-term prediction, the existing models are of limited use.

With no accurate predictive model in the offing, much effort has been spent in attempts to minimize the destruction of life and property. One area of active research is the design of buildings and other structures to withstand moderate-magnitude earthquake shocks, in particular, the horizontal motions that have been shown to be especially destructive. Based on the findings associated with the Mexico City earthquake, engineers and architects are attempting to design buildings whose inherent vibrational frequency will fall outside the normal range of vibrational frequencies induced by seismic waves.

The basic concepts of land-use planning have been applied to restrict the construction of certain kinds of structures such as homes, schools, and hospitals within active fault zones to minimize injury and loss of life. Storage facilities for dangerous materials such as oil, caustic chemicals, and liquefied gas are excluded from zones of potentially high earth movement. In planning future construction, geologic maps are used extensively to delineate areas underlain by unconsolidated materials as well as areas underlain by bedrock. Many cities located in earthquake-prone areas have established disaster relief systems to alert citizens to potential dangers as well as to be responsible for implementing various procedures such as evacuation, fire fighting, and the distribution of medical help and food supplies.

Through it all, the inhabitants of an earthquake-prone area must be educated to the potential dangers. Some people are not aware of the potential threat. Many individuals are aware but not impressed with a model that predicts that a major earthquake will occur within the next half-century. Such people apparently do not consider an event that may occur only once within two or three generations to be a real threat. Also, there will always be a few individuals who *are* aware of the dangers, *know* that a major earthquake will most likely occur within their lifetime, and yet either choose to stay or, unfortunately, have no choice but to stay.

SPOT REVIEW

1. What parameters are required for the true prediction of any future natural disaster? With our present understanding of earthquakes, how precisely can these parameters be evaluated?
2. What kinds of efforts are currently being taken to minimize the death and destruction that may arise from future earthquakes?

CONCEPTS AND TERMS TO REMEMBER

Distribution of earthquakes
- Plate margins

Frequency and location of earthquakes
- seismic waves
- subduction zones
  - Benioff Zone
- Focus or hypocenter
  - shallow, intermediate, and deep
- epicenter

Measurement of earthquakes
- intensity (damage)
  - Modified Mercalli Scale
- magnitude
  - earth movements and energy released
  - Richter Scale
  - seismograph

Earthquake damage
- tsunami
  - Seismic Sea Wave Warning System (SSWWS)

Seismic waves
- kinds of waves
  - shear
  - compressional-extensional (compression)
- body waves
  - S and P waves
- surface waves
  - Love (LQ) waves
  - Rayleigh (LR) waves

Seismology
- seismograph
  - seismogram
    - time-distance (travel-time) curve

World seismic network
- National Earthquake Information Center (NEIC)
- U.S. National Seismograph Network (USNSN)

Prediction
- long-, medium-, and short-term prediction
- seismic gaps

REVIEW QUESTIONS

1. Most earthquakes and nearly all major earthquakes are located
   a. along the oceanic ridges.
   b. along the zones of subduction.
   c. at points where major masses of magma are moving within the underlying crust.
   d. in association with oceanic hot spots.
2. The most intense earthquake in North America within historic time occurred at
   a. San Francisco, California, in 1906.
   b. Los Angeles, California, in 1847.
   c. Boston, Massachusetts, in 1636.
   d. New Madrid, Missouri, in 1811.
3. Each higher step in the Richter Scale represents a ______-fold increase in energy released.
   a. 10 b. 50
   c. 30 d. 100
4. A tsunami is a(an)
   a. surface shock wave large enough in amplitude to result in an earthquake of at least 8.0 on the Richter Scale.
   b. earthquake body wave transmitted totally within the crust.
   c. seawave generated by an earthquake located within or near an ocean basin.
   d. major earth movement such as a landslide or rock fall resulting from an earthquake.
5. The earthquake shock wave that causes the most damage is the
   a. Love surface wave.
   b. Rayleigh surface wave.
   c. P body wave.
   d. S body wave.
6. The distance from an earthquake focus to a seismic station is determined by the
   a. difference in the arrival times of the body waves.
   b. arrival time of the Love surface wave.
   c. difference in the arrival times of the body and surface waves.
   d. arrival time of the P body wave.
7. Why is the amount of earthquake energy released from zones of subduction much greater than that released from oceanic spreading centers (oceanic ridges)?
8. Why don't body waves contribute to the damage caused by an earthquake?
9. An earthquake is the result of what kind of strain?
10. Why can compression-type shock waves be transmitted through any medium while shear-type shock waves can only be conducted through solids?
11. Why must seismic stations have at least two horizontal pendulum instruments to record the arrival of earthquake shock waves?
12. How are the times of arrival of earthquake body waves used to determine the distance from the earthquake focus to the seismic station?

THOUGHT PROBLEMS

1. If any instrument designed to measure the movement of another object requires a part that either remains immobile while the rest of the instrument moves or that is moving in a known direction at a known speed, how does the speedometer in an automobile work? How is the airspeed of an airplane determined by an on-board instrument?
2. Individual faults experience successive movements as the rocks on opposite sides of the fault surface "lock up," allowing stresses to rebuild. If the rocks are prevented from locking up, successive faulting, and subsequent earthquakes, will not occur. Is it feasible to prevent lockup and, if so, how could it be done?

FOR YOUR NOTEBOOK

Since most of us do not live in active earthquake areas, our information about recent earthquakes is obtained from the media. Using information reported in newspapers and weekly periodicals, compile information about the magnitude and intensity of some of the major earthquakes that have occurred during the past 10 or so years. Compare the geological settings, impact on the inhabitants of the region, and extent of damage. For sequences of earthquakes in a given area, establish the frequency of earth movement. Note in particular instances where earthquake prevention methods, or their lack, were involved and especially where particular structures either survived or failed to survive the earthquake.

ESS 101

Lab 11: Earthquakes along San Andreas Fault

Student Name: ____________________

1. What type of fault is the San Andreas Fault?

____

____

2. What type of the plate boundary is the San Andreas Fault?

____

____

3. How old is the fault?

____

____

4. It is the boundary between what plates?

____

____

5. In what direction do plates move relative to each other?

____

____

6. How were the ancient earthquakes that occurred along the fault dated?

____

____

7. How fast is the land moving along the fault? Why is it important to know?

____

____

8. Many earthquakes occur daily and annually along the fault, but not in the town of Hollister which is located along the San Andreas Fault. Why are there no earthquakes, but evidence of strain and deformation of land movement?

____

____

9. How do geologists predict earthquakes?

____

____

10. Why do geologist believe that the Los Angeles area is due for a major earthquake within the next thirty years?

____

____

11. How does local geology (type of rocks) influence the amount of damage expected from future earthquakes in the Los Angeles area?

____

____

12. Why was the damage of the 1906 San Francisco earthquake so much out of proportion?

____

____

13. Do you want to move to Los Angeles?

____

____

CHAPTER 13

Mountain Building

INTRODUCTION No geologic feature captures the imagination quite like mountains. Many people are attracted by their overwhelming beauty. Who would not be impressed by the towering peaks of the Tetons or the majesty of Fujiyama? Others are attracted by the serenity where they can escape from the urban turmoil and the drone of city life. Still others are attracted to the danger that a mountain may present and risk their lives to climb sheer rock faces to feel the exhilaration of triumph as they stand literally on top of the world.

Although most geologists are not willing to risk their lives scaling a sheer rock face, they still are attracted to mountains. According to an old saying, the best geologist is the one who has seen the most rocks. The simple fact is that more rocks are exposed in mountains than anywhere else on Earth. You will also remember from earlier discussions that one of the major objectives of geology is to decipher the history of Earth. Because of the extensive exposures of rock, the greatest segments of Earth's history are to be found recorded in the mountains.

Some geologists are particularly interested in studying and observing basic geologic processes, many of which operate at accelerated rates in the mountains. Relentless freeze-thaw cycles rip at the rocks and transform them into the fragments that commonly litter the ground. Glaciers ply their methodical trade and sculpt mountain peaks and ridges to breathtaking rugged beauty. Turbulent mountain streams crash down steep slopes, carving their channels ever deeper as they attempt to undo the process of mountain building.

Everywhere in the mountains you are surrounded by the feeling of power. Deformed rocks and lofty peaks attest to the enormous forces that brought the mountains into being, while all around the energy of the basic erosional processes works to return the rocks to the sea.

Mountains are also the source of many of the world's mineral deposits.

DEFINITION OF A MOUNTAIN

How does one define a mountain? Someone born and raised in the shadow of the Alps or the Rocky Mountains might wonder how the low ridges of the Appalachians could possibly qualify as mountains. Similarly, someone raised in Appalachia might wonder why a small hill that barely rises a few hundred feet above the flat coastal plain of North Carolina is called "Mount." Nevertheless, that is indeed the definition of a mountain: "a part of Earth's crust sufficiently elevated above the surrounding land surface to be considered worthy of a distinctive name." What "sufficiently elevated" means in terms of feet or meters is not indicated. In like fashion, a mountain chain is defined as "a series of mountains whose bases are continuous." Again, there is no indication of how high the mountain crests must be. Obviously, what constitutes a mountain or mountain chain is in the eye of the beholder. Generally, however, most geologists restrict the term *mountain* to a topographic feature that projects at least 1,000 feet (300 m) above the surrounding land. Some elevated features such as buttes and mesas are usually excluded. It is interesting to note that the government of Nepal restricts the term *mountain* to those topographic features higher than 26,000 feet (8,000 m) above sea level.

Because the definition of mountains is so general, one would expect the classification of mountains to be a difficult task. Actually, mountains can be grouped, at least in general terms, according to their mode of origin and internal structure. Their size or surface appearance is really of little concern. Nevertheless, as we attempt to classify mountains in the discussion that follows, keep in mind that although one mountain range may share some basic qualities with another, each range has a unique geographic and geologic setting and history.

KINDS OF MOUNTAINS

Mountains are the result of the application of enormous amounts of energy acting on and within Earth's crust. Energy sufficient to result in the formation of mountains is available from three different sources: (1) **volcanism** resulting from thermal energy associated with rising asthenospheric plumes and mantle convection cells; (2) **epeirogenic forces** (Greek *epeiros* = continent) involving either (a) local or regional upwarping of Earth's crust *without* faulting or (b) uplift accompanied by extension and normal faulting of segments of Earth's crust; and (3) **orogenic forces** (Greek *oros* = mountains), which involve horizontal compressive forces generated by the convergence of lithospheric plates. The term **diastrophism** (Greek *diastrophe* = twisting or distortion) refers to all tectonically induced crustal movements and includes both epeirogenic and orogenic forces.

The combined effects of these forces result in the creation of four basically different kinds of mountains: (1) *volcanic mountains,* (2) *domal mountains,* (3) *block-fault* (or *fault block*) *mountains,* and (4) *foldbelt mountains.*

Volcanic Mountains

Volcanic mountains are found either in areas where rising mantle convection cells or mantle plumes impinge upon the bottom of the lithosphere or in areas where the thermal energy is being released within zones of subduction. Fujiyama in Japan and Mount Vesuvius in Italy are well-known examples of individual volcanic mountains, while the Cascade Mountains, the Aleutian Islands, and portions of the Andes Mountains are examples of volcanic mountain ranges.

Continental Rift Zones Upwelling mantle convection cells located beneath continental lithosphere initiate the **rifting** of the lithospheric plate. As the fractures develop, volcanism begins, first in the form of hot springs and gaseous emanations; later these are followed by outpourings of lava and finally by the building of cones. An example of rift zone volcanism in the United States is the Rio Grande Rift Zone extending from the Mexico-New Mexico border to Colorado. Cinder cones, small lava cones, and lava flows are common throughout the rift zone. One of the world's most spectacular volcanoes, Mount Kilimanjaro, is located along the East African Rift Valley.

Oceanic Ridges In the past, **oceanic ridges** were not included in discussions of mountains mainly because they are not on land. Since the recognition of the concept of plate tectonics, however, and our increased understanding of the ocean bottom, fewer geologists are making a distinction between elevated features of the continental crust and those of the oceanic crust.

Encompassing more than 40,000 miles (65,000 km) in length, the oceanic ridges are the

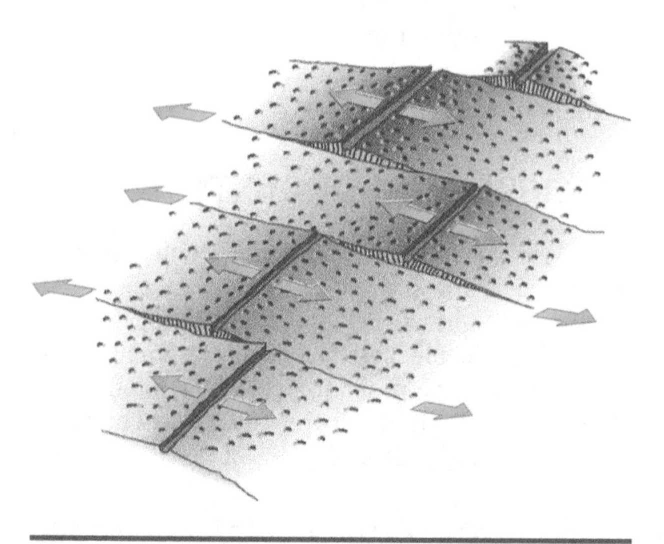

FIGURE 13.1 *Unlike continental mountain ranges with their steep, often precipitous slopes,* oceanic ridges *are very broad structures with gentle outer slopes.*

longest mountain ranges on Earth. They are composed of basalt erupted along oceanic rift zones. Along most oceanic ridges, the summit area is downfaulted forming a long, narrow valley similar in structural form to continental rift valleys. The oceanic ridges are mostly submarine features, but isolated segments do occasionally come to the surface. The best-known example, Iceland, is part of the Atlantic Oceanic Ridge and is emergent only because additional magma is being provided by a hot spot under the oceanic ridge.

For the most part, oceanic ridges do not exhibit the steep slopes characteristic of most terrestrial mountain ranges but are rather broad ranges with gentle outer slopes. In some areas where the height of the ridge is less than 2 miles (3 km), the ridges may be more than 1,000 miles (1,600 km) wide (Figure 13.1).

Hot Spot Volcanoes Localized mantle plumes rising beneath oceanic lithosphere result in the generation of **shield volcanoes** building up from the

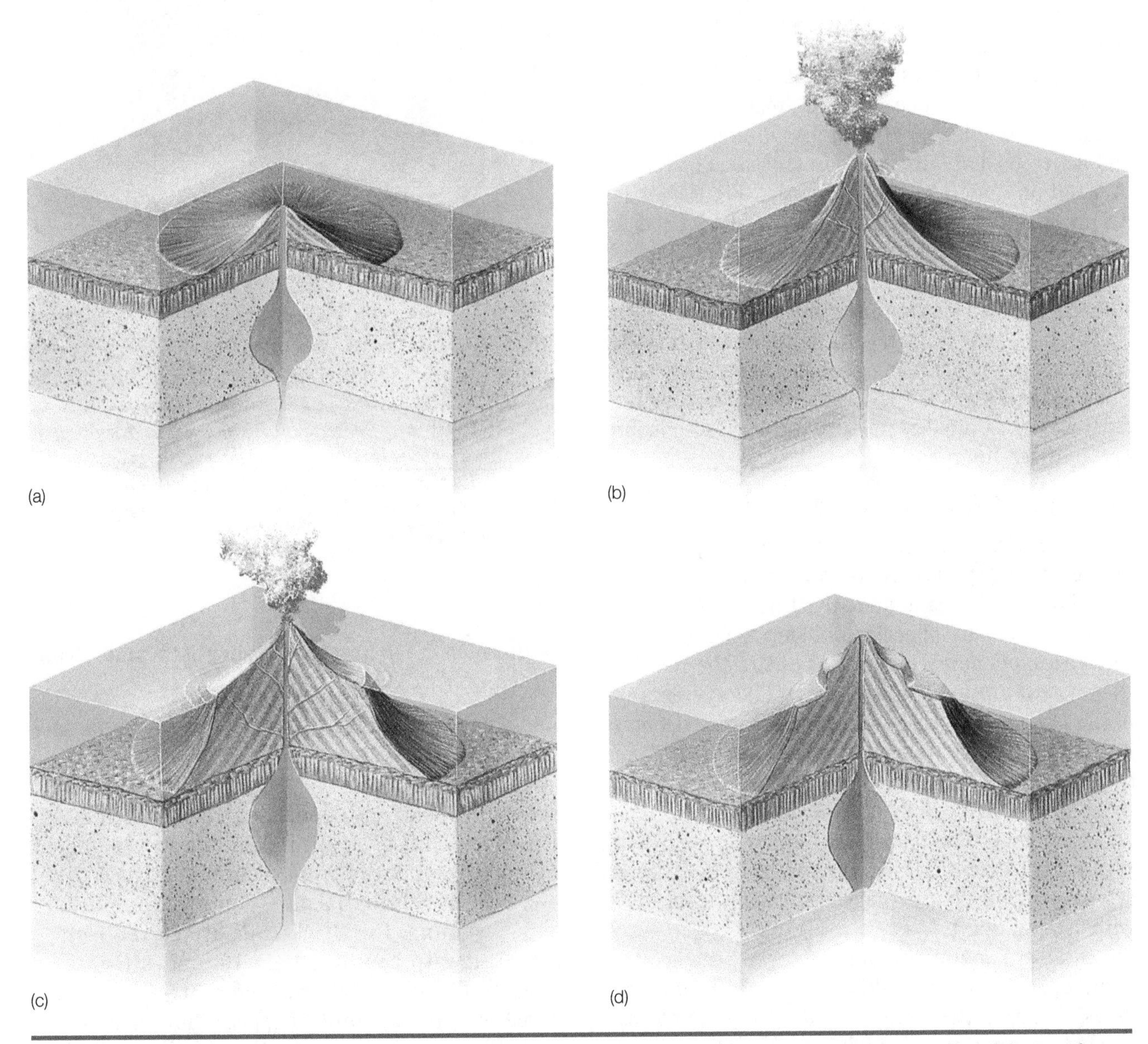

FIGURE 13.2 Shield volcanoes, *formed over stationary hot spots, dot the ocean basins, especially in the Pacific Ocean. Many, called* seamounts, *are submarine while others, such as the* Hawaiian Islands, *build above sea level, only to be eroded and, in time, sink below sea level to once again become seamounts or* guyots.

ocean floor (Figure 13.2). Once the rising magma penetrates the lithosphere, the volcanic cone builds rapidly. In some cases, the peaks remain submarine and are called seamounts. In other cases, the peaks build above sea level to form volcanic islands such as the Hawaiian Islands. As the hot spot goes through as yet unexplained cycles of activity and nonactivity, chains of shield volcanoes—some islands, others seamounts and guyots—form along the ocean floor, aligned in the direction of plate movement.

Subduction Zone Volcanoes The volcanic mountains that form on the edge of the overthrust plate at convergent plate boundaries differ from oceanic ridge and hot spot volcanoes in that they are primarily constructed of andesitic lavas although basalts and rhyolites also occur. Because silicic magmas are more viscous, subduction zone volcanism is often explosive and generates large volumes of pyroclastic materials in addition to outpourings of lava (Figure 13.3). Accumulations of alternating lava and pyroclastic layers construct the stratovolcanoes (composite volcanoes). These volcanoes have steeper slope angles than the shield volcanoes associated with hot spots. Subduction zone volcanoes occur either as **island arcs,** such as the Aleutian or Tonga islands, or as **continental arcs,** such as the Cascade Mountains or the northwestern United States and the western portion of the Andes Mountains.

FIGURE 13.3 *Because of the higher viscosity of the andesitic magmas involved, volcanic eruptions associated with subduction zones are invariably violent and expel large volumes of pyroclastic materials. (Courtesy of T. J. Casaderall/USGS)*

Domal and Block-Fault Mountains

Of the various kinds of mountains, those produced by the epeirogenic forces are in many ways the simplest both geologically and structurally. Mountains forming as a result of epeirogenic forces fall into two categories: (1) **domal mountains** where the relief is the result of erosion following either (a) *local doming* of the continental crust, where the affected area may be a hundred or so miles in diameter, or (b) *regional doming,* where large portions of the continental crust are affected by broad, *regional upwarping;* and (2) **block-fault** (or **fault-block**) **mountains,** which result from *extension* and normal faulting of large segments of the continental crust, usually associated with vertical uplift.

Of all the mountain-building forces, the epeirogenic uplift is the most difficult to explain. In many cases, because the regions undergoing uplift are located far from plate boundaries, the forces do not appear to be directly associated with plate tectonics. Some geologists suggest that the forces originate when heat accumulates under the lithosphere, resulting in the increased buoyancy of the underlying asthenospheric rocks.

Local Doming Except for high-angle faults and monoclinal folds that may develop along the margins of the domal uplift, the upwarping usually occurs without any associated rock deformation. The Black Hills of South Dakota are an example of mountains produced by local doming (Figure 13.4). Located on South Dakota's southwestern border with Wyoming, the Black Hills are carved from a domal upwarp approximately 100 miles (160 km) long by 50 miles (80 km) wide. Streams rejuvenated by the doming stripped the sedimentary rocks from the central portion of the up-arched crystalline continental core, producing monoclinal ridges of sedimentary rocks that dip away under the adjoining Great Plains.

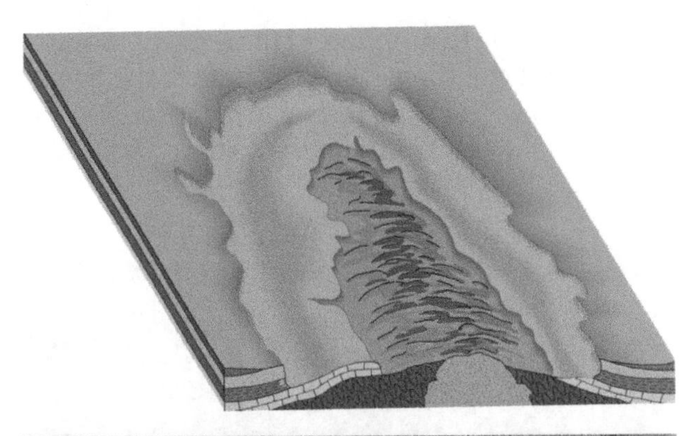

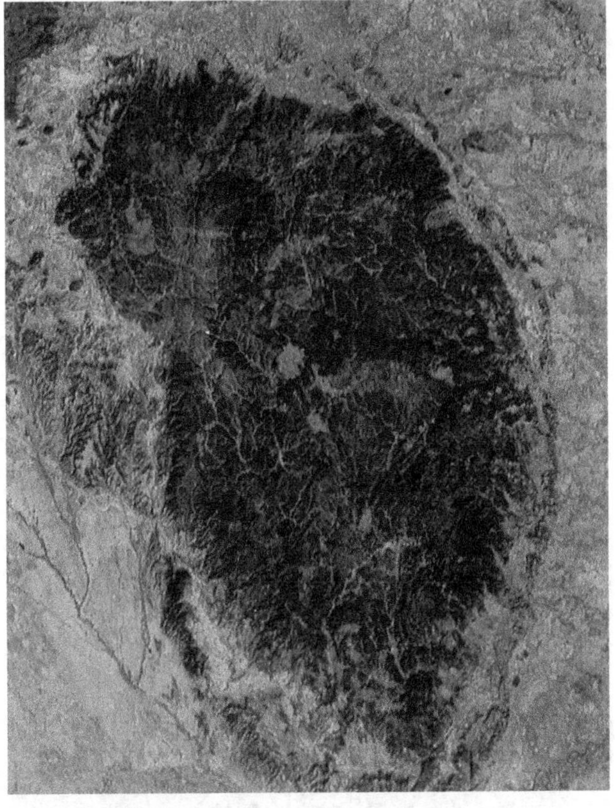

FIGURE 13.4 Local doming *followed by erosion is responsible for the* Black Hills *on the border between South Dakota and Wyoming. The major site of gold production in the United States until the discovery of the California deposits, the Black Hills are now best known for the location of Mount Rushmore on which the faces of several past presidents have been carved. (Photo courtesy of USGS)*

Geologists still do not agree on the source of the energy that created the Black Hills doming. Some argue that the uplift was part of the uplift and later Tertiary rejuvenation of the northern Rocky Mountains to the west.

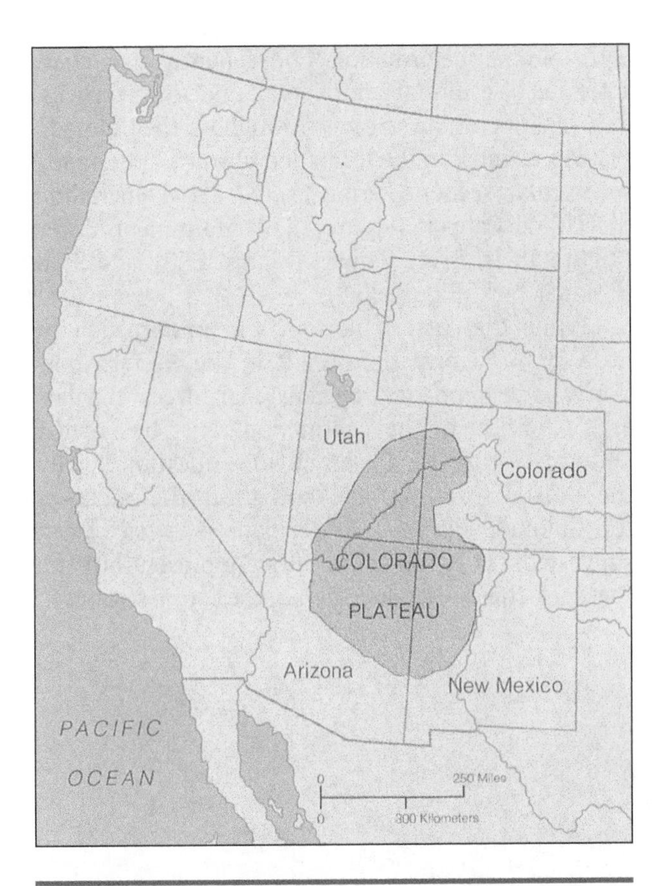

FIGURE 13.5 *One of the most spectacular results of epeirogenic forces in the United States is the* Colorado Plateau, *which has been elevated more than a mile. Erosion of the uplifted rocks has given us many of the awe-inspiring vistas that grace the Southwest.*

Regional Doming Epeirogenic forces are also responsible for the uplift of large portions of lithosphere. An example is the uplift that affected a large area in the southwestern United States centered over the state of Utah and including parts of Colorado, New Mexico, and Arizona (Figure 13.5). Once again, the origin of the epeirogenic forces is not clear. Some geologists believe the upwarping was due to a mantle plume hot spot located beneath the North American lithosphere. Others believe that the uplift was the result of the North American lithosphere overrunning the Pacific spreading center. Whatever the source of the forces, the total uplift of the region amounts to about 1 to 1.2 miles (1.5–2 km) with most of the uplift occurring during the last 15 to 20 million years. Except for some normal faulting and monoclinal folding that developed along the margins, the eastern portion of the uplift, referred to as the Colorado Plateau, was elevated to its present level with minimal deformation. Consequently, it is characterized by essentially horizontal, undeformed rocks. Rejuvenation of the streams throughout the Colorado Plateau resulted in the formation of some of the most spectacular scenery in the United States including the Grand Canyon (Figure 13.6), Monument Valley (Figure 13.7), Bryce Canyon (Figure 13.8), and Zion National Park (Figure 13.9).

While the eastern portion was upwarped with little or no deformation, uplift in the western portion was accompanied by tensional forces and lateral extension of the continental crust by normal faulting (Figure 13.10). An obvious question is why the crustal extension did not affect the rocks of the Colorado Plateau. Two explanations have been suggested: (1) tensional forces were present but the rocks of the Colorado Plateau were too strong to break, or (2) the temperature of the underlying mantle rocks was not high enough to initiate the lateral movement necessary to produce the tensional forces. Whatever the explanation, the crustal extension of the western portion of the uplift resulted in the

FIGURE 13.7 Monument Valley *is truly a monument to the ability of desert streams to erode and remove enormous volumes of rock. The mesas, buttes, and needles that dot the landscape are the remnants of the sandstone body that once extended throughout the area. (Courtesy of E. D. McKee/USGS)*

FIGURE 13.6 *Certainly one of the best-known results of the rejuvenated erosion of the uplifted Colorado Plateau is the* Grand Canyon *of the Colorado River, which has been incised a mile into the rocks of the plateau. The Colorado River is one of the few rivers of the world that flows completely through a desert region to the sea. (Courtesy of E. D. McKee/USGS)*

FIGURE 13.8 *To be truly appreciated, the beauty of* Bryce Canyon *must be experienced firsthand. In the light of the rising or the setting Sun, the colors of the rocks are magnificent.*

development of north-south trending normal faults and the subsequent development of the *block-fault mountains* that characterize the area.

Block-Fault Mountains The faulting associated with block-fault mountains occurs in two different styles that are found together (refer to Figure 13.10). In the first, the strikes of the fault planes are *parallel,* and the fault planes dip in the *same* direction. Because of the listric nature of the faults, the relative motion results in a rotation of the block between each set of parallel faults. As one edge of the block moves *upward* to form the edge of a ridge, the other edge of the block moves *downward* to form a parallel valley. The steeper flank of the ridge is the **scarp** of the exposed fault plane.

In the second tectonic style, the strikes of the fault planes are *parallel,* but the dip directions of the fault planes *alternate.* The relative motion now results in the formation of uplifted blocks called **horsts** and downthrown blocks called **grabens.** Note

FIGURE 13.9 *The sandstones within the soaring cliffs of* Zion National Park *record the presence of an ancient desert and the dunes that swept across its surface.*

that, in this case, both flanks of the ridges are occupied by fault plane scarps.

In western North America, the combination of both faulting styles has given rise to a particularly rugged topography. Because of its unique geology, the region has been designated as a separate geological province called the **Basin and Range.** The Basin and Range Province extends from southernmost Oregon to Mexico with the best development of the structural style visible in Nevada. A quick scan of a Nevada road map reveals one effect of the Basin and Range topography. Because of the northeast-southwest trending block-fault mountains, east-west roadways are scarce.

Difficult though east-west travel within the Basin and Range is today, crossing Nevada was even harder for the early pioneers who found the parallel ridges and adjoining desert valleys to be impenetrable barriers. Westward travel to the Pacific coast followed either a southerly route across Arizona into southern California, roughly the route of old U.S. Route 66, or a more northerly route, used today by the railroad, from Denver, Colorado, to Salt Lake City, Utah, through Reno, Nevada, and across the Sierra Nevada to the Pacific coast. Reno, Nevada, developed as a waiting station for the seasonal crossing of the high Sierra. The crossing is now made through the Donner Pass, named after the Donner party who attempted a late crossing and were stranded by an early snowfall. Most of the party perished in the intense cold of a high Sierra winter.

Foldbelt Mountains

Foldbelt mountains are generated at convergent plate margins. The folds and faults that characterize foldbelt mountains are typically those associated with horizontal compressive stress.

Foldbelt mountains usually consist of two parallel or subparallel linear components: (1) a complex core composed of deformed igneous, metamorphic, and volcanic rocks and (2) an adjoining system of folded and faulted sedimentary rocks (Figure 13.11). A chain of volcanic mountains usually is associated with the complex core either in the form of an island arc separated from the mainland by a shallow sea or as a continental arc paralleling the coast.

Most foldbelt mountains develop near the edges of continents. The two components generally parallel the continental margin with the complex core seaward and the folded mountains landward of the continental margin.

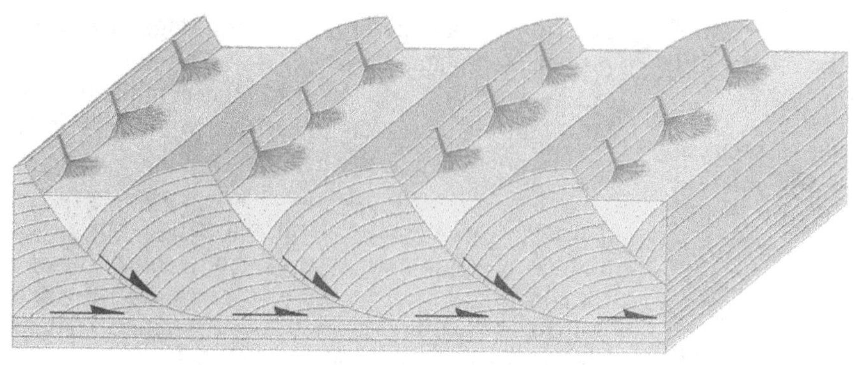

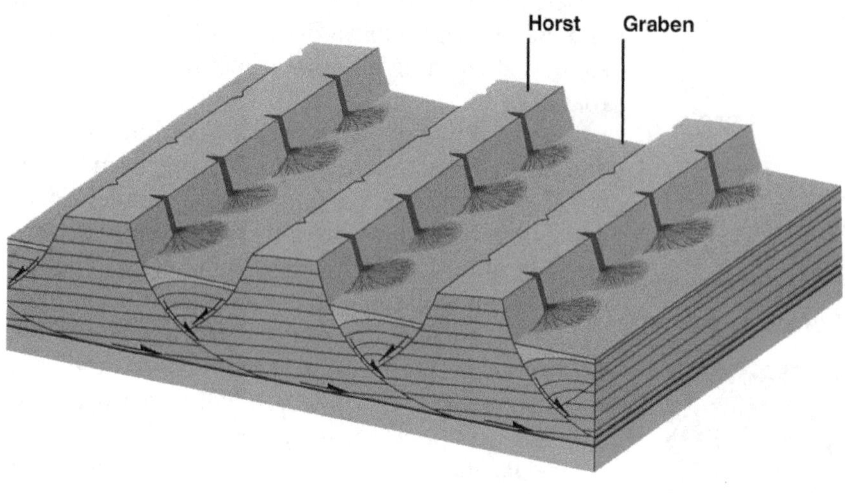

FIGURE 13.10 *The* Basin and Range Province *west of the Colorado Plateau was part of the epeirogenic uplift that affected the entire southwestern portion of the United States. Rather than being uplifted with little deformation, however, the region of the Basin and Range Province was subjected to* extensional forces *that formed two different styles of north-south trending faults. The results is an extremely rugged topography consisting of north-south trending mountain* ranges *separated by* basins *now occupied by deserts.*

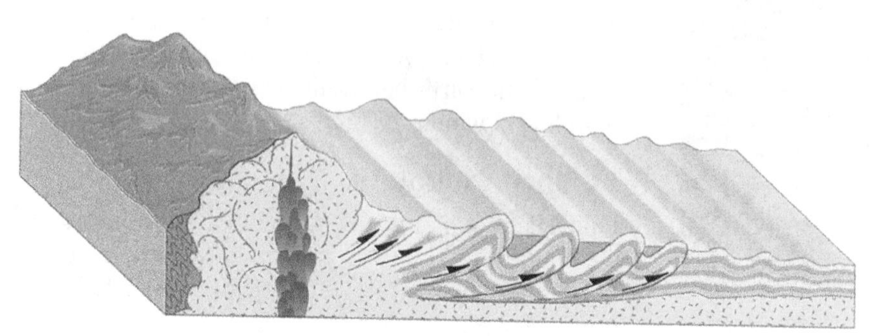

FIGURE 13.11 *Most of the great mountains of the world are* foldbelt mountains, *which consist of two basic components: a* complex core *of igneous, metamorphic, and volcanic rocks and a sequence of* folded and faulted sedimentary rocks. *The mountains are usually located along the leading edge of a continent with the complex core located on the seaward side.*

As the lithospheric plates converge, energy is partly converted to massive compressional forces, which are used to form, deform, and uplift rocks to awe-inspiring heights. Uplift occurs as the result of two processes: (1) crustal thickening due to both igneous intrusion and crustal shortening and (2) a reduction in density resulting from heat and the generation of low-density siliceous magmas. The great mountains of Earth, including the Alps, the Andes, the northern Rockies, and the Himalayas, are examples of foldbelt mountains. The Appalachians are also foldbelt mountains, though greatly subdued due to millions of years of erosion (Figure 13.12).

The Appalachians may be the most studied foldbelt mountains on Earth. Indeed, our present understanding of how the major components of foldbelt mountains are created stems from an early study of the folded sedimentary portion of the Appalachian Mountains.

In the mid-1800s, an American paleontologist, James Hall, made two observations concerning the sedimentary rocks of the folded Appalachians: (1) most of the sedimentary rocks within the folded

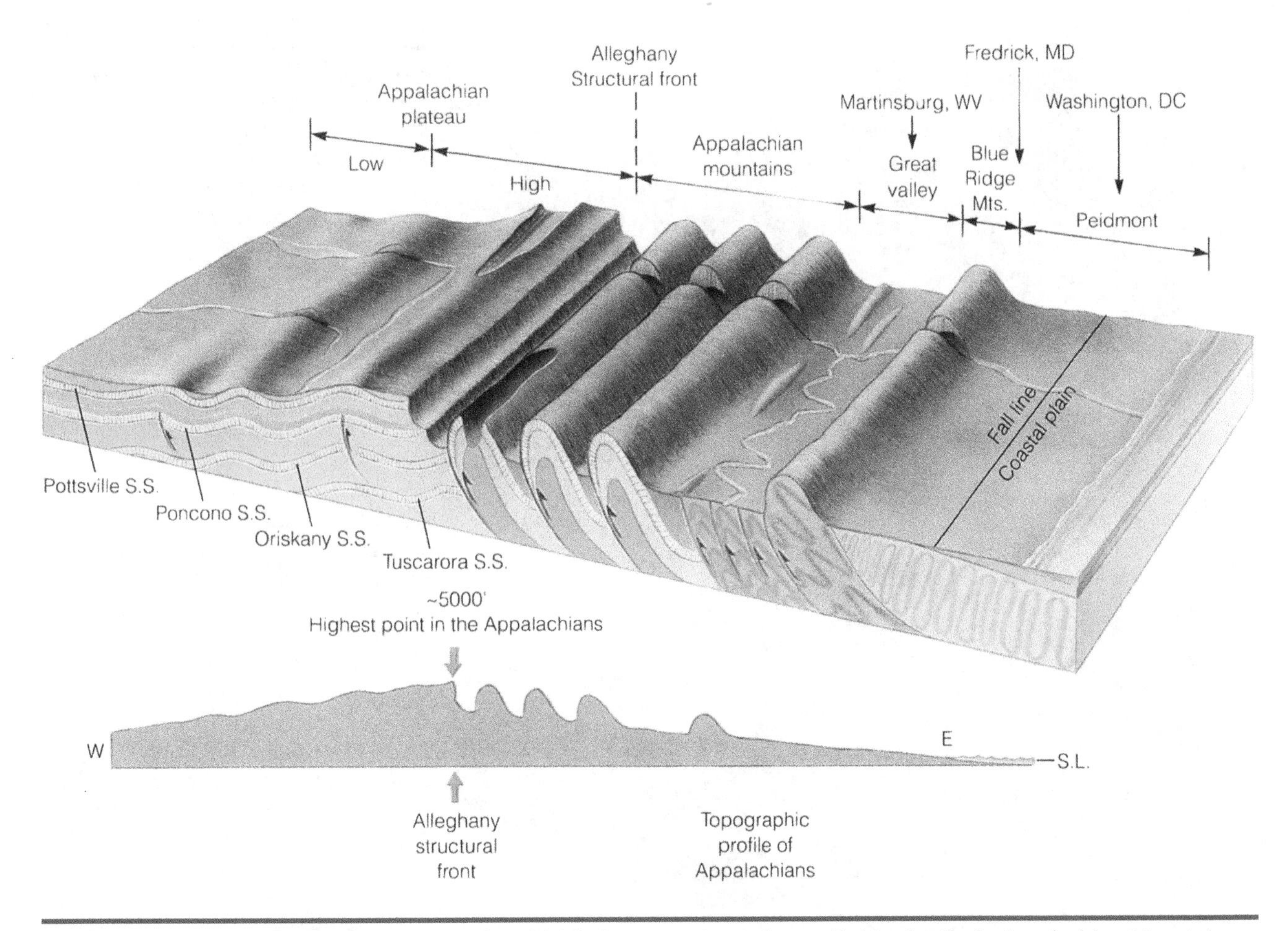

FIGURE 13.12 *One of the first examples of foldbelt mountains to be studied in detail, the* Appalachian Mountains *were formed during the continental collision that created Gondwana. Since their formation, the Appalachians have been reduced by erosion, perhaps to the featureless surface proposed by Davis called a peneplane, and reuplifted by epeirogenic forces that affected the entire eastern portion of North America. The present topography of the Appalachians is the result of erosion by the rejuvenated streams.*

Appalachians were of shallow marine origin and were the same types found within the continental interior, and (2) the sedimentary rocks in the folded Appalachians were many times thicker than the sedimentary rocks that covered the crystalline basement of the continent outside the mountains. The sedimentary rocks over the continental interior were estimated to be a few thousand feet or meters thick, but within the folded Appalachians, Hall found the total thickness of sedimentary rocks to be in the *tens* of thousands of feet. Hall was faced with the problem of explaining how tens of thousands of feet of marine sedimentary rock could accumulate along a continental shelf where the water depth rarely exceeded 500 or 600 feet (150–180 m). He concluded that as the sediments accumulated, the surface of the continental shelf must warp downward into a broad synclinal trough paralleling the continental margin (Figure 13.13). Hall's contemporary, the mineralogist James D. Dana, coined the name **geosyncline** to describe the structure. This term was intended to reflect the great size of these basins and should not be confused with a *structural* syncline. Hall and Dana agreed on the existence of the geosyncline, but developed different hypotheses as to how the structure formed. Hall was of the opinion that the geosyncline formed as the weight of the sediments downwarped the surface of the continental shelf. Dana argued that the accumulating sediments alone could not cause the downwarp of the shelf because the density of the sediments was too low to cause the displacement of the higher-density mantle rocks below. Dana was a proponent of the idea that all the deformation exhibited by Earth's crust, including the formation of geosynclines, could be attributed to the fact that Earth was cooling; as it cooled, the crust supposedly was wrinkling like the skin of a cooling baked

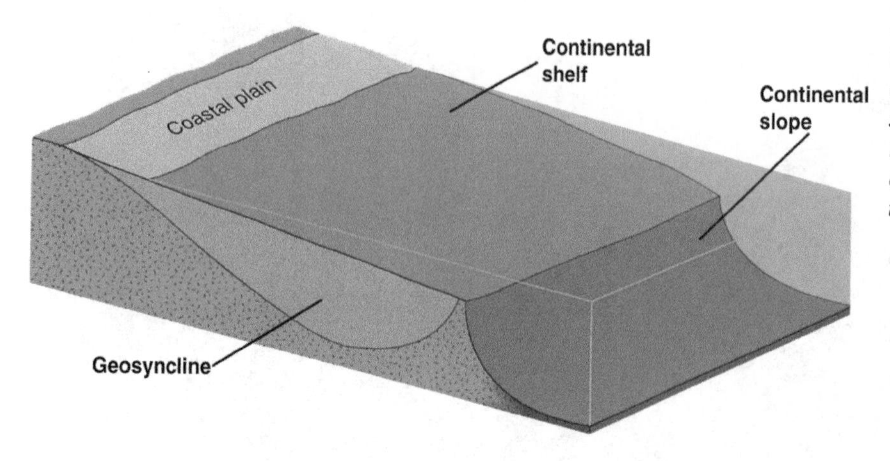

FIGURE 13.13 *As originally conceived, the* geosyncline *was a synclinal basin that formed along the trailing continental margin and continued to downwarp as sediments were introduced from the land.*

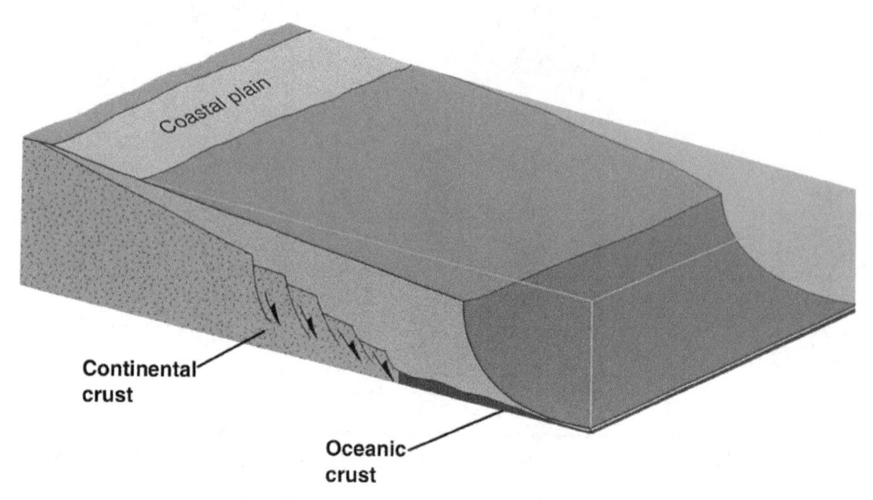

FIGURE 13.14 *With the advent of plate tectonics, combined with seismic data that have allowed geologists to "see" within the rocks and sediments of the continental margin, our structural picture of the passive continental margin has changed. Rather than a synclinal basin, we now recognize a thick wedge of sediment, called a* geocline, *that extends out from the edge of the passive continental margin.*

apple. This "baked apple theory" or "shrinking Earth hypothesis" dominated the thinking of geologists for many years but eventually gave way to the theory of plate tectonics.

The advent of plate tectonics drastically changed our picture of the formation of the thick sedimentary wedge that characterizes continental trailing edges. Because the wedge is not synclinal, some geologists have suggested that it should be called a **geocline** rather than a geosyncline. Although our ideas about the formation of the thick wedge of sediment have changed, the term *geosyncline* is still used today to describe the overly thick accumulations of sediments that are considered to be the precursors to the sedimentary rocks associated with foldbelt mountains.

As the plates move, the continents are carried along as components of a lithospheric plate. Trailing or passive continental margins, such as the eastern margins of North and South America, develop a slowly downwarping continental shelf, which initiates the development of the geocline. The downwarping is thought to be due to crustal thinning and subsidence resulting from continued movement along the high-angle normal faults that formed as a result of extensional forces induced by the former plate movements (Figure 13.14). As the geocline continues to evolve, sediments derived from the weathering and erosion of the adjoining landmass continuously accumulate on the surface of this downwarping margin in the form of a thick sedimentary wedge (refer to Figure 13.14). As the earlier layers of sediment become buried within the downwarping basin, lithifications transforms the sediments into marine sedimentary rocks.

With the advent of plate tectonics, we have come to recognize other settings in which thick sequences of sediments may accumulate. For example, a **back-arc basin** forms between a volcanic island arc and the continental margin (Figure 13.15). In some cases, a mantle convection cell may develop below the back-arc basin, rift the rocks by extension, and pour basalt onto the ocean floor as happens at an oceanic ridge. A

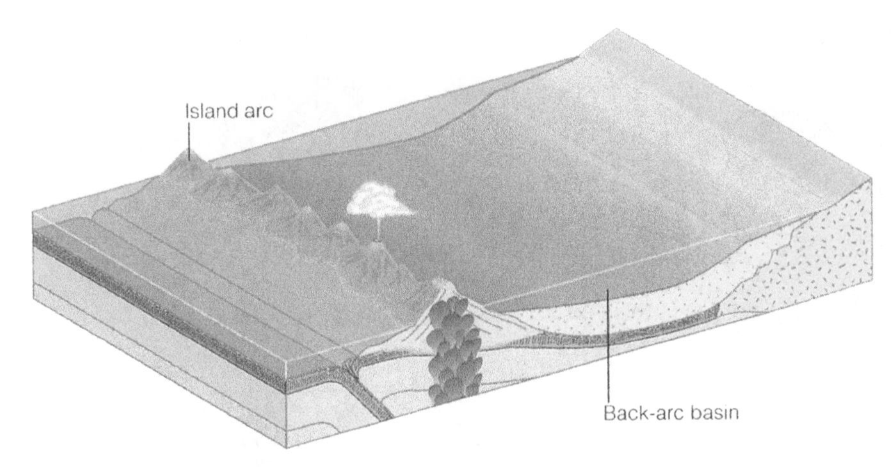

FIGURE 13.15 *The sediments accumulating in geoclines are one source of the sediments from which sedimentary rocks form. Another major source is the* back arc basin *that exists between the island arc and the continental margin. Accumulating sediments from both the continent and the island arc, the back arc basin may develop into a* marginal sea *if a mantle plume develops beneath the underlying oceanic crust. An example of a marginal sea is the* Sea of Japan.

back-arc basin expanded by tensional forces is called a **marginal sea.** The Sea of Japan located between the Japanese Islands and the Asian mainland is an example of a marginal sea.

The back-arc basin and marginal sea differ from the geocline associated with a continental trailing edge in that the accumulated sediment in the back-arc basin is introduced from *both* the mainland and the island arc. With the passage of time, as the mainland is reduced in elevation by weathering and erosion, it becomes less important as a source of sediment. Because the eroded volcanic mountains are "rejuvenated" by fresh eruptions, the island arc eventually becomes the major source of the sediment and will provide sediment more or less continuously for the entire life of the basin.

SPOT REVIEW

1. What is the definition of a mountain?
2. What is the difference between epeirogenic and orogenic forces?
3. Compare and contrast three different scenarios under which volcanic mountains form.
4. What are the basic characteristics of epeirogenic mountains?
5. What are the characteristics of block-fault mountains? Where in the United States are block-fault mountains best developed?
6. What are the basic characteristics of foldbelt mountains? What are some examples of foldbelt mountains?
7. What are the possible depositional sources for the sediments that make up the sedimentary rocks of foldbelt mountains?

OROGENIC STYLES

Three basic convergent plate scenarios can be used to explain most of Earth's foldbelt mountain systems: (1) *ocean-continent,* (2) *ocean-island arc-continent,* and (3) *continent-continent* collisions.

Ocean-Continent Orogenesis

The **ocean-continent** orogenic style, illustrated in Figure 13.16, shows the process of mountain building beginning with a passive continental margin and the associated geoclinal accumulation of sediments. An example is the present west coast of South America or North America. If we go back about 200 million years to the time when our present-day continents were still joined together in Wegener's supercontinent of Pangaea, Figure 13.16a would represent the western part of Gondwana, that portion of the supercontinent that would soon break away to become South America. As South America broke away from Africa and began to move westward against the oncoming Pacific lithospheric plate, a zone of subduction formed seaward of the geoclinal deposits (Figure 13.16b), transforming the western edge of the South American plate from a *passive* to an *active* margin.

As subduction continued, compressional forces developed, initiating the following sequence of events. The downward-moving oceanic lithosphere carried ocean bottom sediments and water with it. The water lowered the melting temperature of rocks within the subduction zone as the system converted from dry to wet, as discussed in Chapter 1. Intense heat and pressure metamorphosed the descending sediments and existing crystalline rocks. Thrust faults formed in the overriding crustal block and brought a mixture of metamorphic rocks, slivers

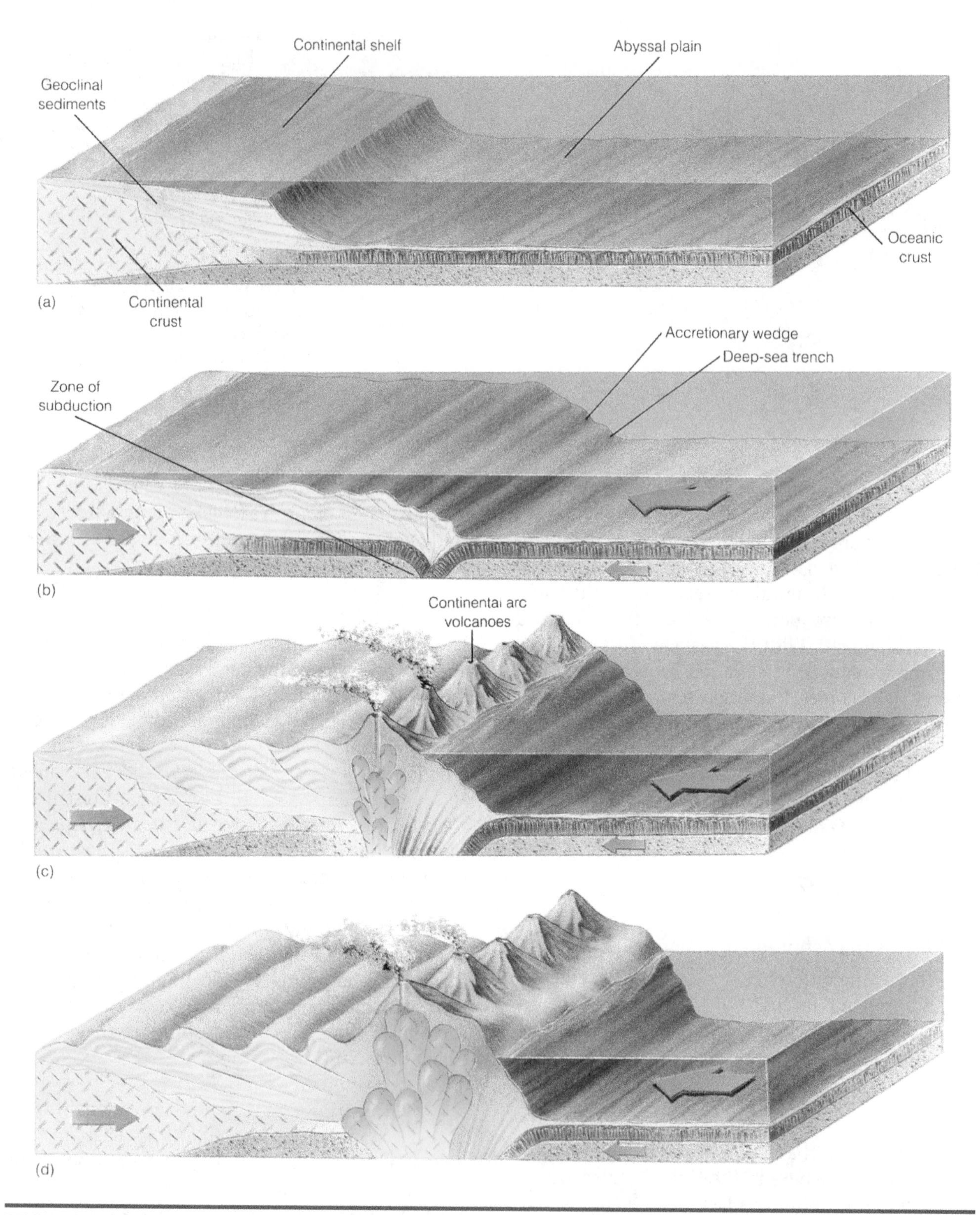

FIGURE 13.16 Opening oceans *form by sea-floor spreading as new oceanic lithosphere is formed at an oceanic ridge (a). At some point in time, the cooling oceanic lithosphere breaks and sinks into the underlying asthenosphere, forming* a zone of subduction *(b). As the oceanic plate continues to subduct, an accretionary wedge forms at the edge of the overriding plate; granitic magmas intrude into the edge of the continent while less viscous andesitic magmas are extruded to form a continental arc; and the sediments of the geocline are folded and thrust onto the continent (c). The result of this orogenic process is a mountain range such as the* Andes Mountains *that form the spine of South America (d).*

of the oceanic crust, igneous rocks, and metasediments to the surface. This assemblage of different rock types, called a **mélange** from the French word for mixture, formed a structure called an **accretionary wedge** (Figure 13.16c). As each new thrust sheet formed and was emplaced *under* the previous sheet, the previously deformed materials were lifted upward as a growing rock mass, eventually forming a coastal mountain range. An example of a mélage formed within an accretionary wedge is the Franciscan Formation, the rocks that make up most of the mountain ranges along the California coast (Figure 13.17).

Massive intrusions of granitic magmas began to be emplaced into the edge of the continental crust while andesitic magmas reached the surface where they created the chain of continental arc stratovolcanoes associated with folded mountain belts. With time, the sediments of the geocline began to respond by plastic deformation as thrust and reverse faults carried the folded sediments eastward onto the continental surface (refer to Figure 13.16d).

FIGURE 13.17 *Most of the rocks seen in the Coastal Range of California, Oregon, and Washington accumulated in an* accretionary wedge *that has been uplifted above sea level. (Courtesy of R. E. Wallace/USGS)*

The deformation produced the structures that we see today along the mountainous backbone of South America, the **Andes Mountains.** The mélage is exposed in the coastal areas of South America and grades eastward into complex exposures of granites and granodiorites capped by the volcanic mountains of the Andes. The crystalline mass makes up the complex core of the Andes mountain chain. The folded portion of the mountain chain is represented by the folded and faulted sedimentary rocks that make up the eastern flanks of the Andes Mountains. The level of present-day volcanism and earthquake activity along the western margin of the South American continent tells us that the Andes are still growing as South America continues to ride over and consume the Pacific ocean floor.

An important aspect of orogenesis is that the batholithic bodies of granite and granodiorite that are generated within the subduction zone through its history are eventually *added* to the continental crust, compensating for losses of continental rocks by weathering, erosion, and sediment transport to the deep sea, followed by subduction. The continental crust that we see today has grown to its present mass and size as a result of repeated orogenic events since the origin of Earth.

Ocean-Island Arc-Continent Orogenesis

The basic scenario of **ocean-island arc-continent** orogenesis is illustrated in Figure 13.18. The events of the early phases of subduction are the same as those for ocean-continent orogenesis except that an island arc develops. Once again, an accretionary wedge, with a mélange, forms along the leading edge of the overriding oceanic plate. Intermediate and granitic magmas are generated and emplaced into the overriding plate, in this case, oceanic lithosphere, which progressively thickens and increases in mass. As a result, the oceanic lithosphere becomes increasingly more continental in character. An example would be the Japanese Islands.

A back-arc basin—and, in some cases, a marginal sea—develops between the growing island arc and the adjacent continent. Sediment derived from both the continent and the island arc pours into the basin (Figure 13.18a). Tensional forces extend the marginal sea, if developed, allowing the continued

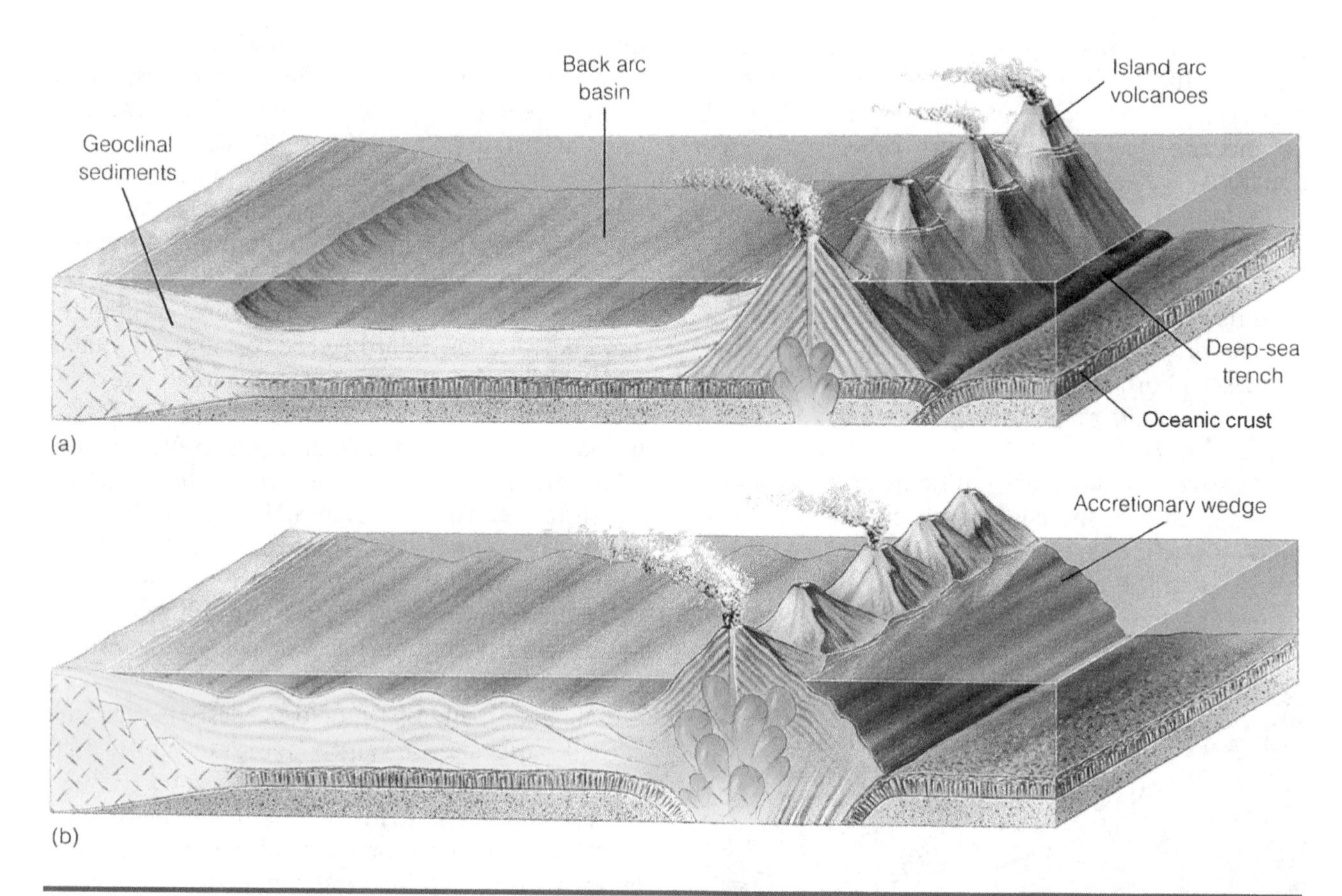

FIGURE 13.18 *An* ocean-island arc-continent orogeny *begins with the formation of a chain of andesitic, volcanic islands, such as the* Aleutian Islands, *along a zone of subduction separated from the mainland by a back arc basin (a). As convergence continues, granitic magmas intrude into the island arc, creating an ever-increasing core of continental-type rocks within the chain of islands. The* Japanese Islands *are an example. At the same time, the sediments within the back-arc basin begin to be deformed (b). Eventually, the sediments of the back-arc basin will be thrusted and folded above sea level and become a part of the continent (c). The orogenic event will end as the original island arc complex is welded to the edge of the continent and the former back-arc basin sediments are stacked onto the surface of the continent along low angle thrust faults (d). The northern Rocky Mountains from Wyoming to Alaska are thought to have formed by such an orogenic event.*

accumulation of sediment. As the plates continue to converge, the sediments in the back-arc basin or the marginal sea begin to be folded and moved landward along low-angle, high-displacement thrust faults (Figure 13.18b). In time, as the entire island arc complex with its core of granitic rocks is thrust and fused to the edge of the continental crust, the deformed sediments of the back-arc basin or marginal sea are thrust onto the continent and stacked, one massive thrust sheet on top of another (Figure 13.18c). Thrust sheets thousands of feet thick are moved laterally with displacements as large as 10 to 20 miles (16–32 km) (Figure 13.18d). An example of a mountain range created by such an orogenic event is the **North American Cordillera** represented by the northern Rocky Mountains extending from Alaska south to Wyoming (Figure 13.19).

Continent-Continent Orogenesis

Some of the most spectacular mountains are the result of **continent-continent** collisions. The continent-continent scenario differs from the ocean-island arc-continent scenario only in that the oceanic lithosphere brings with it a passive continental margin with associated geoclinal sediments.

To demonstrate the mountain-building process, let us return to Gondwana, this time to the eastern side of the continent (Figure 13.20). At the time continental rifting took place, Gondwana had developed a geoclinal basin along its eastern margin. As the supercontinent broke up, the Antarctic continent moved southward, Australia headed eastward, and the fragment that was to become India rifted from the main continental mass and began its journey northward, taking with it a portion of

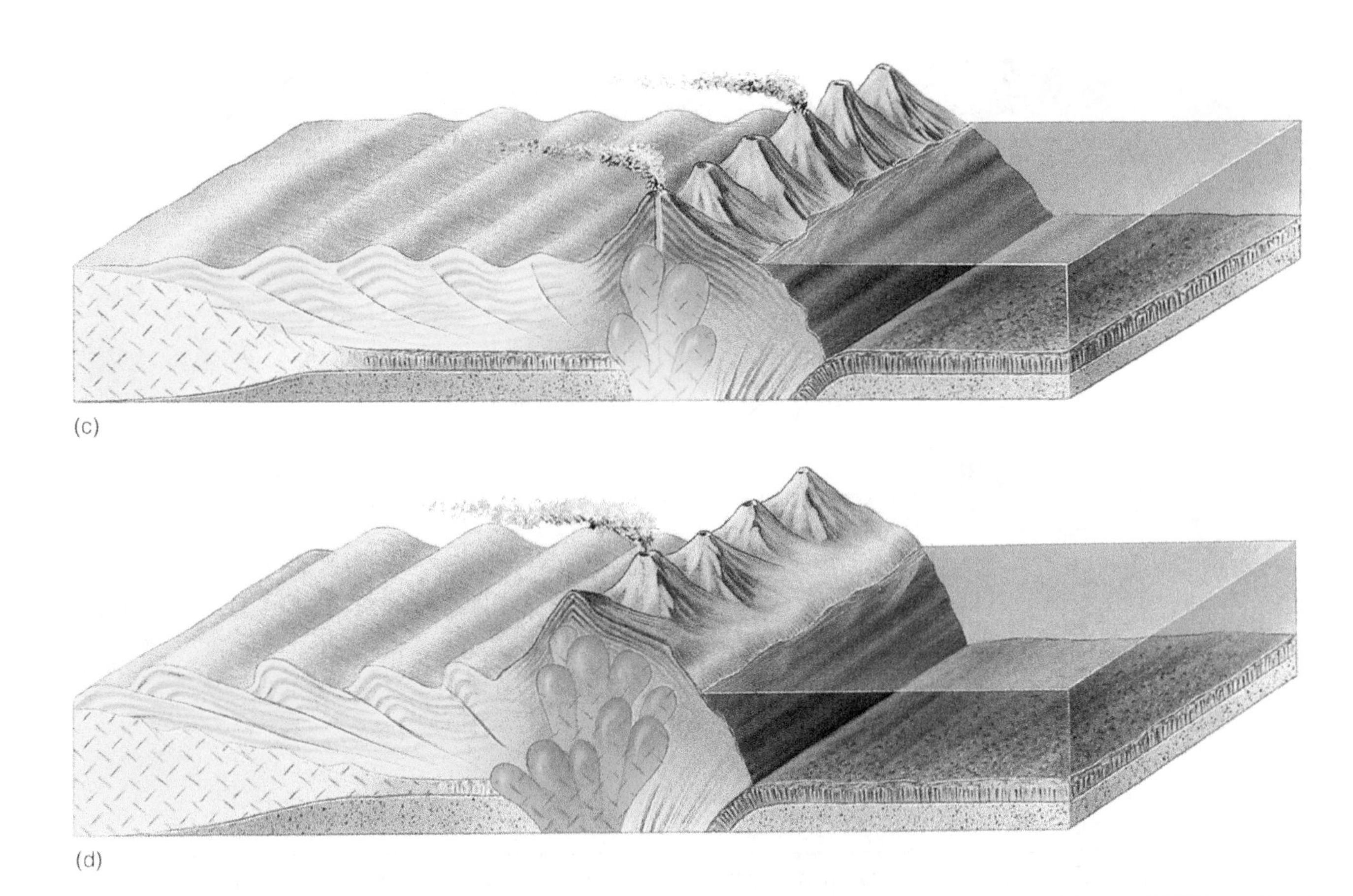

FIGURE 13.18 continued

FIGURE 13.19 *The* northern Rocky Mountains *originally formed along the western margin of North America. Mountain-building events since the formation of the Rocky Mountains have added the portion of the North American continent that extends from the present mountains to the Pacific coast.*

the geoclinal basin with its contained sediments (Figure 13.21).

Across the Tethys Sea, Asia waited with a chain of island arc volcanic mountains and a sediment-filled back-arc basin (Figure 13.21b). As India approached Asia and the Tethys Sea narrowed, the sediments of the Asian back-arc basin began to be folded and thrust northward onto the Asian continent (refer to Figure 13.21). With continued narrowing of the back-arc basin, the island arc complex collided with the continental crust of the Asian mainland and welded to it (Figure 13.21c). As the Tethys Sea narrowed and India approached, the scene was set for the creation of the **Himalayas.**

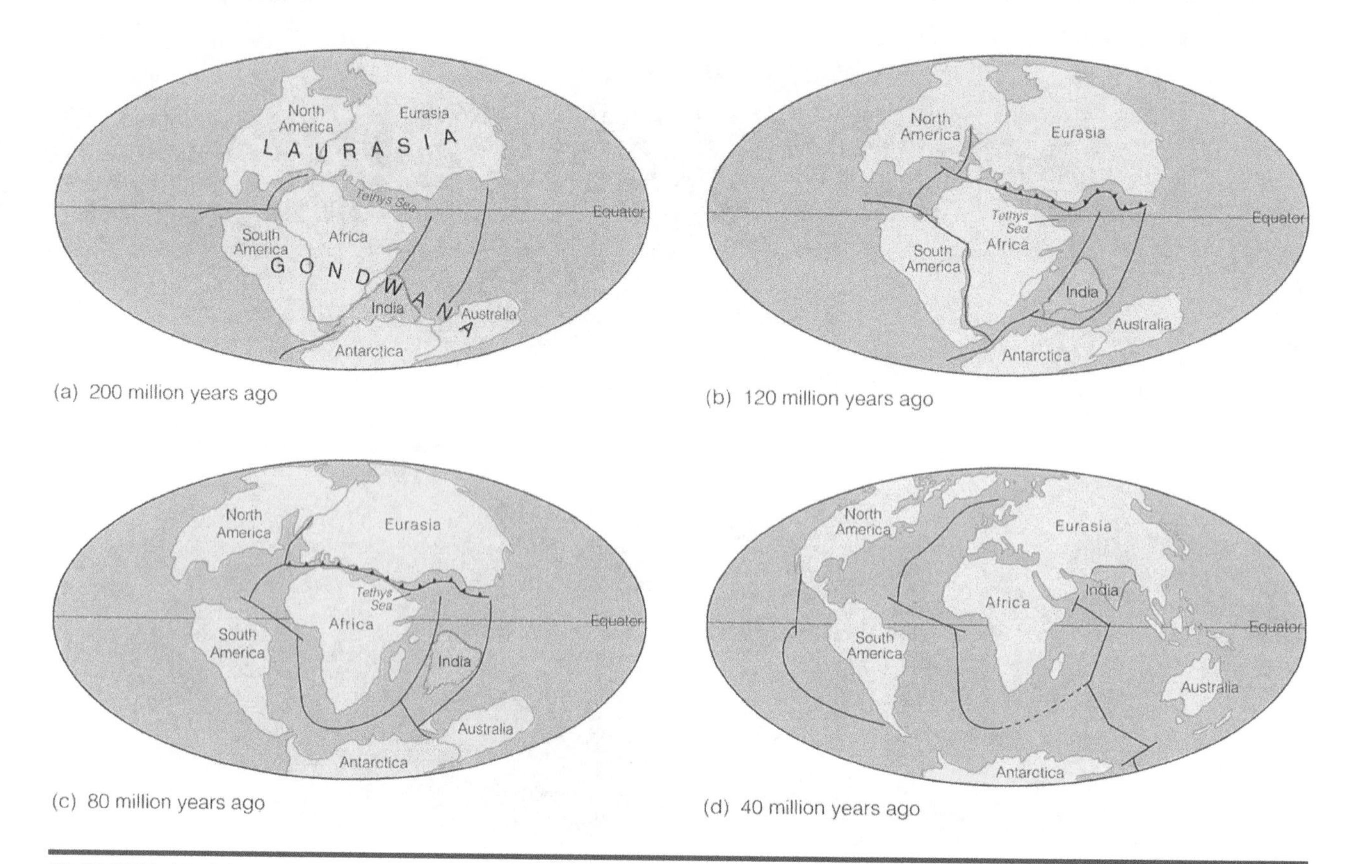

FIGURE 13.20 *The continents that are currently involved in* continent-continent *collisions and those that will be involved in the near geologic future were formed during the breakup of Pangaea about 200 million years ago.*

About 40 million years ago, India collided with Asia. As the continents collided, the geoclinal sediments located along the northern margin of the Indian continental fragment began to deform as the oceanic lithosphere in front of the Indian subcontinent was subducted below the Asian continental lithosphere. As the northern edge of the Indian continental lithosphere plunged downward beneath the southern edge of Asia, sheets of deformed geoclinial sediments and parts of the crystalline crust of India broke away and were thrust *southward* up over the Indian continental crust along major low-angle thrust faults (Figure 13.21d).

Unlike the dense basaltic oceanic lithosphere, the Indian continental lithosphere with its lower-density granitic crustal rocks resisted subduction. The breathtaking heights of the Himalayas that we see today are the result of the buoyancy of the Indian lithosphere as it rams beneath the edge of the Asian plate and lifts the overlying rocks of the Asian margin (refer to Figure 13.21).

Continued movement of the Indian continent northward beneath Asia resulted in a second episode of major low-angle thrusting that drove the Indian geosynclinal sediments farther southward onto the Indian continent (Figure 13.21e). At the same time, some of the folded sediments were thrust back *northward* over the rising edge of the Asian plate. Eventually, these folded and thrust-faulted sediments were uplifted to the very highest elevations in the Himalayas, even above those of the crystalline complex. One result of this uplift is that Mount Everest with its summit at 29,696 feet (9,051 m) is capped by a marine limestone that once lay many thousands of feet below at the bottom of the Tethys Sea!

The formation of the Himalayas is not yet over. Earthquakes indicate that India is still thrusting northward under the Asian plate at an average rate of 2 inches (5 cm) per year. As a result of India's northward movement, the Himalayas are being lifted each year at a rate *greater* than the rate at which the rocks are being removed by erosion. As grand as the Himalayas are, they will become even grander.

Another continent-continent collision that is well underway is that between Africa and Europe. The relationship between Africa and Europe is similar to that between India and Asia before the two continents

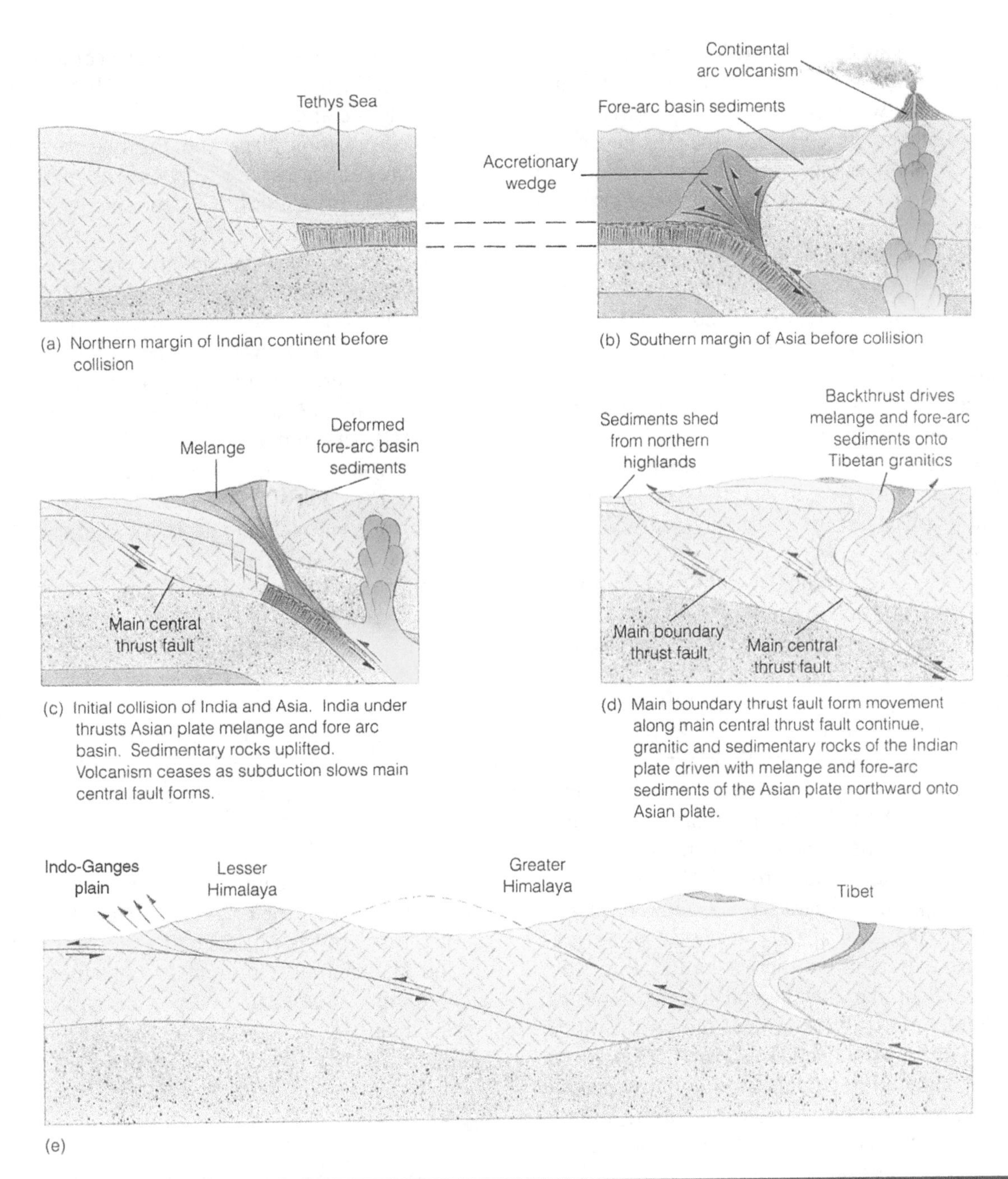

FIGURE 13.21 *The formation of the Himalaya Mountains by the collision of India and Asia about 40 million years ago is an example of a* continent-continent orogeny. *The orogenic event began with the rifting of Gondwana and the closing of the Tethys Sea. India, with its geoclinal wedge of sediments, lay to the south (a). To the north, along the southern margin of Asia, a zone of subduction existed with an accretionary wedge and a continental arc (b). The drawing shows a* fore-arc basin *that commonly develops where an accretionary wedge builds upward fast enough to trap sediment eroded from the continental margin. Eventually, the two continental masses collided. As the Indian continental crust was subducted beneath the edge of Asia, the sediments of the Indian geocline and part of the Asian accretionary wedge were thrusted southward onto the Indian continent. At the same time, the sediments that had accumulated in the Asian fore-arc basin, plus a portion of the accretionary wedge, were thrusted onto the Asian continent (c). Because of its low density, the Indian continental crust resisted subduction. As convergence continued, the buoyancy of the low-density rocks of the Indian continental crust resulted in the uplift of the region. The complex folding of the entire rock complex (d & e) eventually elevated a limestone that had originally formed on the bottom of the Tethys Sea to the top of Mount Everest. Although volcanic activity was terminated as subduction and the subsequent generation of magma slowed, seismic activity indicates that India is still moving northward. No one knows how long it will be before the orogenic episode is concluded.*

collided. The volcanism in the Mediterranean area and the rise of the Alps are the result of the forces being generated as the two continents converge. Eventually, when the collision is complete, the Mediterranean Sea—the last large vestige of the Tethys Sea—will be eliminated, and the Alps will probably be as grand as the present-day Himalayas.

It is important to note that the addition of continental lithosphere to the margin of another continent results in the formation of mountains located *within* the continental mass. The Ural Mountains, which are located well within the Eurasian continent, are an excellent example. The Urals were formed when smaller continents were "sutured" together during a continent-continent collision to form the huge Eurasian continent. In fact, the Ural Mountains are usually considered to be the boundary between Europe and Asia.

Before we leave the topic of continent-continent collisions, let's go back to the time before the supercontinent Gondwana was formed by the collision of whatever continents existed at that time. Referring once again to Figure 13.21, consider this time that the continent to the left is the one that will become North America and the one on the right will become Africa. Picture these two masses of continental lithosphere colliding about 250 million years ago, forming Pangaea and creating a mountain complex that was perhaps every bit as grand as today's Himalayas.

About 20 million years later, Gondwana began to break up. We have already discussed the events that occurred as South America broke away from west Africa. A similar rifting formed North America. When North America broke away from northwestern Africa and Europe, the break occurred somewhere within the crystalline complex of the newly formed mountain range. The northern portion of the Atlantic Ocean was created as the two parts of this once grand mountain range separated.

The western part of the original orogenic system became the rejuvenated Appalachian Mountains of North America. The folded sedimentary rock component of that pre-Atlantic mountain range (comparable to the sediments that were thrust southward onto the Indian continent) now makes up the Appalachian Plateau, the Valley and Ridge, and the Great Valley portions of the Appalachians. A part of the original crystalline core of that early mountain range can be found today in the rocks of the Blue Ridge Mountains and the Piedmont (see Figure 13.12). However, much of the original core complex is buried beneath the modern coastal plain and continental shelf. More extensive exposures of the crystalline complex can be seen today in northern New England, Nova Scotia, and Newfoundland.

As rifting took place and the two parts of this mountain complex separated, streams methodically reduced this once impressive mountain range to a flat erosional surface near sea level. Within the last 20 million years, however, epeirogenic forces in the form of glacio-eustatic rebound upwarped the entire eastern portion of the North American continent. As streams were rejuvenated throughout the area, the increased erosion not only produced the present topography of the eastern United States and southeastern Canada, but also reexposed the structures and rocks of that ancient orogeny. We see them today, perhaps not with the grandeur they once had, but certainly with sufficient clarity to attest to their proud history.

An interesting question might be, what happened to the rocks of the "other half "? On the western side of the Atlantic Ocean, the rocks of the crystalline core complex now lie buried beneath the sediments of the modern Atlantic coast geocline of North America and are exposed as the rocks of Nova Scotia and Newfoundland. On the "other side" of the Atlantic, the rocks of that same crystalline core complex can be seen in the rocks along the western coastlines of Ireland and Scotland and throughout the Scandinavian Peninsula. The folded mountains that existed east of the original central core complex as the "twin" to the Appalachians can now be found in the Mauritanide Mountains of northern Africa, a mountain range structurally quite similar to the Appalachians.

SPOT REVIEW

1. Compare passive and active continental margins. Give an example of each.
2. How can a passive continental margin be converted into an active margin? Give an example.
3. What is a melange and how does it form?
4. Compare the mountain-building scenarios associated with ocean-continent, ocean-island arc-continent, and continent-continent collisions.
5. Where would you look for companion rocks and structures to those exposed in the Appalachian Mountains?

CONCEPTS AND TERMS TO REMEMBER

Mountain-building forces
- volcanic
- epeirogenic
- orogenic
- diastrophism

Kinds of mountains
- volcanic mountains
  - rift zones
  - rift valleys
  - oceanic ridges
  - hot spot shield volcanoes
  - island arc stratovolcanoes
  - continental arc stratovolcanoes
- epeirogenic mountains
  - domal mountains
  - block-fault mountains
    - scarp
    - horst
    - graben
    - Basin and Range Province
- orogenic mountains
  - foldbelt mountains
    - geosyncline or geocline
    - back-arc basin
    - marginal sea

Orogenic styles
- ocean-continent collision
  - mélange
  - accretionary wedge
  - Andes Mountains
- ocean-island arc-continent collision
  - North American Cordillera
- continent-continent collision
  - Himalaya Mountains

REVIEW QUESTIONS

1. The Black Hills of South Dakota are an example of ____ mountains.
   a. block-fault c. erosional
   b. domal d. foldbelt
2. The faults associated with block-fault mountains are
   a. thrust. c. normal.
   b. strike-slip. d. reverse.
3. A modern example of a geocline or geosyncline is now developing
   a. within the Sea of Japan.
   b. along the east coast of North America.
   c. along the west coast of North America.
   d. along the trend of the Aleutian Islands.
4. The forces associated with epeirogenic mountain building are dominantly
   a. horizontally directed nonrotational compression.
   b. volcanic.
   c. vertical uplift with or without tensional extension.
   d. horizontally directed rotational compression.
5. Which of the following features is primarily the result of epeirogenic uplift?
   a. Cascade Mountains
   c. Hawaiian Islands
   b. Grand Canyon
   d. Sierra Nevada
6. Why is there no volcanism associated with the Himalayas?
7. Why do geologists find it difficult to explain the existence of epeirogenic forces?
8. What role do geoclines or geosynclines and back arc-basins play in the formation of foldbelt mountains?
9. What similarities exist between the Appalachian Mountains and the Himalayas?

THOUGHT PROBLEMS

1. Although most foldbelt mountains are located near the margins of continents, the Rocky Mountains of North America and the Ural Mountains of Russia are located inland, but for different reasons. How are the inland locations of these two foldbelt systems explained?

2. Based on what you know about the origin of mountains and about plate tectonics, where do you predict the next ranges of foldbelt mountains will form on Earth? What scenario(s) will bring them to being?

FOR YOUR NOTEBOOK

If you live in an area of deformed rocks, a field trip is definitely in order to describe and perhaps photograph some of the more impressive outcrops. You will also want to investigate the geology of your area in more detail to see how it fits into the larger mountain system.

Another possible exercise, especially if there are no mountains in your area, is to investigate in detail some other mountainous areas that were not treated in the text. In the northeast, there are the White Mountains, the Berkshires, and the Adirondacks of upstate New York. In the mid-continent, there are the Ozark and Arbuckle mountains. In the west Texas, the structure and lithologies of the Guadalupe mountains are unique. Another interesting study would be to compare the structure and stratigraphy of the southern and northern Rocky Mountains.

CPSIA information can be obtained
at www.ICGtesting.com
Printed in the USA
LVHW052017150819
627779LV00002B/2/P

9 781465 260789